PURIFICATION

of

LABORATORY CHEMICALS

THIRD EDITION

PURIFICATION

of

LABORATORY CHEMICALS

THIRD EDITION

D. D. PERRIN
Formerly of the Medical Chemistry Group
Australian National University, Canberra
A.C.T., AUSTRALIA

and

W. L. F. ARMAREGO
Division of Biochemical Sciences (Biochemistry)
Australian National University, Canberra
A.C.T., AUSTRALIA

Butterworth-Heinemann Ltd
Linacre House, Jordan Hill, Oxford OX2 8DP

 A member of the Reed Elsevier plc group

OXFORD LONDON BOSTON
MUNICH NEW DELHI SINGAPORE SYDNEY
TOKYO TORONTO WELLINGTON

First published 1966
Second edition 1980
Third edition 1988
Reprinted 1989, 1992, 1993, 1994

© Butterworth-Heinemann Ltd 1988

British Library Cataloguing in Publication Data
Perrin, D. D. (Douglas Dalzell), *1922-*
    Purification of laboratory chemicals 3rd ed
    1. Chemical compounds. Purification
    I. Title   II. Armarego, W.L.F.
    542

ISBN 0 7506 2332 2

Library of Congress Cataloguing in Publication Data
Perrin, D. D. (Douglas Dalzell), *1922-*
    Purification of laboratory chemicals
    Includes index
    1. Chemicals-purification   I   Armarego, W. L.F.
    II. Title
    TP156.P83P47   1988   542   88-5247

Printed and bound in Great Britain

CONTENTS

Preface to the First Edition

WE BELIEVE that a need exists for a book to help the chemist or biochemist who wishes to purify the reagents she or he uses. This need is emphasised by the previous lack of any satisfactory central source of references dealing with individual substances. Such a lack must undoubtedly have been a great deterrent to many busy research workers who have been left to decide whether to purify at all, to improvise possible methods, or to take a chance on finding, somewhere in the chemical literature, methods used by some previous investigators.

Although commercially available laboratory chemicals are usually satisfactory, as supplied, for most purposes in scientific and technological work, it is also true that for many applications further purification is essential.

With this thought in mind, the present volume sets out, firstly, to tabulate methods, taken from the literature, for purifying some thousands of individual commercially available chemicals. To help in applying this information, two chapters describe the more common processes currently used for purification in chemical laboratories and give fuller details of new methods which appear likely to find increasing application for the same purpose. Finally, for dealing with substances not separately listed, a chapter is included setting out the usual methods for purifying specific classes of compounds.

To keep this book to a convenient size, and bearing in mind that its most likely users will be laboratory-trained, we have omitted manipulative details with which they can be assumed to be familiar, and also detailed theoretical discussion. Both are readily available elsewhere, for example in Vogel's very useful book **Practical Organic Chemistry** (Longmans, London, 3rd ed., 1956), or Fieser's **Experiments in Organic Chemistry** (Heath, Boston, 3rd ed, 1957).

For the same reason, only limited mention is made of the kinds of impurities likely to be present, and of the tests for detecting them. In many cases, this information can be obtained readily from existing monographs.

By its nature, the present treatment is not exhaustive, nor do we claim that any of the methods taken from the literature are the best possible. Nevertheless, we feel that the information contained in this book is likely to be helpful to a wide range of laboratory workers, including physical and inorganic chemists, research students, biochemists, and biologists. We hope that it will also be of use, although perhaps to only a limited extent, to experienced organic chemists.

We are grateful to Professor A. Albert and Dr D.J. Brown for helpful comments on the manuscript.

D.D.P.
W.L.F.A.
D.R.P.

Preface to the Second Edition

SINCE the publication of the first edition of this book there have been major advances in purification procedures. Sensitive methods have been developed for the detection and elimination of progessively lower levels of impurities. Increasingly stringent reqiurements for reagent purity have gone hand-in-hand with developments in semiconductor technology, in the preparation of special alloys and in the isolation of highly biologically active substances. The need to eliminate trace impurities at the micro- and nanogram levels has placed greater emphasis on ultrapurification technique. To meet these demands the range of purities of laboratory chemicals has become correspondingly extended. Purification of individual chemicals thus depends more and more critically on the answers to two questions - Purification from what, and to what permissible level of contamination. Where these questions can be specifically answered, suitable methods of purification can usually be devised.

Several periodicals devoted to ultrapurification and separations have been started. These include "Progress in Separation and Purification" Ed. (vol. 1) E.S. Perry, Wiley-Interscience, New York, vols. 1-4, 1968-1971, and **Separation and Purification Methods** Ed. E.S.Perry and C.J.van Oss, Marcel Dekker, New York, vol. 1 -, 1973 - . Nevertheless, there still remains a broad area in which a general improvement in the level of purity of many compounds can be achieved by applying more or less conventional procedures. The need for a convenient source of information on methods of purifying available laboratory chemicals was indicated by the continuing demand for copies of this book even though it had been out of print for several years.

We have sought to revise and update this volume, deleting sections that have become more familiar or less important, and incorporating more topical material. The number of compounds in Chapters 3 and 4 have been increased appreciably. Also, further details in purification and physical constants are given for many compounds that were listed in the first edition.

We take this opportunity to thank users of the first edition who pointed out errors and omissions, or otherwise suggested improvements or additional material that should be included. We are indebted to Mrs S.Schenk who emerged from retirement to type this manuscript.

D.D.P.
W.L.F.A.
D.R.P.

Preface to the Third Edition

THE CONTINUING demand for this monograph and the publisher's request that we prepare a new edition, are an indication that **Purification of Laboratory Chemicals** fills a gap in many chemists' reference libraries and laboratory shelves. The present volume is an updated edition which contains significantly more detail than the previous edition, as well as an increase in the number of individual entries and a new chapter.

Additions have been made to Chapters 1 and 2 in order to include more recent developments in techniques (e.g. Schlenk-type, *cf* p. 10), and chromatographic methods and materials. Chapter 3 still remains the core of the book, and lists in alphabetical order relevant information on *ca* 4000 organic compounds. Chapter 4 gives a smaller listing of *ca* 750 inorganic and metal-organic substances, and makes a total increase of *ca* 13% of individual entries in these two chapters. Some additions have also been made to Chapter 5.

We are currently witnessing a major development in the use of physical methods for purifying large molecules and macromolecules, especially of biological origin. Considerable developments in molecular biology are apparent in techniques for the isolation and purification of key biochemicals and substances of high molecular weight. In many cases something approaching homogeneity has been achieved, as evidenced by electrophoresis, immunological and other independent criteria. We have consequently included a new section, Chapter 6, where we list upwards of 100 biological substances to illustrate their current methods of purification. In this chapter the details have been kept to a minimum, but the relevant references have been included.

The lists of individual entries in Chapters 3 and 4 range in length from single line entries to *ca* one page or more for solvents such as acetonitrile, benzene, ethanol and methanol. Some entries include information such as likely contaminants and storage conditions. More data referring to physical properties have been inserted for most entries [i.e. melting and boiling points, refractive indexes, densities, specific optical rotations (where applicable) and UV absorption data]. Inclusion of molecular weights should be useful when deciding on the quantities of reagents needed to carry out relevant synthetic reactions, or preparing analytical solutions. The Chemical Abstracts registry numbers have also been inserted for almost all entries, and should assist in the precise identification of the substances.

In the past ten years laboratory workers have become increasingly conscious of safety in the laboratory environment. We have therefore in three places in Chapter 1 (pp. 3 and 33, and bibliography p. 52) stressed more strongly the importance of safety in the laboratory. Also, where possible, in Chapters 3 and 4 we draw attention to the dangers involved with the manipulation of some hazardous substances.

The world wide facilities for retrieving chemical information provided by the Chemical Abstract Service (CAS on-line) have made it a relatively easy matter to obtain CAS registry numbers of substances, and most of the numbers in this monograph were obtained *via* CAS on-line. We should point out that two other available useful files are CSCHEM and CSCORP which provide, respectively, information on chemicals (and chemical products) and addresses and telephone numbers of the main branch offices of chemical suppliers.

The present edition has been produced on an IBM PC and a Laser Jet printer using the **Microsoft Word (4.0)** word-processing program with a set stylesheet. This has allowed the use of a variety of fonts and font sizes which has made the presentation more attractive than in the previous edition. Also, by altering the format and increasing slightly the sizes of the pages, the length of the monograph has been reduced from 568 to 391 pages. The reduction in the number of pages has been achieved in spite of the increase of *ca* 15% of total text.

We extend our gratitude to the readers whose suggestions have helped to improve the monograph, and to those who have told us of their experiences with some of the purifications stated in the previous editions, and in particular with the hazards that they have encountered. We are deeply indebted to Dr M.D. Fenn for the several hours that he has spent on the terminal to provide us with a large number of CAS registry numbers.

This monograph could not have been produced without the expert assistance of Mr David Clarke who has spent many hours to load the necessary fonts in the computer, and for advising one of the authors (W.L.F.A.) on how to use them together with the idiosyncrasies of Microsoft Word.

D.D.P.
W.L.F.A.

CHAPTER 1

COMMON PHYSICAL TECHNIQUES USED IN PURIFICATION

GENERAL REMARKS

Purity is a matter of degree. Quite apart from any adventitious contaminants such as dust, scraps of paper, wax, cork, etc., that may have been incorporated into the sample during manufacture, all commercially available chemical substances are in some measure impure. Methods used in their synthesis may lead to the inclusion of significant amounts of unreacted starting material or intermediates; and isomers and related compounds may be present. Inorganic reagents may deteriorate because of defective packaging (glued liners affected by sulphuric acid, zinc extracted from white rubber stoppers by ammonia), corrosion or prolonged storage. Organic molecules may undergo changes on storage. In extreme cases the container may be incorrectly labelled or, where compositions are given, they may be misleading or inaccurate for the proposed use. Where any doubt exists it is usual to check for impurities by appropriate spot tests, or by recourse to tables of physical or spectral properties such as the extensive infrared and nmr libraries published by the Aldrich Chemical Co.

The important question, then, is not whether a substance is pure but whether a given sample is sufficiently pure for some intended purpose. That is, are the contaminants likely to interfere in the process or measurement that is to be studied. By suitable manipulation it is often possible to reduce levels of impurities to acceptable limits, but absolute purity is an ideal which, no matter how closely approached, can never be shown to be attained. A *negative* physical or chemical test indicates only that the amount of an impurity in a substance lies below a certain level; no test can demonstrate that a specified impurity is entirely absent.

When setting out to purify a laboratory chemical, it is desirable that the starting material is of the best grade commercially available. Particularly among organic solvents there is a range of qualities varying from *laboratory chemical* to *spectroscopic*, *chromatographic* and *electronic* grades. Many of these are suitable for use as received. With many of the commoner reagents it is possible to obtain from the current literature some indications of likely impurities, their probable concentrations and methods for detecting them. However, in many cases complete analyses are not given so that significant concentrations of unspecified impurities may be present. See for example **Reagent Chemicals** (American Chemical Society Specifications, 6th edn., 1981), the American Chemical Society for Testing Materials D56-36, D92-46, and national pharmacopoeias. Other useful sources include **Analar Standards for Laboratory Chemicals**, 7th edn, 1977 (The British Drug Houses Ltd, and Hopkin and Williams Ltd), and **Reagent Chemicals and Standards**, J.Rosin, 5th edn, 1967 (Van Nostrand, New York). For purification of proteins, see **Protein Purification**, R.K.Scopes, Springer-Verlag, New York, 1982.

1

Solvents and substances that are specified as **pure** for a particular purpose may, in fact, be quite impure for other uses. Absolute ethanol may contain traces of benzene, which makes it unsuitable for ultraviolet spectroscopy, or plasticizers which make it unsuitable for use in solvent extraction.

Irrespective of the grade of material to be purified, it is essential that some criteria exist for assessing the degree of purity of the final product. The more common of these include:

1. Examination of physical properties such as:
    (a) Melting point, freezing point, boiling point, and the freezing curve (i.e. the variation, with time, in the freezing point of a substance that is being slowly and continuously frozen).
    (b) Density.
    (c) Refractive index at a specified temperature and wave-length. The sodium D line at 589.26 nm (weighted mean of D_1 and D_2 lines) is the usual standard of wavelength but results from other wavelengths can often be interpolated from a plot of refractive index versus $1/(\text{wavelength})^2$.
    (d) Absorption spectra (ultraviolet, visible, infrared, and nuclear magnetic resonance).
    (e) Specific conductivity. This can be used to detect, for example, water, salts, inorganic and organic acids and bases, in non-electrolytes).
    (f) Optical rotation, optical rotatory dispersion and circular dichroism.
    (g) Mass spectroscopy.

2. Empirical analysis, for C, H, N, ash, etc.

3. Chemical tests for particular types of impurities, e.g. for peroxides in aliphatic ethers (with acidified KI), or for water in solvents (quantitatively by the Karl Fischer method).

4. Physical tests for particular types of impurities:
    (a) Emission and atomic absorption spectroscopy for detecting and determining metal ions.
    (b) Chromatography, including paper, thin layer, liquid (high, medium and normal pressure) and vapour phase.
    (c) Electron spin resonance for detecting free radicals.
    (f) X-ray spectroscopy.
    (e) Mass spectroscopy.
    (f) Fluorimetry.

5. Electrochemical methods.

6. Nuclear methods.

A substance is usually taken to be of an acceptable purity when the measured property is unchanged by further treatment (especially if it agrees with a recorded value). In general, at least two different methods, such as recrystallisation and distillation, should be used in order to ensure maximum purification. Crystallisation may be repeated until the substance has a constant melting point or absorption spectrum, and until it distils repeatedly within a narrow, specified temperature range.

With liquids, the refractive index at a specified temperature and wavelength is a sensitive test of purity. Under favourable conditions, freezing curve studies are sensitive to impurity levels of as little as 0.001 moles per cent. (See, for example, Mair, Glasgow and Rossini *J Res Nat Bur Stand* **26** 591 *1941*). Analogous fusion curve or heat capacity measurements can be up to ten times as sensitive as this (see, for example Aston and Fink *Anal Chem* **19** 218 *1947*). However, with these exceptions, most of the above methods are rather insensitive, especially if the impurities and the substances in which they occur are chemically similar. In some cases, even an impurity comprising many parts per million of a sample may escape detection.

The common methods of purification, discussed below, comprise distillation (including fractional distillation, distillation under reduced pressure, sublimation and steam distillation), crystallisation, extraction, chromatographic and other methods. In some cases, volatile and other impurities can be removed by heating. Impurities can also sometimes be eliminated by the formation of derivatives from which the purified material is recovered.

Safety in the Chemical Laboratory

Although most of the manipulations involved in purifying laboratory chemicals are inherently safe, it remains true that care is necessary if hazards are to be avoided in the chemical laboratory. In particular there are dangers inherent in the inhalation of vapours and absorption of liquids and low melting solids through the skin. To the toxicity of solvents must be added the risk of their flammability and the possibility of eye damage. Chemicals, particularly in admixture, may be explosive. Compounds may be carcinogenic or otherwise deleterious to health. Present day chemical catalogues specifically indicate the particular dangerous properties of the individual chemicals they list and these should be consulted whenever the use of commercially available chemicals is contemplated. Radioisotopic labelled compounds pose special problems of human exposure to them and of disposal of laboratory waste.

The commonest hazards are:
  (1) Explosions due to the presence of peroxides formed by aerial oxidation of ethers and tetrahydrofuran, decahydronaphthalene, acrylonitrile and styrene.
  (2) Compounds with low flash points (below room temperature). Examples are acetaldehyde, acetone, acetonitrile, benzene, carbon disulphide, cyclohexane, diethyl ether, ethyl acetate and n-hexane.
  (3) Contact of oxidising agents ($KMnO_4$, $HClO_4$, chromic acid) with organic liquids.
  (4) Toxic reactions with tissues.

For a fuller discussion, see Bretherick **Handbook of Reactive Chemical Hazards**, Butterworths, London, 3rd edn, 1985.

At least the laboratory should be well ventilated and safety glasses should be worn, particularly during distillation and manipulations carried out under reduced pressure or elevated temperatures. With this in mind we have endeavoured to warn users of this book whenever greater than usual care is needed in handling chemicals. As a general rule, however, **all chemicals which users are unfamiliar with should be treated with extreme care and assumed to be highly flammable and toxic.** The safety of others in a laboratory should always be foremost in mind, with ample warning whenever a potentially hazardous operation is in progress. Also, unwanted solutions or solvents should never be disposed of *via* the laboratory sink. The operator should be aware of the usual means for disposal of chemicals in her/his laboratories and she/he should remove unwanted chemicals accordingly.

Further aspects of safety are detailed on p.33.

Trace Impurities in Solvents

Some of the more obvious sources of contamination of solvents arise from storage in metal drums and plastic containers, and from contact with grease and screw caps. Many solvents contain water. Others have traces of acidic materials such as hydrochloric acid in chloroform. In both cases this leads to corrosion of the drum and contamination of the solvent by traces of metal ions, especially Fe^{3+}. Grease, for example on stopcocks of separating funnels and other apparatus, e.g. greased ground joints, is also likely to contaminate solvents during extractions and chemical manipulation.

A much more general source of contamination that has not received the consideration it merits comes from the use of plastics for tubing and containers. Plasticisers can readily be extracted by organic solvents from PVC and other plastics, so that most solvents, irrespective of their grade (including spectrograde and ultrapure) have been reported to contain 0.1 to 5 ppm of plasticizer [de Zeeuw, Jonkman and van Mansvelt *Anal Biochem* **67** 339 *1975*]. Where large quantities of solvent are used for extraction (particularly of small amounts of compounds), followed by evaporation, this can introduce significant amounts of impurity, even exceeding the weight of the genuine extract and giving rise to spurious peaks in gas chromatography (for example of fatty acid methyl esters, Pascaud, *Anal Biochem* **18** 570 *1967*). Likely contaminants are di(2-ethylhexyl)phthalate and dibutyl phthalate, but upwards of 20 different phthalic esters are listed as plasticisers as well as adipates, azelates, phosphates, epoxides, polyesters, trimellitates, and various heterocyclic compounds. These plasticisers would enter the solvent during passage through plastic tubing or containers or from plastic coatings used in cap liners for bottles. Such contamination could arise at any point in the manufacture or distribution of a solvent. The trouble with cap liners is avoidable by using corks wrapped in aluminium foil, although even in this case care should be taken because aluminium foil can dissolve in some liquids e.g. benzylamine and propionic acid.

Solutions in contact with polyvinyl chloride can become contaminated with trace amounts of lead, titanium, tin, zinc, iron, magnesium or cadmium from additives used in the manufacture and moulding of PVC.

N-Phenyl-2-naphthylamine is a contaminant of solvents and biological materials that have been in contact with black rubber or neoprene (in which it is used as an antioxidant). Although it was only an artefact of the separation procedure it has been isolated as an apparent component of vitamin K preparations, extracts of plant lipids, algae, livers, butter, eye tissue and kidney tissue [Brown *Chem Brit* **3** 524 *1967*].

Most of the above impurities can be removed by prior distillation of the solvent, but care should be taken to avoid plastic or black rubber as much as possible.

On Cleaning Apparatus
Laboratory glassware and Teflon equipment can be cleaned satisfactorily for most purposes by treating initially with a solution of sodium dichromate in concentrated sulphuric acid, draining, and rinsing copiously with distilled water. Where traces of chromium (adsorbed on the glass) must be avoided, a 1:1 mixture of concentrated sulphuric and nitric acid is a useful alternative. (*Used in a fumehood to remove vapour and with adequate face protection.*) Acid washing is also suitable for polyethylene ware but prolonged contact (some weeks) leads to severe deterioration of the plastic. For much glassware, washing with hot detergent solution, using tap water, followed by rinsing with distilled water and acetone, and heating to 200-300° overnight, is adequate. (Volumetric apparatus should not be heated: after washing it is rinsed with acetone, then hexane, and air-dried. Prior to use, equipment can be rinsed with acetone, then with petroleum ether or hexane, to remove the last traces of contaminants.) Teflon equipment should be soaked, first in acetone, then in petroleum ether or hexane for ten minutes prior to use.

For trace metal analyses, prolonged soaking of equipment in 1M nitric acid may be needed to remove adsorbed metal ions.

Soxhlet thimbles and filter papers contain traces of lipid-like material; for manipulations with highly pure materials, as in trace-pesticide analysis, they should be extracted with hexane before use.

Trace impurities in silica gel for TLC can be removed by heating at 300° for 16hr or by Soxhlet extraction for 3hr with redistilled chloroform, followed by 4hr extraction with redistilled hexane.

DISTILLATION

One of the most widely applicable and most commonly used methods of purification (especially of organic chemicals) is fractional distillation at atmospheric, or some lower, pressure. Almost without exception, this method can be assumed to be suitable for all organic liquids and most of the low-melting organic solids. For this reason it has been possible in Chapter 3 to omit many procedures for purification of organic chemicals when only a simple fractional distillation is involved - the suitability of such a procedure is implied from the boiling point.

The boiling point of a liquid varies with the atmospheric pressure to which it is exposed. A liquid boils when its vapour pressure is the same as the external pressure on its surface, its normal boiling point being the temperature at which its vapour pressure is equal to that of a standard atmosphere (760mm

Hg). Lowering the external pressure lowers the boiling point. For most substances, boiling point and vapour pressure are related by an equation of the form,

$$\log p = A + B/(t + 273),$$

where p is the pressure, t is in °C, and A and B are constants. Hence, if the boiling points at two different pressures are known the boiling point at another pressure can be calculated from a simple plot of $\log p$ *versus* $1/(t + 273)$. For organic molecules that are not strongly associated, this equation can be written in the form,

$$\log p = 8.586 - 5.703\,(T + 273)/(t + 273)$$

where T is the boiling point in °C at 760mm Hg. Table 1 gives computed boiling points over a range of pressures. Some examples illustrate its application. Ethyl acetoacetate, **b** 180° (with decomposition) at 760mm Hg has a predicted **b** of 79° at 8mm; the experimental value is 78°. Similarly 2,4-diaminotoluene, **b** 292° at 760mm, has a predicted **b** of 147° at 8mm; the experimental value is 148-150°. For self-associated molecules the predicted **b** are lower than the experimental values. Thus, glycerol, **b** 290° at 760mm, has a predicted **b** of 168° at 8mm: the experimental value is 182°.

For pressures near 760mm, the change in boiling point is given approximately by [Crafts *Ber* **20** 709 *1887*],

$$\delta t = a(760 - p)(t + 273)$$

where $a = 0.00012$ for most substances, but $a = 0.00010$ for water, alcohols, carboxylic acids and other associated liquids, and $a = 0.00014$ for very low-boiling substances such as nitrogen or ammonia.

When all the impurities are non-volatile, simple distillation is an adequate purification. The observed boiling point remains almost constant and approximately equal to that of the pure material. Usually, however, some of the impurities are appreciably volatile, so that the boiling point progressively rises during the distillation because of the progressive enrichment of the higher-boiling components in the distillation flask. In such cases, separation is effected by fractional distillation using an efficient column.

The principle involved in fractional distillation can be seen by considering a system which approximately obeys *Raoult's law*. (This law states that the vapour pressure of a solution at any given temperature is the sum of the vapour pressures of each substance multiplied by its mole fraction in the solution.) If two substances, A and B, having vapour pressures of 600mm Hg and 360mm Hg, respectively, were mixed in a mole ratio of 2:1, the mixture would have (ideally) a vapour pressure of 520mm Hg and the vapour phase would contain 77% of A and 23% of B. If this phase was now condensed, the new liquid phase would, therefore, be richer in the volatile component A. Similarly, the vapour in equilibrium with this phase is still further enriched in A. Each such liquid-vapour equilibrium constitutes a "theoretical plate". The efficiency of a fractionating column is commonly expressed as the number of such plates to which it corresponds in operation. Alternatively, this information may be given in the form of the height equivalent to a theoretical plate, or HETP.

In most cases, systems deviate to a greater or less extent from Raoult's law, and vapour pressures may be greater or less than those calculated from it. In extreme cases, vapour pressure-composition curves pass through maxima or minima, so that attempts at fractional distillation lead finally to the separation of a constant-boiling (azeotropic) mixture and one (but not both) of the pure species if either of the latter is present in excess.

Technique

Distillation apparatus consists basically of a distillation flask, usually fitted with a vertical fractionating column (which may be empty or packed with suitable materials such as glass helices or stainless-steel wool) to which is attached a condenser leading to a receiving flask. The bulb of a thermometer projects into the vapour phase just below the region where the condenser joins the column. The distilling flask is heated so that its contents are steadily vaporised by boiling. The vapour passes up into the column where, initially, it condenses and runs back into the flask. The resulting heat transfer gradually warms the column so that there is a progressive movement of the vapour phase-liquid boundary up the column, with increasing enrichment of the more volatile component. Because of this fractionation, the vapour finally passing into the condenser (where it condenses and flows into the receiver) is commonly that of the lowest-boiling components in the system. The conditions apply until all of the low-boiling material has been distilled, whereupon distillation ceases until the column temperature is high enough to permit the next component to distil. This usually results in a temporary fall in the temperature indicated by the thermometer.

The efficiency of a distillation apparatus used for purification of liquids depends on the difference in boiling points of the pure material and its impurities. For example, if two components of an ideal mixture have vapour pressures in the ratio 2:1, it would be necessary to have a still with an efficiency of at least seven plates (giving an enrichment of $2^7 = 128$) if the concentration of the higher-boiling component in the distillate was to be reduced to less than 1% of its initial value. For a vapour pressure ratio of 5:1, three plates would achieve as much separation.

In a fractional distillation, it is usual to reject the initial and final fractions, which are likely to be richer in the lower-boiling and higher-boiling impurities. The centre fraction can be further purified by repeated fractional distillation.

To achieve maximum separation by fractional distillation:

1. The column must be flooded initially to wet the packing. For this reason it is customary to operate a still at reflux for some time before beginning the distillation.

2. The reflux ratio should be high (i.e the ratio of drops of liquid which return to the distilling flask and the drops which distil over), so that the distillation proceeds slowly and with minimum disturbance of the equilibria in the column.

3. The hold-up of the column should not exceed one-tenth of the volume of any one component to be separated.

4. Heat loss from the column should be prevented but, if the column is heated to offset this, its temperature must not exceed that of the distillate in the column.

5. Heat input to the still-pot should remain constant.

6. For distillation under reduced pressure there must be careful control of the pressure to avoid flooding or cessation of reflux.

Distillation at Atmospheric Pressure

The distilling flask. To minimise superheating of the liquid (due to the absence of minute air bubbles or other suitable nuclei for forming bubbles of vapour), and to prevent bumping, one or more of the following precautions should be taken:

(a) The flask is heated uniformly over a large part of its surface, either by using an electrical heating mantle or, much better, by partial immersion in a bath somewhat above the boiling point of the liquid to be distilled.

(b) Before heating begins, small pieces of unglazed fireclay or porcelain (porous pot, boiling chips), pumice, carborundum, Teflon, diatomaceous earth, or platinum wire are added to the flask. These act as sources of air bubbles.

(c) The flask may contain glass siphons or boiling tubes. The former are inverted J-shaped tubes, the end of the shorter arm being just above the surface of the liquid. The latter comprise long capillary tubes sealed above the lower end.

(d) A steady slow stream of gas is passed through the liquid.

(e) In some cases zinc dust can also be used. It reacts chemically with acidic or strongly alkaline solutions to liberate fine bubbles of hydrogen.

(f) The liquid in the flask is stirred mechanically. This is especially necessary when suspended insoluble material is present.

For simple distillations a Claisen flask (see, for example, Quickfit and Quartz Ltd cataloque of interchangeable laboratory glassware, or Kontes Glass Co, Vineland, New Jersey, cat.no TG-15) is often used. This flask is, essentially, a round-bottomed flask to the neck of which is joined another neck carrying a side arm. This second neck is sometimes extended so as to form a Vigreux column.

For heating baths, see Table 2 (p 37). For distillation on a semi-micro scale, see Linstead, Elvidge and Whalley, 1955 (p 53).

Types of columns and packings. A slow distillation rate is necessary to ensure that equilibrium conditions operate and also that the vapour does not become superheated so that the temperature rises above the boiling point. Efficiency is improved if the column is heat insulated (either by vacuum jacketing or by lagging) and, if necessary, heated to just below the boiling point of the most volatile component. Electrical heating tape is convenient for this purpose.) Efficiency of separation also improves with increase in the heat of vaporisation of the liquids concerned (because fractionation depends on heat equilibration at multiple liquid-gas boundaries). Water and alcohols are more easily purified by distillation for this reason.

Columns used in distillation vary in their shapes and types of packing. Packed columns are intended to give efficient separation by maintaining a large surface of contact between liquid and vapour. Efficiency of separation is further increased by operation under conditions approaching total reflux, i.e. under a high reflux ratio (see p 6). Better control of reflux ratio is achieved by fitting a total condensation, variable take-off still-head (see, for example, catalogues by Quickfit and Quartz, or Kontes) to the top of the fractionating colum. However, great care must be taken to avoid flooding of the column during distillation. The minimum number of theoretical plates for satisfactory separation of two liquids differing in boiling point by δt is approximately $(273 + t)/3\delta t$, where t is the average boiling point in $^\circ$C.

Some of the commonly used columns are:

Bruun column. A type of all-glass bubble-cap column.

Bubble-cap column. A type of plate column in which inverted cups (bubble caps) deflect ascending vapour through reflux liquid lying on each plate. Excess liquid from any plate overflows to the plate lying below it and ultimately returns to the flask. (For further details, see Bruun and Faulconer *Ind Eng Chem (Anal Ed)* 9 247 *1937*). Like most plate columns, it has a high through-put, but a relatively low number of theoretical plates for a given height.

Dufton column. A plain tube, into which fits closely (preferably ground to fit) a solid glass spiral wound round a central rod. It tends to choke at temperatures above 100° unless it is lagged (Dufton *J Soc Chem Ind* (London) 38 45T *1919*).

Hempel column. A plain tube (fitted near the top with a side arm) which is almost filled with a suitable packing, which may be of rings or helices.

Oldershaw column. An all-glass perforated-plate column. The plates are sealed into a tube, each plate being equipped with a baffle to direct the flow of reflux liquid, and a raised outlet which maintains a definite liquid level on the plate and also serves as a drain on to the next lower plate [see Oldershaw *Ind Eng Chem (Anal Ed)* 11 265 *1941*].

Podbielniak column. A plain tube containing "Heli-Grid" Nichrome or Inconel wire packing. This packing provides a number of passage-ways for the reflux liquid, while the capillary spaces ensure very even spreading of the liquid, so that there is a very large area of contact between liquid and vapour while, at the same time, channelling and flooding are minimised. A column 1m high has been stated to have an efficiency of 200-400 theoretical plates (for further details, see Podbielniak *Ind Eng Chem (Anal Ed)* **13** 639 *1941*; Mitchell and O'Gorman *Anal Chem* **20** 315 *1948*).

Stedman column. A plain tube containing a series of wire-gauze discs stamped into flat, truncated cones and welded together, alternatively base-to-base and edge-to-edge, with a flat disc across each base. Each cone has a hole, alternately arranged, near its base, vapour and liquid being brought into intimate contact on the gauze surfaces (Stedman *Canad J Res B* **15** 383 *1937*).

Todd column. A column (which may be a Dufton type, fitted with a Monel metal rod and spiral, or a Hempel type, fitted with glass helices) which is surrounded by an open heating jacket so that the temperature can be adjusted to be close to the distillation temperature (Todd *Ind Eng Chem (Anal Ed)* **17** 175 *1945*).

Vigreux column. A glass tube in which have been made a number of pairs of indentations which almost touch each other and which slope slightly downwards. The pairs of indentations are arranged to form a spiral of glass inside the tube.

Widmer column. A Dufton column, modified by enclosing within two concentric tubes the portion containing the glass spiral. Vapour passes up the outer tube and down the inner tube before entering the centre portion. In this way flooding of the column, especially at high temperatures, is greatly reduced (Widmer *Helv Chim Acta* **7** 59 *1924*).

The packing of a column greatly increases the surface of liquid films in contact with the vapour phase, thereby increasing the efficiency of the column, but reducing its capacity (the quantities of vapour and liquid able to flow in opposite directions in a column without causing flooding). Material for packing should be of uniform size, symmetrical shape, and have a unit diameter less than one eighth that of the column. (Rectification efficiency increases sharply as the size of the packing is reduced but so, also, does the hold-up in the column.) It should also be capable of uniform, reproducible packing.
  The usual packings are:

  (a) **Rings.** These may be hollow glass or porcelain (Raschig rings), of stainless steel gauze (Dixon rings), or hollow rings with a central partition (Lessing rings) which may be of porcelain, aluminium, copper or nickel.

  (b) **Helices.** These may be of metal or glass (Fenske rings), the latter being used where resistance to chemical attack is important (e.g. in distilling acids, organic halides, some sulphur compounds, and phenols). Metal single-turn helices are available in aluminium, nickel or stainless steel. Glass helices are less efficient, because they cannot be tamped to ensure uniform packing.

  (c) **Balls.** These are usually glass.

  (d) **Wire packing.** For use of "Heli-Grid" and "Heli-Pak" packings see references given for Podbielniak column. For Stedman packing, see entry under Stedman column.

Condensers. Some of the more commonly used condensers are:

  Air condenser. A glass tube such as the inner part of a Liebig condenser. Used for liquids with boiling points above 90°. Can be of any length.

Allihn condenser. The inner tube of a Liebig condenser is modified by having a series of bulbs to increase the condensing surface. Further modifications of the bubble shapes give the Julian and Allihn-Kronbitter condensers.

Bailey-Walker condenser. A type of all-metal condenser fitting into the neck of extraction apparatus and being supported by the rim. Used for high-boiling liquids.

Coil condenser. An open tube, into which is sealed a glass coil or spiral through which water circulates. The tube is sometimes also surrounded by an outer cooling jacket.

Double surface condenser. A tube in which the vapour is condensed between an outer and inner water-cooled jacket after impinging on the latter. Very useful for liquids boiling below 40°.

Friedrichs condenser. A "cold-finger" type of condenser sealed into a glass jacket open at the bottom and near the top. The cold finger is formed into glass screw threads.

Graham condenser. A type of coil condenser.

Hopkins condenser. A cold-finger type of condenser resembling that of Friedrichs.

Liebig condenser. An inner glass tube surrounded by a glass jacket through which water is circulated.

Othmer condenser. A large-capacity condenser which has two coils of relatively large bore glass tubing inside it, through which the water flows. The two coils join at their top and bottom.

West condenser. A Liebig condenser with a light-walled inner tube and a heavy-walled outer tube, with only a narrow space between them.

Wiley condenser. A condenser resembling the Bailey-Walker type.

VACUUM DISTILLATION

This expression is commonly used to denote a distillation under reduced pressure lower than that of the normal atmosphere. Because the boiling point of a substance depends on the pressure, it is often possible by sufficiently lowering the pressure to distil materials at a temperature low enough to avoid partial or complete decomposition, even if they are unstable when boiled at atmospheric pressure.
Sensitive or high-boiling liquids should invariably be distilled or fractionally distilled under reduced pressure. The apparatus is essentially as described for distillation except that ground joints connecting the different parts of the apparatus should be greased with the appropriate vacuum grease. For low, moderately high, and very high temperatures Apiezon L, M and T, respectively, are very satisfactory. Alternatively, it is often preferable to avoid grease and to use thin Teflon sleeves in the joints. The distilling flask, must be supplied with a capillary bleed (which allows a fine stream of air, nitrogen or argon into the flask), and the receiver should be of the fraction collector type (e.g. a Perkin triangle, see Quickfit and Quartz Ltd interchangeable glassware catalogue, or Kontes Glass Co, Vineland, New Jersey, cat. no. TG-15). When distilling under vacuum it is very important to place a loose packing of glass wool above the liquid to buffer sudden boiling of the liquid. The flask should be not more than two-thirds full of liquid. The vacuum must have attained a steady state before the heat source is applied, and the temperature of the heat source must be raised *very slowly* until boiling is achieved.
If the pump is a filter pump off a high-pressure water supply, its performance will be limited by the temperature of the water because the vapour pressure of water at 10°, 15°, 20° and 25° is 9.2, 12.8, 17.5 and 23.8mm Hg respectively. The pressure can be measured with an ordinary manometer. For vacuums in the range 10^{-2}mm Hg (10μ) to 10mm Hg, rotary mechanical pumps (oil pumps) are used and the pressure can be measured with a Vacustat McLeod type gauge. If still higher vacuums are required, for example for high vacuum sublimations, a mercury diffusion pump is suitable. In principle,

this pump resembles an ordinary water pump. It has a single, double or triple jet through which the mercury vapour and condensate pass. Such a pump can provide a vacuum up to 10^{-6} mm Hg. Two pumps can be used in series. For better efficiency these pumps are backed by a mechanical pump. The pressure is measured with a Pirani gauge. Where there is fear of contamination with mercury vapour, the mercury in the pumps can be replaced with vacuum oils, e.g. Apiezon type G or Silicone fluid (Dow Corning no. 702 or 703), which produce a vacuum range of 10^{-4} to 10^{-7} mm Hg depending on pump design and system used. These fluids are resistant to oxidation, are non-corrosive and are non-toxic. The gauge should be as close to the distillation apparatus as possible in order to obtain the distillation pressure as accurately as possible, thus minimising the pressure drop between the gauge and the apparatus.

In all cases, the pump is connected to the still through several traps to remove vapours. These traps may operate by chemical action, for example the use of sodium hydroxide pellets to react with acids, or by condensation, in which case empty tubes cooled in solid carbon dioxide-ethanol or liquid nitrogen (contained in wide-mouthed Dewar flasks) are used.

Special oil or mercury traps are available commercially and a liquid-nitrogen trap is the most satisfactory one to use between these and the apparatus. It has an advantage over liquid air or oxygen in that it is non-explosive if it becomes contaminated with organic matter. Air should not be sucked through the apparatus before starting a distillation or sublimation because this will cause liquid air to condense in the liquid nitrogen trap and a good vacuum cannot be readily achieved. Hence, it is advisable to degas the system for a short period before the trap is immersed into the liquid nitrogen (which is kept in a Dewar flask).

Vacuum-lines, Schlenk and glovebox techniques. Manipulations involving materials sensitive to air or water vapour can be carried out by these procedures. Vacuum-line methods make use of quantitative transfers, and P(pressure)-V(volume)-T(temperature) measurements, of gases, and trap-to-trap separations of volatile substances.

It is usually more convenient to work under an inert-gas atmosphere, using **Schlenk** type apparatus. The *principle* of Schlenk methods is the bottle which has a standard ground-glass joint and a sidearm with a tap. The system can be purged by pumping down and flushing with an inert gas (usually nitrogen, or in some cases, argon), repeating the process until the contaminants in the vapour phases have been diminished to acceptable limits. If the bottom of the bottle has sealed into it a tap and a cone, a dropping bottle is produced, while further addition of a sinter disk in the bottle converts it to a filter funnel.

With these, and related pieces of glassware, inert atmospheres can be maintained during crystallisation, filtration, sublimation and transfer. Schlenk-type glassware is commercially available (as *Airless Ware*) from Kontes Glass Co, Vineland, NJ, USA.

Syringe techniques have been worked out for small volumes, while for large volumes or where much manipulation is required, use can be made of dryboxes (*glove boxes*).

For fuller discussion, see Sanderson **Vacuum Manipulation of Volatile Compounds** John Wiley and Sons Ltd, New York 1948; Shriver **The Manipulation of Air-sensitive Compounds** McGraw-Hill Book Co, New York 1969; Brown **Organic Syntheses via Boranes,** Wiley, New York 1975.

Spinning-band columns. Factors which limit the performance of distillation columns include the tendency to flood (which occurs when the returning liquid blocks the pathway taken by the vapour through the column) and the increased hold-up (which decreases the attainable efficiency) in the column that should, theoretically, be highly efficient. To overcome these difficulties, especially for distillation under high vacuum of heat sensitive or high-boiling highly viscous fliuds, spinning band columns have become commercially available.

In such units, the distillation columns contain a rapidly rotating, motor-driven, spiral band, which may be of polymer-coated metal, stainless steel or platinum. The rapid rotation of the band in contact with the walls of the still gives intimate mixing of descending liquid and ascending vapour while the screw-like motion of the band drives the liquid towards the still-pot, helping to reduce hold-up. There is very little pressure drop in such a system, and high throughputs are possible, at high efficiency. For example, a 30-in 10-mm diameter commercial column is reported to have an efficiency of 28 plates and a pressure drop of 0.2mm Hg for a throughput of 330ml/hr. The columns may be either vacuum jacketed or heated externally. The stills can be operated down to 10^{-5} mm Hg. The principle, which was first used commercially in the Podbielniak Centrifugal Superfractionator, has also been embodied in descending-film molecular distillation apparatus.

STEAM DISTILLATION

When two immmiscible liquids distil, the sum of their (independent) partial pressures is equal to the atmospheric pressure. Hence in steam distillation, the distillate has the composition

$$\frac{\text{Moles of substance}}{\text{Moles of water}} = \frac{P_{subst}}{P_{water}} = \frac{760 - P_{water}}{P_{water}}$$

where the P's are vapour pressures in mm Hg) in the boiling mixture. One of the advantages of using water in this way lies in its low molecular weight.

The customary technique consists of heating the substance and water in a flask (to boiling), usually with the passage of steam, followed by condensation and separation of the aqueous and non-aqueous phases. Its advantages are those of selectivity (because only some water-insoluble substances, such as naphthalene, nitrobenzene, phenol and aniline are volatile in steam) and of ability to distil certain high-boiling substances well below their boiling point. It also facilitates the recovery of a non-steam-volatile solid at a relatively low temperature from a high-boiling solvent such as nitrobenzene. The efficiency of steam distillation is increased if superheated steam is used (because the vapour pressure of the organic component is increased relative to water). In this case the flask containing the material is heated (without water) in an oil bath and the steam passing through it is superheated by prior passage through a suitable heating device (such as a copper coil over a bunsen burner or an oil bath). (For further detail, see Krell 1963, p 53).

AZEOTROPIC DISTILLATION

In some cases two or more liquids form constant-boiling mixtures, or azeotropes. Azeotropic mixtures are most likely to be found with components which readily form hydrogen bonds or are otherwise highly associated, especially when the components are dissimilar, for example an alcohol and an aromatic hydrocarbon, but have similar boiling points. (Many systems are summarised in *Azeotropic Data - III*, L.H.Horsley, *Advances in Chemistry Series 116*, American Chemical Society, Washington, 1973.)

Examples where the boiling point of the distillate is a minimum (less than either pure component) include:

Water with ethanol, *n*-propanol and isopropanol, *tert*-butanol, propionic acid, butyric acid, pyridine,
methanol with methyl iodide, methyl acetate, chloroform,
ethanol with ethyl iodide, ethyl acetate, chloroform, benzene, toluene, methyl ethyl ketone,
benzene with cyclohexane,
acetic acid with toluene.

Although less common, azeotropic mixtures are known which have higher boiling points than their components. These include water with most of the mineral acids (hydrofluoric, hydrochloric, hydrobromic, perchloric, nitric and sulphuric) and formic acid. Other examples are acetic acid-pyridine, acetone-chloroform, aniline-phenol, and chloroform-methyl acetate.

The following azeotropes are important commercially for drying ethanol:

ethanol 95.5% (by weight) - water 4.5%	b 78.1°
ethanol 32.4% - benzene 67.6%	b 68.2°
ethanol 18.5% - benzene 74.1% - water 7.4%	b 64.9°

Materials are sometimes added so as to form an azeotropic mixture with the substance to be purified. Because the azeotrope boils at a different temperature, this facilitates separation from substances distilling in the same range as the pure material. (Conversely, the impurity might form the azeotrope and be removed in this way.) This method is often convenient, especially where the impurities are isomers or are otherwise closely related to the desired substance. Formation of low-boiling azeotropes also facilitates distillation.

One or more of the following methods can generally be used for separating the components of an azeotropic mixture:
1. By using a chemical method to remove most of one species prior to distillation. (For example, water can be removed by suitable drying agents; aromatic and unsaturated hydrocarbons can be removed by sulphonation.)
2. By redistillation with an additional substance which can form a ternary azeotropic mixture (as in ethanol-water-benzene example given above).
3. By selective adsorption of one of the components. (For example , of water on to a silica gel or molecular sieve, or of unsaturated hydrocarbons on to alumina.)
4. By fractional crystallisation of the mixture, either by direct freezing or after solution in a suitable solvent.

ISOPIESTIC OR ISOTHERMAL DISTILLATION

This technique can be useful for the preparation of metal-free solutions of volatile acids and bases for use in trace metal studies. The procedure involves placing two beakers, one of distilled water and the other of a solution of the material to be purified, in a desiccator. The desiccator is sealed and left to stand at room temperature for several days. The volatile components distribute themselves between the two beakers whereas the non-volatile contaminants remain in the original beaker. This technique has afforded metal-free pure solutions of ammonia, hydrochloric acid and hydrogen fluoride.

SUBLIMATION

Sublimation differs from ordinary distillation because the vapour condenses to a solid instead of a liquid. Usually, the pressure in the heated system is diminished by pumping, and the vapour is condensed (after travelling a relatively short distance) on to a cold finger or some other cooled surface. This technique, which is applicable to many organic solids, can also be used with inorganic solids such as aluminium chloride, ammonium chloride, arsenious oxide and iodine. In some cases, passage of a stream of inert gas over the heated substance secures adequate vaporisation.

RECRYSTALLISATION

Technique
The most commonly used procedure for the purification of a solid material by recrystallisation from a solution involves the following steps:

(a) The impure material is dissolved in a suitable solvent, by shaking or vigorous stirring, at or near the boiling point, to form a near-saturated solution.
(b) The hot solution is filtered to remove any insoluble particles. To prevent crystallisation during this filtration, a heated (jacketted) filter funnel can be used or the solution can be somewhat diluted with more of the solvent.
(c) It is then allowed to cool so that the dissolved substance crystallises out.
(d) The crystals are separated from the mother liquor, either by centrifuging or by filtering, under suction, through a sintered glass, a Hirsch or a Büchner, funnel. Usually, centrifuging is much preferred because of the much greater ease and efficiency of separating crystals and mother liquor, and also because of the saving of time and effort, particularly when very small crystals are formed or when there is entrainment of solvent.
(e) They are washed free from mother liquor with a little fresh cold solvent, then dried.

If the solution contains extraneous coloured material likely to contaminate the crystals, this can often be removed by adding some activated charcoal (decolorising carbon) to the hot, but not boiling, solution which is then shaken frequently for several minutes before being filtered. (The large active surface of the carbon makes it a good adsorbent for this purpose.) In general, the cooling and crystallisation step should be rapid so as to give small crystals which occlude less of the mother liquor. This is usually satisfactory with inorganic material, so that commonly the filtrate is cooled in an ice-water bath while being vigorously stirred. In many cases, however, organic molecules crystallise much more slowly, so that the filtrate must be set aside to cool to room temperature or left in the refrigerator. It is often desirable to subject material that is very impure to preliminary purification, such as steam distillation, Soxhlet extraction, or sublimation, before recrystallising it. A greater degree of purity is also to be expected if the crystallisation process is repeated several times, especially if different solvents are used. The advantage of several crystallisations from different solvents lies in the fact that the material sought, and its impurities, are unlikely to have similar solubilities as solvents and temperatures are varied.

For the final separation of solid material, sintered-glass discs are preferable to filter paper. Sintered glass is unaffected by strongly acid solutions or by oxidising agents. Also, with filter paper, cellulose fibres are likely to become included in the sample, The sintered-glass discs or funnels can be readily cleaned by washing in freshly prepared *chromic acid cleaning mixture*. This mixture is made by adding 100 ml of concentrated sulphuric acid slowly with stirring to a solution of 5 g of sodium dichromate in 5 ml of water. (The mixture warms to about 70°.)

For materials with melting points below 70° it is sometimes convenient to use dilute solutions in acetone, methanol, pentane, ethyl ether or $CHCl_3$-CCl_4. The solutions are cooled to -78° in Dry-ice, to give a filtrable slurry which is filtered off through a precooled Büchner funnel. Experimental details, as applied to the purification of nitromethane, are given by Parrett and Sun *(J Chem Educ* **54** 448 *1977)*.

Where substances vary little in solubility with temperature, isothermal crystallisation may sometimes be employed. This usually takes the form of a partial evaporation of a saturated solution at room temperature by leaving it under reduced pressure in a desiccator.

However, in rare cases, crystallisation is not a satisfactory method of purification, especially if the impurity forms crystals that are isomorphous with the material being purified. In fact, the impurity content may even be greater in such recrystallised material. For this reason, it still remains necessary to test for impurities and to remove or adequately lower their concentrations by suitable chemical manipulation prior to recrystallisation.

Filtration

Filtration removes particulate impurities rapidly from liquids and is also used to collect insoluble or crystalline solids which separate or crystallise from solution. The usual technique is to pass the solution, cold or hot, through a fluted filter paper in a conical glass funnel (see Vogel's *Textbook of Practical Organic Chemistry*, p 53).

If a solution is hot and needs to be filtered rapidly a Büchner funnel and flask are used and filtration is performed under a slight vacuum (water pump), the filter medium being a circular cellulose filter paper wet with solvent. If filtration is slow, even under high vacuum, a pile of about twenty filter papers, wet as before, are placed in the Büchner funnel and, as the flow of solution slows down, the upper layers of the filter paper are progressively removed. Alternatively, a filter aid, e.g. Celite, Florisil or Hyflo-supercel, is placed on top of a filter paper in the funnel. When the flow of the solution (under suction) slows down the upper surface of the filter aid is scratched gently. Filter papers with various pore sizes are available covering a range of filtration rates. Hardened filter papers are slow filtering but they can withstand acidic and alkaline solutions without appreciable hydrolysis of the cellulose (see Table 3). When using strong acids it is preferable to use glass micro fibre filters which are commercially available (see Table 3).

Freeing a solution from extremely small particles (e.g. for ORD or CD measurements) requires filters with very small pore size. Commercially available (Millipore, Gelman, Nucleopore) filters other than cellulose or glass include nylon, Teflon, and polyvinyl chloride, and the pore diameter may be as small as 0.01micron (see Table 4). Special containers are used to hold the filters, through which the solution is pressed by applying pressure, e.g. from a syringe. Some of these filters can be used to clear strong sulphuric acid solutions .

As an alternative to the Büchner funnel for collecting crystalline solids, a funnel with a sintered glass-plate under suction may be used. Sintered-glass funnels with various porosities are commercially available and can easily cleaned with warm chromic or nitric acid.

When the solid particles are too fine to be collected on a filter funnel because filtration is extremely slow, separation by **centrifugation** should be used. Bench type centrifuges are most convenient for this purpose. The solid is placed in the centrifuge tube, the tubes containing the solutions on opposite sides of the rotor should be balanced accurately (at least within 0.05 to 0.1g), and the solutions are spun at maximum speed for as long as it takes to settle the solid (usually *ca* 3-5 minutes). The solid is washed with cold solvent by centrifugation, and finally twice with a pure volatile solvent in which the solid is insoluble, also by centrifugation. After decanting the supernatant the residue is dried in a vacuum, at elevated temperatures if necessary. In order to avoid "spitting" and contamination with dust while the solid in the centrifuge tube is dried, the mouth of the tube is covered with silver paper and held fast with a tight rubber band near the lip. The flat surface of the silver paper is then perforated in several places with a pin.

Choice of Solvents

The best solvents for recrystallisation have these properties:

(a) The material is much more soluble at higher temperatures than it is at room temperature or below.
(b) Well-formed (but not large) crystals are produced.
(c) Impurities are either very soluble or only sparingly soluble.
(d) The solvent must be readily removed from the purified material.
(e) There must be no reaction between the solvent and the substance being purified.
(f) The solvent must not be inconveniently volatile or too highly flammable. (These are reasons why ethyl ether and carbon disulphide are not commonly used in this way.)

The following generalisations provide a rough guide to the selection of a suitable solvent:

(a) Substances usually dissolve best in solvents to which they are most closely related in chemical and physical characteristics. Thus, hydroxylic compounds are likely to be most soluble in water, methanol, ethanol, acetic acid or acetone. Similarly, petroleum ether might be used with water-insoluble substances. However, if the resemblance is too close, solubilities may become excessive.
(b) Higher members of homologous series approximate more and more closely to their parent hydrocarbon.
(c) Polar substances are more soluble in polar, than in non-polar, solvents.

Although Chapters 3 and 4 provide details of the solvents used for recrystallising a large portion of commercially available laboratory chemicals, they cannot hope to be exhaustive, nor need they necessarily be the best choice. In other cases where it is desirable to use this process, it is necessary to establish whether a given solvent is suitable. This is usually done by taking only a small amount of material in a small test-tube and adding enough solvent to cover it. If it dissolves readily in the cold or on gentle warming, the solvent is unsuitable. Conversely, if it remains insoluble when the solvent is heated to boiling (adding more solvent if necessary), the solvent is again unsuitable. If the material dissolves in the hot solvent but does not crystallise readily within several minutes of cooling in an ice-salt mixture, another solvent should be tried.

Solvents commonly used for recrystallisation, and their boiling points, are given in Table 5.

Mixed Solvents

Where a substance is too soluble in one solvent and too insoluble in another, for either to be used for recrystallisation, it is often possible (provided they are miscible) to use them as a mixed solvent. (In general, however, it is preferable to use a single solvent if this is practicable.) Table 6 comprises many of the common pairs of miscible solvents.

The technique of recrystallisation from a mixed solvent is as follows:

The material is dissolved in the solvent in which it is the more soluble, then the other solvent (heated to near boiling) is added cautiously to the hot solution until a slight turbidity persists or crystallisation begins. This is cleared by adding several drops of the first solvent, and the solution is allowed to cool and crystallise in the usual way.

A variation of this procedure is simply to precipitate the material in a microcrystalline form from solution in one solvent at room temperature, by adding a little more of the second solvent, filtering this off, adding a little more of the second solvent and repeating the process. This ensures, at least in the first or last precipitation, a material which contains as little as possible of the impurities which may also be precipitated in this way. With salts the first solvent is commonly water, and the second solvent is alcohol or acetone.

Solidification from the Melt

A crystalline solid melts when its temperature is raised sufficiently for the thermal agitation of its molecules or ions to overcome the restraints imposed by the crystal lattice. Usually, impurities weaken crystal structures, and hence lower the melting points of solids (or the freezing points of liquids). If an impure material is melted and cooled slowly (with the addition, if necessary, of a trace of solid material near the freezing point to avoid supercooling), the first crystals that form will usually contain less of the impurity, so that fractional solidification by partial freezing can be used as a purification process for solids with melting points lying in a convenient temperature range (or for more readily frozen liquids). In some cases, impurities form higher melting eutectics with substances to be purified, so that the first material to solidify is less pure than the melt. For this reason, it is often desirable to discard the first crystals and also the final portions of the melt. Substances having similar boiling points often differ much more in melting points, so that fractional solidification can offer real advantages, especially where ultrapurity is sought.

The technique of recrystallisation from the melt as a means of purification dates back from its use by Schwab and Wichers (*J Res Nat Bur Stand* **25** 747 *1940*) to purify benzoic acid. It works best if material is already nearly pure, and hence tends to be a final purification step. A simple apparatus for purifying organic compounds by progressive freezing is described by Matthias and Coggeshall (*Anal Chem* **31** 1124 *1959*). In principle, the molten substance is cooled slowly by progressive lowering of the tube containing it into a suitable bath. For temperatures between 0° and 100°, waterbaths are convenient. Where lower temperatures are required, the cooling baths given in Table 7 can be used. Cooling is stopped when part of the melt has solidified, and the liquid phase is drained off. Column crystallisation has been used to purify stearyl alcohol, cetyl alcohol, myristic acid; fluorene, phenanthrene, biphenyl, terphenyls, dibenzyl; phenol, 2-naphthol; benzophenone and 2,4-dinitrotoluene; and many other organic (and inorganic) compounds. [See, for example, *Developments in Separation Science* N.N.Lee (ed), CRC Press, Cleveland, Ohio, 1972].

Thus, an increase in purity from 99.80 to 99.98 mole% was obtained when acetamide was slowly crystallised in an insulated round bottom flask until half the material had solidified and the solid phase was then recrystallised from benzene [Schwab and Wichers *J Res Nat Bur Stand* **32** 253 *1944*].

Fractional solidification and its applications to obtaining ultrapure chemical substances, has been treated in detail in *Fractional Solidification* by M.Zief and W.R.Wilcox eds, Edward Arnold Inc, London 1967, and *Purification of Inorganic and Organic Materials* by M.Zief, Marcel Dekker Inc, New York 1969. These monographs should be consulted for discussion of the basic principles of solid-liquid processes such as zone melting, progressive freezing and column crystallisation, laboratory apparatus and industrial scale equipment, and examples of applications. These include the removal of cyclohexane from benzene, and the purification of aromatic amines, dienes and naphthalene, and inorganic species such as the alkali iodides, potassium chloride, indium antimonide and gallium trichloride. The authors also discuss analytical methods for assessing the purity of the final material.

Zone Refining

Zone refining (or zone melting) is a particular development for fractional solidification and is applicable to all crystalline substances that show differences in soluble impurity concentration in liquid and solid states at solidification. The apparatus used in this technique consists, essentially, of a device by which a narrow molten zone moves slowly down a long tube filled with the material to be purified. The machine can be set to recycle repeatedly. At its advancing side, the zone has a melting interface with the impure material whereas on the upper surface of the zone there is a constantly growing face of higher-melting, resolidified material. This leads to a progressive increase in impurity in the liquid phase which, at the end of the run, is discarded. Also, because of the progressive increase in impurity in the liquid phase, the resolidified material becomes correspondingly less further purified. For this reason, it is usually necessary to make several zone-melting runs before a sample is satisfactorily purified. This is also why the method works most successfully if the material is already fairly pure. In all these operations the zone must travel slowly enough to enable impurities to diffuse or be convected away from the area where resolidification is occurring.

The technique finds commercial application in the production of metals of extremely high purity (impurities down to 10^{-9} ppm), in purifying refractory oxides, and in purifying organic compounds, using commercially available equipment. Criteria for indicating that definite purification is achieved include elevation of melting point, removal of colour, fluorescence or smell, and a lowering of electrical conductivity.

Difficulties likely to be met with in organic compounds, especially those of low melting points and low rates of crystallisation, are supercooling and, because of surface tension and contraction, the tendency of the molten zone to seep back into the recrystallised areas. The method is likely to be useful in cases where fractional distillation is not practicable, either because of unfavourable vapour pressures or ease of decomposition, or where super-pure materials are required. It has been used for the latter purpose with anthracene, benzoic acid, chrysene, morphine and pyrene.

For a description of an apparatus for purifying organic compounds by zone refining, see Herington, Handley and Cook, *Chem & Ind* 292 *1956*; and for a semi-micro version see Handley and Herington, *Chem & Ind* 304 *1956*.

DRYING

Removal of Solvents

Where substances are sufficiently stable, removal of solvent from recrystallised materials presents no problems. The crystals, after filtering at the pump (and perhaps air-drying by suction), are heated in an oven above the boiling point of the solvent (but below their melting point), followed by cooling in a desiccator. Where this treatment is inadvisable, it is still often possible to heat to a lower temperature under reduced pressure, for example in an Abderhalden pistol. This device consists of a small chamber which is heated externally by the vapour of a boiling solvent. Inside this chamber, which can be evacuated by a water pump or some other vacuum pump, is placed a small boat containing the sample to be dried and also a receptacle with a suitable drying agent. Convenient liquids for use as boiling liquids in an Abderhalden pistol, and their temperatures, are given in Table 9. In cases where heating above room temperature cannot be used, drying must be carried out in a vacuum desiccator containing suitable absorbants. For example, hydrocarbons, such as benzene, cyclohexane and petroleum ether, can be removed by using shredded paraffin wax, and acetic acid and other acids can be absorbed by pellets of sodium, or potassium, hydroxide. However, in general, solvent removal is less of a problem than ensuring that the water content of solids and liquids is reduced below an acceptable level.

Removal of Water

Methods for removing water from solids depends on the thermal stability of the solids or the time available. The safest way is to dry in a vacuum desiccator over concentrated sulphuric acid, phosphorus pentoxide, silica gel, calcium chloride, or some other desiccant. Where substances are stable in air and melt above 100° drying in an air oven may be adequate. In other cases, use of an Abderhalden pistol may be satisfactory.

Often, in drying inorganic salts, the final material that is required is a hydrate. In such cases, the purified substance is left in a desiccator to equilibrate above an aqueous solution having a suitable water-vapour pressure. A convenient range of solutions used in this way is given in Table 10.

The choice of desiccants for drying liquids is more restricted because of the need to avoid all substances likely to react with the liquids themselves. In some cases, direct distillation of an organic liquid is a suitable method for drying both solids and liquids, especially if low-boiling azeotropes are formed. Examples include acetone, aniline, benzene, chloroform, carbon tetrachloride, ethylene dichloride, heptane, hexane, methanol, nitrobenzene, petroleum ether, toluene and xylene. Addition of benzene can be used for drying ethanol by distillation. In carrying out distillations intended to yield anhydrous products, the apparatus should be fitted with guard-tubes containing calcium chloride or silica gel to prevent entry of moist air into the system. (Many anhydrous organic liquids are appreciably hygroscopic.)

Traces of water can be removed from solvents such as benzene, 1,2-dimethoxyethane, ethyl ether, CH_2Cl_2, pentane, toluene and tetrahydrofuran by refluxing under nitrogen a solution containing **sodium benzophenone ketyl**, and fractionally distilling. Drying with, and distilling from CaH_2 is applicable to a number of solvents including aniline, benzene, tert-butylamine, tert-butanol, 2,4,6-collidine, diisopropylamine, dimethylformamide, hexamethylphosphoramide, methylenedichloride, pyridine, tetramethylethylenediamine, toluene, triethylamine.

Removal of water from gases may be by physical or chemical means, and is commonly by adsorption on to a drying agent in a low-temperature trap. The effectiveness of drying agents depends on the vapour pressure of the hydrated compound - the lower the vapour pressure the less the remaining moisture in the gas.

The most usually applicable of the specific methods for detecting and determining water in organic liquids is due to Karl Fischer. (See J.Mitchell and D.M.Smith, *Aquametry*, Interscience, New York, 1948.) Other techniques include electrical conductivity measurements and observation of the temperature at which the first cloudiness appears as the liquid is cooled (applicable to liquids in which water is only slightly soluble). Addition of anhydrous cobalt (II) iodide (blue) provides a convenient method (colour change to pink on hydration) for detecting water in alcohols, ketones, nitriles and some esters. Infrared absorption measurements of the broad band for water near 3500 cm^{-1} can also sometimes be used for detecting water in non-hydroxylic substances.

Intensity and Capacity of Common Desiccants
Drying agents can be conveniently be grouped into three classes, depending on whether they combine with water reversibly, they react chemically (irreversibly) with water, or they are molecular sieves. The first group vary in their drying intensity with the temperature at which they are used, depending on the vapour pressure of the hydrate that is formed. This is why, for example, drying agents such as anhydrous sodium sulphate, magnesium sulphate or calcium chloride should be filtered off from the liquids before the latter are heated. The intensities of drying agents belonging to this group fall in the sequence:

$P_2O_5 >>$ BaO $>$ Mg(ClO$_4$)$_2$, CaO, MgO, KOH (fused), conc H$_2$SO$_4$, CaSO$_4$, Al$_2$O$_3$ $>$ KOH (sticks), silica gel, Mg(ClO$_4$)$_2$.3H$_2$O $>$ NaOH (fused), 95% H$_2$SO$_4$, CaBr$_2$, CaCl$_2$ (fused) $>$ NaOH (sticks), Ba(ClO$_4$)$_2$, ZnCl$_2$ (sticks), ZnBr$_2$ $>$ CaCl$_2$ (technical) $>$ CuSO$_4$ $>$ Na$_2$SO$_4$, K$_2$CO$_3$.

Where large amounts of water are to be removed, a preliminary drying of liquids is often possible by shaking with concentrated solutions of calcium chloride or potassium carbonate, or by adding sodium chloride to salt out the organic phase (for example, in the drying of lower alcohols).

Drying agents that combine irreversibly with water include the alkali metals, the metal hydrides (discussed in Chapter 2), and calcium carbide.

Suitability of Individual Desiccants
Alumina. (Preheated to 175° for about 7 hr.) Mainly as a drying agent in a desiccator or as a column through which liquid is percolated.

Aluminium amalgam. Mainly used for removing traces of water from alcohols, which are distilled from it after refluxing.

Barium oxide. Suitable for drying organic bases.

Barium perchlorate. Expensive. Used in desiccators (*covered with a metal guard*). Unsuitable for drying solvents or organic material where contact is necessary, because of the danger of **explosion**

Boric anhydride. (Prepared by melting boric acid in an air oven at a high temperature, cooling in a desiccator, and powdering.) Mainly used for drying formic acid.

Calcium chloride (anhydrous). Cheap. Large capacity for absorption of water, giving the hexahydrate below 30°, but is fairly slow in action and not very efficient. Its main use is for preliminary drying of alkyl and aryl halides, most esters, saturated and aromatic hydrocarbons and ethers. Unsuitable for drying alcohols and amines (which form addition compounds), fatty acids, amides, amino acids, ketones, phenols, or some aldehydes and esters. Calcium chloride is suitable for drying the following gases: hydrogen, hydrogen chloride, carbon monoxide, carbon dioxide, sulphur dioxide, nitrogen, methane, oxygen, also paraffins, ethers, olefines and alkyl chlorides.

Calcium hydride. See Chapter 2.

Calcium oxide. (Preheated to 700-900° before use.) Suitable for alcohols and amines (but does not dry them completely). Need not be removed before distillation, but in that case the head of the distillation column should be packed with glass wool to trap any calcium oxide powder that might be carried over. Unsuitable for acidic compounds and esters. Suitable for drying gaseous amines and ammonia.

Calcium sulphate (anhydrous). (Prepared by heating the dihydrate or the hemihydrate in an oven at 235° for 2-3 hr; it can be regenerated.) Available commercially as Drierite. It forms the hemihydrate, $2CaSO_4.H_2O$, so that its capacity is fairly low (6.6% of its weight of water), and hence is best used on partially dried substances. It is very efficient (being comparable with phosphorus pentoxide and concentrated sulphuric acid). Suitable for most organic compounds. Solvents boiling below 100° can be dried by direct distillation from calcium sulphate.

Copper (II) sulphate (anhydrous). Suitable for esters and alcohols. Preferable to sodium sulphate in cases where solvents are sparingly soluble in water (for example, benzene or toluene).

Lithium aluminium hydride. See Chapter 2.

Magnesium amalgam. Mainly used for removing traces of water from alcohols, which are distilled from it after refluxing.

Magnesium perchlorate (anhydrous). (Available commercially as Dehydrite. Expensive.) Used in desiccators. Unsuitable for drying solvents or any organic material where contact is necessary, because of the **danger of explosion**.

Magnesium sulphate (anhydrous). (Prepared from the heptahydrate by drying at 300° under reduced pressure.) More rapid and effective than sodium sulphate. It has a large capacity, forming $MgSO_4.7H_2O$ below 48°. Suitable for the preliminary drying of most organic compounds.

Molecular sieves. See page 32.

Phosphorus pentoxide. Very rapid and efficient, but difficult to handle and should only be used after the organic material has been partially dried, for example with magnesium sulphate. Suitable for acid anhydrides, alkyl and aryl halides, ethers, esters, hydrocarbons and nitriles, and for use in desiccators. Not suitable with acids, alcohols, amines or ketones, or with organic molecules from which a molecule of water can be fairly readily abstracted by an elimination rection. Suitable for drying the following gases: hydrogen, oxygen, carbon dioxide, carbon monoxide, sulphur dioxide, nitrogen, methane, ethylene and paraffins. It is available with an indicator (cobalt salt, blue when dry and pink when wet) under the name *Sicapent* (from Merck).

Potassium (metal). Properties and applications are similar to those for sodium, and it is a correspondingly hazardous substance.

Potassium carbonate (anhydrous). Has a moderate efficiency and capacity, forming the dihydrate. Suitable for an initial drying of alcohols, bases, esters, ketones and nitriles by shaking with them, then filtering off. Also suitable for salting out water-soluble alcohols, amines and ketones. Unsuitable for acids, phenols and other acidic substances.

Potassium hydroxide. Solid potassium hydroxide is very rapid and efficient. Its use is limited almost entirely to the initial drying of organic bases. Alternatively, sometimes the base is shaken first with a concentrated solution of potassium hydroxide to remove most of the water present. Unsuitable for acids, aldehydes, ketones, phenols, amides and esters. Also used for drying gaseous amines and ammonia.

Silica gel. Granulated silica gel is a commercially available drying agent for use with gases, in desiccators, and (because of its chemical inertness) in physical instruments (pH meters, spectrometers, balances). Its drying action depends on physical adsorption, so that silica gel must be used at room temperature or below. By incorporating cobalt chloride into the material it can be made self indicating, re-drying in an oven at 110° being necessary when the colour changes from blue to pink.

Sodium (metal). Used as a fine wire or as chips, for more completely drying ethers, saturated hydrocarbons and aromatic hydrocarbons which have been partially dried (for example with calcium chloride or magnesium sulphate). Unsuitable for acids, alcohols, alkyl halides, aldehydes, ketones, amines and esters. Reacts violently if much water is present and can cause a fire with highly flammable liquids.

Sodium hydroxide. Properties and applications are similar to those for potassium hydroxide.

Sodium-potassium alloy. Used as lumps. Lower melting than sodium, so that its surface is readily renewed by shaking. Properties and applications are similar to those for sodium.

Sodium sulphate (anhydrous). Has a large capacity for absorption of water, forming the decahydrate below 33°, but drying is slow and inefficient, especially for solvents that are sparingly soluble in water. It is suitable for the preliminary drying of most types of organic compounds.

Sulphuric acid (concentrated). Widely used in desiccators. Suitable for drying bromine, saturated hydrocarbons, alkyl and aryl halides. Also suitable for drying the following gases: hydrogen, nitrogen, carbon dioxide, carbon monoxide, chlorine, methane and paraffins. Unsuitable for alcohols, bases, ketones or phenols. Also available with an indicator (a cobalt salt, blue when dry and pink when wet) under the name *Sicacide* (from Merck) for desiccators.

For convenience, the above drying agents are listed in Table 11 under the classes of organic compounds for which they are commonly used.

FREEZE-PUMP-THAW AND PURGING

Volatile contaminants, e,g, traces of low boiling solvent residue or oxygen, in liquid samples or solutions can be very deleterious to the samples on storage. These contaminants can be removed by repeated freeze-pump-thaw cycles. This involves freezing the liquid material under high vacuum in an appropriate vessel (which should be large enough to avoid contaminating the vacuum line with liquid that has bumped) connected to the vacuum line *via* efficient liquid nitrogen traps. The frozen sample is then thawed until it liquefies, kept in this form for some time (*ca*, 10-15min), refreezing the sample and the cycle repeated several times without interupting the vacuum. This procedure applies equally well to solutions, as well as purified liquids, e.g. as a means of removing oxygen from solutions for NMR and other measurements. If the presence of nitrogen, helium or argon, is not a serious contaminant then solutions can be freed from gases, e.g. oxygen, carbon dioxide, and volatile impurities by purging with N_2, He or Ar at room, or slightly elevated, temperature. The gases used for purging are then removed by freeze-pump-thaw cycles or simply by keeping in a vacuum for several hours.

CHROMATOGRAPHY

Chromatography is often used with advantage for the purification of small amounts of complex organic mixtures, either as liquid chromatography or as vapour phase (gas) chromatography.

The mobile phase in liquid chromatography is a liquid and the stationary phase is of four main types. These are for adsorption, partition, ion-chromatography, and gel filtration. The technique of chromatography which applies to all liquid chromatography at atmospheric pressure comprises the following distinct steps. The material is adsorbed as a level bed onto the column of stationary phase. (It is important that this bed is as narrow as possible because the bands of components in the mixture that is applied widen as they move with the mobile phase down the column.) The column is washed (developed) with a quantity of pure solvent or solvent mixture. The column may be pushed out of the tube so that it can be divided into zones. The desired components are then extracted from the appropriate zones using a suitable solvent. Alternatively, and more commonly, the column is left intact and the bands are progressively eluted by passing more solvent through the column.

The mobile phase in vapour phase chromatography is a gas (e.g. hydrogen, helium, nitrogen or argon) and the stationary phase is a non-volatile liquid impregnated onto a porous material. The mixture to be purified is injected into a heated inlet whereby it is vaporised and taken into the column by the carrier gas. It is separated into its components by partition between the liquid on the porous support and the gas. For this reason vapour-phase chromatography is sometimes referred to as gas-liquid chromatography.

Adsorption Chromatography

Adsorption chromatography is based on the difference in the extent to which substances in solution are adsorbed onto a suitable surface. The substances to be purified are usually placed on the top of the column and the solvent is run down the column. In a more common variation of this method, the column containing the adsorbent is full of solvent before applying the mixture at the top of the column. In another application the mixture is adsorbed onto a small amount of stationary phase and placed at the bottom of the column with the dry stationary phase above it. By applying a slight vacuum at the top of the column, the eluting solvent can be sucked slowly upwards from the bottom of the column. When the solvent has reached the top of the column the separation is complete and the vacuum is released. The packing is pushed gently out of the tube and cut into strips as above. Alternatively the vacuum is kept and the effluent from the top of the column is collected in fractions. The fractions are monitored by UV or visible spectra, colour reactions or other means for identifying the components.

Graded adsorbents and solvents. Materials used in columns for adsorption chromatography are grouped in Table 12 in an approximate order of effectiveness. Other adsorbents sometimes used include barium carbonate, calcium sulphate, charcoal (usually mixed with kieselguhr or other form of diatomaceous earth, for example, the filter aid Celite), cellulose, glucose and lactose. The alumina can be prepared in several grades of activity (see below).

In most cases, adsorption takes place most readily from non-polar solvents, such as petroleum ether or benzene, and least readily from polar solvents such as alcohols, esters, and acetic acid. Common solvents, arranged in approximate order of increasing eluting ability are also given in Table 12.

Eluting power roughly parallels the dielectric constants of solvents. The series also reflects the extent to which the solvent binds to the column material, thereby displacing the substances that are already adsorbed. This preference of alumina and silica gel for polar molecules explains, for example, the use of percolation through a column of silica gel for the following purposes-drying of ethylbenzene, removal of aromatics from 2,4-dimethylpentane and of ultraviolet absorbing substances from cyclohexane.

Mixed solvents are intermediate in strength, and so provide a finely graded series. In choosing a solvent for use as an eluent it is necessary to consider the solubility of the substance in it, and the ease with which it can subsequently be removed.

Preparation and standardisation of alumina. The activity of alumina depends inversely on its water content, and a sample of poorly active material can be rendered more active by leaving for some time in a round-bottomed flask heated up to about 200° in an oil bath or a heating mantle while a slow stream of a dry inert gas is passed through it. Alternatively, it is heated to red heat (380-400°) in an open vessel for 4-6 hr with occasional stirring and then cooled in a vacuum desiccator: this material is then of grade I activity. Conversely, alumina can be rendered less active by adding small amounts of water and thoroughly mixing for several hours. Addition of about 3% (w/w) of water converts grade I alumina to grade II.

Used alumina can be regenerated by repeated extraction, first with boiling methanol, then with boiling water, followed by drying and heating. The degree of activity of the material can be expressed conveniently in terms of the scale due to Brockmann and Schodder (*Ber* B **74** 73 *1941*).

This system is based on the extent of adsorption of five pairs of azo dyestuffs, being adjacent members of the set: azobenzene, *p*-methoxyazobenzene, Sudan yellow, Sudan red, aminoazobenzene, hydroxyazobenzene. In testing the alumina, a tube 10 cm long by 1.5 cm internal diameter is packed with alumina to a depth of 5 cm and covered with a disc of filter paper. The dyestuff solutions are prepared by dissolving 2 mg of each azo dye of the pair in 2 ml of purified benzene (distilled from potassium hydroxide) and 8 ml of petroleum ether. The solution is applied to the column and developed with 20 ml of benzene-petroleum ether mixture (1:4 v/v) at a flow rate of about 20-30 drops per min. The following behaviour is observed:

		POSITION OF ZONES	
GRADE	(a)	(b)	(c)
I	p-Methoxyazobenzene	Azobenzene	
II	p-Methoxyazobenzene(d)	Azobenzene	
II	Sudan Yellow	p-Methoxyazobenzene	
III		Sudan Yellow	p-Methoxy-azobenzene
III	Sudan Red	Sudan Yellow	
IV	Sudan Red		Sudan Yellow
IV	Aminoazobenzene	Sudan Red	
V	Hydroxyazobenzene	Aminoazobenzene	

(a) Near top of column. (b) Near bottom of column. (c) In effluent. (d) 1 to 2 cm from top. Grade I is most active, Grade V is least active.

Alumina is normally slightly alkaline. A (less strongly adsorbing) neutral alumina can be prepared by making a slurry in water and adding 2 M hydrochloric acid until the solution is acid to Congo red. The alumina is then filtered off, washed with distilled water until the wash water gives only a weak violet colour with Congo red paper, and dried.

Alumina used in tlc can be recovered by washing in ethanol for 48 hr with occasional stirring, to remove binder material and then washed with successive portions of ethyl acetate, acetone and finally with distilled water. Fine particles are removed by siphoning. The alumina is first suspended in 0.04M acetic acid, then in distilled water, siphoning off 30 minutes after each wash. The process is repeated 7-8 times. It is then dried and activated at 200° [Vogh and Thomson *Anal Chem* **53** 1365 *1981*].

Preparation of other adsorbents. Silica gel can be prepared from commercial water-glass by diluting it with water to a density of 1.19 and, while keeping it cooled to 5°, adding concentrated hydrochloric acid with stirring until the solution is acid to thymol blue. After standing for 3 hr, the precipitate is filtered off, washed on a Büchner funnel with distilled water, then suspended in 0.2M hydrochloric acid. The suspension is stood for 2-3 days, with occasional stirring, then filtered, washed well with water and dried at 110°. It can be activated by heating up to about 200° as described for alumina.

Powdered commercial silica gel can be purified by suspending and standing overnight in concentrated hydrochloric acid (6 ml/g), decanting the supernatant and repeating with fresh acid until the latter remains colourless. After filtering with suction on a sintered-glass funnel, the residue is suspended in water and washed by decantation until free of chloride ions. It is then filtered, suspended in 95% ethanol, filtered again and washed on the filter with 95% ethanol. The process is repeated with anhydrous ethyl ether before the gel is heated for 24 hr at 100° and stored for another 24 hr in a vacuum desiccator over phosphorus pentoxide.

Commercial silica gel has also been purified by suspension of 200g in 2L of 0.04M ammonia, allowed to stand for 5min before siphoning off the supernatant. The procedure was repeated 3-4 times, before rinsing with distilled water and drying and activating the silica gel in an oven at 110° [Vogh and Thomson, *AC* **53** 1345 *1981*].

Diatomaceous earth. (Celite 535 or 545, Hyflo Super-cel, Dicalite, kieselguhr) is purified before use by washing with 3M hydrochloric acid, then water, or it is made into a slurry with hot water, filtered at the pump and washed with water at 50° until the filtrate is no longer alkaline to litmus. Organic materials can be removed by repeated extraction at 50° with methanol, benzene or chloroform, followed by washing with methanol, filtering and drying at $90\text{-}100^\circ$.

Activation of **charcoal** is generally achieved satisfactorily by heating gently to red heat in a crucible or quartz beaker in a muffle furnace, finally allowing to cool under an inert atmosphere in a desiccator. To improve the porosity, charcoal columns are usually prepared in admixture with diatomaceous earth.

Purification of **cellulose** for chromatography is by sequential washing with chloroform, ethanol, water, ethanol, chloroform and acetone. More extensive purification uses aqueous ammonia, water, hydrochloric acid, water, acetone and ethyl ether, followed by drying in a vacuum. Trace metals can be removed from filter paper by washing for several hours with 0.1M oxalic or citric acid, followed by repeated washing with distilled water.

Cellex CM ion-exchange cellulose can be purified by treatment of 30-40g (dry weight) with 500ml of 1mM cystein hydrochloride. It is then filtered through a Büchner funnel and the filter cake is suspended in 500ml of 0.05M NaCl/0.5M NaOH. This is filtered and the filter cake is resuspended in 500ml of distd water and filtered again. The process is repeated until the washings are free from chloride ions. The filter cake is again suspended in 500ml of 0.01M buffer at the desired pH for chromatography, filtered, and the last step repeated several times.

Cellex D and other anionic celluloses are washed with 0.25M NaCl/0.25M NaOH solution, then twice with deionised water. This is followed with 0.25M NaCl and then washed with water until chloride-free. The Cellex is then equilibrated with the desired buffer as above.

Crystalline **hydroxylapatite** is a structurally organised, highly polar material which, in aqueous solution (in buffers) strongly adsorbs macromolecules such as proteins and nucleic acids, permitting their separation by virtue of the interaction with charged phosphate groups and calcium ions, as well by physical adsorption. The procedure therefore is not entirely ion-exchange in nature. Chromatographic separations of singly and doubly stranded DNA are readily achievable whereas there is negligible adsorption of low molecular weight species.

Partition Chromatography

Partition chromatography is concerned with the distribution of substances between a mobile phase and a non-volatile liquid which is itself adsorbed onto an inert supporting stationary phase. The mobile phase may be a gas (see vapour phase chromatography) or a liquid. Paper chromatography, and reverse-phase thin layer chromatography are other applications of partition chromatography. Yet another application is paired-ion chromatography which is used for the separation of substances by virtue of their ionic properties. In principle, the separation of components of a mixture depends on the differences in their distribution ratios between the mobile phase and the liquid stationary phase. The more the distribution of a substance favours the stationary phase, the more slowly it progresses through the column.

When cellulose is used as a stationary phase, with water or aqueous organic solvents as eluents, the separation of substances is by partition between the eluting mixture and the water adsorbed on the column. This is similar to the cellulose in paper chromatography.

For chromatography on dextran gels see page 26.

Paired-ion Chromatography

Mixtures containing ionic compounds (e.g. acids and/or bases), non-ionisable compounds, and zwitterions, can be separated successfully by paired-ion chromatography (PIC). It utilises the 'reverse-phase' technique (Eksberg and Schill *Anal Chem* **45** 2092 *1973*). The stationary phase is lipophilic, such as μ-BONDAPAK C$_{18}$ (Waters Assoc) or any other adsorbent that is compatible with water. The mobile phase is water or aqueous methanol containing the acidic or basic counter ion. Thus the mobile phase consists of dilute solutions of strong acids (e.g. 5 mM 1-heptanesulphonic acid) or strong bases (e.g. 5 mM tetrabutylammonium phosphate) that are completely ionised at the operating pH values which are usually between 2 and 8. An equilibrium is set up between the neutral species of a mixture in the stationary phase and the respective ionised (anion or cation) species which dissolve in the mobile phase containing the counter ions. The extent of the equilibrium will depend on the ionisation constants of the respective components of the mixture, and the solubility of the unionised species in the stationary phse. Since the ionisation constants and the solubility in the stationary phase will vary with the water-methanol ratio of the mobile phase, the separation may be improved by altering this ratio gradually (gradient elution) or stepwise. If the compounds are eluted too rapidly the water content of the mobile phase should be increased, e.g. by steps of 10%. Conversely, if components do not move, or move slowly, the methanol content of the mobile phase should be increased by steps of 10%.

The application of pressure to the liquid phase in liquid chromatography generally increases the separation (see HPLC). Also in PIC improved efficiency of the column is observed if pressure is applied to the mobile phase (Wittmer, Nuessle and Haney *Anal Chem* **47** 1422 *1975*).

Flash Chromatography

A faster method of separating components of a mixture has come into use in recent years. In flash chromatography (see Still *et al, JOC* **43** 2923 *1978*) the eluent flows through the column under a pressure of *ca* 1 to 4 atmospheres. The lower end of the chromatographic column has a relatively long taper closed with a tap. The upper end of the column is connected through a ball joint to a tap. The tapered portion is plugged with cotton, or quartz, wool and *ca* 1 cm of fine washed sand. The adsorbant is then placed in the column as a dry powder or as a slurry in a solvent and allowed to fill about one third of the column. A fine grade of adsorbant is required in order to slow the flow rate at the higher pressure, e.g. Silica 60, 230 to 400 mesh (ASTM) with particle size 0.040-0.063 mm (from Merck). The top of the adsorbant is layered with *ca* 1 cm of fine washed sand. The mixture in the smallest volume of solvent is applied at the top of the column and allowed to flow into the adsorbant under gravity by opening the lower tap momentarily. The top of the column is filled with eluent, the ball joint assembled, clipped together, the upper tap is connected by a tube to a nitrogen supply from a cylinder, or to compressed air, and turned on to the desired pressure (monitor with a gauge). The lower tap is turned on and fractions are collected rapidly until the level of eluent has reached the top of the adsorbant (do not allow the column to run dry). If further elution is desired then both taps are turned off, the column is filled with more eluting solvent and the process repeated. The top of the column can be modified so that gradient elution can be performed. Alternatively, an apparatus for producing the gradient is connected to the upper tap by a long tube and placed high above the column in order to produce the required hydrostatic pressure. Flash chromatography is more efficient and gives higher resolution than conventional chromatography at atmospheric pressure and is completed in a relatively shorter time. A successful separation of components of a mixture by TLC using the same adsorbant is a good indication that flash chromatography will give the desired separation on a larger scale.

Ion-exchange Chromatography

Ion-exchange chromatography involves an electrostatic process which depends on the relative affinities of various types of ions for an immobilised assembly of ions of opposite charge. The stationary phase is an aqueous buffer with a fixed *p*H or an aqueous mixture of buffers in which the *p*H is continuously increased or decreased as the separation may require. This form of liquid chromatography can also be performed at high inlet pressures of liquid with increased column performances.

Ion-exchange resins. An ion-exchange resin is made up of particles of an insoluble elastic hydrocarbon network to which is attached a large number of ionisable groups. Materials commonly used comprise synthetic ion-exchange resins made, for example, by crosslinking polystyrene to which has been attached non-diffusible ionised or ionisable groups. Resins with relatively high crosslinkage (8-12%) are suitable for the chromatography of small ions, whereas those with low crosslinkage (2-4%) are suitable for larger molecules. Applications to hydrophobic systems are possible using aqueous gels with phenyls bound to the rigid matrix (Phenyl-Superose, Pharmacia) or neopentyl chains (Alkyl-Superose, Pharmacia). (Superose is a cross-linked agarose-based medium with an almost uniform bead size.) These groups are further distinguishable as strong ($-SO_2OH$, $-NR_3^+$) or weak ($-OH$, $-CO_2H$, $-PO(OH)_2$, $-NH_2$). (For an extensive collection listing most of the commercially available materials and their characteristics, see *Ion Exchangers in Organic Chemistry and Biochemistry*, C.Calmon and T.R.E.Kressman, Interscience, New York, 1957, *pp* 116-129.) Their charges are counterbalanced by diffusible ions, and the operation of a column depends on its ability and selectivity to replace these ions. The exchange that takes place is primarily an electrostatic process but adsorptive forces and hydrogen bonding can also be important. A typical sequence for the relative affinities of some common anions (and hence the inverse order in which they pass through such a column), is the following, obtained using a quaternary ammonium (strong base) anion-exchange column:

Fluoride < acetate < bicarbonate < hydroxide < formate < chloride < bromate < nitrite < cyanide < bromide < chromate < nitrate < iodide < thiocyanate < oxalate < sulphate < citrate.

For an amine (weak base) anion-exchange column in its chloride form, the following order has been observed:

Fluoride < chloride < bromide = iodide = acetate < molybdate < phosphate < arsenate < nitrate < tartrate < citrate < chromate < sulphate < hydroxide.

With strong cation-exchangers, the usual sequence is that polyvalent ions bind more firmly than mono- or di- valent ones, a typical series being as follows:

$$Th^{4+} > Fe^{3+} > Al^{3+} > Ba^{2+} > Pb^{2+} > Sr^{2+} > Ca^{2+} > Co^{2+} > Ni^{2+} = Cu^{2+} > Zn^{2+} =$$
$$Mg^{2+} > UO_2^+ = Mn^{2+} > Ag^+ > Tl^+ > Cs^+ > Rb^+ > NH_4^+ = K^+ > Na^+ > H^+ > Li^+.$$

Thus, if an aqueous solution of a sodium salt contaminated with heavy metals is passed through the sodium form of such a column, the heavy metal ions will be removed from the solution and will be replaced by sodium ions from the column. This effect is greatest in dilute solution. Passage of sufficiently strong solutions of alkali metal salts or mineral acids readily displaces all other cations from ion-exchange columns. (The regeneration of columns depends on this property.) However, when the cations lie well to the left in the above series it is often advantageous to use a complex-forming species to facilitate removal. For example, iron can be displaced from ion-exchange columns by passage of sodium citrate or sodium ethylenediaminetetraacetate.

Some of the more common commercially available resins are listed in Table 13.

Ion-exchange resins swell in water to an extent which depends on the amount of crosslinking in the polymer, so that columns should be prepared from the wet material by adding it as a suspension in water to a tube already partially filled with water. (This also avoids trapping air bubbles,) The exchange capacity of a resin is commonly expressed as mg equiv./ml of wet resin. This quantity is pH-dependent for weak-acid or weak-base resins but is constant at about 0.6-2 for most strong-acid or strong-base types.

Apart from their obvious applications to inorganic species, sulphonic acid resins have been used in purifying amino acids, aminosugars, organic acids, peptides, purines, pyrimidines, nucleosides, nucleotides and polynucleotides. Thus, organic bases can be applied to the H^+ form of such resins by adsorbing them from neutral solution and, after washing with water, they are eluted sequentially with suitable buffer solutions or dilute acids. Alternatively, by passing alkali solution through the column, the bases will be displaced in an order that is governed by their pK values. Similarly, strong-base anion exchangers have been used for aldehydes and ketones (as bisulphite addition compounds), carbohydrates (as their borate complexes), nucleosides, nucleotides, organic acids, phosphate esters and uronic acids. Weakly acidic and weakly basic exchange resins have also found extensive applications, mainly in resolving weakly basic and acidic species. For demineralisation of solutions without large changes in pH, mixed-bed resins can be prepared by mixing a cation-exchange resin in its H^+ form with an anion-exchange resin in its OH^- form. Commercial examples include Amberlite MB-1 (IR-120 + IRA-400) and Bio-Deminrolit (Zeo-Karb 225 and Zerolit FF). The latter is also available in a self-indicating form.

Ion-exchange Cellulose and Sephadex. A different type of ion-exchange column that is finding extensive application in biochemistry for the purification of proteins, nucleic acids and acidic polysaccharides derives from cellulose by incorporating acidic and basic groups to give ion-exchangers of controlled acid and basic strengths. Commercially available cellulose-type resins are given in Tables 14 and 15. AG 501 x 8 (Bio-Rad) is a mixed-bed resin containing equivalents of AG 50W-x8 H^+ form and AG 1-x8 OH⁻ form, and Bio-Rex MSZ 501 resin. A dye marker indicates when the resin is exhausted. Removal of unwanted cations, particularly of the transition metals, from amino acids and buffer can be achieved by passage of the solution through a column of Chelex 20 or Chelex 100. The metal-chelating abilities of the resin reside in the bonded iminodiacetate groups. Chelex can be regenerated by washing in two bed volumes of 1M HCl, two bed volumes of 1M NaOH and five bed volumes of water.

Ion-exchange celluloses are available in different particle sizes. It is important that the amounts of 'fines' are kept to a minimum otherwise the flow of liquid through the column can be extremely slow and almost stop. Celluloses with a large range of particle sizes should be freed from 'fines' before use. This is done by suspending the powder in the required buffer and allowing it to settle for one hour and then decanting the 'fines'. This separation appears to be wasteful but it is necessary for reasonable flow rates without applying high pressures at the top of the column. Good flow rates can be obtained if the cellulose column is packed dry whereby the 'fines' are evenly distributed throughout the column. Wet packing causes the fines to rise to the top of the column, which thus becomes clogged.

Several ion-exchange celluloses require recycling before use, a process which must be applied for recovered celluloses. Recycling is done by stirring the cellulose with 0.1M aqueous sodium hydroxide, washing with water until neutral, then suspending in 0.1M hydrochloric acid and finally washing with water until neutral. When regenerating a column it is advisable to wash with a salt solution (containing the required counter ions) of increasing ionic strength up to 2M. The cellulose is then washed with water and recycled if necessary. Recycling can be carried out more than once if there are doubts about the purity of the cellulose and when it had been used previously for a different purification procedure than the one to be used. The basic matrix of these ion-exchangers is cellulose and it is important not to subject them to strong acid (> 1M) and strongly basic (> 1M) solutions.

When storing ion-exchange celluloses, or during prolonged usage, it is important to avoid growth of microorganisms or moulds which slowly destroy the cellulose. Good inhibitors of microorganisms are phenyl mercuric salts (0.001%, effective in weakly alkaline solutions), chlorohexidine (Hibitane at 0.002% for anion exchangers), 0.02% aqueous sodium azide or 0.005% of ethyl mercuric thiosalicylate (Merthiolate) are most effective in weakly acidic solutions for cation exchangers. Trichlorobutanol (Chloretone, at 0.05% is only effective in weakly acidic solutions) can be used for both anion and cation exchangers. Most organic solvents (e.g. methanol) are effective antimicrobial agents but only at high concentrations. These inhibitors must be removed by washing the columns thoroughly before use because they may have adverse effects on the material to be purified (e.g. inactivation of enzymes or other active preparations).

In recent years other carbohydrate matrices such as *Sephadex* have been developed which have more uniform particle sizes. Their advantages over the celluloses include faster and more reproducible flow rates and they can be used directly without removal of 'fines'.

Sephadex, which can also be obtained in a variety of ion-exchange forms, consists of beads of a cross-linked dextran gel which swells in water and aqueous salt solutions. The smaller the bead size the higher the resolution that is possible. Typical applications of Sephadex gels are the fractionation of mixtures of polypeptides, proteins, and in nucleic acids and polysaccharides, desalting solutions.

Sephadex is a bead form of cross-linked dextran gel. *Sepharose CL* and *Bio-Gel A* are derived from agarose. Sephadex ion-exchangers, unlike celluloses, are available in narrow ranges of particle sizes. These are of two medium types, the G-25 and G-50, and their dry bead diameter sizes are 50 to 150 microns. They are available as cation and anion exchange Sephadex. One of the disadvantages of using Sephadex ion-exchangers is that the bed volume can change considerably with alteration of *p*H. *Ultragels* also suffer from this disadvantage to a varying extent, but ion-exchangers of the bead type have been developed e.g. *Fractogels, Toyopearl*, which do not suffer from this disadvantage.

Sepharose is a bead form of agarose gel which is useful for the fractionation of high molecular weight subatances, for the molecular weight determination of large molecules (molecular weight > 5000), and the immobilisation of enzymes, antibodies, hormones and receptors usually by affinity chromatography. In preparing any of the above for use in columns, the dry powder is evacuated, then mixed under reduced pressure with water or the appropriate buffer solution. Alternatively it is stirred gently with the solution until all air bubbles are removed. Because some of the wet powders change volumes reversibly with alteration of *p*H or ionic strength, it is imperative to make allowances when packing columns (see above) in order to avoid overflowing of packing when the pH or salt concentrations are altered.

Gel Filtration

The corresponding CM, DEAE and SE resins based on *dextran* are sold in bead form as Sephadex ion exchangers. The gel-like nature of wet Sephadex (a modified dextran) enables small molecules such as inorganic salts to diffuse freely into it while, at the same time, protein molecules are unable to do so. Hence, passage through a Sephadex column can be used for complete removal of salts from protein solutions. Polysaccharides can be freed from monosaccharides and other small molecules because of their differential retardation. Similarly, amino acids can be separated from proteins and large peptides. Gel filtration using Sephadex G-types (50 to 200, from Pharmacia, Uppsala, Sweden) is essentially useful for fractionation of large molecules with molecular weights above 1000. For Superose (Pharmacia) the range is given as 5000 to 5×10^6. Fractionation of lower molecular weight solutes (e,g, ethylene glycols, benzyl alcohols) can now be achieved with Sephadex G-10 (up to Mol.Wt 700) and G-25 (up to Mol.Wt 1500). These dextrans are used only in aqueous solutions. More recently, however, Sephadex LH-20 and LH-60 (prepared by hydroxypropylation of Sephadex) have become available and are used for the separation of small molecules (Mol.Wt less than 500) using most of the common organic solvents as well as water.

Sephasorb HP (ultrafine, prepared by hydroxypropylation of crossed-linked dextran) can also be used for the separation of small molecules in organic solvents and water, and in addition it can withstand pressures up to 1400 psi making it useful in HPLC. Because solutions with high and low pH values slowly decompose, these gels are best operated at pH values between 2 and 12.

High Performance Liquid Chromatography (HPLC)

When pressure is applied at the inlet of a liquid chromatographic column the performance of the column can be increased by several orders of magnitude. This is partly because of the increased speed at which the liquid flows through the column and partly because fine column packings can be used which have larger surface areas. Because of the improved efficiency of the columns this technique has been referred to as high performance, high pressure, or high speed liquid chromatography.

Equipment consists of a hydraulic system to provide the pressure at the inlet of the column, a column, a detector and a recorder. The pressures used in HPLC vary from a few psi to 4000-5000 psi. The most convenient pressures are, however, between 500 and 1800 psi. The plumbing is made of stainless steel or non-corrosive metal tubing to withstand high pressures. Increase of temperature has a very small effect on the performance of a column in liquid chromatography. Small variations in temperatures, however, do upset the equilibrium of the column, hence it is advisable to place the column in an oven at ambient temperature in order to achieve reproducibility. The packing (stationary phase) is specially prepared for withstanding high pressures. It may be an adsorbent (for adsorption or solid-liquid HPLC), a material impregnated with a high boiling liquid (e.g. octadecyl sulphate, in *reverse-phase* or *liquid-liquid or paired-ion* HPLC), an ion-exchange material (in *ion-exchange* HPLC), or a highly porous non-ionic gel (for high performance *gel filtration*). The mobile phase is water, aqueous buffers, salt solutions, organic solvents or mixtures of these. The more commonly used detectors have UV, visible, or fluorescence monitoring for light absorbing substances, and refractive index monitoring for transparent compounds. The sensitivity of the refractive index monitoring is usually lower than the light absorpting monitoring by a factor of ten or more. The cells of the monitoring devices are very small (*ca* 5 μl) and the detection is very good. The volumes of the analytical columns are quite small (*ca* 2ml for a 1 metre column) hence the result of an analysis is achieved very quickly. Larger columns have been used for preparative work and can be used with the same equipment. Most modern machines have solvent mixing chambers for solvent gradient or ion gradient elution. The solvent gradient (for two solvents) or pH or ion gradient can be adjusted in a linear, increasing or decreasing exponential manner. Some of the more common column packings are listed in Table 14.

Purification of stereoisomers has been achieved by applying HPLC using a chiral stationary phase such as (R)-N-3,5-dinitrobenzoylphenylglycine or (S)-3,5-dinitrobenzoylleucine. Examples covering a wide range of compounds are given in references by Pirkle *et al* in *JACS* 103 3964 *(1981)*, and ACS

Symposium Series no 185, *"Asymmetric Reactions and Processes in Chemistry"* Eliel and Otsuka eds (*Amer Chem Soc*, Washington DC, *pp* 245-260, *1982*).

Other Types of Liquid Chromatography

New stationary phases for specific purposes in chromatographic separation are being continually proposed. *Charge transfer adsorption chromatography* makes use of a stationary phase which contains immobilised aromatic compounds and permits the separation of aromatic compounds by virtue of the ability to form charge transfer complexes (sometimes coloured) with the stationary phase. The separation is caused by the differences in stability of these complexes (Porath and Dahlgren-Caldwell *J Chromatog* **133** 180 *1977*).

In *metal chelate adsorption chromatography* a metal is immobilised by partial chelation on a column which contains bi- or tri- dentate ligands. Its application is in the separation of substances which can complex with the bound metals and depends on the stability constants of the various ligands (Porath, Carlsson, Olsson and Belfrage *Nature* **258** 598 *1975*; Loennerdal, Carlsson and Porath *Febs Letts* **75** 89 *1977*).

An application of chromatography which has found extensive use in biochemistry and has brought a new dimension in the purification of enzymes is *affinity chromatography*. A specific enzyme inhibitor is attached by covalent bonding to a stationary phase (e.g. AH-Sepharose 4B for acidic inhibitors and CH-Sepharose 4B for basic inhibitors), and will strongly adsorb only the specific enzyme which is inhibited, allowing all other proteins to flow through the column. The enzyme is then eluted with a solution of high ionic strength (e.g. 1M sodium chloride) or a solution containing a substrate or reversible inhibitor of the specific enzyme. (The ionic medium can be removed by gel filtration using a mixed-bed gel.) Similarly, an immobilised lectin may interact with the carbohydrate moiety of a glycoprotein. The most frequently used matrixes are cross-linked (4-6%) agarose and polyacrylamide gel. Many adsorbents are commercially available for nucleotides, coenzymes and vitamins, amino acids, peptides and lectins. Considerable purification can be achieved by one passage through the column and the column can be reused several times.

The affinity method may be *biospecific*, for example as an antibody-antigen interaction, chemical as in the chelation of boronate by *cis*-diols, or of unknown origin as in the binding of certain dyes to albumin.

Hydrophobic adsorption chromatography takes advantage of the hydrophobic properties of substances to be separated and has also found use in biochemistry (Hoftsee *Biochem Biophys Res Comm* **50** 751 *1973*; Jennissen and Heilmayer Jr *Biochemistry* **14** 754 *1975*). Specific covalent binding with the stationary phase, a procedure that was called *covalent chromatography*, has been used for separation of compounds and for immobilising enzymes on a support: the column was then used to carry out specific bioorganic reactions (Mosbach **Methods in Enzymology** vol 44, 1976).

Vapour phase chromatography

Although this technique was first used for analytical purposes in 1952, its application to the purification of chemicals at a preparative level is much more recent and commercial instruments for this purpose are currently in a state of rapid develpoment. This type of partition chromatography uses a tubular column packed with an inert material which is impregnated with a liquid. This liquid separates components of gases or vapours as they flow through the column. On a preparative scale, use of a large column heated slightly above the boiling point of the material to be processed makes it possible to purify in this way small quantities of many volatile organic substances. For example, if the impurities have a greater affinity for the liquid in the column than the desired component has, the latter will emerge first and in a substantially pure form.

In operation, the organic material is carried as a vapour in a *carrier* gas such as hydrogen, helium, carbon dioxide, nitrogen or argon (in a manner analogous to a solution in a suitable solvent in liquid chromatography). The technique that is almost invariably used is to inject the substance (for example, by means of a hypodermic syringe) over a relatively short time on to the surface of the column through which is maintained a slow continuous passage of the chemically inert carrier gas. This leads to the progressive elution of individual components from the column in a manner analogous to the movement of bands in conventional chromatography. As substances emerge from the column they can be condensed in suitable traps. The carrier gas blows the vapour through these traps hence these traps have to be very efficient. Improved collection of the effluent vaporised fractions in preparative work is attained by strong cooling, increasing the surface of the traps by packing them with glass wool, and by applying an electrical potential which neutralises the charged vapour and causes it to condense.

The choice of carrier gas is largely determined by the type of detection system that is available (see below). Column efficiency is greater in argon, nitrogen or carbon dioxide than it is in helium or hydrogen, but the latter are less impeded by flowing through packed columns so that lower pressure differentials exist between inlet and outlet. The packing in the column is usually an inert supporting material such as powdered firebrick, or a firebrick-Celite mixture coated with a high-boiling organic liquid as the stationary phase. These liquids include Apiezon oils and greases, di-esters (such as dibutylphthalate or di-2-ethylhexyl sebacate), polyesters (such as diethyleneglycol sebacate), polyethylene glycols, hydrocarbons (such as Nujol or squalene), silicone oils and tricresyl phosphate.

The coating material (about 75 ml per 100 ml of column packing) is applied as a solution in a suitable solvent such as methylene chloride, acetone, methanol or pentane, which is then allowed to evaporate in air, over a steam-bath, or in a vacuum oven (provided the adsorbed substance is sufficiently non-volatile). The order in which a mixture of substances travels through such columns depends on their relative solubilities in the materials making up the stationary phases.

Stationary Phase	Mixture
Benzyl diphenyl	Aromatic molecules
Benzyl ether	Saturated hydrocarbons and olefines
Bis(2-n-butoxyethyl)phthalate	Saturated hydrocarbons and olefines
Diethylene glycol adipate	Methyl esters of fatty acids up to C_{24}
Dimethyl sulpholane (below 40°)	Saturated and unsaturated hydrocarbons
Dinonyl phthalate	Paraffins, olefines, low molecular weight aromatics, alcohols (up to amyl alcohol), lower ethers, esters and carbonyl compounds.
Hexadecane	Low-boiling hydrocarbons
Mineral oil	Aliphatic and aromatic amines
2,2'-Oxydipropionitrile	Paraffins, cycloalkanes, olefines, ethers, alkylbenzenes, acetates, aldehydes, alcohols, acetals and ketones
Polyethylene glycols	Aromatic molecules from paraffins
Silicone oil	Aromatic hydrocarbons, alcohols, esters
Silicone-stearic acid	Fatty acids
Squalane	Saturated hydrocarbons
Tricresyl phosphate	Hexanes, heptanes, aromatics, organic sulphur compounds and aliphatic chlorides

The three main requirements of a liquid for use in a gas chromatograph column are that it must have a high boiling point, a low vapour pressure, and at the same time permit adequate separation of components fairly rapidly. As a rough guide, the boiling point should be at least 250° above the temperature of the column, and, at column temperatures, the liquid should not be too viscous, nor should it react chemically with the sample. Liquids suitable for use as stationary phases in gas chromatography are given in Table 17 and above.

Where the stationary phase is chemically similar to the material to be separated, the main factors governing the separation will be the molecular weight and the shape. Otherwise, polar interactions must also be considered, for example hydroxylated compounds used for stationary phases are likely to retard the movement through the column of substances with hydrogen accepting groups. A useful guide to the selection of a suitable stationary phase is to compare, on the basis of polarity, possible materials with the components to be separated. This means that, in general, solute and solvent will be members of the same, or of adjacent, classes in the following groupings:

A. Water, polyhydric alcohols, aminoalcohols, oxyacids, polyphenols, di- and tri-carboxylic acids.
B. Alcohols, fatty acids, phenols, primary and secondary amines, oximes, nitro compounds, nitriles with α-H atoms.
C. Ethers, ketones, aldehydes, esters, tertiary amines, nitriles without α-H atoms.
D. Chlorinated aromatic or olefinic hydrocarbons.
E. Saturated hydrocarbons, carbon disulphide, tetrachloromethane.

Material emerging from the column is detected by a thermal-conductivity cell, an ionisation method, or a gas-density balance.

The first of these methods, which is applicable when hydrogen or helium is used as carrier gas, depends on the differences in heat conductivities between these gases and most others, including organic substances. The resistance of a tungsten or platinum wire heated by a constant electric current will vary with its temperature which, in turn, is a function of the thermal conductivity through the gas. These devices, also known as catharometers, can detect about 10^{-8} moles of substance. When argon is used as carrier gas, an ionisation method is practicable. It is based on measurement of the current between two electrodes at different voltages in the presence of a suitable emitter of β-radiation. The gas-density balance method depends on measurement of the difference in thermal e.m.f. between two equally warmed copper-constantan thermocouples located in the cross-channel of what constitutes a mechanical equivalent to the Wheatstone bridge. Any increase in density of the effluent gas relative to the reference gas will cause movement of gas along the cross-channel, and hence cool one of the thermocouples relative to the other. The technique is comparable in sensitivity with the thermal-conductivity method.

More recently *glass capillary columns* have been used. These columns can be several metres long. The glass capillary wall acts as the support onto which is coated the liquid phase. These columns have much superior separating powers than the conventional columns. In some cases the resolution is so good that enantiomeric and diastereomeric compounds have been separated. When these columns are attached to a mass spectrometer a very powerful analytical tool (*gas chromatography-mass spectrometry*; GC-MS) is produced. Because of the relatively small amounts of material required for mass spectrometry, a splitting system is inserted between the column and the mass spectrometer. This enables only a small fraction of the effluent to enter the spectrometer, the rest of the effluent is usually vented to the air. Even more recently a liquid chromatogrphic column has replaced the gas chromatographic column in the chromatography-mass spectrometry analyses

Paper Chromatography

Paper chromatography is basically a type of partition chromatography between water adsorbed onto the cellulose fibre of the paper and a liquid mobile phase in a closed tank. The most common application is the ascending solvent technique. The paper is hung by means of clips or string and the lower end is made to dip into the eluting solvent. The material under test is applied as a spot 2.5 cm or so above the lower end of the paper and marked with a pencil. It is important that the spots are above the eluting solvent before it begins to rise up the paper by capillarity. Eluents are normally aqueous mixtures of organic solvents, acids or bases. (For solvent systems see Lederer and Lederer, p 54.) The descending technique has also been used, and in this case the top of the paper dips into a trough containing the eluent which travels downwards, also by capillarity. The spots are applied at the top of the paper close to the solvent trough. A closed tank is necessary for these operations because better reproducibility is achieved if the solvent and vapour in the tank are in equilibrium. The tanks have to be kept away from draughts. Elution times vary from several hours to a day depending on the solvent system and paper. For more efficient separations the dried paper is eluted with a different solvent along a direction which is 90° from that of the first elution. This is referred as *two dimensional paper chromatography*. In a third application (*circular paper chromatography*) ordinary circular filter papers are used. The filter paper is placed between two glass plates. The upper plate has a hole in the centre which is coincident with the centre of the paper. A strong solution of the mixture is then separated radially by the eluting solvent. A strong solution of the mixture is placed in this hole followed by the eluting solvent. After the solvents have travelled the required distances in the above separations, the papers are air dried and the spots are revealed by their natural colours or, by spraying with a reagent that forms a coloured product with the spots. In many cases, the positions of the spots can be seen as light fluorescing or absorbing spots when viewed under UV light.

The use of *thick paper* such as Whatman nos 3 or 31 (0.3-0.5 mm) increases the amounts that can be handled (up to about 100 mg per sheet). Larger quantities require multiple sheets or cardboard, e.g. Scheicher and Schüll nos 2071 (0.65 mm), 2230 (0.9 mm) or 2181 (4 mm). For even larger amounts recourse may be had to *chromatopack* or *chromatopile* procedures. The latter use a large number (200-500) of identical filter papers stacked and compressed in a column, the material to be purified being adsorbed onto a small number of these discs which, after drying, are placed almost at the top of the column. The column is then subjected to descendiing development, and bands are separated mechanically by disassembling the filter papers. Instead of filter papers, *cellulose powder* may be suitable, the column being packed by first suspending the powder in the solvent to be used for development. Yet another variation employs tightly wound *paper roll columns* contained in thin polythene skins. (These are unsuitable for such solvents as benzene, chloroform, collidine, ethyl ether, pyridine and toluene.)

The technique of paper chromatography has been almost entirely superseded by thin- or thick-layer chromatography.

Thin or Thick Layer Chromatography (TLC)

Thin layer chromatography is in principle similar to paper chromatography when used in the ascending method, i.e. the solvent creeps up the stationary phase by capillarity. The adsorbent (e.g. silica, alumina, cellulose) is spread on a rectangular glass plate (or solid inert plastic sheet). Some adsorbents (e.g. silica) are mixed with a setting material (e.g. $CaSO_4$) by the manufacturers which causes the film to set on drying. The adsorbent can be activated by heating at 100-110° for a few hours. Other adsorbents (e.g. celluloses) adhere on glass plates without a setting agent. The materials to be purified are spotted in the solvent close to the lower end of the plate and allowed to dry. The spots will need to be placed at such a distance as to ensure that when the lower end of the plate is immersed in the solvent, the spots are a few mm above the eluting solvent. The plate is placed upright in a tank containing the eluting solvent. Elution is carried out in a closed tank as in paper chromatography to ensure equilibrium. It requires less than three hours for the solvent to reach the top of the plate. Good separations can be achieved with square plates if a second elution is performed at right angles to the first as in two dimensional paper chromatography. For rapid work plates of the size of microscopic slides or even smaller are used which can decrease the elution time to as little as fifteen minutes without loss of resolution. The advantage of plastic backed plates is that the size of the plate can be made as required by cutting the sheet with scissors.

The thickness of the plates could be between 0.2 mm to 2 mm or more. The thicker plates are used for preparative work in which hundreds of milligrams of mixtures can be purified conveniently and quickly. The spots or areas are easily scraped off the plates and eluted with the required solvent. These can be revealed on the plates by UV light if they are UV absorbing or fluorescing substances, by spraying with a reagent that gives coloured products with the spot (e.g. iodine solution or vapour gives brown colours with amines), or with dilute sulphuric acid (organic compounds become coloured or black when the plates are heated at 100°) if the plates are of alumina or silica, but not cellulose. Some alumina and silica powders are available with fluorescent materials in them, in which case the whole plate fluoresces under UV light. Non-fluorescing spots are thus clearly visible, and fluorescent spots invariably fluoresce with a different colour. The colour of the spots can be different under UV light at 254 nm and at 365 nm. Another useful way of showing up non-UV absorbing spots is to spray the plate with a 1-2% solution of Rhodamine 6G in acetone. Under UV light the dye fluoresces and reveals the non-fluorescing spots. If the material in the spot is soluble in ether, benzene or light petroleum, the spots can be extracted from the powder with these solvents which leave the water soluble dye behind.

Thin and thick layer chromatography have been used successfully with ion-exchange celluloses as stationary phases and various aqueous buffers as mobile phases. Also, gels (e.g. Sephadex G-50 to G-200 superfine) have been adsorbed on glass plates and are good for fractionating substances of high molecular weights (1500 to 250,000). With this technique, which is called *thin layer gel filtration* (TLG), molecular weights of proteins can be determined when suitable markers of known molecular weights are run alongside.

Commercially available precoated plates with a variety of adsorbents are generally very good for quantitative work because they are of a standard quality. More recently plates of a standardised silica gel 60 (as medium porosity silica gel with a mean porosity of 6 mm) were released by Merck. These have a specific surface of 500 m^2/g and a specific pore volume of 0.75 ml/g. They are so efficient that they have been called *high performance thin layer chromatography* (**HPTLC**) plates (Ropphahn and Halpap *J.Chromatog* **112** 81 *1975*). In another variant of thin layer chromatography the adsorbent is coated with an oil as in gas chromatography thus producing *reverse-phase thin layer chromatography*.

A thin layer form of circular paper chromatography makes use of a circular glass disc coated with an adsorbent (silica, alumina or cellulose). The disc is allowed to rotate and the sample followed by the eluting solvent are allowed to drip onto a central position on the plate. As the plate rotates the solvent elutes the mixture, centrifugally, while separating the components in the form of circles radiating from the central point. When the elution is complete the revolving circular plate is stopped and the circular bands are scraped off and extracted with a suitable solvent.

SOLVENT EXTRACTION AND DISTRIBUTION

Extraction of a substance from suspension or solution into another solvent can sometimes be used as a purification process. Thus, organic substances can often be separated from inorganic impurities by shaking an aqueous solution or suspension with suitable immiscible solvents such as benzene, carbon tetrachloride, chloroform, ethyl ether, isopropyl ether or petroleum ether. After several such extractions the combined organic phase is dried and the solvent is evaporated. Grease from the glass taps of conventional separating funnels is invariably soluble in the solvents used. Contamination with grease can be very troublesome particularly when the amounts of material to be extracted are very small. Instead, the glass taps should be lubricated with the extraction solvent; or better, the taps of the extraction funnels should be made of the more expensive material *Teflon*. Immiscible solvents suitable for extractions are given in Table 18. Addition of electrolytes (such as ammonium sulphate, calcium chloride or sodium chloride) to the aqueous phase helps to ensure that the organic layer separates cleanly and also decreases the extent of extraction into the latter. Emulsions can also be broken up by filtration (with suction) through Celite, or by adding a little octyl alcohol or some other paraffinic alcohol. The main factor in selecting a suitable immiscible solvent is to find one in which the material to be extracted is readily soluble, whereas the substance from which it is being extracted is not. The same considerations apply irrespective of whether it is the substance being purified, or one of its contaminants, that is taken into the new phase. (The second of these processes is described as washing.)

Common examples of washing with aqueous solutions include the following:

> Removal of acids from water-immiscible solvents by washing with aqueous alkali, sodium carbonate or sodium bicarbonate.
> Removal of phenols from similar solutions by washing with aqueous alkali.
> Removal of organic bases by washing with dilute hydrochloric or sulphuric acids.
> Removal of unsaturated hydrocarbons, of alcohols and of ethers from saturated hydrocarbons or alkyl halides by washing with cold concentrated sulphuric acid.

This process can also be applied to purification of the substance if it is an acid, a phenol or a base, by extracting into the appropriate aqueous solution to form the salt which, after washing with pure solvent, is again converted to the free species and re-extracted. Paraffin hydrocarbons can be purified by extracting them with phenol (in which aromatic hydrocarbons are highly soluble) prior to fractional distillation.

For extraction of solid materials with a solvent, a *Soxhlet* extractor is commonly used. This technique is applied, for example, in the alcohol extraction of dyes to free them from insoluble contaminants such as sodium chloride or sodium sulphate.

Acids, bases and amphoteric substances can be purified by taking advantage of their ionisation constants. Thus an acid can be separated from other acidic impurities which have different pK_a values and from basic and neutral impurities, by extracting a solution of the organic acid into an organic solvent (e.g. benzene or amyl alcohol) with a set of inorganic buffers of increasing pH (see Table 19). The acid will dissolve to form its salt in a set of buffers of pH greater than the pK_a value. It can then be isolated by adding excess mineral acid to the buffer and extracting the free acid with an organic solvent. On a large scale, a *countercurrent distribution* machine (e.g. Craig type, see Quickfit and Quartz catalogue) can be used. In this way a very large number of liquid-liquid extractions can be carried out automatically. The closer the ionisation constants of the impurities are to those of the required material, the larger should the be the number of extractions to effect a good separation. A detailed discussion is available in review articles such as that by L.L.Craig, D.Craig and E.G.Scheibel, p 53. Applications are summarised in C.G.Casinovi's review, "A Comprehensive Bibliography of Separations of Organic Substances by Countercurrent Distribution" *Chromatographic Reviews* **5** 161 *1963*. This technique, however, appears to have been displaced almost completely by chromatographic methods.

MOLECULAR SIEVES

Molecular sieves are types of adsorbents composed of crystalline zeolites (sodium and calcium aluminosilicates). By heating them, water of hydration is removed, leaving holes of molecular dimensions in the crystal lattices. These holes are of uniform size and allow the passage into the crystals of small molecules, but not of large ones. This *sieving* action explains their use as very efficient drying agents for gases and liquids. The pore size of these sieves can be modified (within limits) by varying the cations built into the lattices. The three types of Linde (Union Carbide) molecular sieves currently available are:

> **Type 4A**, a crystalline sodium aluminosilicate.
> **Type 5A**, a crystalline calcium aluminosilicate.
> **Type 13X**, a crystalline sodium aluminosilicate.

They are unsuitable for use with strong acids but are stable over the pH range 5-11.

> **Type 4A sieves.** The pore size is about 4 Angstroms, so that, besides water, the ethane molecules (but not butane) can be adsorbed. Other molecules removed from mixtures include carbon dioxide, hydrogen sulphide, sulphur dioxide, ammonia, methanol, ethanol, ethylene, acetylene, propylene, n-propyl alcohol, ethylene oxide and (below -30°) nitrogen, oxygen and methane. The material is supplied as beads, pellets or powder.
> **Type 5A sieves.** Because the pore size is about 5 Angstroms, these sieves adsorb larger molecules than type 4A. For example, as well as the substances listed above, propane, butane, hexane, butene, higher n-olefines, n-butyl alcohol and higher n-alcohols, and cyclopropane can be adsorbed, but not branched-chain C_6 hydrocarbons, cyclic hydrocarbons such as benzene and cyclohexane, or secondary and tertiary alcohols, carbon tetrachloride or boron trifluoride. This is the type generally used for drying gases.
> **Type 13X sieves.** Their pore size of about 10 Angstroms enables many branched-chain and cyclic materials to be adsorbed, in addition to all the substances taken out by type 5A sieves.

Because of their selectivity, molecular sieves offer advantages over silica gel, alumina or activated charcoal, especially in their very high affinity for water, polar molecules and unsaturated organic compounds. Their relative efficiency is greatest when the impurity to be removed is present at low concentrations. Thus, at 25° and a relative humidity of 2%, type 5A molecular sieves adsorb 18% by weight of water, whereas for silica gel and alumina the figures are 3.5 and 2.5% respectively. Even at 100° and a relative humidity of 1.3% molecular sieves adsorb about 15% by weight of water.

The much greater preference of molecular sieves for combining with water molecules explains why this material can be used for drying ethanol and why molecular sieves are probably the most universally useful and efficient drying agent. Percolation of ethanol with an initial water content of 0.5% through a

57-in long column of type 4A molecular sieves reduced the water content to 10 ppm. Similar results have been obtained with pyridine.

The main applications of molecular sieves to purification comprise:

1. Drying of gases and liquids containing traces of water.
2. Drying of gases at elevated temperatures.
3. Selective removal of impurities (including water) from gas streams.

(For example, carbon dioxide from air or ethylene; nitrogen oxides from nitrogen; methanol from ethyl ether. In general, carbon dioxide, carbon monoxide, ammonia, hydrogen sulphide, mercaptans, ethane, ethylene, acetylene, propane and propylene are readily removed at 25°. In mixtures of gases, the more polar ones are preferentially adsorbed.)

The following applications include the removal of straight-chain from branched-chain or cyclic molecules. For example, type 5A sieves will adsorb n-butyl alcohol but not its branched-chain isomers. Similarly, it separates n-tetradecane from benzene, or n-heptane from methylcyclohexane. A logical development is the use of molecular sieves as chromatographic columns for particular preparations.

The following liquids have been dried with molecular sieves: acetone, acetonitrile, acrylonitrile, allyl chloride, amyl acetate, benzene, butadiene, n-butane, butene, butyl acetate, n-butylamine, n-butyl chloride, carbon tetrachloride, chloroethane, 1-chloro-2-ethylhexane, cyclohexane, , dichloromethane, dichloroethane, 1,2-dichloropropane, 1,1-dimethoxyethane, dimethyl ether, 2-ethylhexanol, 2-ethylhexylamine, n-heptane, n-hexane, isoprene, isopropyl alcohol, isopropyl ether, methanol, methyl ethyl ketone, oxygen, n-pentane, phenol, propane, n-propyl alcohol, propylene, pyridine, styrene, tetrachloroethylene, toluene, trichloroethylene and xylene. In addition, the following gases have been dried: acetylene, air, argon, carbon dioxide, chlorine, ethylene, helium, hydrogen, hydrogen chloride, hydrogen sulphide, nitrogen, oxygen and sulphur hexafluoride.

After use, molecular sieves can be regenerated by heating at between 150° and 300° for several hours, preferably in a stream of dry air, then cooling in a desiccator.

However, care must be exercised in using molecular sieves for drying organic liquids. Appreciable amounts of impurities were *formed* when samples of acetone, 1,1,1-trichloroethane and methyl-*t*-butyl ether were dried in the liquid phase by contact with molecular sieves 4A (Connett *Lab.Practice* **21** 545 *1972*). Other, less reactive types of sieves may be more suitable but, in general, it seems desirable to make a preliminary test to establish that no unwanted reaction takes place.

SOME HAZARDS OF CHEMICAL MANIPULATION IN PURIFICATION AND RECOVERY FROM RESIDUES

Performing chemical manipulations calls for some practical knowledge if danger is to be avoided. However, with care, hazards can be kept to an acceptable minimum. A good general approach is to consider every operation as potentially perilous and then to adjust one's attitude as the operation proceeds. A few of the commonest dangers are set out below. For a larger coverage of the following sections, and of the literature, the bibliography at the end of this chapter should be consulted. Several precautions on **Safety in the chemical laboratory** have been emphasised earlier in this monograph on page 3.

1. Perchlorates and perchloric acid. At 160° perchloric acid is an exceedingly strong oxidising acid and a strong dehydrating agent. Organic perchlorates, such as methyl and ethyl perchlorates, are unstable and are violently explosive compounds. A number of heavy-metal perchlorates are extremely prone to explode. The use of anhydrous magnesium perchlorate *Anhydrone* as a drying agent for organic vapours is **not** recommended. Desiccators which contain this drying agent should be adequately shielded at all times and kept in a cool place, i.e **never** on a window sill where sunlight can fall on it.

No attempt should be made to purify perchlorates, except for ammonium, alkali metal and alkaline earth salts which, in water or aqueous alcoholic solutions are insensitive to heat or shock. Note that perchlorates react relatively slowly in aqueous organic solvents, but as the water is removed there is an increased possibility of an explosion. Perchlorates, often used in non-aqueous solvents, are explosive in the presence of even small amounts of organic compounds when heated. Hence stringent care should be taken when purifying perchlorates, and direct flame and infrared lamps should be avoided. Tetra-alkylammonium perchlorates should be dried below 50° under vacuum (and protection). Only very small amounts of such materials should be prepared, and stored, at any one time.

2. Peroxides. These are formed by aerial oxidation or by autoxidation of a wide range of organic compounds, including ethyl ether, allyl ethyl ether, allyl phenyl ether, benzyl ether, benzyl butyl ether, *n*-butyl ether, *iso*-butyl ether, *t*-butyl ether, dioxane, tetrahydrofuran, olefines, and aromatic and saturated aliphatic hydrocarbons. They accumulate during distillation and can detonate violently on evaporation or distillation when their concentration becomes high. If peroxides are likely to be present materials should be tested for peroxides before distillation (for tests see entry under "Ethers", in Chapter 5). Also, distillation should be discontinued when at least one quarter of the residue is left in the distilling flask.

3. Heavy-metal-containing-explosives. Ammoniacal silver nitrate, on storage or treating, will eventually deposit the highly explosive silver nitride *"fulminating silver"*. Silver nitrate and ethanol may give silver fulminate (see Chapter 4), and in contact with azides or hydrazine and hydrazides may form silver azide. Mercury can form such compounds. Similarly, ammonia or ammonium ions can react with gold salts to form *"fulminating gold"*. Metal fulminates of cadmium, copper, mercury and thallium are powerfully explosive, and some are detonators [Luchs, *Photog Sci Eng* **10** 334 *1966*]. Heavy metal containing solutions, particularly when organic material is present should be treated with great respect and precautions towards possible explosion should be taken.

4. Strong acids. In addition to perchloric acid (see above), extra care should be taken when using strong mineral acids. Although the effects of concentrated sulphuric acid are well known these cannot be stressed strongly enough. Contact with tissues will leave irreparable damage. **ALWAYS DILUTE THE CONCENTRATED ACID BY CAREFULLY ADDING THE ACID DOWN THE SIDE OF THE FLASK WHICH CONTAINS WATER, AND THE PROCESS SHOULD BE CARRIED OUT UNDER COOLING. THIS SOLUTION IS NOT SAFE TO HANDLE UNTIL THE ACID HAS BEEN THOROUGHLY MIXED WITH THE WATER. PROTECTIVE FACE AND BODY COVERAGE SHOULD BE USED AT ALL TIMES.** Fuming sulphuric acid and chlorosulphonic acid are even more dangerous than concentrated sulphuric acid and adequate precautions should be taken. Chromic acid cleaning mixture (hot and cold, see p.4) contains strong sulphuric acid and should be treated in the same way; and in addition the mixture is potentially *carcinogenic*.
Concentrated and fuming nitric acids are also dangerous because of their severe deleterious effects on tissues.

5. Solvents. The flammability of low-boiling organic liquids cannot be emphasised strongly enough. These invariably have very low flash points and can ignite spontaneously. Special precautions against explosive flammability should be taken when recovering such liquids. Care should be taken with small volumes (*ca* 250ml) as well as large volumes (> 1L), and the location of all the fire extinguishers, and fire blankets, in the immediate vicinity of the apparatus should be checked. The fire extinguisher should be operational. The following flammable liquids (in alphabetical order) are common fire hazards in the laboratory: acetaldehyde, acetone, acrylonitrile, acetonitrile, benzene, carbon disulphide, cyclohexane, diethyl ether, ethyl acetate, hexane, low-boiling petroleum ethers, tetrahydrofuran and toluene. Toluene should always be used in place of benzene due to the potential *carcinogenic* effects of the liquid and vapour of the latter.
The drying of flammable solvents with sodium or potassium metal and metal hydrides poses serious potential fire hazards and adequate precautions should be stressed.

6. Salts. In addition to the dangers of perchlorate salts, other salts such as nitrates and diazo salts can be hazardous and care should be taken when these are dried. Large quantities should never be prepared or stored for long periods.

TABLE 1A. PREDICTED EFFECT OF PRESSURE ON BOILING POINT[*]

760 mm Hg	\multicolumn{10}{c}{Temperature in degrees Centigrade}									
	0	20	40	60	80	100	120	140	160	180
0.1	-111	-99	-87	-75	-63	-51	-39	-27	-15	-4
0.2	-105	-93	-81	-69	-56	-44	-32	-19	-7	5
0.4	-100	-87	-74	-62	-49	-36	-24	-11	2	15
0.6	-96	-83	-70	-57	-44	-32	-19	-6	7	20
0.8	-94	-81	-67	-54	-41	-28	-15	-2	11	24
1.0	-92	-78	-65	-52	-39	-25	-12	1	15	28
2.0	-85	-71	-58	-44	-30	-16	-3	11	25	39
4.0	-78	-64	-49	-35	-21	-7	8	22	36	51
6.0	-74	-59	-44	-30	-15	-1	14	29	43	58
8.0	-70	-56	-41	-26	-11	4	19	34	48	63
10.0	-68	-53	-38	-23	-8	7	22	37	53	68
14.0	-64	-48	-33	-23	-2	13	28	44	59	74
16.0	-61	-45	-29	-14	2	17	33	48	64	79
20.0	-59	-44	-28	-12	3	19	35	50	66	82
30.0	-54	-38	-22	-6	10	26	42	58	74	90
40.0	-50	-34	-17	-1	15	32	48	64	81	97
50.0	-47	-30	-14	3	19	36	52	69	86	102
60.0	-44	-28	-11	6	23	40	56	73	86	107
80.0	-40	-23	-6	11	28	45	62	79	97	114
100.0	-37	-19	-2	15	33	50	67	85	102	119
150.0	-30	-12	6	23	41	59	77	95	112	130
200.0	-25	-7	11	29	47	66	84	102	120	138
300.0	-18	1	19	38	57	75	94	113	131	150
400.0	-13	6	25	44	64	83	102	121	140	159
500.0	-8	11	30	50	69	88	108	127	147	166
600.0	-5	15	34	54	74	93	113	133	152	172
700.0	-2	18	38	58	78	98	118	137	157	177
750.0	0	20	40	60	80	100	120	140	160	180
770.0	0	20	40	60	80	100	120	140	160	180
800.0	1	21	41	61	81	101	122	142	162	182

[*] *How to use the Table*: Take as an example a liquid with a b.p. of 80°C at 760mm Hg. The Table gives values of the b.ps of this liquid at pressures from 0.1 to 800mm Hg. Thus at 50mm Hg this liquid has a b.p. of 19°C, and at 2mm Hg its b.p. would be -30°C.

TABLE 1B. PREDICTED EFFECT OF PRESSURE ON BOILING POINT[*]

760 mm Hg	Temperature in degrees Centigrade										
	200	220	240	260	280	300	320	340	360	380	400
0.1	8	20	32	44	56	68	80	92	104	115	127
0.2	17	30	42	54	67	79	91	103	116	128	140
0.4	27	40	53	65	78	91	103	116	129	141	154
0.6	33	40	59	72	85	98	111	124	137	150	163
0.8	38	51	64	77	90	103	116	130	143	156	169
1.0	41	54	68	81	94	108	121	134	147	161	174
2.0	53	66	80	94	108	121	135	149	163	176	190
4.0	65	79	93	108	122	136	151	156	179	193	208
6.0	72	87	102	116	131	146	160	175	189	204	219
8.0	78	93	108	123	137	152	167	182	197	212	227
10.0	83	98	113	128	143	158	173	188	203	218	233
14.0	90	105	120	136	151	166	182	197	212	228	243
18.0	95	111	126	142	157	173	188	204	219	235	251
20.0	97	113	129	144	160	176	191	207	223	238	254
30.0	106	123	139	155	171	187	203	219	235	251	267
40.0	113	130	146	162	179	195	211	228	244	260	277
50.0	119	135	152	168	185	202	218	235	251	268	284
60.0	123	140	157	174	190	207	224	241	257	274	291
80.0	131	148	165	182	199	216	233	250	267	284	301
100.0	137	154	171	189	206	223	241	258	275	293	310
150.0	148	166	184	201	219	237	255	273	290	308	326
200.0	156	174	193	211	229	247	265	283	302	320	338
300.0	169	187	206	225	243	262	281	299	318	337	355
400.0	178	197	216	235	254	273	292	311	330	350	369
500.0	185	205	224	244	263	282	302	321	340	360	379
600.0	192	211	231	251	270	290	310	329	349	368	388
700.0	197	217	237	257	277	296	316	336	356	376	396
750.0	200	220	239	259	279	299	319	339	359	279	399
770.0	200	220	241	261	281	301	321	341	361	381	401
800.0	202	222	242	262	282	302	322	342	262	382	403

[*] *How to use the Table*: Taking as an example a liquid with a b.p. of 340°C at 760mm Hg, the column headed 340°C gives values of the b.ps of this liquid at each value of pressures from 0.1 to 800mm Hg. Thus, at 100mm Hg its b.p. is 258°C, and at 0.8mm Hg its b.p. will be 130°C.

TABLE 2. HEATING BATHS

Up to 100°	Water baths
-20 to 200°	Glycerol or di-*n*-butyl phthalate
Up to about 200°	Medicinal paraffin
Up to about 250°	Hard hydrogenated cotton-seed oil (**m** 40- 60°) or a 1:1 mixture of cotton-seed oil and castor oil containing about 1% of hydroquinone.
-40 to 250°	(to 400° under nitrogen); D.C. 550 silicone fluid
Up to about 260°	A mixture of 85% orthophosphoric acid (4 parts) and metaphosphoric acid (1 part)
Up to 340°	A mixture of 85% orthophosphoric acid (2 parts) and metaphosphoric acid (1 part)
60 to 500°	Fisher bath wax (highly unsaturated)
150 to 500°	A mixture of $NaNO_2$ (40%), $NaNO_3$ (7%) and KNO_3 (53%)
73 to 350°	Wood's Metal[*]
250 to 800°	Solder[*]
350 to 800°	Lead[*]

[*] In using metal baths, the container (usually a metal crucible) should be removed while the metal is still molten.

TABLE 3. WHATMAN FILTER PAPERS

Grade No.	1	2	3	4	5	6	113
Particle size retained in microns	11	8	5	12	2.4	2.8	28
Filtration speed[*] sec/100ml	40	55	155	20	<300	125	9

Whatman routine ashless filters

Grade No.	40	41	42	43	44
Particle size retained in microns	7.5	12	3	12	4
Filtration speed[*] sec/100ml	68	19	200	38	125

(continued)

TABLE 3 (continued). WHATMAN FILTERS

Whatman	Hardened			Hardened ashless		
Grade No.	50	52	54	540	541	542
Particle size retained in microns	3	8	20	9	20	3
Filtration speed* sec/100ml	250	55	10	55	12	250

Whatman Glass Micro Filters

Grade No	GF/A	GF/B	GF/C	GF/D	GF/F
Particle size retained in microns	1.6	1.0	1.1	2.2	0.8
Filtration speed sec/100ml*	8.3	20.0	8.7	5.5	17.2

*Filtration speeds are rough estimates of initial flowrates and should be considered on a relative basis.

TABLE 4.   MICRO FILTERS*

Nucleopore (polycarbonate) Filters

Mean Pore Size (microns)	8.0	2.0	1.0	0.1	0.03	0.015
Av. pores/cm^2	10^5	2×10^6	2×10^7	3×10^8	6×10^8	$1\text{-}6 \times 10^9$
Water flowrate ml/min/cm^2	2000	2000	300	8	0.03	0.1-0.5

Millipore Filters

Type	Cellulose ester		Teflon		Microweb#	
	MF/SC	MF/VF	LC	LS	WS	WH
Mean Pore Size (microns)	8	0.01	10	5	3	0.45
Water flowrate ml/min/cm^2	850	0.2	170	70	155	55

(continued)

TABLE 4 (Continued).   MICRO FILTERS*

Gelman Membranes

Type	Cellulose ester				Copolymer	
	GA-1	TCM-450	VM-1	DM-800	AN-200	Tuffryn-450
Mean Pore Size(microns)	5	0.45	5	0.8	0.2	0.45
Water flow-rate(ml/min/cm^2)	320	50	700	200	17	50

Sartorius Membrane Filters (SM)

Application	Gravi-metric	Biological clarificatn.	Sterili-zation	Particle count in H$_2$O	For acids & bases
Type No.	11003	11004	11006	11011	12801
Mean Pore Size (microns)	1.2	0.6	0.45	0.01	8
Water flowrate ml/min/cm^2	300	150	65	0.6	1100

* Only a few representative filters are tabulated (available ranges are more extensive).
# Reinforced nylon.

TABLE 5.   COMMON SOLVENTS USED IN RECRYSTALLISATION
(and their boiling points)

Acetic acid (118°) (glacial)	*Ethyl acetate (78°)
*Acetone (56°)	Ethyl benzoate (98°/19mm)
Acetylacetone (139°)	Ethylene glycol (68°/4mm)
*Benzene (80°)	Formamide (110°/10mm)
Benzyl alcohol (93°/10mm)	Glycerol (126°/11mm)
n-Butanol (118°)	Isoamyl alcohol (131°)
Butyl acetate (126.5°)	*Methanol (64.5°)
n-Butyl ether (142°)	Methyl cyanide (82°)
γ-Butyrolactone (206°)	Methylene chloride (41°)
Carbon tetrachloride (77°)	Methyl ethyl ketone (80°)
Cellosolve (135°)	Methyl isobutyl ketone (116°)
Chlorobenzene (132°)	Nitrobenzene (210°)
Chloroform (61°)	Nitromethane (101°)
*Cyclohexane (81°)	*Petroleum ether (various)
Diethyl cellosolve (121°)	Pyridine (115.5°)
Diethyl ether (34.5°)	Pyridine trihydrate (93°)
Dimethyl formamide (76°/39mm)	*Tetrahydrofuran (64-66°)
*Dioxane (101°)	Toluene (110°)
*Ethanol (78°)	Trimethylene glycol (59°/11mm)
	Water (100°)

* Highly flammable, should be heated or evaporated on steam or electrically heated water baths only (preferably in a nitrogen atmosphere).

TABLE 6. PAIRS OF MISCIBLE SOLVENTS

Acetic acid: with chloroform, ethanol, ethyl acetate, methyl cyanide, petroleum ether, or water.

Acetone: with benzene, butyl acetate, butyl alcohol, carbon tetrachloride, chloroform, cyclohexane, ethanol, ethyl acetate, methyl acetate, methyl cyanide, petroleum ether or water.

Ammonia: with ethanol, methanol, pyridine.

Aniline: with acetone, benzene, carbon tetrachloride, ethyl ether, *n*-heptane, methanol, methyl cyanide or nitrobenzene.

Benzene: with acetone, butyl alcohol, carbon tetrachloride, chloroform, cyclohexane, ethanol, methyl cyanide, petroleum ether or pyridine.

Butyl alcohol: with acetone or ethyl acetate.

Carbon disulphide: with petroleum ether.

Carbon tetrachloride: with cyclohexane.

Chloroform: with acetic acid, acetone, benzene, ethanol, ethyl acetate, hexane, methanol or pyridine.

Cyclohexane: with acetone, benzene, carbon tetrachloride, ethanol or ethyl ether.

Dimethyl formamide: with benzene, ethanol or ether.

Dimethyl sulphoxide: with acetone, benzene, chloroform, ethanol, ethyl ether or water.

Dioxane: with benzene, carbon tetrachloride, chloroform, ethanol, ethyl ether, petroleum ether, pyridine or water.

Ethanol: with acetic acid, acetone, benzene, chloroform, cyclohexane, dioxane, ethyl ether, pentane, toluene, water or xylene.

Ethyl acetate: with acetic acid, acetone, butyl alcohol, chloroform, or methanol.

Ethyl ether: with acetone, cyclohexane, ethanol, methanol, methylal, methyl cyanide, pentane or petroleum ether.

Glycerol: with ethanol, methanol or water.

Hexane: with benzene, chloroform or ethanol.

Methanol: with chloroform, ethyl ether, glycerol or water.

Methylal: with ethyl ether.

Methyl ethyl ketone: with acetic acid, benzene, ethanol or methanol.

Nitrobenzene: with aniline, methanol or methyl cyanide.

Pentane: with ethanol or ethyl ether.

Petroleum ether: with acetic acid, acetone, benzene, carbon disulphide or ethyl ether.

Phenol: with carbon tetrachloride, ethanol, ethyl ether or xylene.

Pyridine: with acetone, ammonia, benzene, chloroform, dioxane, petroleum ether, toluene or water.

Toluene: with ethanol, ethyl ether or pyridine.

Water: with acetic acid, acetone, ethanol, methanol, or pyridine.

Xylene: with ethanol or phenol.

TABLE 7. MATERIALS FOR COOLING BATHS

Temperature	Composition
0°	Crushed ice
-5° to -20°	Ice-salt mixtures
-33°	Liquid ammonia
-40° to -50°	Ice (3.5-4 parts) - $CaCl_2 6H_2O$ (5 parts)
-72°	Solid CO_2 with ethanol
-77°	Solid CO_2 with chloroform or acetone
-78°	Solid CO_2 (powdered)
-100°	Solid CO_2 with ethyl ether
-192°	liquid air
-196°	liquid nitrogen

Alternatively, the following liquids can be used, partially frozen, as cryostats, by adding solid CO_2 from time to time to the material in a Dewar-type container and stirring to make a slush:

13°	p-Xylene
12°	Dioxane
6°	Cyclohexane
5°	Benzene
2°	Formamide
-8.6°	Methyl salicylate
-9°	Hexane-2,5-dione
-10.5°	Ethylene glycol
-11.9°	tert-Amyl alcohol
-12°	Cycloheptane or methyl benzoate
-15°	Benzyl alcohol
-16.3°	n-Octanol
-18°	1,2-Dichlorobenzene
-22°	Tetrachloroethylene
-22.4°	Butyl benzoate
-22.8°	Carbon tetrachloride
-24.5°	Diethyl sulphate
-25°	1,3-Dichlorobenzene
-29°	o-Xylene or pentachloroethane
-30°	Bromobenzene
-32°	m-Toluidine
-32.6°	Dipropyl ketone
-38°	Thiophen
-41°	Methyl cyanide
-42°	Pyridine or diethyl ketone
-44°	Cyclohexyl chloride
-45°	Chlorobenzene
-47°	m-Xylene
-50°	Ethyl malonate or n-butylamine

(continued)

TABLE 7 (continued).   MATERIALS FOR COOLING BATHS

-52°	Benzyl acetate or diethylcarbitol
-55°	Diacetone
-56°	n-Octane
-60°	Isopropyl ether
-73°	Trichloroethylene or isopropyl acetate
-74°	o-Cymene or p-cymene
-77°	Butyl acetate
-79°	Isoamyl acetate
-83°	Propylamine

By using liquid nitrogen[*] instead of solid CO_2, this range can be extended.

-83.6°	Ethyl acetate
-86°	Methyl ethyl ketone
-89°	n-Butanol
-90°	Nitroethane
-91°	Heptane
-92°	n-Propyl acetate
-93°	2-Nitropropane or cyclopentane
-94°	Ethyl benzene or hexane
-94.6°	Acetone
-95.1°	Toluene
-97°	Cumene
-98°	Methanol or methyl acetate
-99°	Isobutyl acetate
-104°	Cyclohexene
-107°	Isooctane
-108°	1-Nitropropane
-116°	Ethanol or ethyl ether
-117°	Isoamyl alcohol
-126°	Methylcyclohexane
-131°	n-Pentane
-160°	Isopentane

For other organic materials used in low temperature slush-baths with liquid nitrogen see R.E.Rondeau [*J.Chem.Eng.Data* **11** 124 *1966*]. **NOTE** that the liquid nitrogen should be oxygen-free. Liquid nitrogen that has been in contact with air will contain oxygen (see Table 8 for boiling points) and should not be used.

[*] **Use high quality pure nitrogen, do not use liquid air or liquid nitrogen that has been in contact with air for a long period (due to the dissolution of oxygen in it) which could EXPLODE in contact with organic matter. If the quality of the liquid nitrogen is not known, or is uncertain then it should NOT be used.**

TABLE 8.   BOILING POINTS OF SOME USEFUL GASES AT ONE ATMOSPHERE PRESSURE

Argon	-185.6°	Krypton	-152.3°
Carbon dioxide	-78.5° (sublimes)	Methane	-164.0°
		Neon	-246.0°
Carbon monoxide	-191.3°	Nitrogen	-209.9°
Ethane	-88.6°	Nitrous oxide	-88.5°
Helium	-268.6°	Nitric oxide	-195.8°
Hydrogen	-252.6°	Oxygen	-182.96°

TABLE 9.   LIQUIDS FOR DRYING PISTOLS

	Boiling points (760mm)
Ethyl chloride	12.2°
Methylene dichloride	39.8°
Acetone	56.1°
Chloroform	62.0°
Methanol	64.5°
Carbon tetrachloride	76.5°
Ethanol	78.3°
Benzene	79.8°
Trichloroethylene	86.0°
Water	100.0°
Toluene	110.5°
Tetrachloroethylene	121.2°
Chlorobenzene	132.0°
m-Xylene	139.3°
Isoamyl acetate	142.5°
Tetrachloroethane	146.3°
Bromobenzene	155.0°
p-Cymene	176.0°
Tetralin	207.0°

TABLE 10. VAPOUR PRESSURES (mm Hg) OF SATURATED AQUEOUS SOLUTIONS IN EQUILIBRIUM WITH SOLID SALTS

| Salt | Temperature | | | | | % Humidity at 20° |
	10°	15°	20°	25°	30°	
LiCl.H$_2$O			2.6			15
CaBr$_2$.6H$_2$O	2.1	2.7	3.3	4.0	4.8	19
KOAc			3.5			20
CaCl$_2$.6H$_2$O	3.5	4.5	5.6	6.9	8.3	20
CrO$_3$			6.1			32
Zn(NO$_3$)$_2$.6H$_2$O			7.4			42
K$_2$CO$_3$.2H$_2$O			7.7	10.7		44
KCNS			8.2			47
Na$_2$Cr$_2$O$_7$.2H$_2$O			9.1			52
Ca(NO$_3$)$_2$.4H$_2$O	6.0	7.7	9.6	11.9	14.2	55
Mg(NO$_3$)$_2$.6H$_2$O			9.8			56
NaBr.2H$_2$O	5.8	7.8	10.3	13.5	17.5	58
NaNO$_2$			11.6			66
NaClO$_3$			13.1			75
NaCl	6.9	9.6	13.2	17.8	21.4	75
NaOAc			13.3			76
NH$_4$Cl			13.8			79
(NH$_4$)$_2$SO$_4$			14.2			81
KBr			14.7			84
KHSO$_4$			15.1			86
KCl			15.1	20.2	27.0	86
K$_2$CrO$_4$			15.4			88
ZnSO$_4$.7H$_2$O			15.8			90
NH$_4$.H$_2$PO$_4$			16.3			93
Na$_2$HPO$_4$.12H$_2$O			16.7			95
KNO$_3$			16.7	22.3	29.8	95
Pb(NO$_3$)$_2$			17.2			98
(Water)	9.21	12.79	17.53	23.76	31.82	100

TABLE 11. DRYING AGENTS FOR CLASSES OF COMPOUNDS

Class	Dried with
Acetals	Potassium carbonate
Acids (organic)	Calcium sulphate, magnesium sulphate, sodium sulphate.
Acyl halides	Magnesium sulphate, sodium sulphate.
Alcohols	Calcium oxide, calcium sulphate, magnesium sulphate, potassium carbonate, followed by magnesium and iodine.
Aldehydes	Calcium sulphate, magnesium sulphate, sodium sulphate.
Alkyl halides	Calcium chloride, calcium sulphate, magnesium sulphate, phosphorus pentoxide, sodium sulphate.
Amines	Barium oxide, calcium oxide, potassium hydroxide, sodium carbonate, sodium hydroxide.
Aryl halides	Calcium chloride, calcium sulphate, magnesium sulphate, phosphorus pentoxide, sodium sulphate.
Esters	Magnesium sulphate, potassium carbonate, sodium sulphate.
Ethers	Calcium chloride, calcium sulphate, magnesium sulphate, sodium, lithium aluminium hydride.
Heterocyclic bases	Magnesium sulphate, potassium carbonate, sodium hydroxide.
Hydrocarbons	Calcium chloride, calcium sulphate, magnesium sulphate, phosphorus pentoxide, sodium (not for olefines).
Ketones	Calcium sulphate, magnesium sulphate, potassium carbonate, sodium sulphate.
Mercaptans	Magnesium sulphate, sodium sulphate.
Nitro compounds and nitriles	Calcium chloride, magnesium sulphate, sodium sulphate.
Sulphides	Calcium chloride, calcium sulphate.

TABLE 12.   GRADED ADSORBENTS AND SOLVENTS

Adsorbents (decreasing effectiveness)	Solvents (inreasing eluting ability)
Fuller's earth (hydrated aluminosilicate)	Petroleum ether, b 40-60°.
Magnesium oxide	Petroleum ether, b 60-80°.
Charcoal	Carbon tetrachloride.
Alumina	Cyclohexane.
Magnesium trisilicate	Benzene.
Silica gel	Ethyl ether.
Calcium hydroxide	Chloroform.
Magnesium carbonate	Ethyl acetate.
Calcium phosphate	Acetone.
Calcium carbonate	Ethanol.
Sodium carbonate	Methanol.
Talc	Pyridine.
Inulin	Acetic acid.
Sucrose = starch	

TABLE 13.   **REPRESENTATIVE ION-EXCHANGE RESINS**

Sulphonated polystyrene
Strong-acid cation exchanger   **Distributor**

AG 50W-x8   (Bio-Rad, USA)
Amberlite IR-120   (Rohm and Haas, USA)
Dowex 50W-x8   (Dow Chemical Co., USA)
Duolite 225   (Dia-Prosim Ltd)
Permutit RS   (Permutit AG, Germany)
Permutite C50D   (Phillips and Pain-Vermorel,
                              France)

Carboxylic acid-type
Weak acid cation exchangers

Amberlite IRC-50   (Rohm and Haas, USA)
Bio-Rex 70   (Bio-Rad, USA)
Chelex 100   (Bio-Rad, USA)
Duolite 436   (Dia-Prosim Ltd)
Permutit C   (Permutit AG, Germany)
Permutits H and H-70   (Permutit Co, USA)

Aliphatic amine-type
weak base anion exchangers

Amberlites IR-45 and IRA-67 (Rohm and Haas, USA)
Dowex 3-x4A   (Dow Chemical Co, USA)
Permutit E   (Permutit AG, Germany)
Permutit A 240A   (Phillips and Pain-Vermorel,
                              France)

Strong Base, anion exchangers

AG 2x8   (Bio-Rad, USA)
Amberlite IRA-400   (Rohm and Haas, USA)
Dowex 2-x8   (Dow Chemical Co, USA)
Duolite 113   (Dia-Prosim Ltd)
Permutit ESB   (Permutit AG, Germany)
Permutite 330D   (Phillips and Pain-Vermorel,
                              France)

TABLE 14.   MODIFIED FIBROUS CELLULOSES FOR ION-EXCHANGE

Cation exchange	Anion exchange
CM cellulose (carboxymethyl)	DEAE cellulose (diethylaminoethyl)
CM 22, 23 cellulose	DE 22, 23 cellulose
P cellulose (phosphate)	PAB cellulose (p-aminobenzyl)
SE cellulose (sulphoethyl)	TEAE cellulose (triethylaminoethyl)
SM cellulose (sulphomethyl)	ECTEOLA cellulose

SE and SM are much stronger acids than CM, whereas P has two ionisable groups (pK 2-3, 6-7), one of which is stronger, the other weaker, than for CM (3.5 - 4.5).

For basic strengths, the sequence is:
TEAE » DEAE (pK 8 - 9.5) > ECTEOLA (pK 5.5 - 7) > PAB.
Their exchange capacities lie in ths range 0.3 to 1.0 mg equiv./g.

TABLE 15.   BEAD FORM ION-EXCHANGE PACKAGINGS[1]

Cation exchange	Capacity (meq/g)	Anion exchange	Capacity (meq/g)
CM-Sephadex C-25, C-50.[2] (weak acid)	4.5±0.5	DEAE-Sephadex A-25, A-50.[7] (weak base)	3.5±0.5
SP-Sephadex C-25, C-50.[3] (strong acid)	2.3±0.3	QAE-Sephadex A-25, A-50.[8] (strong base)	3.0±0.4
CM-Sepharose CL-6B.[4]	0.12±0.02	DEAE-Sepharose CL-6B.[4]	0.13±0.02
Fractogel TSK CM-650.[5]		Fractogel TSK DEAE-650.[5]	
		DEAE-Sephacel.[9]	1.4±0.1
CM-32 Cellulose.		DE-32 Cellulose.	
CM-52 Cellulose.[6]		DE-52 Cellulose	

[1] May be sterilised by autoclaving at pH 7 and below 120°.
[2] Carboxymethyl. [3] Sulphopropyl. [4] Crosslinked agarose gel, no precycling required, pH range 3-10.
[5] Hydrophilic methacrylate polymer with very little volume change on change of pH; equivalent to *Toyopearl*. [6] Microgranular, pre-swollen, does not require precycling. [7] Diethylaminoethyl.
[8] Diethyl(2-hydroxypropyl)aminoethyl. [9] Bead form cellulose, pH range 2-12, no precycling.

TABLE 16.	COLUMNS FOR HPLC[1,2]	
Column	Mobile Phase[3]	Application
DUPONT		
ODS"permaphase" (octadecyl silane).	Most solvents, not strong acids and bases, for gradient elution.	Aromatic compds, sterols, drugs, natural products.
HCP (hydrocarbon polymer).	Aqueous alcohols up to 50% isopropanol.	Aromatic compds. quinones.
CWT (carbowax 4000).	Hydrocarbons only.	Steroids and polar organic compounds.
TMG (trimethylene methylene glycol)	Hydrocarbons, $CHCl_3$, dioxane and tetra-hydrofuran. Not alcohols, Phase must be saturated with tri-methylene glycol.	Hydroxy and amino compds, pesticides, polymer inter-mediates.
BOP (2,2'-oxydi-propionitrile).	Hydrocarbons, butyl ether, up to 15% of THF. Phase must be satd with 2,2'-oxydi-propionitrile.	Alkaloids, pest-icides, polymer additives, steroids
WAX (weak anion exchange).	Water only, retention and resolution are modified by pH and ionic strength	Ionic compounds.
SAX (strong anion exchange).	Water only, as above.	As above.
SCX (strong cation exchange).	Water only, as above.	As above.

(continued)

TABLE 16 (cont.).	COLUMNS FOR HPLC[1,2]	
Column	Mobile Phase[3]	Application
MERCK		
Silica Gel 60-Kieselgel 60.	EtOH, $CHCl_3$, CH_2Cl_2, n-C_7H_{16}, EtOAc, acetic acid.[2]	Vitamins, alkaloids esters, steroids, drugs, aromatic, compds.
LiChromosorb SI60 SI 100 and Altex T.	Hydrocarbons, ether aliphatic acids, Me_2SO, $CHCl_3$, CH_2Cl_2 t-BuOH.	As above, phthal-imido-acids, anti-oxidants.
Perisorb A.	Hexane, acetic acid, isooctane, EtOAc.	Acids, esters, aromatic amines and hydrocarbons.
Perisorb PA6.	MeOH, H_2O, AcOH.	As above.
Perisorb KAT.	Aq. Buffers to pH 11.	Heterocycles, nucleosides, acids and bases.
BAKER		
Bakerbond Chiral DNBPG (ionic or covalent).[4]	t-BuOH, 2-PrOH, Butyl methyl ether, hexane, $CHCl_3$, and phases below.	Chiral mixts of alcohols, acids, amines, variety of enantiomeric compds
DNBLeu (cov-alent).[5]	Chiral phases: S-Asp-artyl-S-phenylalanine methyl ester, N,N-Di-propyl-S-alanine cupric acetate.	Phosphonates, aryl-sulphoxides, nitro-gen heterocycles, di-β-naphthols.

[1] Only a few representative columns are tabulated.
[2] Waters Assoc., T.J.Baker and Altex also have a wide range of columns.
[3] Not to be used above 50°, halide acids and salts are corrosive and must be avoided.
[4] $R-N-$3,5-dinitrobenzoylphenylglycine.
[5] $R-N-$3,5-dinitrobenzoylleucine.

TABLE 17. LIQUIDS FOR STATIONARY PHASES IN GAS CHROMATOGRAPHY

Material	Temp.	Retards
Dimethylsulpholane	0-40°	Olefines and aromatic hydrocarbons
Di-*n*-butyl phthalate	0-40°	General purposes
Squalane	0-150°	Volatile hydrocarbons and polar molecules
Silicone oil or grease	0-250°	General purposes
Diglycerol	20-120°	Water, alcohols, amines, esters, and aromatic hydrocarbons
Dinonyl phthalate	20-130°	General purposes
Polydiethylene glycol succinate	50-200°	Aromatic hydrocarbons, alcohols, ketones, esters.
Polyethylene glycol	50-200°	Water, alcohols, amines, esters and aromatic hydrocarbons
Apiezon grease	50-200°	Volatile hydrocarbons and polar molecules
Tricresyl phosphate	50-250°	General purposes

TABLE 18. SOME COMMON IMMISCIBLE OR SLIGHTLY MISCIBLE PAIRS OF SOLVENTS

Carbon tetrachloride with ethanolamine, ethylene glycol, formamide or water.

Dimethyl formamide with cyclohexane or petroleum ether.

Dimethyl sulphoxide with cyclohexane or petroleum ether.

Ethyl ether with ethanolamine, ethylene glycol or water.

Methanol with carbon disulphide, cyclohexane or petroleum ether.

Petroleum ether with aniline, benzyl alcohol, dimethyl formamide, dimethyl sulphoxide, formamide, furfuryl alcohol, phenol or water.

Water with aniline, benzene, benzyl alcohol, carbon disulphide, carbon tetrachloride, chloroform, cyclohexane, cyclohexanol, cyclohexanone, ethyl acetate, isoamyl alcohol, methyl ethyl ketone, nitromethane, tributyl phosphate or toluene.

TABLE 19.	AQUEOUS BUFFERS
Approx. pH	Composition

0	$2N$ sulphuric acid or N hydrochloric acid
1	$0.1N$ hydrochloric acid or $0.18N$ sulphuric acid
2	**Either** $0.01N$ hydrochloric acid or $0.013N$ sulphuric acid
	Or 50 ml of $0.1M$ glycine (also $0.1M$ NaCl) + 50 ml of $0.1N$ hydrochloric acid
3	**Either** 20 ml of the $0.2M$ Na_2HPO_4 + 80 ml of $0.1M$ citric acid
	Or 50 ml of $0.1M$ glycine + 22.8 ml of $0.1N$ hydrochloric acid in 100 ml
4	**Either** 38.5 ml of $0.2M$ Na_2HPO_4 + 61.5 ml of $0.1M$ citric acid
	Or 18 ml of $0.2M$ NaOAc + 82 ml of $0.2M$ acetic acid
5	**Either** 70 ml of $0.2M$ NaOAc + 30 ml of $0.2M$ acetic acid
	Or 51.5 ml of $0.2M$ Na_2HPO_4 + 48.5 ml of $0.1M$ citric acid
6	63 ml of $0.2M$ Na_2HPO_4 + 37 ml of $0.1M$ citric acid
7	82 ml of M Na_2HPO_4 + 18 ml of $0.1M$ citric acid
8	**Either** 50 ml of $0.1M$ Tris buffer + 29 ml of $0.1N$ hydrochloric acid, in 100 ml
	Or 30 ml of $0.05M$ borax + 70 ml of $0.2M$ boric acid
9	80 ml of $0.05M$ borax + 20 ml of $0.2M$ boric acid
10	**Either** 25 ml of $0.05M$ borax + 43 ml of $0.1N$ NaOH, in 100 ml
	Or 50 ml of $0.1M$ glycine + 32 ml of $0.1N$ NaOH, in 100 ml
11	50 ml of $0.15M$ Na_2HPO_4 + 15 ml of $0.1N$ NaOH
12	50 ml of $0.15M$ Na_2HPO_4 + 75 ml of $0.1N$ NaOH
13	$0.1N$ NaOH or KOH
14	N NaOH or KOH

These buffers are suitable for use in obtaining ultraviolet spectra. Alternatively, for a set of accurate buffers of low, but constant, ionic strength ($I = 0.01$) covering a pH range 2.2 to 11.6 at 20°, see Perrin *Aust J Chem* **16** 572 *1963.*

BIBLIOGRAPHY

The following books and reviews provide fuller details of the topics indicated.

Safety in the Chemical Laboratory

L.Bretherick, *Handbook of Reactive Chemical Hazards*, 3rd edn, Butterworths, London, 1985.
College Safety Committee, *Safety in the Chemical Laboratory and in the Use of Chemicals*, 3th edn, Imperial College, London, 1971.
College Safety Committee, *Code of Practice against Radiation Hazards*, 6th edn, Imperial College, London, 1973.
M.J.Lefevre, *First Aid Manual for Chemical Accidents*, Dowden, Hutchinson and Ross, PA, 1980.
R.E.Lenga (ed.), *The Sigma-Aldrich Library of Chemical Safety Data*, Sigma-Aldrich Coporation, Milwaukee, Wl, 1986.
G.M.Muir (ed.), *Hazards in the Chemical Industry*, 2nd edn, The Royal Society of Chemistry, London, 1977.
D.A.Pipitone, *Safe Storage of Laboratory Chemicals*, J.Wiley & Sons Inc., New York, 1984.

Prudent Practices for Handling Hazardous Chemicals in Laboratories, National Academy Press, Washington, D.C., 1983.

N.I.Sax, *Dangerous Properties of Industrial Materials*, 6th edn, Van Nostrand Reinhold, New York, 1984.

N.V.Steere (ed.), *CRC Handbook of Laboratory Safety*, CRC Press, Florida, 1971.

Laboratory Technique and Theoretical Discussion

E.W.Berg, *Physical and Chemical Methods of Separation*, McGraw-Hill, New York, 1963.

K.L.Cheng, K.Ueno and T.Imamura, *CRC Handbook of Organic Analytical Reagents*, CRC Press Inc, Florida, 1982.

N.D.Cheronis and J.B.Entrikin, *Identification of Organic Compounds*, Interscience, New York, 1963.

P.S.Diamond and R.F.Denman, *Laboratory Techniques in Chemistry and Biochemistry*, 2nd edn, Butterworths, 1973.

Fieser and Fieser's *Reagents for Organic Chemistry*, Vols 1 to 11, Wiley-Interscience, New York, 1967 to 1984.

B.S.Furniss, A.J.Hannaford, V.Rogers, P.W.G.Smith and A.R.Tatchell, Vogel's *Textbook of Practical Organic Chemistry*, 4th edn, Longmans, London, 1978.

B.Keil, *Laboratoriumstechnik der Organischen Chemie*, Academie Verlag, Berlin, 1961 (German edn, translated and revised by H.Fürst).

R.P.Linstead, J.A.Elvidge and M.Whalley, *A Course in Modern Techniques of Organic Chemistry*, Butterworths, London, 1955.

C.J.O.R.Morris and P.Morris, *Separation Methods in Biochemistry*, 2nd edn, Interscience, New York, 1976.

P.A.Ongley (ed), Organicum, *Practical Handbook of Organic Chemistry*, Pergamon Press, Oxford (English edn), 1973.

R.K.Scopes, *Protein Purification. Principles and Practice*, Springer-Verlag, New York, 1982.

F.J.Wolf, *Separation Methods in Chemistry and Biochemistry*, Academic Press, New York, 1969.

Organic solvents

J.A.Riddick and W.B.Bunger, *Organic Solvents: Physical Properties and Methods of Purification, Techniques of Chemistry*, Vol II, Wiley-Interscience, New York, 1970.

J.F.Coetzee (ed), *Purification of Solvents,* Pergamon Press, Oxford, 1982.

Solvent Extraction and Distribution

L.C.Craig, D.Craig and E.G.Scheibel, in A. Weissberger's (ed) *Techniques of Organic Chemistry*, vol III, pt I, 2nd edn, Interscience, New York, 1956.

F.A.von Metzsch, in W.G.Berl's (ed), *Physical Methods in Chemical Analysis*, vol IV, Academic Press, New York, 1961.

Distillation

T.P.Carney, *Laboratory Fractional Distillation*, Macmillan, New York, 1949.

E.Krell, *Handbook of Laboratory Distillation*, Elsevier, Amsterdam, 1963.

A.Weissberger (ed), *Techniques of Organic Chemistry*, vol IV, *Distillation*, Interscience, New York, 1951.

Crystallisation

R.S.Tipson, in A.Weissberger (ed), *Techniques of Organic Chemistry*, vol III, pt I, 2nd edn, Interscience, New York, 1956.

Zone Refining

E.F.G.Herington, *Zone Melting of Organic Compounds*, Wiley, New York, 1963.
W.Pfann, *Zone Melting*, 2nd edn, Wiley, New York, 1966.
H.Schildknecht, *Zonenschmelzen*, Verlag Chemie, Weiheim, 1964.
W.R.Wilcox, R.Friedenberg *et al*, *Chem.Rev.*, **64** 187 *1964*.
M.Zief and W.R.Wilcox (eds), *Fractional Solidification*, vol I, Marcel Dekker, New York, 1967.

Drying

G.Broughton, in A.Weissberger's (ed) *Techniques of Organic Chemistry*, vol III, pt I, 2nd edn, Interscience, New York, 1956.

Molecular Sieves

P.Andrews, *Molecular Sieve Chromatography, Brit.Med.Bull.*, **22** 109 *1966*.
D.W.Breck, *Zeolite Molecular Sieves. Structure, Chemistry and Use*, Wiley & Sons, New York, 1974.
C.K.Hersh, *Molecular Sieves*, Reinhold, New York, 1961.
G.R.Landolt and G.T.Kerr, *Sepn.Purif.Meth.*, **2** 283 *1973*.
Union Carbide Molecular Sieves for Selective Adsorption, 2nd edn, British Drug Houses, Poole, England, 1961.

Chromatography

J.N.Balston, B.G.Talbot and T.S.G.Jones, *A Guide to Filter Paper and Cellulose Powder Chromatography*, Reeve Angel, London, and Balston, Maidstone, 1952.
R.J.Block, E.L.Durrum and G.Zweig, *A Manual of Paper Chromatography and Paper Electrophoresis*, Academic Press, New York, 1955.
H.G.Cassidy, *Fundamentals of Chromatography, Techniques of Organic Chemistry*, Vol X, Interscience, New York, 1957.
J.C.Giddings, E.Grushka and P.R.Brown (eds), *Advances in Chromatography*, (M.Dekker), Vols 1-27, 1955-1987.
E.Heftman (ed), *Chromatography*, 2nd Edn, Reinhold, New York, 1967.
E.Lederer and M.Lederer, *Chromatography, A Review of Principles and Applications*, 2nd edn, Elsevier, Amsterdam, 1957.
W.L.Hinze and D.W.Armstrong, eds, *Ordered Media in Chemical Separations*, ACS Symposia Series 342, ACS, Washington DC, 1987.
R.Stock and C.B.F.Rice, *Chromatographic Methods*, 3rd edn, Chapman and Hall, London, 1974.
T.I.Williams, *An Introduction to Chromatography*, Blackie, London, 1947.

Ion Exchange

C.Calmon and T.R.E.Kressman (eds), *Ion Exchangers in Organic and Biochemistry*, Interscience, New York, 1957.
Dowex: Ion Exchange, Dow Chemical Co., Midland, Michigan, 1959.
C.G.Horvath, *Ion Exchangers*, Vol. 3, Dekker, New York, 1972.
Ion Exchange Resins, British Drug Houses, 5th edn, Poole, England, 1977.
J.Khym, *Analytical Ion Exchange Procedures in Chemistry and Biology*, Prentice Hall, New Jersey, 1974.
R.Kunin, *Ion Exchange Resins*, 2nd edn, Wiley & Sons, New York, 1958.
E.A.Peterson in *Laboratory Techniques in Biochemistry and Molecular Biology*, Vol 2, Pt II, T.S.Work and E.Work (eds), North Holland, Amsterdam, 1970.
E.A.Peterson and H.A.Sober, *J.Amer.Chem.Soc.*, **78** 751 *1956*.
J.Porath, *Ark.Kemi Min.Geol.*, **11** 97 *1957*.

H.A.Sober *et al, J.Amer.Chem.Soc.*, **78** 756 *1956*.
H.F.Walton (ed), *Ion-exchange Chromatography*, Dowden, Hutchinson and Ross Inc, Stroudsburg, Pa., distributed by Halstead Press, 1976.

Ionic Equilibria

G.Kortüm, W.Vogel and K.Andrussow, *Dissociation Constants of Organic Acids in Aqueous Solution*, Butterworths, London, 1961.
D.D.Perrin, *Dissociation Constants of Organic Bases in Aqueous Solution*, Butterworths, London, 1965; and Supplement 1972.
D.D.Perrin, *Dissociation Constants of Inorganic Acids and Bases in Aqueous Solution*, Butterworths, London, 1969.
D.D.Perrin and B.Dempsey, *Buffers for pH and Metal Ion Control*, Chapman and Hall, London, 1974.
E.P.Serjeant and B.Dempsey, *Ionization Constants of Organic Acids in Aqueous Solution*, Pergamon Press, Oxford, 1979.

Gas Chromatography

D.Ambrose, *Gas Chromatography*, 2nd edn, Butterworths, London, 1971.
V.J.Coates, H.J.Noebels and I.S.Fagerson (eds), *Gas Chromatography*, Academic Press, New York, 1958.
D.H.Desty (ed), *Vapour Phase Chromatography*, Butterworths, London, 1957.
D.H.Desty (ed), *Gas Chromatography*, Butterworths, London, 1958.
W.Hemes, *Capillary Gas Chromatography-Fourier Transform Infra Red Spectroscopy-Applications*, Huethig, 1987.
H.Jaeger, *Capillary Gas Chromatography-Mass Spectrometry in Medicine and Pharmacology*, Huethig, 1987.
R.A.Jones, *An Introduction to Gas-Liquid Chromatography*, Academic Press, London, 1970.
R.Kaiser, *Gas Chromatography*, vol.3, Butterworths, London, 1963 (practical details of liquid phases, adsorbents, etc., with references).
A.I.M.Keulemans, *Gas Chromatography*, 2nd edn, Reinhold, New York, 1959. Materials for Gas Chromatography, May & Baker, 1958.
W.A.König, *The Practice of Enantiomeric Separation By Capillary Gas Chromatography*, Huethig, 1987.
A.B.Littlewood, *Gas Chromatography: Principles, Techniques and Applications*, 2nd edn, Academic Press, New York, 1970.
H.J.Noebels, R.F.Wall and Brenner (eds), *Gas Chromatography*, Academic Press, New York, 1961.
R.L.Pecsok, *Principles and Practice of Gas Chromatography*, Wiley & Sons, New York, 1959.
R.W.Zumwalt, K.C.Kuo and C.W.Gehrke, *Amino Acid Analysis by Gas Chromatography*, CRC Press Inc, Florida, 1987.

High Pressure Liquid Chromatography

B.A.Bidlingmeyer, *Preparative Liquid Chromatography*, Elsevier, Amsterdam, 1987.
P.R.Brown, *High Pressure Liquid Chromatography*, Academic Press, New York, 1973.
Z.Deyl, K.Macek and J.Janak (eds), *Liquid Column Chromatography. A Survey of Modern Techniques and Applications*, Elsevier, Amsterdam, 1975.
P.F.Dixon, C.H.Gray, C.K.Lim and M.S.Stoll (eds), *High Pressure Liquid Chromatography in Clinical Chemistry (Symposium)*, Academic Press, New York, 1976.
H.Engelhardt, *High Performance Liquid Chromatography*, (translated from the German by G.Gutnikov), Springer-Verlag, Berlin, 1979.
J.J.Kirkland, Columns for HPLC, *Anal.Chem.*, **43(12)** 36A *1971*.
A.M.Krstulovic and P.R.Brown, *Reversed Phase HPLC. Theory, Practical and Biomedical Applications*, Wiley & Sons, New York, 1982.
E.Molner (ed), *Practical Aspects of Modern HPLC*, W.deGruyter, Berlin, 1983.

N.A.Parris, *Instrumental Liquid Chromatography.   A Manual of High Performance Liquid Chromatography Methods,* Elsevier, Amsterdam, 1976.
A.P.Pryde and M.T.Gilbert, *Applications of High Performance Liquid Chromatography*, Chapman and Hall, London, 1979.
A.Zlatkis and R.E.Kaiser (eds), *High Performance Thin Layer Chromatography*, Elsevier, Amsterdam, 1977.

Affinity Chromatography

I.A.Chaiken, *Analytical Affinity Chromatography*, CRC Press Inc, Florida, 1987.
E.V.Gorman and M.Wilchek, *Recent Developments in Affinity Chromatography Effects, Trends in Biotechnology,* **5** 220 *1987.*
W.B.Jakoby and M.Wilchek, *Affinity Chromatography* in *Methods in Enzymology*, vol.34, 1975.
C.R.Lowe and P.D.G.Dean, *Affinity Chromatography*, Wiley & Sons, New York, 1974.
J.Turkova, *Affinity Chromatography*, Elsevier, Amsterdam, 1978,

Gel Filtration

H.Determann, *Gel Filtration*, Springer-Verlag, Berlin, 1969.
L.Fischer, *Introduction to Gel Chromatography*, North-Holland, Amsterdam, 1968.
L.Fischer, *Gel Filtration*, 2nd edn, Elsevier/North-Holland, Amsterdam, 1980.
P.Flodin, *Dextran Gels and Their Applications in Gel Filtration*, Pharmacia, Uppsala, Sweden, 1962.

CHAPTER 2

CHEMICAL METHODS
USED
IN PURIFICATION

GENERAL REMARKS

Greater selectivity in purification can often be achieved by making use of differences in chemical properties between the substance to be purified and its contaminants. Unwanted metal ions may be removed by precipitation in the presence of a *collector* (see p 58). Sodium borohydride and other metal hydrides transform organic peroxides and carbonyl-containing impurities such as aldehydes and ketones in alcohols and ethers. Many classes of organic chemicals can be purified by conversion into suitable derivatives, followed by regeneration. This Chapter describes relevant procedures.

REMOVAL OF TRACES OF METALS FROM REAGENTS

It is necessary to purify the reagents used for determinations of the more common heavy metals. Also, there should be very little if any metallic contamination of many of the materials required for biochemical studies. The main methods for removing impurities of this type are as follows.

Distillation. Reagents such as water, ammonia, hydrochloric acid, nitric acid, perchloric acid (under reduced pressure), and sulphuric acid can be purified in this way using all-glass stills. Isothermal distillation is convenient for ammonia: a beaker containing concentrated ammonia is left alongside a beaker of distilled water for several days in an empty desiccator so that some of the ammonia distils over into the water. Hydrochloric acid can be purified in the same way. The redistilled ammonia should be kept in polyethylene or paraffin-waxed bottles. In some cases, instead of attempting to purify a salt it is simpler to synthesise it from distilled components. Ammonium acetate is an example.

Use of ion-exchange resin. Application of ion-exchange columns has greatly facilitated the removal of heavy metal ions such as Cu^{2+}, Zn^{2+} and Pb^{2+} from aqueous solutions of many reagents. Thus, sodium salts and sodium hydroxide can be purified by passage through a column of a cation-exchange resin in its sodium form. Similarly, for acids, a resin in its H^+ form is used. In some cases, where metals form anionic complexes, they can be removed by passage through an anion-exchange resin. Iron in hydrochloric acid solution is an example.

Ion exchange resins are also useful for demineralising biochemical preparations such as proteins. Removal of metal ions from protein solutions using polystyrene-based resins, however, may lead to protein denaturation. This difficulty may be avoided by using a weakly acidic cation exchanger such as Bio-Rex 70 (which is a carboxylic acid exchange resin based on a polyacrylic lattice).

Heavy metal contamination of *p*H buffers can be removed by passage of the solutions through a Chelex 100 column. For recent examples of biological studies, see Crouch and Klee *Biochem* **19** 3692 *1980* (HEPES, calmodulin); Bruger *et al*, *Biochem* **23** 1966 *1984* (TES, calmodulin); Ma and Ray *Biochem* **19** 751 *1980* (TRIS, phosphoglucomutase); Reiss *et al*, *Biochem* **140(1)** 62 *1984* (sodium phosphate, nucleotides). For the removal of metal ions from enzyme solutions, see Dunn *et al*, *Biochem* **19** 718 *1980*. For examples of the use of ion-exchange resins, see Viola *et al*, *Anal Biochem* **96** 334 *1979* (reduced nicotinamide adenine dinucleotide); Němecek *et al*, *JBC* **254** 598 *1979* (cyclic AMP);

MacManus *Anal Biochem* **96** 407 *1979* (cyclic GMP); Kawakita and Yamazaki *Biochem* **17** 3546 *1978* (IDP); Krishnan and Krishnan *Anal Biochem* **70** 18 *1978* (cyclic GMP).

Water, with very low concentrations of ionic impurities (and approaching conductivity standards), is very readily obtained by percolation through alternate columns of cation- and anion-exchange resins, or through a mixed-bed resin, and many commercial devices are available for this purpose. For some applications, this method is unsatisfactory because the final water may contain traces of organic material after passage through the columns. However, organic matter can also be removed by using yet another special column in series for this purpose (see Milli Q water preparation, Millipore Corp.).

Precipitation.   In removing traces of impurities by precipitation it is necessary to include a material to act as a *collector* of the precipitated substance so as to facilitate its removal by filtration or decantation. Aqueous hydrofluoric acid can be freed from lead by adding 1ml of 10% strontium chloride per 100ml of acid, lead being co-precipitated as lead fluoride with the strontium fluoride. If the acid is decanted from the precipitate and the process repeated, the final lead content in the acid is less than 0.003 ppm. Similarly, lead can be precipitated from a nearly saturated sodium carbonate solution by adding 10% strontium chloride dropwise (1-2ml per 100ml), then filtering. (If the sodium carbonate is required as a solid, the solution can be evaporated to dryness in a platinum dish.) Removal of lead from potassium chloride uses precipitation as lead sulphide, followed, after filtration, by evaporation and recrystallisation of the potassium chloride.

Several precipitation methods are available for iron. It has been removed from potassium thiocyanate solutions by adding a few milligrams of an aluminium salt, then precipitating aluminum and iron as their hydroxides by adding a few drops of ammonia. Iron is also carried down on the hydrated manganese dioxide precipitate formed in cadmium chloride or cadmium sulphate solutions by adding 0.5% aqueous potassium permanganate (0.5ml per 100ml of solution), sufficient ammonia to give a slight precipitate, and 1 ml of ethanol. The solution is heated to boiling to coagulate the precipitate, then filtered. For the removal of iron from sodium potassium tartrate, a small amount of cadmium chloride solution and a slight excess of ammonium sulphide are added, the solution is stood for 1 hour, and the sulphide precipitate is filtered off. Ferrous iron can be removed from copper solutions by adding some hydrogen peroxide to the solution to oxidise the iron, followed by precipitation of ferric hydroxide by adding a small amount of sodium hydroxide.

Traces of calcium can be removed from solutions of sodium salts by precipitation at pH 9.5-10 as its 8-hydroxyquinolinate. The excess of 8-hydroxyquinoline acts as a *collector*. The magnesium content of calcium chloride solutions can be reduced by making them about 0.1M in sodium hydroxide and filtering.

Extraction.   In some cases, a simple solvent extraction is sufficient to remove a particular impurity. For example, traces of gallium can be removed from titanous chloride in hydrochloric acid by extraction with isopropyl ether. Similarly, ferric chloride can be removed from aluminium chloride solutions containing hydrochloric acid by extraction with ethyl ether. Usually, however, it is necessary to extract with an organic solvent in the presence of a suitable complexing agent such as dithizone or sodium diethyl dithiocarbamate. When the former is used, weakly alkaline solutions are extracted with dithizone in chloroform (at about 25mg/L of chloroform) or carbon tetrachloride until the colour of some fresh dithizone solution remains unchanged after shaking. Excess dithizone is taken out by extracting with the pure solvent, the last traces of which, in turn, are removed by aeration. This method has been used with aqueous solutions of ammonium hydrogen citrate, potassium bromide, potassium cyanide, sodium acetate and sodium citrate. The advantage of dithizone for such a purpose lies in the wide range of metals with which it combines under these conditions. 8-Hydroxyquinoline (oxine) can also be used in this way. Sodium diethyl dithiocarbamate has been used to purify aqueous hydroxylamine hydrochloride (made just alkaline to thymol blue by adding ammonia) from copper and other heavy metals by repeated extraction with chloroform until no more diethyl dithiocarbamate remained in the solution (which was then acidified to thymol blue by adding hydrochloric acid).

Complexation.   Although not strictly a removal of an impurity, addition of a suitable complexing agent such as ethylenediaminetetra-acetic acid often overcomes the undesirable effects of contaminating metal ions by reducing the concentrations of the free metal species to very low levels. For a detailed discussion of this *masking*, see **Masking and Demasking of Chemical Reactions**, D.D.Perrin, Wiley-Interscience, New York, 1970.

USE OF METAL HYDRIDES

This group of reagents has become commercially available in large quantities; some of its members - notably lithium aluminium hydride ($LiAlH_4$), calcium hydride (CaH_2), sodium borohydride ($NaBH_4$) and potassium borohydride (KBH_4) - have found widespread use in the purification of chemicals.

Lithium aluminium hydride. This solid is stable at room temperature, and is soluble in ether-type solvents. It reacts violently with water, liberating hydrogen, and is a powerful drying and reducing agent for organic compounds. It reduces aldehydes, ketones, esters, carboxylic acids, peroxides, acid anhydrides and acid chlorides to the corresponding alcohols. Similarly, amides, nitriles, aldimines and aliphatic nitro compounds yield amines, while aromatic nitro compounds are converted to azo compounds. For this reason it finds extensive application in purifying organic chemical substances by the removal of water and carbonyl containing impurities as well as peroxides formed by autoxidation. Reactions can generally be carried out at room temperature, or in refluxing ethyl ether, at atmospheric pressure. *When drying organic liquids with this reagent it is important that the concentration of water in the liquid is below 0.1% otherwise a violent reaction or explosion may occur. The mixing of the liquid with the reagent should be performed at ice bath temperature and under a reflux condenser.*

Calcium hydride. This powerful drying agent is suitable for use with hydrogen, argon, helium, nitrogen, hydrocarbons, chlorinated hydrocarbons, esters and higher alcohols.

Sodium borohydride. This solid which is stable in dry air up to 300° like potassium borohydride, is a less powerful reducing agent than lithium aluminium hydride, from which it differs also by being soluble in hydroxylic solvents and to a lesser extent in ether-type solvents. Sodium borohydride forms a dihydrate melting at 36-37°, and its aqueous solutions decompose slowly unless stabilised to above pH 9 by alkali. (For example, a useful solution is one nearly saturated at 30-40° and containing 0.2% sodium hydroxide.) Its solubility in water is 25, 55 and 88g per 100ml of water at 0°, 25° and 60°, respectively. Its aqueous solutions are rapidly decomposed by boiling or acidification. The reagent, available either as a hygroscopic solid or as an aqueous sodium hydroxide solution, is useful as a water soluble reducing agent for aldehydes, ketones and organic peroxides. This explains its use for the removal of carbonyl-containing impurities and peroxides from alcohols, polyols, esters, polyesters, amino-alcohols, olefines, chlorinated hydrocarbons, ethers, polyethers, amines (including aniline), polyamines and aliphatic sulphonates.
Purifications can be carried out conveniently using alkaline aqueous or methanolic solutions, allowing the reaction mixture to stand at room temperature for several hours. Other solvents that can be used with this reagent include isopropyl alcohol (without alkali), amines (including liquid ammonia, in which its solubility is 104g per 100g of ammonia at 25°, and ethylenediamine), diglyme, formamide, dimethylformamide and tetrahydrofurfuryl alcohol. Alternatively, the material to be purified can be percolated through a column of the borohydride. In the absence of water, sodium borohydride solutions in organic solvents such as dioxane or amines decompose only very slowly at room temperature. Treatment of ethers with sodium borohydride appears to inhibit peroxide formation.

Potassium borohydride. Potassium borohydride is similar in properties and reactions to sodium borohydride, and, like it, is used as a reducing agent for removing aldehydes, ketones and organic peroxides. It is non-hygroscopic and can be used in water, ethanol, methanol or water-alcohol mixtures, provided some alkali is added to minimise decomposition, but it is somewhat less soluble than sodium borohydride in most solvents. For example, its solubility in water at 25° is 19g per 100ml of water (compare sodium borohydride, 55g).

PURIFICATION *via* DERIVATIVES

Relatively few derivatives of organic substances are suitable for use as aids to purification. This is because of the difficuly in regenerating the starting material. For this reason, we list below, the common methods of preparation of derivatives that can be used in this way.

Whether or not any of these derivatives is likely to be satisfactory for the use of any particular case will depend on the degree of difference in properties, such as solubility, volatility or melting point, between the starting material, its derivative and likely impurities, as well as on the ease with which the substance can be recovered. Purification *via* a derivative is likely to be of most use when the quantity of pure material that is required is not too large. Where large quantities (for example, more than 50 g) are available, it is usually more economical to purify the material directly and discard larger fractions (for example, in distillations and recrystallisations).

The most generally useful purifications *via* derivatives are as follows:

Alcohols. Aliphatic or aromatic alcohols are converted to solid esters. *p*-Nitrobenzoates are the most convenient esters to form because of their sharp melting points, and the ease with which they can be recrystallised and the alcohol recovered. The *p*-nitrobenzoyl chloride used in the esterification is prepared by refluxing dry *p*-nitrobenzoic acid with a 3mole excess of thionyl chloride for 30min on a steam bath (*in a fume cupboard*). The solution is cooled slightly and the excess thionyl chloride is distilled off under (water-pump) vacuum, keeping the temperature below 40°. Dry toluene is added to the residue in the flask, then distilled off under vacuum, the process being repeated two or three times to ensure complete removal of thionyl chloride, hydrogen chloride and sulphur dioxide. (This freshly prepared *p*-nitrobenzoyl chloride cannot be stored without decomposition; it should be used directly.) A solution of the acid chloride (1mol) in dry toluene or alcohol-free chloroform (distilled from P_2O_5 or by passage through an activated Al_2O_3 column) under a reflux condenser is cooled in an ice bath while the alcohol (1mol), with or without a solvent (preferably miscible with toluene or alcohol-free chloroform), is added dropwise to it. When addition is over and the reaction subsides, the mixture is refluxed for 30min and the solvent is removed under reduced pressure. The solid ester is then recrystallised to constant melting point from toluene, acetone, light petroleum or mixtures of these, but not from alcohols.

Hydrolysis of the ester is achieved by refluxing in aqueous N or $2N$ NaOH solution until the insoluble ester dissolves. The solution is then cooled, and the alcohol is extracted into a suitable solvent, e.g. ether, toluene or alcohol-free chloroform. The extract is dried ($CaSO_4$, $MgSO_4$) and distilled, then fractionally distilled if liquid or recrystallised if solid. (The nitro acid can be recovered by acidification of the aqueous layer.) In most cases where the alcohol to be purified is readily freed from ethanol, the hydrolysis of the ester is best achieved with N or $2N$ ethanolic NaOH or 85% aqueous ethanolic N NaOH. The former is prepared by dissolving the necessary alkali in a minimum volume of water and diluting with absolute alcohol. The ethanolic solution is refluxed for one to two hours and hydrolysis is complete when an aliquot gives a clear solution on dilution with four or five times its volume of water. The bulk of the ethanol is distilled off and the residue is extracted as above. Alternatively, use can be made of ester formation with benzoic acid, toluic acid or 3,5-dinitrobenzoic acid, by the above method. Other derivtives can be prepared by reaction of the alcohol with an acid anhydride. For example, phthalic or 3-nitrophthalic anhydride (1 mol) and the alcohol (1mol) are refluxed for half to one hour in a non-hydroxylic solvent, e.g. toluene or alcohol-free chloroform, and then cooled. The phthalate ester crystallises out, is precipitated by the addition of light petroleum or is isolated by evaporation of the solvent. It is recrystallised from water, 50% aqueous ethanol, toluene or light petroleum. Such an ester has a characteristic melting point and the alcohol can be recovered by acid or alkaline hydrolysis.

Aldehydes (and ketones). The best derivative from which an aldehyde can be recovered readily is its bisulphite addition compound, the main disadvantage being the lack of a sharp melting point. The aldehyde (sometimes in ethanol) is shaken with a cold saturated solution of sodium bisulphite until no more solid adduct separates. The adduct is filtered off, washed with a little water, then alcohol. A better reagent is freshly prepared saturated aqueous sodium bisulphite solution to which 75% ethanol is added to near-saturation. (Water may have to be added dropwise to render this solution clear.) With this reagent the aldehyde need not be dissolved separately in alcohol and the adduct is finally washed with alcohol. The aldehyde is recovered by dissolving the adduct in the least volume of water and adding an equivalent quantity of sodium carbonate (not sodium hydroxide) or concentrated hydrochloric acid to react with the bisulphite, followed by steam distillation or solvent extraction.

Other derivatives that can be prepared are the Schiff bases and semicarbazones. Condensation of the aldehyde with an equivalent of primary aromatic amine yields the Schiff base, for example aniline at 100° for 10-30min.

Semicarbazones are prepared by dissolving semicarbazide hydrochloride (*ca* 1g) and sodium acetate (*ca* 1.5g) in water (8-10ml) and adding the aldehyde or ketone (0.5-1g) and stirring. The semicarbazone crystallises out and is recrystallised from ethanol or aqueous ethanol. These are hydrolysed by steam distillation in the presence of oxalic acid or better by exchange with pyruvic acid (Hershberg *JOC* **13** 542 *1948*).

Amines. (a) Picrates: The most versatile derivative from which the free base can be readily recovered is the picrate. This is very satisfactory for primary and secondary aliphatic amines and aromatic amines and is particularly so for heterocyclic bases. The amine, dissolved in water, alcohol or benzene, is treated with excess of a saturated solution of picric acid in water, alcohol or benzene, respectively, until separation of the picrate is complete. If separation does not occur, the solution is stirred vigorously and warmed for a few minutes, or diluted with a solvent in which the picrate is insoluble. Thus, a solution of the amine and picric acid in ethanol or benzene can be treated with benzene or light petroleum, repectively, to precipitate the picrate. Alternatively, the amine can be dissolved in alcohol and aqueous picric acid added. The picrate is filtered off, washed with water, ethanol or benzene, and recrystallised from boiling water, ethanol, methanol, aqueous ethanol or methanol, chloroform or benzene. The solubility of picric acid in water, ethanol and benzene is 1.4, 6.23 and 5.27% respectively at 20°.

It is not advisable to store large quantities of picrates for long periods, *particularly when they are dry due to their potential explosive nature*. The free base should be recovered as soon as possible. The picrate is suspended in an excess of $2N$ aqueous NaOH and warmed a little. Because of the limited solubility of sodium picrate, excess hot water must be added. Alternatively, because of the greater solubility of lithium picrate, aqueous 10% lithium hydroxide solution can be used. The solution is cooled, the amine is extracted with a suitable solvent such as ethyl ether or toluene, washed with $5N$ NaOH until the alkaline solution remains colourless, then with water, and the extract is dried with anhydrous sodium carbonate. The solvent is distilled off and the amine is fractionally distilled (under reduced pressure if necessary) or recrystallised.

If the amines are required as their hydrochlorides, picrates can often be decomposed by suspending them in much acetone and adding two equivalents of $10N$ HCl. The hydrochloride of the base is filtered off, leaving the picric acid in the acetone. Dowex No 1 anion-exchange resin in the chloride form is useful for changing solutions of the more soluble picrates (for example, of adenosine) into solutions of their hydrochlorides, from which sodium hydroxide precipitates the free base (Davoll and Lowy *JACS* **73** 1650 *1951*).

(b) Salts: Amines can also be purified *via* their salts, e.g. hydrochlorides. A solution of the amine in dry toluene, ether, methylene chloride or chloroform is saturated with dry hydrogen chloride (generated by addition of concentrated sulphuric acid to dry sodium chloride, or to concentrated HCl followed by drying the gas through sulphuric acid, or from a hydrogen chloride cylinder) and the insoluble hydrochloride is filtered off and dissolved in water. The solution is made alkaline and the amine is extracted, as above. Hydrochlorides can also be prepared by dissolving the amine in ethanolic HCl and adding ether or light petroleum. Where hydrochlorides are too hygroscopic or too soluble for satisfactory isolation, other salts, e.g. nitrate, sulphate, bisulphate or oxalate, can be used.

(c) As double salts: The amine (1mol) is added to a solution of anhydrous zinc chloride (1mol) in concentrated hydrochloric acid (42ml) in ethanol (200ml, or less depending on the solubility of the double salt). The solution is stirred for 1hr and the precipitated salt is filtered off and recrystallised from ethanol. The free base is recovered by adding excess of 5-10N NaOH (to dissolve the zinc hydroxide that separates) and is steam distilled. Mercuric chloride in hot water can be used instead of zinc chloride and the salt is crystallised from 1% hydrochloric acid. Other double salts have been used, e.g. cuprous salts, but are not as convenient as the above salts.

(d) **N-Acetyl derivatives:** Purification as their *N*-acetyl derivatives is satisfactory for primary, and to a limited extent secondary, amines. The base is refluxed with slightly more than one equivalent of acetic anhydride for half to one hour, cooled and poured into ice-cold water. The insoluble derivative is filtered off, dried, and recrystallised from water, ethanol, aqueous ethanol, benzene or benzene-light petroleum. The derivative is then hydrolysed by refluxing with 70% sulphuric acid for a half to one hour. The solution is cooled, poured onto ice, and made alkaline. The amine is steam distilled or extracted as above. Alkaline hydrolysis is very slow.

(e) **N-Tosyl derivative:** Primary and secondary amines are converted into their tosyl derivatives by mixing equimolar amounts of amine and toluene-*p*-sulphonyl chloride in dry pyridine (*ca* 5-10mols) and allowing to stand at room temperature overnight. The solution is poured into ice-water and the *p*H adjusted to 2 with HCl. The solid derivative is filtered off, washed with water, dried (vac. desiccator) and recrystallised from an alcohol or aqueous alcohol solution to a sharp melting point. The derivative is decomposed by dissolving in liquid ammonia (*fume cupboard*) and adding sodium metal (in small pieces with stirring) until the blue colour persists for 10-15min. Ammonia is allowed to evaporate (*fume cupboard*), the residue treated with water and the solution checked that the *p*H is above 10. If the *p*H is below 10 then the solution has to be basified with 2*N* NaOH. The mixture is extracted with ether or toluene, the extract is dried (K_2CO_3), evaporated and the residual amine recrystallised if solid or distilled if liquid.

Aromatic hydrocarbons. (a) Adducts: Aromatic hydrocarbons can be purified as their picrates using the procedures described for amines. Instead of picric acid, 1,3,5-trinitrobenzene or 2,4,7-trinitrofluorenone can also be used. In all these cases, following recrystallisation, the hydrocarbon can be isolated either as described for amines or by passing a solution of the adduct through an activated alumina column and eluting with toluene or light petroleum. The picric acid and nitro compounds are more strongly adsorbed on the column.

(b) **By sulphonation:** Naphthalene, xylenes and alkyl benzene can be purified by sulphonation with concentrated sulphuric acid and crystallisation of the sodium sulphonates. The hydrocarbon is distilled out of the mixture with superheated steam.

Carboxylic acids (a) 4-Bromophenacyl esters: (See Lee, Judefind and Reid *JACS* **42** 1043 *1920*.) A solution of the sodium salt of the acid is prepared. If the salt is not available, the acid is dissolved in an equivalent of aqueous NaOH and the *p*H adjusted to 8-9 with this base. A solution of one equivalent of 4-bromophenacyl bromide (for a monobasic acid, two equivalents for a dibasic acid, etc) in ten times its volume of ethanol is then added. The mixture is heated to boiling, and, if necessary, enough ethanol is added to clarify the solution which is then refluxed for half to three hours depending on the number of carboxylic groups that have to be esterified. (One hour is generally sufficient for monocarboxylic acids.) On cooling, the ester should crystallise out. If it does not do so, the solution is heated to boiling, and enough water is added to produce a slight turbidity. The solution is again cooled. The ester is collected, and recrystallised or fractionally distilled.
The ester is hydrolysed by refluxing for 1-2hr with 1-5% of barium carbonate suspended in water or with aqueous sodium carbonate solution. The solution is cooled and extracted with ether, toluene or chloroform. It is then acidified and the acid is collected by filtration or extraction, and recrystallised or fractionally distilled.
p-Nitrobenzyl esters can be prepared in an analogous manner using the sodium salt of the acid and *p*-nitrobenzyl bromide. They are readily hydrolysed.

(b) **Alkyl esters:** Of the alkyl esters, methyl esters are the most useful because of their rapid hydrolysis. The acid is refluxed with one or two equivalents of methanol in excess alcohol-free chloroform (or methylene chloride) containing about 0.1g of toluene-*p*-sulphonic acid (as catalyst), using a Dean and Stark trap. (The water formed by the esterification is carried away into the trap.) When the theoretical amount of water is collected in the trap, esterification is complete. The chloroform solution in the flask is washed with 5% aqueous sodium carbonate solution, then water, and dried over sodium sulphate or magnesium sulphate. The chloroform is distilled off and the ester is fractionally distilled through an efficient column. The ester is hydrolysed by refluxing with 5-10%

aqueous NaOH solution until the insoluble ester has completely dissolved. The aqueous solution is concentrated a little by distillation to remove all of the methanol. It is then cooled and acidified. The acid is either extracted with ether, toluene or chloroform, or filtered off and isolated as above. Other methods for preparing esters are available.

(c) **Salts:** The most useful salt derivatives for carboxylic acids are the isothiouronium salts. These are prepared by mixing almost saturated solutions containing the acid (carefully neutralised with N NaOH using phenolphthalein indicator) then adding two drops of N HCl and an equimolar amount of S-benzylisothiouronium chloride in ethanol and filtering off the salt that crystallises out. After recrystallisation from water, alcohol or aqueous alcohol the salt is decomposed by suspending or dissolving in $2N$ HCl and extracting the carboxylic acid in ether, chloroform or toluene.

Hydroperoxides. These can be converted to their sodium salts by precipitation below 30° with aqueous 25% NaOH. The salt is then decomposed by addition of solid (powdered) carbon dioxide and extracted with low-boiling petroleum ether. The solvent should be removed under reduced pressure below 20°. **The apparatus should be adequately shielded at all times for the safety of the operator from explosions.**

Ketones. (a)Bisulphite adduct: The adduct can be prepared and decomposed as described for aldehydes. Alternatively, because no Cannizzaro reaction is possible, it can also be decomposed with $0.5N$ NaOH.

(b) **Semicarbazones:** A powdered mixture of semicarbazide hydrochloride (1 mol) and anhydrous sodium acetate (1.3 mol) is dissolved in water by gentle warming. A solution of the ketone (1 mol) in the least volume of ethanol needed to dissolve it is then added. The mixture is warmed on a water bath until separation of the semicarbazone is complete. The solution is cooled, and the solid is filtered off. After washing with a little ethanol followed by water, it is recrystallised from ethanol or dilute aqueous ethanol. The derivative should have a characteristic melting point. The semicarbazone is decomposed by refluxing with excess of oxalic acid or with aqueous sodium carbonate solution. The ketone (which steam distils) is distilled off. It is extracted or separated from the distillate (after saturating with NaCl), dried with $CaSO_4$ or $MgSO_4$ and fractionally distilled using an efficient column (under vacuum if necessary).

Phenols. The most satisfactory derivatives for phenols that are of low molecular weight or monohydric are the benzoate esters. (Their acetate esters are generally liquids or low-melting solids.) Acetates are more useful for high molecular weight and polyhydric phenols.

(a) **Benzoates:** The phenol (1mol) in 5% aqueous NaOH is treated (while cooling) with benzoyl chloride (1mol) and the mixture is stirred in an ice bath until separation of the solid benzoyl derivative is complete. The derivative is filtered off, washed with alkali, then water, and dried (in a vacuum desiccator over NaOH). It is recrystallised from ethanol or dilute aqueous ethanol. The benzoylation can also be carried out in dry pyridine at low temperature (*ca* 0°) instead of in NaOH solution, finally pouring the mixture into water and collecting the solid above. The ester is hydrolysed by refluxing in an alcohol (for example, ethanol, n-butanol) containing two or three equivalents of the alkoxide of the corresponding alcohol (for example sodium ethoxide or sodium n-butoxide) and a few (*ca* 5-10) millilitres of water, for half to three hours. When hydrolysis is complete, an aliquot will remain clear on dilution with four to five times its volume of water. Most of the solvent is distilled off. The residue is diluted with cold water and acidified, and the phenol is steam distilled. The latter is collected from the distillate, dried and either fractionally distilled or recrystallised.

(b) **Acetates:** These can be prepared as for the benzoates using either acetic anhydride with $3N$ NaOH or acetyl chloride in pyridine. They are hydrolysed as described for the benzoates. This hydrolysis can also be carried out with aqueous 10% NaOH solution, completion of hydrolysis being indicated by the complete dissolution of the acetate in the aqueous alkaline solution. On steam distillation, acetic acid also distils off but in these cases the phenols (see above) are invariably solids which can be filtered off and recrystallised.

Phosphate and phosphonate esters. These can be converted to their nitrate addition compounds. The crude or partially purified ester is saturated with uranyl nitrate solution and the adduct filtered off. It is recrystallised from *n*-hexane, toluene or ethanol. For the more soluble members crystallisation from hexane using low temperatures (-40°) has been successful. The adduct is decomposed by shaking with sodium carbonate solution and water, the solvent is steam distilled (if hexane or toluene is used) and the es⸱⸱r is collected by filtration. Alternatively, after decomposition, the organic layer is separated, dried with CaCl$_2$ or BaO, filtered, and fractionally distilled at high vacuum.

Alternatively, impurities can sometimes be removed by conversion to derivatives under conditions where the major component does not react. For example, normal (straight-chain) paraffins can be freed from unsaturated and branched-chain components by taking advantage of the greater reactivity of the latter with chlorosulphonic acid or bromine. Similarly, the preferential nitration of aromatic hydrocarbons can be used to remove e.g. benzene or toluene from cyclohexane by shaking for some hours with a mixture of concentrated nitric acid (25%), sulphuric acid (58%), and water (17%).

BIBLIOGRAPHY

Trace Metal Analysis

N.Zief and J.W.Mitchell, *Contamination Control in Trace Analysis*, Wiley, New York, 1979.

F.Feigl and V.Anger, *Spot Tests in Inorganic Analysis*, Elsevier, Amsterdam, 1972.

Metal Hydrides

Sodium Borohydride and Potassium Borohydride: A Manual of Techniques, Metal Hydrides, Beverly, Massachusettes, 1958.

N.G.Gaylord, *Reductions with Complex Metal Hydrides*, Interscience, New York, 1956.

Characterization of Organic and Inorganic Compounds

N.D.Cheronis and J.B.Entrikin, *Identification of Organic Compounds*, Interscience, New York, 1963.

W.J.Hickinbottom, *Reactions of Organic Compounds*, Longmans, London, 3rd edn, 1958.

S.M.McElvain, *The Characterisation of Organic Compounds*, Macmillan, New York, 2nd edn, 1958.

R.L.Shriner and R.C.Fuson, *The Systematic Identification of Organic Compounds*, Wiley, New York, 3rd edn, 1948.

B.S.Furniss, A.J.Hannaford, V.Rogers, P.W.G.Smith and A.R.Tatchell, *Vogel's Textbook of Practical Organic Chemistry*, Longmans, London, 4th edn, 1978.

R.C.Weast, *Physical Constants of Inorganic Compounds*, in *CRC Handbook of Chemistry and Physics*, CRC Press, Cleveland, Ohio, 58th edn, 1978.

CHAPTER 3

PURIFICATION OF ORGANIC CHEMICALS

Most organic liquids and a number of solids, can readily be purified by fractional distillation, usually at atmospheric pressure: sometimes, especially with higher boiling materials, distillation under reduced pressure is desirable. The general principles and techniques are described in Chapter 1. For this reason, and to save space, the present chapter omits most of the substances for which the published purification methods involve only distillation. Where boiling points are given, purification by distillation is another means of separating impurities. Similarly, references are omitted for methods which require simple recrystallisation from solution if the correct solvent can be guessed readily. Otherwise, substances are listed alphabetically, usually with some criteria of purity, giving brief details of how they can be purified; occasionally some relevant literature references are included. Also noted, when available, are the molecular weights (to the first decimal place), melting points and boiling points together with the respective refractive indexes and densities in the cases of liquids, and optical rotations when the compounds are chiral. When the temperatures and/or the wavelengths are not given for the last three named properties then it should be assumed that the temperature is 20°C and the wavelength is that of the sodium D line; and densities are relative to water at 4°.

In Chapters 3 and 4, *drying with* and *distillation from* are to be taken as implying physical contact between the substances concerned, whereas *drying over (or above)* indicate remoteness, for example as in drying of a solid in a desiccator containing sulphuric acid.

To save space the following abbreviations are used: abs (absolute), aq (aqueous), crystd (crystallised), crystn (crystallisation), dec (decomposes), dil (dilute), distd (distilled), distn (distillation), pet ether (petroleum ether), pptd (precipitated), ppte (precipitate), pptn (precipitation), satd (saturated), soln (solution), tlc (thin layer chromatography) and vac (vacuum).

In addition, the following journals are designated by their initials:

Anal.Chem.	*AC*
Biochem.J.	*BJ*
Ind.Eng.Chem.(Anal.Ed.)	*IECAE*
J.Am.Chem.Soc.	*JACS*
J.Biol.Chem.	*JBC*
J.Chem.Phys.	*JCP*
J.Chem.Soc.	*JCS*
J,Chem.Soc.Dalton Trans.	*JCSDT*
J.Chem.Soc.Farad.Trans.	*JCSFT*
J.Org.Chem.	*JOC*
J.Phys.Chem.	*JPC*
Pure Appl.Chem.	*PAC*
Trans.Faraday Soc.	*TFS*

As a good general rule all low boiling (<110°) organic liquids should be treated as highly flammable and the necessary precautions should be taken.

Abietic acid *[514-10-3]* M 302.5, m 172-175°, $[\alpha]^{25}$ -116° (c 1; EtOH). Crystd from aq EtOH.

Abscisic acid *[21293-39-8]* M 264.3, m 160-161° (sublimation), $[\alpha]_{287}$ + 24,000°, $[\alpha]_{245}$ -69,000° (c 1-50 μg/ml in acidified MeOH or EtOH). Crystd from CCl_4-pet.ether.

Acenaphthalene *[208-96-8]* M 152.2, m 92-93°. Dissolved in warm redistd MeOH, filtered through a sintered glass funnel and cooled to -78° to ppte the material as yellow plates [Dainton, Ivin and Walmsley *TFS* **56** 1784 1960]. Alternatively can be sublimed *in vacuo*.

Acenaphthaquinone *[82-86-0]* M 182.2, m 260-261°. Extracted with, then recrystd twice from benzene [LeFevre, Sundaram and Sundaram *JCS* 974 1963].

Acenaphthene *[83-32-9]* M 154.2, m 94.0°. Crystd from EtOH. Purified by chromatography from CCl_4 on alumina with benzene as eluent [McLaughlin and Zainal *JCS* 2485 1960].

Acetal *[105-57-7]* M 118.2, b 103.7-104°, n 1.38054, n^{25} 1.3682, d 0.831. Dried over Na to remove alcohols and water, and to polymerise aldehydes, then fractionally distd. Or, treat with alkaline H_2O_2 soln at 40-45° to remove aldehydes, then the soln is satd with NaCl, separated, dried with K_2CO_3 and distd from Na [Vogel *JCS* 616 *1948*].

Acetaldehyde *[75-07-0]* M 44.1, b 20.2°, n 1.33113, d 0.788. Usually purified by fractional distn in a glass helices-packed column under dry N_2, discarding the first portion of distillate. Or, shaken for 30 min with NaHCO₃, dried with $CaSO_4$ and fractionally distd at 760 mm through a 70 cm Vigreux column. The middle fraction was taken and further purified by standing for 2 hr at 0° with a small amount of hydroquinone, followed by distn [Longfield and Walters *JACS* **77** 810 *1955*].

Acetamide *[60-35-5]* M 59.1, m 81°. Crystd by soln in hot MeOH (0.8 ml/g), diltd with Et_2O and allowed to stand [Wagner *J Chem Ed* **7** 1135 *1930*]. Alternate crystns are from acetone, benzene, chloroform, dioxane, methyl acetate or from benzene-ethyl acetate mixture (3:1 and 1:1). It has also been recrystd from hot water after treating with HCl-washed activated charcoal (which had been repeatedly washed with water until free from chloride ions), then crystd again from hot 50% aq. EtOH and finally twice from hot 95% EtOH [Christoffers and Kegeles *JACS* **85** 2562 *1963*]. Final drying is in a vac. desiccator over P_2O_5. Acetamide is also purified by distn (b 221-223°) or by sublimation *in vacuo*. Also purified by twice recrystn from cyclohexane containing 5% (v/v) of benzene. Needle-like crystals separated by filtn, washed with distd H_2O and dried with a flow of dry N_2. [Slebocka-Tilk *et al*, *JACS* **109** 4620 *1987*.]

Acetamidine hydrochloride *[124-42-5]* M 94.5, m 174°. Recrystd from EtOH.

N-(2-Acetamido)-2-aminoethanolsulphonic acid (ACES) *[7365-82-4]* M 182.2, m > 220°(dec). Recrystd from hot aq EtOH.

4-Acetamidobenzaldehyde *[122-85-0]* M 163.2, m 156°. Recrystd from water.

p-Acetamidobenzenesulphonyl chloride *[121-60-8]* M 233.7, m 149°(dec). Crystd from toluene, $CHCl_3$, or ethylene dichloride.

N-(2-Acetamido)iminodiacetic acid (ADA) *[26239-55-4]* M 190.2, m 219° (dec). Dissolved in water by adding one equivalent of NaOH soln (to final pH of 8-9), then acidified with HCl to ppte the free acid. Filtered and washed with water.

2-Acetamido-5-nitrothiazole *[140-40-9]* M 187.2, m 264-265°. Recrystd from EtOH or glacial acetic acid.

2-Acetamidophenol *[614-80-2]* M 151.2, m. 209°. Recrystd from water or aq EtOH.

3-Acetamidophenol *[621-42-1]* **M 151.2, m 148-149°.** Recrystd from water.

4-Acetamidophenol *[103-90-2]* **M 151.2, m 169-170.5°.** Recrystd from water or EtOH.

5-Acetamido-1,3,4-thiadiazole-2-sulphonamide *[59-66-5]* **M 222.3, m 256-259° (dec).** Recrystd from water.

Acetanilide *[103-84-4]* **M 135.2, m 114°.** Recrystd from water, aq EtOH, benzene or toluene.

Acetic acid (glacial) *[64-19-7]* **M 60.1, m 16.6°, b 118°, n 1.37171, n^{25} 1.36995, d 1.049.** Usual impurities are traces of acetaldehyde and other oxidisable substances and water. (Glacial acetic acid is very hygroscopic. The presence of 0.1% water lowers its **m.** by 0.2°.) Purified by adding some acetic anhydride to react with water present, heating for 1 hr to just below boiling in the presence of 2 g CrO_3 per 100 ml and then fractionally distilling [Orton and Bradfield *JCS* 960 *1924*, 983 *1927*]. Instead of CrO_3, 2-5% (w/w) of $KMnO_4$, with boiling under reflux for 2-6 hr, has been used.
Traces of water have been removed by refluxing with tetraacetyl diborate (prepared by warming 1 part of boric acid with 5 parts (w/w) of acetic anhydride at 60°, cooling, and filtering off), followed by distn [Eichelberger and La Mer *JACS* 55 3633 *1933*].
Refluxing with acetic anhydride in the presence of 0.2 g % of 2-naphthalenesulphonic acid as catalyst has also been used [Orton and Bradfield *JCS* 983 *1927*]. Other suitable drying agents include $CuSO_4$ and chromium triacetate: P_2O_5 converts some acetic acid to the anhydride. Azeotropic removal of water by distn with thiophene-free benzene or with butyl acetate has been used [Birdwhistell and Griswold *JACS* 77 873 *1955*]. An alternative purification uses fractional freezing.

Acetic anhydride *[108-24-7]* **M 102.1, b 138°, n 1.3904, d 1.082.** Adequate purification can usually be obtained by fractional distn through an efficient column. Acetic acid can be removed by prior refluxing with CaC_2 or with coarse Mg filings at 80-90° for 5 days, or by distn from synthetic quinoline (1% of total charge) at 75mm pressure. Acetic anhydride can also be dried by standing with Na wire for up to a week, and distilling from it under vac. (Na reacts vigorously with acetic anhydride at 65-70°.) Dippy and Evans [*JOC* 15 451 *1950*] let the anhydride (500g) stand over P_2O_5 (50g) for 3h, then decanted it and stood it with ignited K_2CO_3 for a further 3h. The supernatant liquid was distd and the fraction **b** 136-138°, was further dried with P_2O_5 for 12h, followed by shaking with ignited K_2CO_3, before two further distns through a five-section Young and Thomas fractionating column. The final material distd at 137.8-138.0°. Can also be purified by azeotropic distn with toluene: the azeotrope boils at 100.6°. After removal of the remaining toluene, the anhydride is distd [sample had a specific conductivity of 5 x 10^{-9} ohm^{-1}cm^{-1}].

Acetin Blue Crystd from 1:3 benzene-methanol.

Acetoacetanilide *[102-01-2]* **M 177.2, m 86°.** Crystd from H_2O, aq EtOH or pet ether (b 60-80°).

Acetoacetylpiperidide *[1128-87-6]* **M 169.2, b 88.9°/0.1mm, n^{25} 1.4983.** Dissolved in benzene, extracted with 0.5M HCl to remove basic impurities, washed with water, dried, and distd at 0.1mm [Wilson *JOC* 28 314 *1963*].

α-Acetobromoglucose *[572-09-8]* **M 411.2, m 88-89°, $[\alpha]_D^{25}$ +199.3° (c 3, $CHCl_3$).** Crystd from isopropyl ether or pet ether (b 40-60°).

Acetoin see **3-Hydroxy-2-butanone.**

2-Acetonaphthalene *[93-08-3]* **M 170.2, m 55-56°.** Crystd from pet ether, EtOH or acetic acid. [Gorman and Rodgers *JACS* 108 5074 *1986*.]

2-Acetonaphthenone, see **2-acetonaphthalene.**

β-Acetonaphthone *[93-08-3]* **M 170.2, m 54-55°.** Recrystd from EtOH [Levanon *et al, JPC* **91** 14 *1987*].

Acetone *[67-64-1]* **M 58.1, b 56.2°, n 1.35880, d 0.791.** The commercial preparationn of acetone by catalytic dehydrogenation of isopropyl alcohol gives relatively pure material. Analytical reagent quality generally contains less than 1% organic impurities but may have up to about 1% H_2O. Dry acetone is appreciably *hygroscopic*. The main organic impurity in acetone is mesityl oxide, formed by the aldol condensation. It can be dried with anhydrous $CaSO_4$, K_2CO_3 or type 4A Linde molecular sieves, and then distd. Silica gel and alumina cause acetone to undergo the aldol condensation, so that its water content is increased by passage through these reagents, and mildly acidic or basic desiccants. This also occurs to some extent when P_2O_5 or sodium amalgam is used. Anhydrous $MgSO_4$ is an inefficient drying agent, and $CaCl_2$ forms an addition compound. Drierite (anhydrous $CaSO_4$) offers the minimum acid and base catalysis of aldol formation and is the recommended drying agent for this solvent (Coetzee and Siao *Inorg Chem* **14v** 2 *1987*; Riddick and Bunger *Organic Solvents* Wiley-Interscience, N.Y., 3rd edn, 1970). Acetone was shaken with Drierite (25g/L) for several hours before it was decanted and distd from fresh Drierite (10g/L) through an efficient column, maintaining atmospheric contact through a Drierite drying tube. The equilibrium water content is about 10^{-2}M. Anhydrous $Mg(ClO_4)_2$ **should not be used as drying agent because of the risk of explosion with acetone vapour.**

Organic impurities have been removed from acetone by adding 4g of $AgNO_3$ in 30ml of water to 1L of acetone, followed by 10ml of M NaOH, shaking for 10min, filtering, drying with anhydrous $CaSO_4$ and distilling [Werner *Analyst* **58** 335 *1933*]. Alternatively, successive small portions of $KMnO_4$ have been added to acetone at reflux, until the violet colour persists, followed by drying and distn. Refluxing with chromic anhydride has also been used. Methanol has been removed from acetone by azeotropic distn (at 35°) with methyl bromide, and treatment with acetyl chloride.

Small amounts of acetone can be purified as the NaI addition compound, by dissolving 100g of finely powdered NaI in 400g of boiling acetone, then cooling in ice and salt to -8°. Crystals of $NaI.3Me_2CO$ are filtered off and, on warming in a flask, acetone distils off readily. [This method is more convenient than the one using the bisulphite addition compound]. Also purified by gas chromatography on a 20% free fatty acid phthalate (on Chromosorb P) column at 100°.

For efficiency of desiccants in drying acetone see Burfield and Smithers [*JOC* **43** 3966 *1978*]. The water content of acetone can be determined by a modified Karl Fischer titration (Koupparis and Malmstadt *AC* **54** 1914 *1982*]

Acetonedicarboxylic acid *[542-05-2]* **M 146.1, m 138° (dec).** Crystd from ethyl acetate and stored over P_2O_5.

Acetone semicarbazone *[110-20-3]* **M 115.1, m 187°.** Crystd from water or from aq EtOH.

Acetonitrile *[75-05-8]* **M 41.1, b 81.6°, n 1.3441, n^{25} 1.34163, d^{25} 0.77683.** Commercial acetonitrile is a byproduct of the reaction of propylene and ammonia to acrylonitrile. The procedure that significantly reduces the levels of acrylonitrile, allyl alcohol, acetone and benzene was used by Kiesel [*AC* **52** 2230 *1988*]. Methanol (300ml) is added to 3L of acetonitrile fractionated at high reflux ratio until the boiling temperature rises from 64° to 80°, and the distillate becomes optically clear down to 240nm. Add sodium hydride (1g) free from paraffin, to the liquid, reflux for 10min, and then distil rapidly until about 100ml of residue remains. Immediately pass the distillate through a column of acidic alumina, discarding the first 150ml of percolate. Add 5g of CaH_2 and distil the first 50ml at a high reflux ratio. Discard this fraction, and collect the following main fraction. The best way of detecting impurities is by gas chromatography.

Usual contaminants in commercial acetonitrile include H_2O, acetamide, NH_4OAc and NH_3. Anhydrous $CaSO_4$ and $CaCl_2$ are inefficient drying agents. Preliminary treatment of acetonitrile with cold, satd aq KOH is undesirable because of base-catalysed hydrolysis and the introduction of water. Drying by shaking with silica gel or Linde 4A molecular sieves removes most of the water in acetonitrile. Subsequent stirring with CaH_2 until no further hydrogen is evolved leaves only traces of water and removes acetic acid. The acetonitrile is then fractionally distd at high reflux, taking precaution to exclude moisture by refluxing over CaH_2 [Coetzee *PAC* 13 429 *1966*]. Alternatively, 0.5-1% (w/v) P_2O_5 is often added to the distilling flask to remove most of the remaining water. Excess P_2O_5 should be avoided because it leads to the formation of an orange polymer. Traces of P_2O_5 can be removed by distilling from anhydrous K_2CO_3.

Kolthoff, Bruckenstein and Chantooni [*JACS* 83 3297 *1961*] removed acetic acid from 3L of acetonitrile by shaking for 24h with 200g of freshly activated alumina (which had been reactivated by heating at 250° for 4h). The decanted solvent was again shaken with activated alumina, followed by five batches of 100-150g of anhydrous $CaCl_2$. (Water content of the solvent was then less than 0.2%). It was shaken for 1h with 10g of P_2O_5, twice, and distd in a 1m x 2cm column, packed with stainless steel wool and protected from atmospheric moisture by $CaCl_2$ tubes. The middle fraction had a water content of 0.7 to 2mM.

Traces of unsaturated nitriles can be removed by an initial refluxing with a small amount of aq KOH (1ml of 1% solution per L). Acetonitrile can be dried by azeotropic distn with dichloromethane, benzene or trichloroethylene. Isonitrile impurities can be removed by treatment with conc HCl until the odour of isonitrile has gone, followed by drying with K_2CO_3 and distn.

Acetonitrile was refluxed with, and distd from alk. $KMnO_4$ and $KHSO_4$, followed by fractional distn from CaH_2. (This was better than fractionation from molecular sieves or passage through a type H activated alumina column, or refluxing with KBH_4 for 24h and fractional distn)[Bell, Rodgers and Burrows *JCSFT 1* 73 315 *1977*; Moore *et al*, *JACS* 108 2257 *1986*].

Material suitable for polarography was obtained by refluxing over anhydrous $AlCl_3$ (15g/L) for 1h, distg, refluxing over Li_2CO_3 (10g/L) for 1h and redistg. It was then refluxed over CaH_2 (2g/L) for 1h and fractionally distd., retaining the middle portion. The product was not suitable for UV spectroscopy use. A better purification used refluxing over anhydrous $AlCl_3$ (15g/L) for 1h, distg, refluxing over alk. $KMnO_4$ (10g $KMnO_4$, 10g Li_2CO_3/L) for 15min, and distg. A further reflux for 1h over $KHSO_4$ (15g/L), then distn, was followed by refluxing over CaH_2 (2g/L) for 1h, and fractional distn. The product was protected from atmospheric moisture and stored under nitrogen [Walter and Ramalay *AC* 45 165 *1973*].

Acetonitrile has been distd from $AgNO_3$, collecting the middle fraction over freshly activated Al_2O_3. After standing for two days, the liquid was distd from the activated Al_2O_3. Specific conductivity 0.8-1.0 x 10^{-8} mhos [Harkness and Daggett *Can J Chem* 43 1215 *1965*].

4-Acetophenetidine *[62-44-2]* M 179.2, m 136°. Crystd from H_2O or purified by soln in cold dil alkali and repptd by addn of acid to neutralisation point. Air-dried.

Acetophenone *[98-86-2]* M 120.2, m 19.6°, b 54°/2.5mm, 202°/760mm, n ^{25} 1.5322, d^{25} 1.0238. Dried by fractional distn or by standing with anhydrous $CaSO_4$ or $CaCl_2$ for several days, followed by fractional distn under reduced pressure (from P_2O_5, optional), and careful, slow and repeated partial crystns from the liquid at 0° excluding light and moisture. It can also be crystd at low temperatures from isopentane. Distn can be followed by purification using gas-liquid chromatography [Earls and Jones *JCSFT 1* 71 2186 *1975*].

Acetoxime *[127-06-0]* M 73.1, m 63°, b 135°/760mm. Crystd from pet ether (b 40-60°). Can be sublimed.

21-Acetoxypregnenolone M 374.5, m 184-185°. Crystd from acetone.

Aceto-*o*-toluidide *[120-66-1]* M 149.2, m 110°, b 296°/760mm,
Aceto-*m*-toluidide *[537-92-8]* m 65.5°, b 182-183°/14mm, 307°/760mm. Crystd from H_2O, EtOH or aq EtOH.

Aceto-*p*-toluidide *[103-89-9]* M 149.2, m 146°, b 307°/760mm. Crystd from aq EtOH.

Acetylacetone *[123-54-6]* **M 100.1, b 45°/30mm, d$^{30.2}$ 0.9630, n$^{18.5}$ 1.45178.** Small amounts of acetic acid were removed by shaking with small portions of 2M NaOH until the aqueous phase remained faintly alkaline. The sample, after washing with water, was dried with anhydrous Na$_2$SO$_4$, and distd through a modified Vigreux column [Cartledge *JACS* **73** 4416 *1951*]. An additional purification step is fractional crystn from the liquid. Alternatively, there is less loss of acetylacetone if it is dissolved in four volumes of benzene and the soln is shaken three times with an equal volume of distd water (to extract acetic acid): the benzene is then removed by distn at 43-53° and 20-30mm through a helices-packed column. It is then refluxed over P$_2$O$_5$ (10g/L) and fractionally distd under reduced pressure. The distillate (sp conductivity 4 x 10^{-8} ohm^{-1}cm^{-1}) was suitable for polarography [Fujinaga and Lee *Talanta* **24** 395 *1977*]. To recover used acetylacetone, metal ions were stripped from the soln at *p*H 1 (using 100ml 0.1M H$_2$SO$_4$/L of acetylacetone). The acetylacetone was washed (1:10) ammonia soln (100ml/L) and with distd water (100ml/L, twice), and treated as above.

N-**Acetyl-L-alaninamide** *[15062-47-7]* **M 130.2, m 162°.** Crystd repeatedly from EtOH-ethyl ether.

N-**Acetyl-β-alanine** *[3025-95-4]* **M 127.2, m 78.3-80.3°.** Crystd from acetone.

N-**Acetyl-L-alanyl-L-alaninamide** *[30802-37-0]* **M 201.2, m 250-251°.** Crystd repeatedly from EtOH/ethyl ether.

N-**Acetyl-L-alanyl-L-alanyl-L-alaninamide** *[29428-34-0]* **M 272.3, m 295-300°.** Crystd from MeOH/ether.

N-**Acetyl-L-alaninylglycinamide** *[76571-64-7]* **M 187.2, m 148-149°.** Crystd repeatedly from EtOH/ethyl ether.

Acetyl-α-amino-*n*-butyric acid *[34271-24-4]* **M 145.2.** Crystd twice from water (charcoal) and air dried [King and King *JACS* **78** 1089 *1956*].

2-Acetylaminofluorene see *N*-2-fluorenylacetamide.

9-Acetylanthracene *[784-04-3]* **M 220.3, m 75-76°.** Crystd from EtOH. [Masnori *et al*, *JACS* **108** 1126 *1986*.]

N-**Acetylanthranilic acid** *[89-52-1]* **M 179.2, m 185°.** Crystd from water, EtOH or acetic acid.

Acetylbenzonitrile *[1443-80-7]* **M 145.2, m 57-58°.** Recrystd from EtOH [Wagner *et al*, *JACS* **108** 7727 *1986*.]

4-Acetylbiphenyl *[92-91-1]* **M 196.3, m 120-121°, b 325-327°/760mm.** Crystd from EtOH or acetone.

Acetyl-5-bromosalicylic acid *[1503-53-3]* **M 168-169°.** Crystd from EtOH.

Acetylcarnitine chloride *[R:5080-50-2][S:5061-35-8][RS:2504-11-2]* **M 239.7.** Recrystd from isopropanol. Dried over P$_2$O$_5$ under high vac.

Acetyl chloride *[75-36-5]* **M 78.5, b 52°, n 1.38976, d 1.1051.** Refluxed with PCl$_5$ for several hours to remove traces of acetic acid, then distd. Redistd from one-tenth volume of dimethylaniline or quinoline to remove free HCl. A.R. quality is freed from HCl by pumping it for 1hr at -78° and distg into a trap at -196°.

Acetylcholine bromide *[66-23-9]* **M 226.1, m 146°.** Crystd from EtOH.

Acetyldigitoxin-α **M 807.0, m 217-221°.** Crystd from MeOH.

Acetylene *[74-86-2]* **M 26.0, m -80.8°, b -84°.** Purified by successive passage through spiral wash bottles containing, in this order, satd aq NaHSO$_4$, water, 0.2 M iodine in aq KI (two bottles), sodium thiosulphate soln (two bottles), alkaline sodium hydrosulphite with sodium anthraquinone-2-sulphonate as indicator (two bottles), and 10% aq, KOH soln (two bottles). The gas was then passed through a Dry-ice trap and two drying tubes, the first containing CaCl$_2$, and the second, Dehydrite [Conn, Kistiakowsky and Smith *JACS* **61** 1868 *1939*]. Acetone vapour can be removed from acetylene by passage through two traps at -65°.

Sometimes contains acetone and air. These can be removed by a series of bulb-to-bulb distns, e.g. a train consisting of a conc H$_2$SO$_4$ trap and a cold EtOH trap (-73°), or passage through H$_2$O and H$_2$SO$_4$, then over KOH and CaCl$_2$.

Acetylenedicarboxamide *[543-21-5]* **M 112.1, m 294°(dec).** Crystd from MeOH.

Acetylenedicarboxylic acid *[142-45-0]* **M 114.1, m 179°(anhydrous).** Crystd from aq ether as dipicrate.

2-Acetylfluorene *[781-73-7]* **M 208.3, m 132°.** Crystd from EtOH.

Acetyl fluoride *[557-99-3]* **M 62.0, b 20.5°/760mm, d 1.032.** Purified by fractional distn.

N-**Acetyl-D-galactosamine** *[14215-68-0]* **M 221.2, m 160-161°, [α]$_{546}$ +102° (c 1, H$_2$O),**
N-**Acetyl-D-glucosamine** *[7512-17-6]* **M 221.2, m ca 215°, [α]$_{546}$ +49° after 2hr (c 2,H$_2$O).** Crystd from MeOH/Et$_2$O.

N-**Acetylglutamic acid** *[1188-37-0]* **M 189.2, m 185° (RS); 201° (S), [α]^{25} -16.6° (in H$_2$O).** Likely impurity is glutamic acid. Crystd from boiling water.

N-**Acetylglycine** *[543-24-8]* **M 117.1, m 206-208°.** Treated with acid-washed charcoal and recrystd three times from water or EtOH/Et$_2$O and dried in vac over KOH [King and King *JACS* **78** 1089 *1956*].

N-**Acetylglycyl-L-alaninamide** *[34017-20-4]* **M 175.2,**
N-**Acetylglycinamide** *[2620-63-5]* **M 116.1, m 139-139.5°,**
N-**Acetylglycylglycinamide** *[27440-00-2]* **M 173.2, m 207-208°,**
N-**Acetylglycylglycylglycinamide** *[35455-24-4]* **M 230.2, m 253-255°.** Repeated crystn from EtOH/Et$_2$O. Dried in a vac desiccator over KOH.

N-**Acetylhistidine** (H$_2$O) *[39145-52-3]* **M 171.2, m 148° (RS); 169° (S) [α]^{25} +46.2° (H$_2$O).** Likely impurity is histidine. Crystd from water, then 4:1 acetone:water.

N-**Acetylimidazole** *[2466-76-4]* **M 110.1, m 101.5-102.5°.** Crystd from isopropenyl acetate. Dried in vac over P$_2$O$_5$.

Acetyl iodide *[507-02-8]* **M 170.0, b 108°/760mm.** Purified by fractional distn.

N-**Acetyl-L-leucinamide** *[28529-34-2]* **M 177.2, m 133-134°.** Recrystd from CHCl$_3$ and pet ether (b 40-60°).

Acetyl mandelic acid (R-) *[51019-43-3]* **M 194.2, m 98-99° [α]$_D$ -152.4° (c 2, acetone); (S+)** *[7322-88-5]* **m 97-99° [α]$_D$ +150.4° (c 2, acetone).** Crystd from benzene or toluene.

N-**Acetyl-L-methionine** *[65-82-7]* **M 191.3, m 104°, [α]$_{546}$ -24.5° (c 1, in H$_2$O).** Crystd from water or ethyl acetate. Dried in vac over P$_2$O$_5$.

Acetylmethionine nitrile *[538-14-7]* **M 172.3, m 44-46°.** Crystd from ethyl ether.

5-Acetyl-2-methoxybenzaldehyde *[531-99-7]* M 166.2 , m 144°. Crystd from EtOH or Et$_2$O.

N-**Acetyl-*N*'-methyl-L-alanimide** *[1901-83-8]* M 144.2. Crystd from EtOAc/Et$_2$O, then from EtOH and Et$_2$O.

Acetylmethylcarbinol see **3-hydroxy-2-butanone.**

N-**Acetyl-6*N*'-methylglycinamide** *[7606-79-3]* M 130.2. Recrystd from EtOH/Et$_2$O mixt.

N-**Acetyl-6*N*'-methyl-L-leucine amide** *[32483-15-1]* M 186.3. Recrystd from EtOH/hexane mixt.

N-**Acetyl-D-penicillamine** *[15537-71-0]* M 191.3, m 189-190° (dec), [α]$_D$ +18° (c 1, in 50% EtOH). Crystd from water.

N-**Acetyl-L-phenylalanine** *[2018-61-3]* M 207.2, m 170-171°, [α]$_D$ +49.3, (DL) m 152.5-153°. Crystd from CHCl$_3$ and stored in a desiccator at 4°. (DL)-isomer crystd from water or acetone.

N-**Acetyl-L-phenylalanine ethyl ester** *[2361-96-8]* M 235.3. Crystd from water.

1-Acetyl-2-phenylhydrazine *[114-83-0]* M 150.2, m 128.5°. Crystd from aq EtOH.

Acetylsalicylic acid *[50-78-2]* M 180.2, m 133.5-135°. Crystd twice from toluene, washed with cyclohexane and dried at 60° under vac for several hr [Davis and Hetzer *J Res Nat Bur Stand* **60** 569 *1958*]. Has also been recrystd from isopropanol and from ethyl ether/pet ether (b 40-60°).

N-**Acetylsalicylsalicylic acid** *[530-75-6]* M 300.3, m 159°. Crystd from dil acetic acid.

N-**(4)-Acetylsulphanilamide** *[144-80-9]* M 214.2, m 216°. Crystd from aq EtOH.

N-**Acetylsulphanilyl chloride** see *p*-acetamidobenzenesulphonyl chloride.

N-**Acetyltryptophan** M 246.3, *[87-32-1]* m 206° (RS); *[1218-34-4]* m 188° (S), [α]^{25} +30.1° (aq NaOH). Likely impurity is tryptophan. Crystd from EtOH by adding water.

N-**Acetyl-L-valine amide** *[37933-88-3]* M.158.2, m 275°. Recrystd from CH$_3$OH/ether.

cis-**Aconitic acid** *[585-84-2]* M 174.1, m 126-129°(dec). Crystd from water by cooling (sol: 1g in 2ml of water at 25°). Dried in a vac desiccator.

Aconitine *[302-27-2]* M 645.8, m 204°, [α]$_{546}$ +20° (c 1, CHCl$_3$). Crystd from EtOH, CHCl$_3$ or toluene.

Aconitine hydrobromide M 726.7, m 207°. Crystd from water or EtOH/ether.

Acraldehyde see **acrolein**

Acridine *[260-94-6]* M 179.2, m.111°, b 346°. Crystd twice from benzene/cyclohexane, or from aq EtOH, then sublimed, removing and discarding the first 25% of the sublimate. The remainder was again crystd and sublimed, discarding the first 10-15% [Wolf and Anderson *JACS* **77** 1608 *1955*].
Acridine can also be purified by crystn. from *n*-heptane and then from ethanol/water after pre-treatment with activated charcoal, or by chromatography on alumina with pet ether in a darkened room. Alternatively, acridine can be pptd as the hydrochloride from benzene soln by adding HCl, after which the base is regenerated, dried at 110°/50mm, and crystd to constant melting point from pet ether [Cumper, Ginman and Vogel *JCS* 4518 *1962*]. The regenerated free base may be recrystd,

chromatographed on basic alumina, then vac-sublimed and zone-refined. [Williams and Clarke, *JCSFT 1* 73 514 *1977*.]

Acridine Orange *[494-38-2]* **M 349.94, m 181-182° (free base).** The double salt with $ZnCl_2$ (6g) was dissolved in water (200ml) and stirred with four successive portions (12g each) of Dowex-50 ion-exchange resin (K^+ form) to remove the zinc. The soln was then concentrated in vac to 20ml, and 100ml of ethanol was added to ppte KCl which was removed. Ether (160 ml) was added to the soln from which, on chilling, the dye cryst as its chloride. It was separated by centrifuging, washed with chilled ethanol and ether, and dried under vac, before being recryst from ethanol (100ml) by adding ether (50ml), and chilling. Yield 1g. [Pal and Schubert *JACS* 84 4384 *1962*.]
It was recrystd twice as the free base from ethanol or methanol/water by dropwise addition of NaOH (less than 0.1M). The ppte was washed with water and dried under vac. It was dissolved in $CHCl_3$ and chromatographed on alumina: the main sharp band was collected, concentrated and cooled to -20°. The ppte was filtered, dried in air, then dried for 2hr under vac at 70°. [Stone and Bradley *JACS* 83 3627 *1961*; Blauer and Linschitz *JPC* 66 453 *1962*.]

Acridine Yellow *[135-49-9]* **M 237.8, m 325°.** Crystd from 1:1 benzene/methanol.

Acridinol see **4-hydroxyacridine.**

Acriflavine *[8048-52-0]* **M 196.2.** Treated twice with freshly pptd AgOH to remove proflavine, then recrystd from absolute methanol [Wen and Hsu *JPC* 66 1353 *1962*].

Acrolein *[107-02-8]* **M 56.1, b 52.1°, n 1.3992, d 0.839.** Purified by fractionl distn. under nitrogen, drying with anhydrous $CaSO_4$ and then distilling under vac. Blacet, Young and Roof [*JACS* 59 608 *1937*] distd under nitrogen through a 90cm column packed with glass rings. To avoid formation of diacryl, the vapour was passed through an ice-cooled condenser into a receiver cooled in an ice-salt mixture and containing 0.5g catechol. The acrolein was then distd twice from anhydrous $CuSO_4$ at low pressure, catechol being placed in the distilling flask and the receiver to avoid polymerization. [Alternatively, hydroquinone (1% of the final soln) can be used.]

Acrolein semicarbazone *[6055-71-6]* **M 113.1, m 171°.** Crystd from water.

Acrylamide *[79-06-1]* **M 71.1, m 84°, b 125°/25mm.** Crystd from acetone, chloroform, ethyl acetate, methanol or benzene/chloroform mixture, then vac dried and kept in the dark under vac. Recryst from $CHCl_3$ (200g dissolved in 1L heated to boiling and filtered without suction in a warmed funnel through Whatman 541 filter paper. Allowed to cool to room temp and kept at -15° overnight). Crystals were collected with suction in a cooled funnel and washed with 300ml of cold MeOH. Crystals were air-dried in a warm oven. [Dawson *et al*, **Data for Biochemical Research**, Oxford Press 1986 p 449.]
CAUTION: *Acrylamide is extremely toxic and precautions must be taken to avoid skin contact or inhalation. Use gloves and handle in a well ventilated fume cupboard.*

Acrylic acid *[79-10-7]* **M 72.1, m 13°, b 30°/3mm, d 1.051.** Can be purified by steam distn, or vac distn through a column packed with copper gauze to inhibit polymerisation. (This treatment also removes inhibitors such as methylene blue that may be present.) Azeotropic distn of the water with benzene converts aq acrylic acid to the anhydrous material.

Acrylonitrile *[107-13-1]* **M 53.1, b 78°, n^{25} 1.3886, d 0.806.** Washed with dil H_2SO_4 or dil H_3PO_4, then with dilute Na_2CO_3 and water. Dried with Na_2SO_4, $CaCl_2$ or (better) by shaking with molecular sieves. Fractionally distd under nitrogen. Can be stabilised by adding 10ppm *tert*-butyl catechol. Immediately before use, the stabilizer can be removed by passage through a column of activated alumina (or by washing with 1% NaOH soln if traces of water are permissible in the final material), followed by distn. Alternatively, shaken with 10% (w/v) NaOH to extract inhibitor, and then washed in turn with 10% H_2SO_4, 20% Na_2CO_3 and distd water. Dried for 24hr over $CaCl_2$ and fractionally distd under N_2

taking the fraction boiling at 75.0 to 75.5°C (at 734mm Hg). Stored with 10ppm *tert*-butyl catechol. Acrylonitrile is distilled off as required. [Burton *et al*, *JCSFT 1* **75** 1050 *1979*.]

Actidione see **cycloheximide.**

Actinomycin D *[50-76-0]* **M 1255.5.** Crystd from ethyl acetate or from MeOH.

Adamantane *[281-23-2]* **M 136.2, m 269.6-270.8°** (sublimes). Crystd from acetone or cyclohexane, sublimed in vac below its melting point. [Butler *et al*, *JCSFT 1* **82** 535 *1986*.] Adamantane was also purified by dissolving in *n*-heptane (*ca* 10ml/g of adamantane) on a hot plate, adding activated charcoal (2g/100g of adamantane), and boiling for 30min, filtering the hot soln through a filter paper, concentrating the filtrate until crystn just starts, adding one quarter of the original volume *n*-heptane and allowing to cool slowly over a period of hours. The supernatant was decanted off and the crystals were dried on a vacuum line at room temperature. [Walter *et al*, *JACS* **107** 793 *1985*.]

Adamantane-1-carboxylic acid *[828-51-3]* **M 180.3, m 177°.** Crystd from absolute EtOH and dried under vac at 100°.

2-Adamantanone *[700-58-3]* **M 150.2, m 256-258°**(sublimes). Purified by repeated sublimation *in vacuo*. [Butler *et al*, *JCSFT 1* **82** 535 *1986*.]

1-Adamantyl bromide *[768-90-1]* **M 215.1, m 117-118°,**
1-Adamantyl fluoride *[768-92-3]* **M 154.2,**
1-Adamantyl iodide *[768-93-4]* **M 262.1.** Purified by recryst from pet ether (40-60°C) followed by rigorous drying and repeated sublimation.

Adenine *[73-24-5]* **M 135.1, m 360-365°** (dec rapid heating). Crystd from distd water.

Adenosine *[58-61-7]* **M 267.3, m 234-236°, [α]$_{546}$ -85°** (c 2, 5% NaOH). Crystd from distd water.

Adenosine-3'-phosphoric acid *[84-21-9]* **M 365.2, m 210°**(dec), **[α]$_{546}$ -50°** (c 0.5, 0.5M Na$_2$HPO$_4$). Crystd from a large vol of distd water, as the monohydrate.

Adenosine-5'-phosphoric acid monohydrate *[18422-05-4]* **M 365.2, m 196-200°**(dec), **[α]$_{546}$ -56°** (c 2, 2% NaOH). Crystd from water by addition of acetone. Purified by chromatography on Dowex 1 (in formate form), eluting with 0.25M formic acid. It was then adsorbed onto charcoal (which had been boiled for 15 min with M HCl, washed free of chloride and dried at 100°), and recovered by stirring three times with isoamyl alcohol/water (1:9 v/v). The aq layer from the combined extracts was evaporated to dryness under reduced pressure, and the product was cryst twice from hot water. [Morrison and Doherty *BJ* **79** 433 *1961*.]

Adenosine-5'-triphosphate *[56-65-5]* **M 507.2, [α]$_{546}$ -35.5** (c 1, 0.5 M Na$_2$HPO$_4$). Pptd as its barium salt when excess barium acetate soln was added to a 5% soln of ATP in water. After filtering off, the ppte was washed with distd water, redissolved in 0.2M HNO$_3$, and again pptd with barium acetate. The ppte, after several washings with distd water, was dissolved in 0.2M HNO$_3$ and slightly more 0.2M H$_2$SO$_4$ than was needed to ppte all the barium as BaSO$_4$, was added. After filtering off the BaSO$_4$, the ATP was pptd by addition of a large excess of 95% ethanol, filtered off, washed several times with 100% EtOH and finally with dry ethyl ether. [Kashiwagi and Rabinovitch *JPC* **59** 498 *1955*.]

Adenylic acid see **adenosinephosphoric acid**

Adipic acid *[124-04-9]* **M 146.1, m. 154°.** For use as a volumetric standard, adipic acid was crystd once from hot water with the addition of a little animal charcoal, dried at 120° for 2 hr, then recrystd from acetone and again dried at 120° for 2 hr. Other purification procedures include crystn from ethyl acetate and from acetone/petroleum ether, fusion followed by filtration and crystn from the melt, and preliminary distn under vac.

Adonitol (Ribitol) *[488-81-3]* M 152.2, m 102°. Cryst. from EtOH by addition of ethyl ether.

Adrenalin see **epinephrine.**

Adrenochrome *[382-45-6]* M 179.2, m 125-130°. Crystd from MeOH/formic acid, as hemihydrate, and stored in a vac desiccator.

Adrenosterone (Reichstein's G) *[382-45-6]* M 300.4, m 220-224°. Crystd from EtOH. Can be sublimed under high vac.

Agaricic acid *[666-99-9]* M 416.6, m 142°(dec), $[\alpha]_D$ -9.8° (in NaOH). Crystd from EtOH.

Agmatine sulphate *[2482-00-0]* M 228.3, m 231°. Crystd from aq MeOH.

Agroclavin *[548-42-5]* M 238.3, m 198-203°(dec), $[\alpha]^{30}$ -242°. Crystd from ethyl ether.

Ajmalicine *[483-04-5]* M 352.4, m 250-252°(dec), $[\alpha]_{546}$ -76° (c 0.5, CHCl$_3$). Crystd from MeOH.

Ajmalicine hydrochloride *[4373-34-6]* M 388.9, m 290°(dec), $[\alpha]_D$ -17° (c 0.5, MeOH). Crystd from EtOH.

Ajmaline *[4360-12-7]* M 326.4, m 160°. Crystd from MeOH.

Ajmaline hydrochloride *[4373-34-6]* M 388.9, m 140°. Crystd from water.

Alanine (RS) *[302-72-7]* M 89.1, m 295-296°, **(S)** *[56-41-7]* m 297°(dec), $[\alpha]_D^{15}$ +14.7° (in 1M HCl). Crystd from water or aq EtOH, e.g. crystd from 25% EtOH in water, recrystd from 62.5% EtOH, washed with EtOH and dried to constant weight in a vac desiccator over P$_2$O$_5$. [Gutter and Kegeles *JACS* **75** 3893 *1953*.] 2,2'-Iminodipropionic acid is a likely impurity.

β-Alanine *[107-95-9]* M 89.1, m 205°(dec). Crystd from filtered hot satd aq soln by adding four volumes of absolute EtOH and cooling in an ice-bath. Recrystd in the same way and then finally, crystd from a warm satd soln in 50% EtOH by adding four volumes of absolute EtOH cooled in an ice bath. Crystals were dried in a vac desiccator over P$_2$O$_5$. [Donovan and Kegeles *JACS* **83** 255 *1961*.]

Albumin (bovine serum) see Chapter 6.

Aldol *[107-89-1]* M 88.1, b 80-81°/20mm. An ethereal soln was washed with a satd aq soln of NaHCO$_3$, then with water. The non-aqueous layer was dried with anhydrous CaCl$_2$ and distd immediately before use. The fraction, b 80-81°/20mm, was taken, [Mason, Wade and Pouncy *JACS* **76** 2255 *1954*].

Aldosterone *[52-39-1]* 360.5, m 108-112°(hydrate), 164°(anhydr). Crystd from aq acetone.

Aldrin *[309-00-2]* M 354.9, m 103-104.5°. Crystd from MeOH.

Aleuritic acid *[533-87-3]* M 304.4, m 100-101°. Crystd from aq EtOH.

Alginic acid *[9005-32-7]* M 48,000-186000. To 5g in 550ml water containing 2.8g KHCO$_3$, were added 0.3ml acetic acid and 5g potassium acetate. EtOH to make the soln 25% (v/v) in EtOH was added and any insoluble material was discarded. Further addition of EtOH, to 37% (v/v), pptd alginic acid. [Pal and Schubert *JACS* **84** 4384 *1962*.]

Aliquat 336 *[5137-55-3]* **M 404.2, d 0.884.** A 30% (v/v) soln in benzene was washed twice with an equal volume of 1.5M HBr. [Petrow and Allen, *AC* **33** 1303 *1961*.] Purified by dissolving 50g in CHCl₃ (100ml) and shaking with 20% NaOH solution (200ml) for 10min, and then with 20% NaCl (200ml) for 10min. Washed with small amount of water and filtered through a dry filter paper (Adam and Pribil *Talanta* **18** 733 *1971*).

Alizarin *[72-48-0]* **M 240.2, d 0.884.** Crystd from glacial acetic acid or 95% EtOH. Can also be sublimed.

Alizarin Complexone (2H₂O) *[3952-78-1]* **M 421.4, m 189°(dec).** Purified by suspension in 0.1M NaOH (1g in 50ml), filtering the solution and extracting alizarin with 5 successive portions of CH₂Cl₂. Then added HCl dropwise to precipitate the reagent, stirring the solution in a bath. Filtered ppte on glass filter, washed with cold water and dried in a vac desiccator over KOH (Ingman *Talanta* **20** 135 *1973*).

Alizarin Orange see **3-nitroalizarin.**

n-**Alkylammonium chloride** **n=2,4,6.** Recrystd from EtOH or an EtOH/Et₂O mixture. [Hashimoto and Thomas *JACS* **107** 4655 *1985;* Chu and Thomas *JACS* **108** 6270 *1986*.]

n-**Alkyltrimethylammonium bromide** **n=10,12,16.** Recrystd from an EtOH/Et₂O mixture. [Hashimoto and Thomas *JACS* **107** 4655 *1985*.]

Allantoin *[97-59-6]* **M158.1, m 238°(dec).** Crystd from water or EtOH.

Allene *[463-49-0]* **M 40.1, m -146°, b -32°.** Frozen in liquid nitrogen, pumped on, then thawed out. This cycle was repeated several times, then the allene was frozen in a methyl cyclohexane-liquid nitrogen bath and pumped for some time. Also purified by hplc.

Allopregnane-3α, 20α-diol *[566-58-5]* **M 320.5, m 248-248.5°, [α]_D +17°** (c 0.15, EtOH). Crystd from EtOH.

Alloxan *[50-71-5]* **M 142.0, m ~170°(dec).** Crystn from water gives the tetrahydrate. Anhydrous crystals are obtained by crystn from acetone, glacial acetic acid or by sublimation *in vacuo*.

Alloxantin *[76-24-4]* **M 286.2, m 253-255°(dec) (yellow at 225°).** Crystd from water or EtOH and kept under nitrogen. Turns red in air.

Allyl acetate *[591-87-7]* **M 100.1, b 103°, n₄ 1.40488, n_D²⁷ 1.4004, d 0.928.** Freed from peroxides by standing with crystalline ferrous ammonium sulphate, then washed with 5% NaHCO₃, followed by satd CaCl₂ soln. Dried with Na₂SO₄ and fractionally distd in an all-glass apparatus.

Allyl alcohol *[107-18-6]* **M 58.1, b 98°, d₄ 0.857, n_D 1.4134.** Can be dried with K₂CO₃ or CaSO₄, or by azeotropic distn with benzene followed by distn under nitrogen. It is difficult to obtain peroxide free. Also reflux with magnesium and fractionally distd [Hands and Norman *Ind Chemist* **21** 307 *1945*].

Allylamine *[107-11-9]* **M 57.1, b 52.9°, n 1.42051, d 0.761.** Purified by fractional distn from calcium chloride.

1-Allyl-6-amino-3-ethyluracil *[642-44-4]* **M 195.2, m 143-144°** (anhydr). Crystd from water (as monohydrate).

Allyl bromide *[106-95-6]* **M 121, b 70°, n 1.46924, d 1.398.** Washed with NaHCO₃ soln then distd water. Dried with CaCl₂ or MgSO₄, and fractionally distd. Protect from strong light.

Allyl chloride *[107-05-1]* **M 76.5, b 45.1°, n 1.4130, d 0.939.** Likely impurities include 2-chloropropene, propyl chloride, iso-propyl chloride, 3,3-dichloropropane, 1,2-dichloropropane and 1,3-dichloropropane. Purified by washing with conc HCl, then with Na_2CO_3 soln, drying with $CaCl_2$, and distn through an efficient column [Oae and Vanderwerf *JACS* **75** 2724 *1953*].

1-*N*-Allyl-3-hydroxymorphinan *[152-02-3]* **M 283.4, m 180-182°.** Crystd from aq EtOH.

Allyl iodide *[556-56-9]* **M 167.7, b 103°, d^{12} 1.848.** Purified in a dark room by washing with aq Na_2SO_3 to remove free iodine, then drying with $MgSO_4$ and distilling at 21mm pressure, to give a very pale yellow liquid. (This material, dissolved in hexane, was stored in a light-tight container at -5° for up to three months before free iodine could be detected, by its colour, in the soln) [Sibbett and Noyes *JACS* **75** 761 *1953*].

Allyl thiourea *[109-57-9]* **M 116.2, m 78°.** Crystd from acetone, EtOH or ethyl acetate, after decolorizing with charcoal.

***N*-Allylurea** *[557-11-9]* **M 100.1, m 85°.** Crystd from EtOH, EtOH/ether, EtOH/chloroform or EtOH/toluene.

D-Altrose *[1990-29-0]* **M 180.2, m 103-105°, $[\alpha]_{546}$ +35° (c 7.6, H_2O).** Crystd fom aq EtOH.

Amberlite IRA-904 Anion-exchange resin (Rohm and Haas). Washed with 1M HCl, CH_3OH (1:10) and then rinsed with distd water until the washings were neutral to litmus paper. Finally extracted successively for 24hr in a Soxhlet apparatus with MeOH, benzene and cyclohexane (Shue and Yan *AC* **53** 2081 *1981*).

Amethopterin (H_2O) *[59-05-2]* **M 454.5, m 185-204° (dec), $[\alpha]_D$ -19.4° (c 2, 0.1 M NaOH).** Crystd from water.

***p*-Aminoacetanilide** *[122-80-5]* **M 150.2, m 162-163°.** Crystd from water.

Aminoacetic acid (Glycine) *[56-40-6]* **M 75.1, m 262° (dec, goes brown at 226°, sublimes at 200°/0.1mm).** Crystd from distd water by dissolving at 90-95°, filtering, cooling to about -5°, and draining the crystals centrifugally. Alternatively, crystd from distd water by addition of MeOH or EtOH (e.g. 50g dissolved in 100ml of warm water, and 400ml of MeOH added). The crystals can be washed with MeOH or EtOH, then with ethyl ether. Likely impurities are ammonium glycinate, iminodiacetic acid, nitrilotriacetic acid, ammonium chloride.

Aminoacetonitrile bisulphate *[151-63-3]* **M 154.1, m 188°(dec)** Crystd from aq EtOH.

α-Aminoacetophenone hydrochloride *[5488-37-1]* **M 171.6, m 188°(dec).** Crystd from acetone/EtOH.

***m*-Aminoacetophenone** *[99-03-6]* **M 135.2, m 98-99°.** Crystd from EtOH.

***p*-Aminoacetophenone** *[99-92-3]* **M 135.2, m 106-107°.** Crystd from water.

9-Aminoacridine *[90-45-9]* **M 194.2, m 241°.** Crystd from EtOH or acetone.

***dl*-α-Aminoadipic acid** (hydrate) *[542-32-5]* **M 161.2, m 196-198°.** Crystd from water.

2-Amino-4-anilino-*s*-triazine *[537-17-7]* **M 168.2, m 235-236°.** Crystd from dioxane or 50% aq EtOH.

1-Aminoanthraquinone-2-carboxylic acid *[82-24-6]* **M 276.2, m 295-296°.** Crystd from nitrobenzene.

4-Aminoantipyrine *[83-07-8]* **M 203.3, m 109°.** Crystd from EtOH or EtOH/ether.

p-**Aminoazobenzene** *[60-09-3]* **M 197.2, m 126°.** Crystd from EtOH, CCl$_4$, pet ether/benzene, or a MeOH/water mixture.

o-**Aminoazotoluene** **(Fast Garnet GBC base)** *[614-63-1]* **M 225.3, m 101.4-102.6°.** Crystd twice from EtOH, once from benzene, then dried in an Abderhalden drying apparatus [Cilento *JACS* **74** 968 *1952*]. **CARCINOGENIC.**

5-Aminobarbituric acid see **uramil.**

2-Aminobenzaldehyde *[529-23-7]* **M 121.1, m 39-40°.** Distd in steam and crystd from water or EtOH/ether.

3-Aminobenzaldehyde *[29159-23-7]* **M 121.1, m 28-30°.** Crystd from ethyl acetate.

p-**Aminobenzeneazodimethylaniline** *[539-17-3]* **M 240.3, m 182-183°.** Crystd from aq EtOH.

p-**Aminobenzenesulphonamide** see **sulphanilamide.**

m-**Aminobenzenesulphonic acid** see **metanilic acid.**

p-**Aminobenzenesulphonic acid** see **sulphanilic acid.**

o-**Aminobenzoic acid (anthranilic acid)** *[118-92-3]* **M 137.1, m 145°.** Crystd from water (charcoal). Has also been crystd from 50% aq acetic acid. Can be vac sublimed.

m-**Aminobenzoic acid** *[99-05-8]* **M.137.1, m 174°.** Crystd from water.

p-**Aminobenzoic acid** *[150-13-0]* **M 137.1, m 187-188°.** Purified by dissolving in 4-5% aq HCl at 50-60°, decolorizing with charcoal and carefully pptg with 30% Na$_2$CO$_3$ to pH 3.5-4 in the presence of ascorbic acid. It can be cryst. from water, EtOH or EtOH/water mixtures.

p-**Aminobenzonitrile** *[873-74-5]* **M 118.1, m 86-86.5°.** Crystd from water, 5% aq EtOH or EtOH and dried over P$_2$O$_5$ or dried *in vacuo* for 6hr at 40°. [Moore *et al, JACS* **108** 2257 *1986*.]

4-Aminobenzophenone *[1137-41-3]* **M 197.2, m 123-124°.** Dissolved in aq acetic acid, filtered and pptd with ammonia. Process repeated several times, then recrystd from aq EtOH.

2-Aminobenzothiazole *[136-95-8]* **M 150.2, m 132°,**
6-Aminobenzothiazole *[533-30-2]* **M 150.2, m 87°.** Crystd from aq EtOH.

N-**(*p*-Aminobenzoyl)-L-glutamic acid** *[4271-30-1]* **M 266.3, m 173°ₗ (L-form), [α]$_{546}$ -17.5° (c 2, 0.1m HCl); 197° (DL).** Crystd from water.

3-*o*-Aminobenzyl-4-methylthiazolium chloride hydrochloride *[534-94-1]* **M 277.4, m 213°(dec).** Crystd from aq EtOH

o-**Aminobiphenyl** *[90-41-5]* **M 169.2, m 49.0°.** Crystd from aq EtOH (charcoal).

p-**Aminobiphenyl** *[92-67-1]* **M 169.2, m 53°, b 191°/16mm.** Crystd from water or EtOH. **CARCINOGENIC.**

2-Amino-5-bromotoluene *[583-75-5]* **M 186.1, m 59°.** Steam distd, and crystd from EtOH.

S-α-Aminobutyric acid *[1492-24-6]* **M 103.1, m 292°(dec), [α]$_D$ + 20.4° (c 2, 2.5N HCl).** Crystd from aq EtOH.

RS-α-Aminobutyric acid *[2835-81-6]* **M 103.1, m 303°(dec).** Crystd from water.

β-Aminobutyric acid *[2835-82-7]* **M 103.1, m 193-194°,**
γ-Aminobutyric acid *[56-12-2]* **M 103.1, m 202°(dec).** Crystd form aq EtOH.

RS-α-Amino-*n*-caproic acid see **RS-norleucine**

2-Amino-5-chlorobenzoic acid *[635-21-1]* **M 171.6, m 100°.** Crystd from water, EtOH or chloroform.

3-Amino-4-chlorobenzoic acid *[2840-28-0]* **M 171.6, m 216-217°.** Crystd from water.

4-Amino-4'-chlorobiphenyl *[135-68-2].***M 203.5, m 134°.** Crystd from pet ether.

2-Amino-4-chloro-6-methylpyrimidine *[5600-21-5]* **M 143.6, m 184-186°.** Crystd from EtOH.

2-Amino -5-chloropyridine *[1072-98-6]* **M 128.6, m 135-136°.** Crystd from pet ether, sublimes at 50°/0.5mm.

1-Amino-1-cyclopentanecarboxylic acid *[52-52-8]* **M 129.2, m 330°(dec).** Crystd from aq EtOH.

2-Amino-3,5-dibromopyridine *[35486-42-1]* **M 251.9, m 103-104°.** Steam distd and crystd from aq EtOH or pet ether.

2-Amino-4,6-dichlorophenol *[527-62-8]* **M 175.0, m 95-96°.** Crystd from CS_2 or benzene.

3-Amino-2,6-dichloropyridine *[62476-56-6]* **M 164.0, m 119°, b 110°/0.3mm.** Crystd from water.

4-Amino-*N,N*-diethylaniline hydrochloride *[16713-15-8]* **M 200.7, m 233.5°.** Crystd from EtOH.

4-Amino-3,5-diiodobenzoic acid *[2122-61-4]* **M 388.9, m >350°.** Purified by soln in dil NaOH and pptn with dil HCl. Air dried.

2-Amino-4,6-dimethylpyridine *[5407-87-4]* **M 122.2, m 69-70.5°.** Crystd from hexane, ether/pet ether or benzene. Residual benzene was removed from crystals over paraffin-wax chips in an evacuated desiccator.

2-Amino-4,6-dimethylpyrimidine *[767-15-7]* **M 123.2, m 152-153°.** Crystn from water gives **m 197°**, and crystn from acetone gives **m 153°**.

2-Aminodiphenylamine *[534-85-0]* **M 184.2, m 79-80°.** Crystd from water.

4-Aminodiphenylamine *[101-54-2]* **M 184.2, b 155°/0.026mm.** Crystn from EtOH gives **m 66°**, and crystn from ligroin gives **m 75°**.

2-Amino-1,2-diphenylethanol *[530-36-9]* **M 213.3, m 165°.** Crystd from EtOH.

2-Aminodiphenylmethane *[28059-64-5]* **M 183.3, m 52°, b 172°/12mm and 190°/22mm.** Crystd from ether.

2-Aminoethanethiol *[60-23-1]* **M 77.2, m 97-98.5°.** Sublimed under vac. [Barkowski and Hedberg *JACS* **109** 6989 *1987*].

2-Aminoethanol *[141-43-5]* **M 61.1, f 10.5°, b 72-73°/12mm, 171.1°/760mm, n 1.14539, d 1.012.** Decomposes slightly when distd at atmospheric pressure, with the formation of conducting impurities. Fractional distn at about 12mm pressure is satisfactory. After distn, 2-aminoethanol was further purified by repeated washing with ether and crystn from EtOH (at low temp). After fractional distn in the absence of CO_2, it was twice crystd by cooling, followed by distn. *Hygroscopic.* [Reitmeier, Silvertz and Tartar *JACS* **62** 1943 *1940*.] Can be dried by azeotropic distn with dry benzene.

2-Aminoethanol hydrochloride *[2002-24-6]* **M 97.6, m 75-77°.** Crystd from EtOH. It is deliquescent.

2-Aminoethanol hydrogen sulphate *[926-39-6]* **M 125.2, m 285-287° (chars at 275°).** Crystd from water or dissolved in water and EtOH added.

2-(2-Aminoethylamino)ethanol see **hydroxyethyl-ethylenediamine.**

S-(2-Aminoethyl)isothiouronium bromide hydrobromide *[56-10-0]* **M 281.0, m 194-195°.** Crystd from abs EtOH/ethyl acetate. It is hygroscopic.

2-Amino-4-(ethylthio)butyric acid see **ethionine.**

(2-Aminoethyl)trimethylammonium chloride hydrochloride (chloramine chloride hydrochloride) *[3399-67-5]* **M 175.1, m 260°(dec).** Crystd from EtOH. (Material is very soluble in water).

RS-α-Aminohexanoic acid see **RS-methionine.**

4-Amino hippuric acid *[61-78-9]* **M 194.2, m 198-199°.** Crystd from water.

1-Amino-4-hydroxyanthraquinone *[116-85-8]* **M 293.2, m 207-208°.** Purified by tlc on SiO_2 gel plates using toluene/acetone (9:1) as eluent. The main band was scraped off and extracted with MeOH. The solvent was evapd and the dye was dried in a drying pistol [Land, McAlpine, Sinclair and Truscott *JCSFT 1* **72** 2091 *1976*]. Crystd from aq EtOH.

2-Amino-4-hydroxybutyric acid see **homoserine.**

dl-**4-Amino-3-hydroxybutyric acid** *[924-49-2]* **M 119.1, m 225°(dec).** Crystd from water or aq EtOH.

2-Amino-2-hydroxymethyl-1,3-propanediol see **tris(hydroxymethyl)aminomethane.**

5-Amino-8-hydroxyquinoline hydrochloride *[3881-33-2]* **M 196.7.** Dissolved in minimum of MeOH, then Et_2O was added to initiate pptn. Ppte was filtered off and dried [Lovell *et al JPC* **88** 1885 *1984*].

3-Amino-4-hydroxytoluene *[95-84-1]* **M 123.2, m 137-138°.** Crystd from water or toluene.

4-Amino-5-hydroxytoluene *[2835-98-5]* **M 123.2, m 159°,**
6-Amino-3-hydroxytoluene *[2835-99-6]* **M 123.2, m 162°(dec).** Crystd from 50% EtOH.

4-Aminoimidazole-5-carboxamide hydrochloride (AICAR HCl) *[72-40-2]* **M 162.6, m 255-256°(dec).** Recrystd from EtOH.

5-Aminoindane *[24425-40-9]* **M 133.2, m 37-38°, b 131°/15mm, 146-147°/25mm, 247-249°/745mm.** Distd and then crystd from pet ether.

6-Aminoindazole *[6967-12-0]* **M 133.2, m 210°.** Crystd from water or EtOH and sublimed in vac.

2-Amino-5-iodotoluene *[13194-68-8]* **M 233.0, m 87°.** Crystd from 50% EtOH.

α-Aminoisobutyric acid *[62-57-7]* **M 103.1, sublimes at 280°.** Crystd from aq EtOH and dried at 110°.

D-4-Amino-3-isoxazolidone (D-cycloserine) *[68-14-7]* **M 102.1, m 154-155°(dec), [α]$_{546}$ +139° (c 2, H$_2$O).** Crystd from aq ammoniacal soln at *p*H 10.5 (100mg/ml) by diluting with 5 vols of isopropanol and then adjusting to *p*H 6 with acetic acid.

5-Aminolaevulinic acid hydrochloride *[5451-09-2]* **M 167.6, m 156-158°(dec).** Dried in a vac desiccator over P$_2$O$_5$ overnight then crystd by dissolving in cold EtOH and adding dry Et$_2$O.

8-Amino-6-methoxyquinoline *[90-52-8]* **M 174.1, m 41-42°.** Distd under N$_2$ at *ca* 50 microns, then recrystd several times from MeOH (0.4ml/g).

4-Aminomethylbenzenesulphonamide hydrochloride *[138-37-4]* **M 222.3, m 265-267°.** Crystd from dil HCl and dried in vac at 100°.

1-Amino-4-methylaminoanthraquinone *[1220-94-6]* **M 252.3.** Purified by tlc on silica gel plates using toluene/acetone (3:1) as eluent. The main band was scraped off and extracted with MeOH. The solvent was evapd and the residue dried in a drying pistol [Land, McAlpine, Sinclair and Truscott *JCSFT 1* **72** 2091 *1976*].

7-Amino-4-methylcoumarin *[26093-31-2]* **M 175.2, 221-442°(dec).** Dissolved in 5% HCl, filtered and basified with 2M ammonia. The ppte is dried in vac, and crystd from dil EtOH. It yields a blue soln and is light sensitive.

4-Amino-2-methyl-1-naphthol hydrochloride *[130-24-5]* **M 209.6, m 283°(dec).** Crystd from dil HCl.

2-Amino-2-methyl-1,3-propanediol *[115-69-5]* **M 105.1, m 111°, b 151-152°/10mm.** Crystd three times from MeOH, dried in a stream of dry N$_2$ at room temp, then in a vac oven at 55°. Stored over CaCl$_2$ [Hetzer and Bates *JPC* **66** 308 *1962*].

2-Amino-2-methyl-1-propanol *[124-68-5]* **M 89.4, m 31°, b 164-166°/760mm, d 0.935.** Purified by distn and fractional freezing.

2-Amino-3-methylpyridine *[1603-40-3]* **M 108.1, m 33.2°, b 221-222°.** Crystd three times from benzene, most of the residual benzene being removed from the crystals over paraffin wax chips in an evacuated desiccator. The amine, transferred to a separating funnel under N$_2$, was left in contact with NaOH pellets for 3hr with occasional shaking. It was then placed in a vac distilling flask where it was refluxed gently in a stream of dry N$_2$ before being fractionally distd [Mod, Magne and Skau *JPC* **60** 1651 *1956*].

2-Amino-4-methylpyridine *[695-34-1]* **M 108.1, m 99.2°, b 230°.** Crystd from EtOH or a 2:1 benzene/acetone mixture, and dried under vac.

2-Amino-5-methylpyridine *[1603-41-4]* **M 108.1, m 76.5°, b 227°.** Crystd from acetone.

2-Amino-6-methylpyridine *[1824-81-3]* **M 108.1, m 44.2°, b 208-209°.** Crystd three times from acetone, dried under vac at *ca* 45°. After leaving in contact with NaOH pellets for 3hr, with occasional shaking, it was decanted and fractionally distd [Mod, Magne and Skau *JPC* **60** 1651 *1956*]. Also recrystd from CH$_2$Cl$_2$ by addition of pet ether. [Marzilli *et al*, *JACS* **108** 4830 *1986*.]

2-Amino-5-methylpyrimidine *[50840-23-8]* **M 109.1, m 193.5°.** Crystd from water and benzene. Sublimes at 50°/0.5mm.

4-Amino-2-methylquinoline *[6628-04-2]* **M 158.2, m 168°, b 333°/760mm.** Crystd from benzene/pet ether.

2-Amino-4-(methylsulphoxyl)butyric acid see **methionine sulphoxide.**

2-Aminonaphthalene (β-naphthylamine)   [91-59-8] **M 143.2, m 111-113°.**   Crystd from water (charcoal). **CARCINOGENIC.**

3-Amino-2-naphthoic acid [5959-52-4] **M 187.2, m 214°(dec).** Crystd from aq EtOH.

4-Amino-5-naphthol-2,7-disulphonic acid [90-20-0] **M 320.3.** Sufficient Na_2CO_3 (ca 22g) to make the soln slightly alkaline to litmus was added to a soln of 100g of the dry acid in 750ml of hot distd water, followed by 5g of activated charcoal and 5g of Celite. The suspension was stirred for 10min and filtered by suction. The acid was pptd by adding ca 40ml of conc HCl (soln blue to Congo Red), then filtered by suction through sharkskin filter paper and washed with 100ml of distd water. The purification process was repeated. The acid was dried overnight in an oven at 60° and stored in a dark bottle [Post and Moore AC **31** 1872 1959].

1-Amino-2-naphthol hydrochloride [1198-27-2] **M 195.7, m 250°(dec).** Crystd from the minimum volume of hot water containing a few drops of stannous chloride in an equal weight of hydrochloric acid (to reduce atmospheric oxidation).

1-Amino-2-naphthol-4-sulphonic acid [116-63-2] **M 239.3, m 295°(dec).** Purified by warming 15g of the acid, 150g of $NaHSO_3$ and 5g of Na_2SO_3 (anhydrous) with 1L of water to ca 90°, shaking until most of the solid had dissolved, then filtering hot. The ppte obtained by adding 10ml of conc HCl to the cooled filtrate was collected, washed with 95% EtOH until the washings were colourless, and dried under vac over $CaCl_2$. It was stored in a dark coloured bottle, in the cold [Chanley, Gindler and Sobotka JACS **74** 4347 1952].

6-Aminonicotinic acid [3167-49-5] **M 138.1, m 312°(dec).** Crystd from aq acetic acid.

2-Amino-4-nitrobenzoic acid [619-17-0] **M 182.1, m 269°(dec).** Crystd from water or aq EtOH.

5-Amino-2-nitrobenzoic acid [13280-60-9] **M 182.1, m 235°(dec).** Crystd from water.

1-Amino-4-nitronaphthalene [776-34-1] **M 188.2, m 195°.** Crystd from EtOH or ethyl acetate.

2-Amino-4-nitrophenol [99-57-0] **M 154.1, m 80-90° (hydrate), 142-143° (anhydr),**
2-Amino-5-nitrophenol [121-88-0] **M 154.1, m 207-208°,**
6-Aminopenicillanic acid [551-16-6] **M 216.2, m 208-209°, $[\alpha]_{546}$ +327°** (in 0.1M HCl). Crystd from water.

2-Aminoperimidine hydrobromide [40835-96-9] **M 264.1, m 299°.** Purified by boiling a saturated aqueous soln with charcoal, filtering and leaving the salt to crystallise. Stored in a cool, dark place.

2-Aminophenol [95-55-6] **M 109.1, m 175-176°.** Purified by soln in hot water, decolorised with activated charcoal, filtered and cooled to induce crystn. It was necessary to maintain an atmosphere of N_2 over the hot phenol soln to prevent its oxidation [Charles and Freiser JACS **74** 1385 1952]. Can also be crystd from EtOH.

3-Aminophenol [591-27-5] **M 109.1, m 122-123°.** Crystd from hot water or toluene.

4-Aminophenol [123-30-8] **M 109.1, m 190° (under N_2).** Crystd from EtOH, then water, excluding oxygen. Can be sublimed at 110°/0.3mm. Has been purified by chromatography on alumina with a 1:4 (v/v) mixture of abs EtOH/benzene as eluent.

4-Aminophenol hydrochloride *[51-78-5]* **M 145.6, m 306°(dec).** Purified by treating an aq soln with satd Na$_2$S$_2$O$_3$, filtering under an inert atmosphere, then recrystd from 50% EtOH twice and once from abs EtOH [Livingston and Ke *JACS* **72** 909 *1950*].

4-Aminophenylacetic acid *[1197-55-3]* **M 151.2, m 199-200°(dec).** Crystd from hot water (60-70ml/g).

S(-)-2-Amino-3-phenyl-1-propanol (L-phenylalaninol) *[3182-95-4]* **M 151.2, m 95°, [α]$_D^{24}$ -25.7° (c 3.3, EtOH).** Crystd from benzene or toluene.

N-**Aminophthalimide** *[1875-48-5]* **M 162.2, m 200-202°.** Recrystd from 96% EtOH (1 part in 44 at b.p.) to form a yellow soln. Sublimes in vac at *ca* 150°. Resolidifies after melting, and remelts at 338-341°.

4-Aminopropiophenone *[70-69-9]* **M 163.1, m 140°.** Crystd from water or EtOH.

α-(α-Aminopropyl)benzyl alcohol *[5053-63-4]* **M 165.1, m 79-80°.** Crystd from benzene/pet ether.

4-(2-Aminopropyl)phenol *[103-86-6]* **M 151.2, m 125-126°.** Crystd from benzene.

1-Aminopyrene *[1606-67-3]* **M 217.3, m 117-118°.** Crystd from hexane.

2-Aminopyridine *[504-29-0]* **M 94.1, m 58°, b 204-210°.** Crystd from benzene/pet ether (b 40-60°) or CHCl$_3$/pet ether.

3-Aminopyridine *[462-08-8]* **M 94.1, m 64°, b 248°.** Crystd from benzene, CHCl$_3$/pet ether (b 60-70°), or benzene/pet ether (4:1).

4-Aminopyridine *[504-24-5]* **M 94.1, m 160°, b 180°/12-13mm.** Crystd from benzene/EtOH, then recrystd twice from water, crushed and dried for 4hr at 105° [Bates and Hetzer *J Res Nat Bur Stand* **64A** 427 *1960*]. Has also been crystd from EtOH, benzene, benzene/pet ether, toluene and sublimes in vac.

2-Aminopyrimidine *[109-12-6]* **M 95.1, m 126-127.5°.** Crystd from benzene. EtOH or water.

Aminopyrine (4-dimethylaminoantipyrene) *[58-15-1]* **M 231.3, m 107-109°.** Crystd from pet ether.

3-Aminoquinoline *[580-17-6]* **M 144.2, m 93.5°.** Crystd from benzene.

4-Aminoquinoline *[578-68-7]* **M 144.2, m 158°.** Purified by zone refining.

5-Aminoquinoline *[611-34-7]* **M 144.2, m 110°, b 184°/10mm, 310°/760mm.** Crystd from pentane, then from benzene or EtOH.

6-Aminoquinoline *[580-15-4]* **M 144.2, m 117-119°.** Purified by column chromatography on a SiO$_2$ column using CHCl$_3$/MeOH (4:1) as eluent. It is an irritant.

8-Aminoquinoline *[578-66-5]* **M 144.2, m 70°.** Crystd from EtOH or ligroin.

p-**Aminosalicylic acid** *[65-49-6]* **M 153.1, m 150-151°(dec),**
2-Amino-5-sulphanilylthiazole *[473-30-3]* **M 238.3, m 219-221°(dec).** Crystd from EtOH.

4-Amino-2-sulphobenzoic acid *[527-76-4]* **M 217.1.** Crystd from water.

2-Aminothiazole *[96-50-4]* **M 108.1, m 93°, b 140°/11mm.** Crystd from pet ether (b 100-120°), or EtOH.

2-Amino-1,2,4-triazole *[244994-60-3]* **M 84.1, m 91-93°.** Crystd from water. [Barszez *et al, JCSDT* 2025 *1986*].

3-Amino-1,2,4-triazole *[61-82-5]* **M 84.1, m 159°.** Crystd from EtOH (charcoal), then three times from dioxane [Williams, McEwan and Henry *JPC* 61 261 *1957*].

4-Amino-1,2,4-triazole *[584-13-4]* **M 84.1,m 80-81°.** Crystd from water. [Barszez *et al, JCSDT* 2025 *1986*.]

7-Amino-4-(trifluoromethyl)coumarin, m 222°. Purified by column chromatography on a C18 column, eluted with acetonitrile/0.01M aq HCl (1:1), and crystd from isopropanol. Alternatively, it is eluted from a silica gel column with CH_2Cl_2, or by extracting a CH_2Cl_2 solution (4g/L) with 1M aq NaOH (3 x 0.1L), followed by drying ($MgSO_4$), filtration and evapn. [Bissell *JOC* 45 2283 *1980*].

9-Aminotriptycene *[793-41-9]* **M 269.3, m 223.5-224.5°.** Recrystd from ligroin [Imashiro *et al, JACS* 109 729 *1987*].

DL-α-Amino-*n*-valeric acid see **norvaline.**

5-Amino-*n*-valeric acid *[660-88-8]* **M 117.2, m 157-158°.** Crystd by dissolving in water and adding EtOH.

5-Amino-*n*-valeric acid hydrochloride *[627-95-2]* **M 153.6, m 103-104°.** Crystd from $CHCl_3$.

Ammonium benzoate *[1863-63-4]* **M 139.2, m 200°(dec).** Crystd from EtOH.

Ammonium *d*-α-bromocamphor-π-sulphonate *[14575-84-9]* **M 328.2, m 284-285°(dec), $[\alpha]_D^{25}$ +84.8° (c 4, H$_2$O).** Passage of a hot aq soln through an alumina column removed water-soluble coloured impurities which remained on the column when the ammonium salt was eluted with hot water. The salt was crystd from water and dried over $CaCl_2$ [Craddock and Jones *JACS* 84 1098 *1962*].

Ammonium dodecylsulphate *[2235-54-3]* **M 283.4.** Recrystd first from 90% EtOH and then twice from abs EtOH, finally dried in vac.

Ammonium nitrosophenylhydroxylamine see **cupferron.**

Ammonium peroxydisulphate *[7727-54-0]* **M 228.2.** Recrystd at room temp from EtOH/water.

Ammonium picrate *[131-74-8]* **M 246.1, EXPLODES above 200°.** Crystd from EtOH and acetone.

Amodiaquin [4-(3-aminomethyl-4-hydroxyanilino)-7-chloroquinoline) *[86-42-0]* **M 287.5, m 208°.** Crystd from 2-ethoxyethanol.

D-Amygdalin *[29883-15-6]* **M 457.4, m 214-216°, $[\alpha]_D^{22}$ -38° (c 1.2, H$_2$O).** Crystd from water.

***n*-Amyl acetate** *[628-63-7]* **M 130.2, b 149.2°, n 1.40228, d 0.876.** Shaken with satd $NaHCO_3$ soln until neutral, washed with water, dried with $MgSO_4$ and distd.

***n*-Amyl alcohol** *[71-41-0]* **M 88.2, b 138.1°, n 1.4100, d^{15} 0.818.** Dried with anhydrous K_2CO_3 or $CaSO_4$, filtered and fractionally distd. Has also been treated with 1-2% of sodium and heated at reflux for 15hr to remove water and chlorides. Traces of water can be removed from the near-dry alcohol by refluxing with a small amount of sodium in the presence of 2-3% *n*-amyl phthalate or succinate followed by distn (see *ethanol*).

Small amounts of amyl alcohol have been purified by esterifying with *p*-hydroxybenzoic acid, recrystallising the ester with CS_2, saponifying with ethanolic-KOH, drying with $CaSO_4$ and fractionally distilling [Olivier *Rec Trav chim Pays-Bas* **55** 1027 *1936*].

tert-**Amyl alcohol** *[75-85-4]* **M 88.2, b 102.3°, n 1.4058, d^{15} 0.8135.** Refluxed with anhydrous K_2CO_3, CaH_2, CaO or sodium, then fractionally distd. Near-dry alcohol can be further dried by refluxing with magnesium activated with iodine, as described for *ethanol*. Further purification is possible using fractional crystn, zone refining or preparative gas chromatography.

n-**Amylamine** *[110-58-7]* **M 87.2, b 105°, d 0.752.** Dried by prolonged shaking with NaOH pellets, then distd.

n-**Amyl bromide** (*n*-pentylbromide) *[110-53-2]* **M 151.1, b 129.7°, n 1.445, d 1.218.** Washed with conc H_2SO_4, then water, 10% Na_2CO_3 soln, again with water, dried with $CaCl_2$ or K_2CO_3, and fractionally distd just before use.

n-**Amyl chloride** *[543-59-9]* **M 106.6, b 107.8°, n 1.41177, d 0.882,**
sec-**Amyl chloride** (1-chloro-2-methylbutane) *[616-13-7]* **M 106.6, b 96-97°.** Purified by stirring vigorously with 95% H_2SO_4, replacing the acid when it became coloured, until the layer remained colourless after 12hr stirring. The amyl chloride was then washed with satd Na_2CO_3 soln, then distd water, and dried with anhydrous $MgSO_4$, followed by filtration, and distn through a 10-in Vigreux column. Alternatively a stream of oxygen containing 5% ozone was passed through the amyl chloride for three times as long as it took to cause the first coloration of starch iodide paper by the exit gas. Washing the liquid with $NaHCO_3$ soln hydrolyzed ozonides and removed organic acids prior to drying and fractional distn [Chien and Willard *JACS* **75** 6160 *1953*].

tert-**Amyl chloride** *[594-36-5]* **M 106.6, b 86°, d 0.866.** Methods of purification commonly used for other alkyl chlorides lead to decomposition. Unsatd materials were removed by chlorination with a small amount of chlorine in bright light, followed by distn [Chien and Willard *JACS* **75** 6160 *1953*].

Amyl ether *[693-65-2]* **M 158.3, b 186.8°, n 1.41195, d 0.785.** Repeatedly refluxed over sodium and distd.

n-**Amyl mercaptan** see **1-pentanethiol**.

Amylose see Chapter 6.

p-*tert*-**Amylphenol** *[80-46-6]* **M 146.3, m 93.5-94.2°.** Purified *via* its benzoate, as for phenol. After evaporating the solvent from its soln in ether, the material was crystd (from the melt) to constant melting point [Berliner, Berliner and Nelidow *JACS* **76** 507 *1954*].

2-*n*-Amylpyridine *[2294-76-0]* **M 149.2, b 63.0°/2mm, n^{26} 1.4861,**
4-*n*-Amylpyridine *[2961-50-4]* **M 149.2, b 78.0°/2.5mm, n 1.4908.** Dried with NaOH for several days, then distd from CaO under reduced pressure, taking the middle fraction and redistilling it.

α-Amyrin *[638-95-9]* **M 426.7, m 186°.** Crystd from EtOH.

β-Amyrin *[508-04-3]* **M 426.7, m 197-197.5°.** Crystd from pet ether or EtOH.

Androstane *[24887-75-0]* **M 260.5, m 50-50.5°.** Crystd from acetone/MeOH.

epi-**Androsterone** *[481-29-8]* **M 290.4, m 172-173°, [α]$_{546}$ +115° (c 1, MeOH).** Crystd from aq EtOH.

cis-**Androsterone** *[53-41-8]* **M 290.4. m 185-185.5°.** Crystd from acetone/Et_2O.

Angelic acid *[565-63-9]* **M 100.1, m 45°.** Steam distd, then crystd from water.

Aniline *[62-53-3]* **M 93.1, f.p. -6.0°, b 68.3/10mm, 184.4°/760mm, n 1.585, n^{25} 1.5832, d 1.0220.** Aniline is hygroscopic. It can be dried with KOH or CaH$_2$, and distd at reduced pressure. Treatment with stannous chloride removes sulphur-containing impurities, reducing the tendency to become coloured by aerial oxidn. Can be crystd from Et$_2$O at low temps. More extensive purifications involve preparation of derivatives, such as the double salt of aniline hydrochloride and cuprous chloride or zinc chloride, or N-acetylaniline (m 114°) which can be recrystd from water.

Recrystd aniline was dropped slowly into an aq soln of recrystd oxalic acid. Aniline oxalate was filtered off, washed several times with water and recrystd three times from 95% EtOH. Treatment with satd Na$_2$CO$_3$ soln, regenerated aniline which was distd from the soln, dried and redistd under reduced pressure [Knowles *Ind Eng Chem* **12** 881 *1920*].

After refluxing with 10% acetone for 10hr, aniline was acidified with HCl (Congo Red as indicator) and extracted with Et$_2$O until colourless. The hydrochloride was purified by repeated crystn before aniline was liberated by addition of alkali, then dried with solid KOH, and distd. The product was sulphur-free and remained colourless in air [Hantzsch and Freese *Ber* **27** 2529, 2966 *1894*].

Non-basic materials, including nitro compounds were removed from aniline in 40% H$_2$SO$_4$ by passing steam through the soln for 1hr. Pellets of KOH were added to liberate the aniline which was steam distd, dried with KOH, distd twice from zinc dust at 20mm, dried with freshly prepared BaO, and finally distd from BaO in an all-glass apparatus [Few and Smith *JCS* 753 *1949*].

Aniline hydrobromide *[542-11-0]* **M 174.0, m 286°,**
Aniline hydrochloride *[142-04-1]* **M 129.6, m 200.5-201°,**
Aniline hydriodide *[45497-73-2]* **M 220.0.** Crystd from water or EtOH and dried at 5mm over P$_2$O$_5$. Crystd four times from MeOH containing a few drops of conc HCl by addition of pet ether (b 60-70°), then dried to constant wieght over paraffin chips, under vac [Gutbezahl and Grunwald *JACS* **75** 559 *1953*]. It was pptd from EtOH soln by addition of Et$_2$O, and the filtered solid was recrystd from EtOH and dried in vac. [Buchanan *et al, JACS* **108** 1537 *1986*.]

p-**Anilinophenol** see *p*-hydroxydiphenylamine.

m-**Anisaldehyde** *[591-31-1]* **M 136.2, b 143°/50mm, d 1.119.** Washed with NaHCO$_3$, then water, dried with anhydrous MgSO$_4$ and distd under reduced pressure under N$_2$. Stored under N$_2$ in sealed glass ampoules.

Anisic acid see *p*-methoxybenzoic acid.

p-**Anisidine** *[104-94-9]* **M 123.2, m 57°.** Crystd from water or aq EtOH. Dried in vac oven at 40° for 6hr and stored in a dry box. [More *et al, JACS* **108** 2257 *1986*.] Purified by vac sublimation [Guarr *et al, JACS* **107** 5104 *1985*.]

Anisole *[100-66-3]* **M 108.1, f.p. -37.5°, b 43°/11mm, 153.8°/760mm, n^{25} 1.5143, d^{15} 0.9988.** Shaken with half volume of 2M NaOH, and emulsion allowed to separate. Repeated 3 times, then washed twice with water, dried over CaCl$_2$, filtered, dried over sodium wire and finally distd from fresh sodium under N$_2$, using a Dean-Stark trap, samples in the trap being rejected until free from turbidity [Caldin, Parbov, Walker and Wilson *JCSFT 1* **72** 1856 *1976*].

Dried with CaSO$_4$ or CaCl$_2$, or by refluxing with sodium or BaO with crystalline FeSO$_4$ or by passage through an alumina column. Traces of phenols have been removed by prior shaking with 2M NaOH, followed by washing with water. Can be purified by zone refining.

2-*p*-Anisyl-1,3-indanone *[117-37-3]* **M 252.3, m 156-157°.** Crystd from acetic acid or EtOH.

Anserine *[584-85-0]* **M 240.3, m 238-239°, [α]$_D$ +11.3° (H$_2$O).** Crystd from aq EtOH. It is hygroscopic.

S-Anserine nitrate *[5937-77-9]* **M 303.3, m 225°(dec), [α]$_D^{30}$ +12.2°.** Likely impurities: 1-methylimidazole-5-alanine, histidine. Crystd from aq MeOH.

Antheraxanthin *[68831-78-7]* **M 584.8, m 205°, λ$_{max}$ 460.5, 490.5nm, in CHCl$_3$.** Likely impurities: violaxanthin and mutatoxanthin. Purified by chromatography on columns of Ca(OH)$_2$ and of ZnCO$_3$. Crystd from benzene/MeOH as needles or thin plates. Stored in the dark, in an inert atmosphere, at -20°.

Anthracene *[120-12-7]* **M 178.2, m 218°.** Likely impurities are anthraquinone, anthrone, carbazole, fluorene, 9,10-dihydroanthracene, tetracene and bianthryl. Carbazole is removed by continuous-adsorption chromatography [see Sangster and Irvine *JPC* **24** 670 *1956*] using a neutral alumina column and passing *n*-hexane. [Sherwood in **Purification of Inorganic and Organic Materials**, Zief (ed), Marcel Dekker, New York, 1969.] The solvent is evapd and anthracene is sublimed under vac, then purified by zone refining, under N$_2$ in darkness or non-actinic light.
Has been purified by co-distillation with ethylene glycol (boils at 197.5°), from which it can be recovered by additn of water, followed by crystn from 95% EtOH, benzene, toluene, a mixture of benzene/xylene (4:1), or Et$_2$O. It has also been chromatographed on alumina with pet ether in a dark room (to avoid photo-oxidation of adsorbed anthracene to anthraquinone). Other purification methods include sublimation in a N$_2$ atmosphere (in some cases after refluxing with sodium), and recrystd from toluene [Gorman *et al, JACS* **107** 4404 *1985*].
Anthracene has also been crystd from EtOH, chromatographed through alumina in hot benzene (*fume hood*) and then vac sublimed in a pyrex tube that has been cleaned and baked at 100°. (For further details see Craig and Rajikan *JCSFT 1* **74** 292 *1978*; and Williams and Zboinski *JCSFT 1* **74** 611 *1978*.) More recently it has been chromatographed on alumina, recrystd from *n*-hexane and sublimed under reduced pressure. [Saltiel *JACS* **108** 2674 *1986*; Masnori *et al, JACS* **108** 1126 *1986*.] Alternatively, it was recrystd from cyclohexane, chromatographed on alumina with *n*-hexane as eluent, and recrystd two more times [Saltiel *et al, JACS* **109** 1209 *1987*].

Anthracene-9-carbonitrile see **9-cyanoanthracene.**

Anthracene-9-carboxylic acid *[723-62-6]* **M 222.2, m 214°(dec).** Crystd from EtOH.

9-Anthraldehyde *[642-31-9]* **M 206.2, m 104-105°.** Crystd from acetic acid or EtOH. [Masnori *et al, JACS* **108** 1126 *1986*.]

Anthranilic acid see *o*-aminobenzoic acid.

Anthranol *[529-86-2]* **M 196.2, m 160-170°(dec).** Crystd from glacial acetic acid or aq EtOH.

Anthranthrone *[641-13-4]* **M 306.3, m 300°.** Crystd from chlorobenzene or nitrobenzene.

Anthraquinone *[84-65-1]* **M 208.2, m 286°.** Crystd from CHCl$_3$ (38ml/g), benzene, or boiling acetic acid, washing with a little EtOH and drying under vac over P$_2$O$_5$.

Anthraquinone Blue B *[2861-02-1]* **M 476.4,**
Anthraquinone Blue RXO *[4403-89-8]* **M 445.5,**
Anthraquinone Green G *[4403-90-1]* **M 624.6.** Purified by salting out three times with sodium acetate, followed by repeated extraction with EtOH [McGrew and Schneider *JACS* **72** 2547 *1950*].

Anthrarufin *[117-12-4]* **M 240.1, m 280°(dec).** Purified by column chromatography on silica gel with CHCl$_3$/Et$_2$O as eluent, followed by recrystn from acetone. Alternatively recrystd from glacial acetic acid [Flom and Barbara *JPC* **89** 4489 *1985*].

1,8,9-Anthratriol *[480-22-8]* **M 226.2, m 176-181°.** Crystd from pet ether.

Anthrimide *[82-22-4]* **M 429.4.** Crystd from chlorobenzene or nitrobenzene.

Anthrone *[90-44-8]* **M 194.2, m 155°.** Crystd from a 3:1 mixture of benzene/pet ether (b 60-80°) (10-12ml/g), or successively from benzene then EtOH. Dried under vac.

Antipyrine *[60-80-0]* **M 188.2, m 114°, b 319°.** Crystd from EtOH/water mixture, benzene, benzene/pet ether or hot water (charcoal), and dried under vac.

β-Apo-4'-carotenal *[12676-20-9]* **M 414.7, m 139°,** $\varepsilon_{1cm}^{1\%}$ **2640 at 461nm,**
β-Apo-8'-carotenal *[1107-26-2]* **M 414.7.** Recrystd from $CHCl_3$/EtOH mixture or *n*-hexane. [Bobrowski and Das *JOC* **91** 1210 *1987*].

β-Apo-8'-carotenoic acid ethyl ester *[1109-11-1]* **M 526.8, m 134-138°,** $\varepsilon_{1cm}^{1\%}$ **2550 at 449nm,**
β-Apo-8'-carotenoic acid methyl ester *[16266-99-2]* **M 512.7, m 136-137°,** $\varepsilon_{1cm}^{1\%}$ **2575 at 446nm and 2160 at 471nm, in pet ether..** Crystd from pet ether or pet ether/ethyl acetate. Stored in the dark in an inert atmosphere at -20°.

Apocodeine *[641-36-1]* **M 281.3, m 124°.** Crystd from MeOH and dried at 80°/2mm.

Apomorphine *[50-00-4]* **M 267.3, m 195°(dec).** Crystd from $CHCl_3$ and pet ether.

β-L-Arabinose (natural) *[87-72-9]* **M 150.1, m 158°, $[\alpha]_D$ +104° (c 4, H_2O after 24hr).** Crystd slowly twice from 80% aq EtOH, then dried under vac over P_2O_5.

D-Arabinose *[28697-53-6]* **M 150.1, m 164°, $[\alpha]_{546}$ -123° (c 10, H_2O after 24hr).** Crystd three times from EtOH, vac dried at 60° for 24hr and stored in a vac desiccator.

L-Arabitol *[7643-75-6]* **M 152.2, m 102°, $[\alpha]_{546}$ -16° (c 5, 8% borax soln),**
DL-Arabitol *[2152-56-9]* **M 152.2, m 105-106°.** Crystd from 90% EtOH.

Araboascorbic acid see isoascorbic acid.

Arachidic acid *[506-30-9]* **M 312.5, m 77°.** Crystd from abs EtOH.

Arachidic alcohol (1-eicosanol) *[629-96-9]* **M 298.6, m 65.5° (71°), b 200°/3mm.** Crystd from benzene or benzene/pet ether.

p-**Arbutin** *[497-76-7]* **M 272.3, m 163-164°.** Crystd from water.

S-Arginine *[74-79-3]* **M 174.2, m 207°(dec), $[\alpha]_D$ +26.5° (c 5, in 5M HCl), $[\alpha]_{546}$ +32° (c 5, in 5M HCl).** Crystd from 66% EtOH.

S-Arginine hydrochloride *[1119-34-2]* **M 210.7, m 217°(dec), $[\alpha]_D^{25}$ +26.9° (c 6, M HCl).** Likely impurity is ornithine. Crystd from water at *p*H 5-7, by adding EtOH to 80% (v/v).

S-Argininosuccinic acid *[2387-71-5]* **M 290.3, $[\alpha]_D^{24}$ +16.4° (H_2O).** Likely impurity is fumaric acid. In neutral or alkaline soln it readily undergoes ring closure. Crystd from water by adding 1.5 vols of EtOH. Barium salt is stable at 0-5° if dry.

S-Argininosuccinic anhydride *[28643-94-9]* **M 272.3, $[\alpha]_D^{23}$ -10° (H_2O for anhydride formed at neutral pH).** Crystd from water by adding two vols of EtOH. An isomeric anhydride is formed if the free acid is allowed to stand at acid pH. In soln, the mixture of anhydrides and free acid is formed.

Ascorbic acid *[50-81-7]* **M 176.1, m 193°(dec), $[\alpha]_{546}$ +23° (c 10, H_2O).** Crystd from MeOH/Et_2O/pet ether [Herbert *et al JCS* 1270 *1933*].

S-Asparagine *[70-47-3]* **M 150.1, m 234-235⁰, (monohydrate** *[5794-13-8]*), [α]$_D$ **+32.6⁰ (0.1M HCl).** Likely impurities are aspartic acid and tyrosine. Crystd from water or aq EtOH. Slowly effloresces in dry air.

Aspartic acid M 133.1, m 338-339⁰ (RS, *[617-45-8]*)**; m 271⁰ (S, requires heating in a sealed tube** *[56-84-8]*)**, [α]$_D$²⁵ +25.4⁰ (3M HCl).** Likely impurities are glutamic acid, cystine and asparagine. Crystd from water by adding 4 vols of EtOH and dried at 110⁰.

L-Aspartic acid β-methyl ester hydrochloride *[16856-13-6]* **M 183.6, m 194⁰.** Recrystd from MeOH by using anhydrous ethyl ether [Bach *et al, Biochem Preps* **13** 20 *1971*].

DL-Aspartic acid dimethyl ester hydrochloride *[14358-33-9]* **M 197.7.** Crystd from abs MeOH. [Kovach *et al, JACS* **107** 7360 *1985*.]

Aspergillic acid *[490-02-8]* **M 224.3, m 97-99⁰.** Sublimed at 80⁰/10⁻³mm. Crystd from MeOH.

Astacin *[514-76-1]* **M 592.8, ε$_{1cm}$¹% 10⁵·⁵ at 498mm (pyridine).** Probable impurity is astaxanthin. Purified by chromatography on alumina/fibrous clay (1:4) or sucrose, or by partition between pet ether and MeOH (alkaline). Crystd from pyridine/water. Stored in the dark under N$_2$ at -20⁰.

Atrolactic acid (½H$_2$O) *[515-30-0]* **M 166.2, m 94.5⁰ (anhydr), 88-91⁰ (½H$_2$O).** Crystd from water and dried at 55⁰/0.5mm.

Atropine *[51-55-8]* **M 289.4, m 114-116⁰.** Crystd from acetone or hot water.

Aureomycin and hydrochloride see Chapter 6.

8-Azaadenine *[1123-54-2]* **M 136.1, m 345⁰(dec).** Crystd from water.

2-Azacyclotridecanone *[947-04-6]* **M 197.3, m 152⁰.** Crystd from CHCl$_3$, stored over P$_2$O$_5$ in a vac desiccator.

8-Azaguanine *[134-58-7]* **M 152.1, m >300⁰.** Dissolved in hot M NH$_4$OH, filtered, and cooled; recrystd, and washed with water.

7-Azaindole *[271-63-6]* **M 118.1, m 105-106⁰.** Repeatedly recrystd from EtOH, then vac sublimed [Tokumura *et al, JACS* **109** 1346 *1987*].

1-Azaindolizine *[274-76-0]* **M 118.1, b 72-73o/1mm.** Purified by distn or gas chromatography.

Azaserine *[115-02-6]* **M 173.1, m 146-162⁰(dec), [α]$_D$²⁷·⁵ -0.5⁰ (c 8.5, H$_2$O, pH 5.2).** Crystd from 90% EtOH.

Azelaic acid *[123-99-9]* **M 188.2, m 105-106⁰.** Crystd from water (charcoal) or thiophen-free benzene. The material cryst from water was dried by azeotropic distn in toluene: the residual toluene soln was cooled and filtered, the ppte being dried in a vac oven. Also purified by zone refining or by sublimation onto a cold finger at 10⁻³ torr.

Azobenzene *[103-33-3]* **M 182.2, m 68⁰.** Ordinary azobenzene is nearly all in the *trans*-form. It is partly converted into the *cis*-form on exposure to light [for isolation see Hartley *JCS* 633 *1938*, and for spectra of *cis*- and *trans*-azobenzenes, see Winkel and Siebert *Ber* **74B** 670 *1941*]. *trans*-Azobenzene is obtained by chromatography on alumina using 1:4 benzene/heptane or pet ether, and crystd from EtOH (after refluxing for several hours) or hexane. All operations should be carried out in diffuse red light or in the dark.

1,1'-Azobis(cyclohexane carbonitrile) *[2094-98-6]* **M 244.3, m 114-114.5°,** ε_{350nm} **16.0.** Crystd from EtOH.

Azobis(isobutyramidinium) chloride M 179.7. Crystd from water.

α,α'-Azobis(isobutyronitrile) *[78-61-1]* **M 164.2, m 103°(dec).** Crystd from acetone, Et$_2$O, CHCl$_3$, aq EtOH or MeOH. Has also been crystd from abs EtOH below 40° in subdued light. Dried under vac at room temp over P$_2$O$_5$ and stored under vac in the dark at <-10° until used. Also crystd from CHCl$_3$ soln by addn of pet ether (b <40°). [Askham *et al, JACS* **107** 7423 *1985*; Ennis *et al, JCSDT* 2485 *1986*; Inoue and Anson *JPC* **91** 1519 *1987; Tanner JOC* **52** 2142 *1987.*]

Azolitmin B *[1395-18-2].* Crystd from water,

Azomethane *[503-28-6]* **M 58.1, m -78°, b 1.5°.** Purified by vac distn and stored in the dark at -80°. Can be **EXPLOSIVE.**

p,p'-Azoxyanisole *[1562-94-3]* **M 258.3, transition temps: 118.1-118.8°, 135.6-136.0°.** Crystd from abs EtOH or acetone, and dried by heating under vac.

Azoxybenzene *[495-48-7]* **M 198.2, m 36°.** Crystd from EtOH or MeOH, and dried for 4hr at 25° and 10^{-3}torr. Sublimed before use.

p,p-Azoxyphenetole *[1562-94-3]* **M 258.3, m 137-138° (turbid liquid clarifies at 167°).** Crystd from toluene or EtOH.

Azulene *[275-51-4]* **M 128.2, m 98.5-99°.** Crystd from EtOH.

Azuleno(1,2-b)thiophene *[25043-00-9]* **M 184.2,**
Azuleno(2,1-b)thiophene *[248-13-5]* **M 184.2.** Crystd from cyclohexane, then sublimed *in vacuo*.

Azure A *[531-53-3]* **M 291.8, CI 52005, m > 290°(dec),** λ_{max} **633nm,**
Azure B *[531-55-3]* **M 305.8, CI 52010, m > 201°(dec),** λ_{max} **648nm,**
Azure C *[531-57-7]* **M 277.8,** λ_{max} **616nm.** Twice recrystd from water, and dried at 100° in an oven for 1hr.

B.A.L. see 1,2-dimercapto-3-propanol.

Barbituric acid *[67-52-7]* **M 128.1, m 250°(dec).** Crystd twice from water, then dried for 2 days at 100°.

Batyl alcohol *[544-62-7]* **M 344.6, m 70.5-71°.** Crystd from aq acetone.

Behenic acid see **docosanoic acid.**

Behenyl alcohol see **1-docosanol.**

Benzalacetone *[122-57-6]* **M 146.2, m 42°.** Crystd from b 40-60° pet ether, or dist (b 137-142° /16mm).

Benzalacetophenone (Chalcone) *[94-41-6]* **M 208.3, m 56-58°.** Crystd from EtOH warmed to 50° (about 5ml/g), iso-octane, or toluene/pet ether, or recrystd from MeOH, and then twice from hexane.

Benzaldehyde *[100-52-7]*.**M 106.1, f -26°, b 62°/10mm, 179.0°/760mm, n 1.5455, d 1.044.** To diminish its rate of oxidation, benzaldehyde usually contains additives such as hydroquinone or catechol. It can be purified *via* its bisulphite addition compound but usually distn (under nitrogen at reduced pressure) is sufficient. Prior to distn it is washed with NaOH or 10% Na_2CO_3 (until no more CO_2 is evolved), then with satd Na_2SO_3 and water, followed by drying with $CaSO_4$, $MgSO_4$ or $CaCl_2$.

anti-**Benzaldoxime** *[932-90-1]* **M 121.1, m 130°.** Crystd from ethyl ether by adding pet ether (b 60-80°).

Benzamide *[55-21-0]* **M 121.1, m 129.5°.** Crystd from hot water (about 5ml/g), EtOH or 1,2-dichloroethane, and air dried. Crystd from dil aq ammonia, water, acetone and then benzene (using a Soxhlet extractor). Dried in an oven at 110° for 8hr and stored in a desiccator over 99% H_2SO_4. [Bates and Hobbs *JACS* **73** 2151 *1951*.]

Benzamidine *[618-39-3]* **M 120.2, m 64-66°.** Liberated from chloride by treatment with 5M NaOH. Extracted into ethyl ether. Sublimed *in vacuo.*

Benzanilide *[93-98-11]* **M 197.2, m, 164°.** Crystd from pet ether (b 70-90°) using a Soxhlet extractor, and dried overnight at 120°. Also crystd from EtOH.

Benz[α]anthracene *[56-55-3]* **M 228.3, m 159-160°.** Crystd from MeOH, EtOH or benzene (charcoal), then chromatographed on alumina from sodium-dried benzene (twice), using vac distn to remove benzene. Final purification was by vac sublimation.

Benz[α]anthracene-7,12-dione *[2498-66-0]* **M 258.3, m 169.5-170.5°.** Crystd from MeOH (charcoal).

Benzanthrone *[82-05-3]* **M 230.3, m 170°.** Crystd from EtOH or xylene.

Benzene *[71-43-2]* **M 78.1, f 5.5°,b 80.1°,d^{20} 0.874, n 1.50110, n^{25} 1.49790.** For most purposes, benzene can be purified sufficiently by shaking with conc H_2SO_4 until free from thiophen, then with water, dil NaOH and water, followed by drying (with P_2O_5, sodium, $LiAlH_4$, CaH_2, 4X Linde molecular sieve, or $CaSO_4$, or by passage through a column of silica gel, for a preliminary drying, $CaCl_2$ is suitable), and distn. A further purification step to remove thiophen, acetic acid and propionic acid, is crystn by partial freezing. The usual contaminants in dry thiophen-free benzene are non-benzenoid hydrocarbons such as cyclohexane, methylcyclohexane, and heptanes, together with naphthenic hydrocarbons and traces of toluene. Carbonyl-containing impurities can be removed by percolation through a Celite column impregnated with 2,4-dinitrophenylhydrazine, phosphoric acid and water. (Prepared by dissolving 0.5g DNPH in 6ml of 85% H_3PO_4 by grinding together, then adding and mixing 4ml of distd water and 10g Celite.) [Schwartz and Parker *AC* 33 1396 *1961*.] Benzene has been freed from thiophen by refluxing with 10% (w/v) of Raney nickel for 15min, after which the nickel was removed by filtn or centrifugation.

Mair *et al* [*J Res Nat Bur Stand* 37 229 *1946*] cooled a mixture of 200ml of benzene and 50ml of EtOH in a cylindrical brass container (5cm dia x 20cm) in an ice-salt cooling bath at *ca* -10° The slurry which was produced on vigorous stirring was transferred to a centrifuge cooled to near -10°. After 5min the benzene crystals were removed from the basket of the centrifuge, allowed to melt, washed three times with distd water, and filtered through silica gel to remove any alcohol and water before distn.

When dry benzene was obtained by doubly distilling high purity benzene from a soln containing the blue ketyl formed by the reaction of sodium-potassium alloy with a small amount of benzophenone.

Thiophen has been removed from benzene (absence of bluish-green colorattion when 3ml of benzene is shaken with a soln of 10mg of isatin in 10ml of conc H_2SO_4) by refluxing the benzene (1Kg) for several hours with 40g HgO (freshly pptd) dissolved in 40ml glacial acetic acid and 300ml of water. The ppte was filtered off, the aq phase was removed and the benzene was washed twice with water, dried and distd. Alternatively, benzene dried with $CaCl_2$ has been shaken vigorously for half an hour with anhydrous $AlCl_3$ (12g/L at 25-35°, then decanted, washed with 10% NaOH, and water, dried and distd. The process was repeated, giving thiophen-free benzene. [Holmes and Beeman *Ind Eng Chem* 26 172 *1934*.]

After shaking successively for about an hour with concnd H_2SO_4, distd water (twice), 6M NaOH, and distd water (twice), benzene was distd through a 3-ft glass column to remove most of the water. Abs EtOH was added and the benzene-alcohol azeotrope was distd. (This low-boiling distn leaves any non-azeotrope-forming impurities behind.) The middle fraction was shaken with distd water to remove EtOH, and again redistd. Final slow and very careful fractional distn from sodium, then $LiAlH_4$ under N_2, removed traces of water and peroxides. [Peebles, Clarke and Stockmayer *JACS* 82 2780 *1960*.] *Benzene liquid and vapour are very* **TOXIC** *and* **HIGHLY FLAMMABLE**, *and all operations should be carried out in an efficient fumecupboard and in the absence of naked flames in the vicinity.*

[$^2H_6$]Benzene (benzene-d_6) *[1076-43-3]* M 84.2, b 80°/773.6mm, 70°/562mm, 60°/399mm, 40°/186.3mm, 20°/77.1mm, 10°/49.9mm, 0°/27.5mm, n 1.4991, n^{40} 1.4865, d 0.9488, d^{40} 0.9257. Hexadeuteriobenzene of 99.5% purity is refluxed over and distd from CaH_2 onto Linde type 5A sieves under N_2.

Benzeneazodiphenylamine *[28110-26-1]* M 273.3, m 82°. Purified by chromatography on neutral alumina using anhydrous benzene with 1% anhydrous MeOH. The major component, which gave a stationary band, was cut out and eluted with EtOH or MeOH. [Högfeldt and Bigeleisen *JACS* 82 15 *1960*.] Crystd from pet ether or EtOH.

1-Benzeneazo-2-naphthol *[842-07-9]* M 248.3, m 134°. Crystd from EtOH.

1-Benzeneazo-2-naphthylamine *[85-84-7]* M 247.3, m 102-104°. Crystd from acetic acid/water.

m-**Benzenedisulphonic acid** *[98-48-6]* M 238.2. Freed from H_2SO_4 by conversion to the calcium or barium salts (using $Ca(OH)_2$ or $Ba(OH)_2$, and filtering). The calcium salt was then converted to the potassium salt, using K_2CO_3. Both the potassium and the barium salts were recrystd from water, and the acid was regenerated by passing through the H^+ form of a cation exchange resin. The acid was recrystd twice from conductivity water and dried over $CaCl_2$ at 25°. [Atkinson, Yokoi and Hallada *JACS* 83 1570 *1961*.] It has also been crystd from Et_2O and dried in a vac oven.

m-**Benzenedisulphonyl chloride** *[585-47-7]* M 275.1, m 63°. Crystd from $CHCl_3$ and dried at 20mm pressure.

Benzenephosphinic acid *[1779-48-2]* **M 142.1, m** *ca* **70°.** Purified by allowing to stand for several days under ethyl ether, with intermittent shaking and several changes of solvent. After filtn, the excess ether was removed in vac.

Benzeneseleninic acid *[6996-92-5]* **M 189.1, m 123-124°.** Recrystd twice from water [Kice and Purkiss *JOC* **52** 3448 *1987*].

Benzenesulphonic anhydride *[512-35-6]* **M 298.3, m 88-91°.** Crystd from ethyl ether.

Benzenesulphonyl chloride *[98-09-9]* **M 176.6, m 14.5°, b 120°/10mm, 251.2°/760mm(dec), d 1.384.** Distd, then treated with 3mole % each of toluene and AlCl$_3$, and allowed to stand overnight. The free benzenesulphonyl chloride was distd off at 1mm pressure, and then carefully fractionally distd at 10mm in an all-glass column. [Jensen and Brown *JACS* **80** 4042 *1958*.]

Benzene-1,2,4,5-tetracarboxylic acid *[89-05-4]* **M 254.2, m 281-284°.** Crystd from water.

Benzenethiol (thiophenol) *[108-98-5]* **M 110.2, f.p. -14.9°, b 46.4°/10mm, 168.0°/760mm, n 1.58973, d 1.073.** Dried with CaCl$_2$ or CaSO$_4$, and distd at 10mm pressure or at 100mm (b 103.5°) in a steam of N$_2$.

Benzene-1,2,3-tricarboxylic acid (H$_2$O) *[36362-97-7]* **M 210.1, m 190°(dec),**
Benzene-1,3,5-tricarboxylic acid (trimesic acid) *[554-95-0]* **M 210.1, m 360°(dec).** Crystd from water.

1,2,4-Benzenetriol *[533-73-3]* **M 126.1, m 141°.** Crystd from ethyl ether.

Benzethonium chloride *[121-54-0]* **M 448.1, m 164-166°.** Crystd from 1:9 MeOH/ethyl ether mixture.

Benzhydrol *[91-01-0]* **M 184.2, m 69°, b 297°/748mm, 180°/20mm.** Crystd from hot water or pet ether (b 60-70°), pet ether sometimes contains a little benzene, from CCl$_4$, or EtOH (1ml/g). An additional purification step is passage of a benzene soln through an activated alumina column. Sublimes in vac. Also crystd three times from MeOH/water [Naguib *JACS* **108** 128 *1986*].

Benzidine *[92-87-5]* **M 184.2, m 128-129°.** Its soln in benzene was decolorized by percolation through two 2-cm columns of activated alumina, then concentrated until benzidine crystd on cooling. Recrystd alternatively from EtOH and benzene to constant absorption spectrum [Carlin, Nelb and Odioso *JACS* **73** 1002 *1951*]. Has also been crystd from hot water (charcoal) and from ethyl ether. Dried under vac in an Abderhalden pistol. Stored in the dark in a stoppered container. **CARCINOGENIC.**

Benzidine dihydrochloride *[531-85-1]* **M 257.2.** Crystd by soln in hot water, with addn of conc HCl to the slightly cooled soln.

Benzil *[134-81-6]* **M 210.2, m 96-96.5°.** Crystd from benzene after washing with alkali. (Crystn from EtOH did not free benzil from material reacting with alkali.) [Hine and Howarth *JACS* **80** 2274 *1958*.] Has also been crystd from CCl$_4$, diethyl ether or EtOH [Inoue *et al*, *JCSFT 1* **82** 523 *1986*].

Benzilic acid *[76-93-7]* **M 228.3, m 150°.** Crystd from benzene (*ca* 6ml/g), or hot water.

Benzil monohydrazone *[5433-88-7]* **M 224.3, m 151°.** Crystd from EtOH.

α-Benzil monoxime *[14090-77-8], [E, 574-15-2], [Z, 574-16-3]* **M 105.1, m 140°.** Crystd from benzene (must not use animal charcoal).

Benzimidazole *[51-17-2]* **M 118.1, m 172-173°.** Crystd from water or aq EtOH (charcoal), and dried at 100° for 12hr.

Benzo[*b*]biphenylene *[259-56-3]* **M 202.2.** Purified by sublimation under reduced pressure.

Benzo-15-crown-5 *[14098-44-3]* **M 268.3.** Recrystd from *n*-heptane.

Benzo[3,4]cyclobuta[1,2-*b*]quinoxaline *[259-57-3]* **M 204.2.** Purified by sublimation under reduced pressure.

2-Benzofurancarboxylic acid *[496-41-3]* **M 162.1, m 192-193°.** Crystd from water.

Benzofurazan *[273-09-6]* **M 20.1,m 55°.** Purified by crystn from EtOH and sublimed.

Benzoic acid *[65-85-0]* **M 122.1, m 122.6-123.1°.** For use as a volumetric standard, analytical reagent grade benzoic acid should be carefully fused to *ca* 130° (to dry it) in a platinum crucible, and then powdered in an agate mortar. Benzoic acid has been crystd from boiling water (charcoal), aq acetic acid, glacial acetic acid, benzene, aq EtOH, pet ether (b 60-80°), and from EtOH soln by adding water. It is readily purified by fractional crystn from its melt and by sublimation in vac at 80°.

Benzoic anhydride *[93-97-0]* **M 226.2, m 42°.** Freed from benzoic acid by washing with $NaHCO_3$, then water, and drying. Crystd from benzene (0.5ml/g) by adding just enough pet ether (b 40-60°), to cause cloudiness, then cooling in ice. Can be distd at 210-220°/20mm.

(±)-Benzoin *[119-53-9]* **M 212.3, m 137°.** Crystd from CCl_4, hot EtOH (8ml/g), or 50% acetic acid. Crystd from high purity benzene, then twice from high purity MeOH, to remove fluorescent impurities [Elliott and Radley *AC* **33** 1623 *1961*].

(±)-α-Benzoinoxime *[441-38-3]* **M 227.3, m 151°.** Crystd from ethyl ether.

Benzonitrile *[100-47-0]* **M 103.1, f.p. -12.9°, b 191.1°, n 1.52823, d 1.010.** Dried with $CaSO_4$, $CaCl_2$, $MgSO_4$ or K_2CO_3, and distd from P_2O_5 in an all-glass apparatus, under reduced pressure (b 69°/10mm), collecting the middle fraction. Distn from CaH_2 causes some decomposition of solvent. Isonitriles can be removed by preliminary treatment with concnd HCl until the smell of isonitrile has gone, followed by preliminary drying with K_2CO_3. (This treatment also removes amines).
Steam distd (to remove small quantities of carbylamine). The distillate was extracted into ether, washed with dil Na_2CO_3, dried overnight with $CaCl_2$, and the ether removed by evaporation. The residue was distd at 40mm (b 96°) [Kice, Perham and Simons *JACS* **82** 834 *1960*].
Conductivity grade benzonitrile (specific conductance 2 x 10^{-8} mho) was obtained by treatment with anhydrous $AlCl_3$, followed by rapid distn at 40-50° under vac. After washing with alkali and drying with $CaCl_2$, the distillate was vac distd several times at 35° before being fractionally crystd several times by partial freezing. It was dried over finely divided activated alumina from which it was withdrawn as required [Van Dyke and Harrison *JACS* **73** 402 *1951*].

3,4-Benzophenanthrene *[195-19-7]* **M 228.3, m 68°.** Crystd from EtOH, pet ether, or EtOH/acetone.

Benzophenone *[119-61-9]* **M 182.2, m 48.5-49°.** Crystd from MeOH, EtOH, cyclohexane, benzene or pet ether, then dried in a current of warm air and stored over BaO or P_2O_5. Also purified by zone melting and by sublimation [Itoh *JPC* **89** 3949 *1985*; Naguib *et al, JACS* **108** 128 *1986*; Gorman and Rodgers *JACS* **108** 5074 *1986*; Ohamoto and Teranishi *JACS* **108** 6378 *1986*; Naguib *et al, JPC* **91** 3033 *1987*].

Benzophenone oxime *[574-66-3]* **M 197.2, m 142°.** Crystd from MeOH (4ml/g).

Benzopinacol *[464-72-2]* **M 366.5, m 170-180° (depends on heating rate).** Crystd from EtOH

Benzopurpurin 4B *[992-59-6]* **M 724.7.** Crystd from water.

Benzo[α]pyrene *[50-32-8]* **M 252.3, m 179.0-179.5°**. Chromatographed on activated alumina, eluted with a cyclohexane-benzene mixture containing up to 8% benzene, and the solvent evapd under reduced pressure [Cahnmann *AC* **27** 1235 *1955*]. It can be recrystd from EtOH [Nithipatikom and McGown *AC* **58** 3145 *1986*].

2,3-Benzoquinoline (acridine) *[260-94-6]* **M 179.2, m 111°** (sublimes),
3,4-Benzoquinoline (phenanthridines) *[229-87-8]* **M 179.2, m 102.5-103.5°**,
5,6-Benzoquinoline *[85-02-9]* **M 179.0, m 85.5-86°**,
7,8-Benzoquinoline *[230-27-3]* **M 179.0, m 52.0-52.5°**. Chromatographed on activated alumina from benzene soln, with ethyl ether as eluent. Evapn of ether gave crystalline material which was freed from residual solvent under vac, then further purified by fractional crystn under N_2, from its melt [Slough and Ubbelhode *JCS* 911 *1957*].

p-Benzoquinone *[106-51-4]* **M 108.1, m 115.7°**. Usually purified in one or more of the following ways: steam distn, followed by filtration and drying (e.g. in a desiccator over $CaCl_2$); crystn from pet ether (b 80-100°), benzene (with , then without, charcoal), water or 95% EtOH; sublimation under vac (e.g. from room temperature to liquid N_2. It slowly decomposes, and should be stored, refrigerated, in an evacuated or sealed glass vessel in the dark. It should be resublimed before use. [Wolfenden *et al*, *JACS* **109** 463 *1987*.]

1,2,3-Benzothiadiazole *[273-77-8]* **M 136.2, m 35°**,
2,1,3-Benzothiadiazole *[272-13-2]* **M 136.2, m 44°, b 206°/760mm**. Crystd from pet ether.

1,2,3-Benzotriazole *[95-14-7]* **M 119.1, m 100°**. Crystd from toluene, $CHCl_3$ or satd aq soln, and dried at room temperature or in a vac oven at 65°.

Benzotrifluoride see α,α,α-trifluorotoluene.

Benzoylacetone *[93-91-4]* **M 162.2, m 58.5-59.0°**. Crystd from ethyl ether or MeOH and dried under vac at 40°.

Benzoylauramine G m 178-179°. Crystd from chlorobenzene.

Benzoyl chloride *[98-88-4]* **M 140.6, b 56°/4mm, 196.8°/745mm, n^{10} 1.5537, d 1.2120.**
A soln of benzoyl chloride (300ml) in benzene (200ml) was washed with two 100ml portions of cold 5% $NaHCO_3$ soln, separated, dried with $CaCl_2$ and distd [Oakwood and Weisgerber *Org Synth* **III** 113 *1955*]. Repeated fractional distn at 4mm through a glass helices-packed column (avoiding porous porcelain or silicon-carbide boiling chips, and hydrocarbon or silicon greases on the ground joints) gave benzoyl chloride that did not darken on addn of $AlCl_3$. Further purification was achieved by adding 3 mole% each of $AlCl_3$ and toluene, standing overnight, and distilling off the benzoyl chloride at 1-2mm [Brown and Jenzen *JACS* **80** 2291 *1958*]. Refluxing for 2hr with an equal weight of thionyl chloride before distn, has also been used. *Strong* **IRRITANT**. *Use in a fume cupboard.*

Benzoyl glycine *[495-69-2]* **M 179.2, m 188°**. Crystd from boiling water.

Benzoyl peroxide *[94-36-0]* **M 242.2, m 95°(dec)**. Dissolved in $CHCl_3$ at room temp and pptd by adding an equal vol of MeOH or pet ether. Similarly pptd from acetone by adding two vols of distd water. Has also been crystd from 50% MeOH, and from ethyl ether. Dried under vac at room temperature for 24hr. Stored in a desiccator in the dark at 0°. When purifying in the absence of water it can be **EXPLOSIVE** and it should be done on a very small scale with adequate protection. Large amounts should be kept moist with water and stored in a refrigerator. [Kim *et al*, *JOC* **52** 3691 *1987*.]

p-Benzoylphenol *[1137-42-4]* **M 198.2, m 133.4-134.8°**. Dissolved in hot EtOH (charcoal), crystd once from EtOH/H_2O and twice from benzene [Grunwald *JACS* **73** 4934 *1951*].

N-Benzoyl-N-phenylhydroxylamine *[304-88-1]* M 213.2, m 121-122°. Recrystd from hot water, benzene or acetic acid.

Benzoyl sulphide *[644-32-6]* M 174.4, m 131.2-132.3°. About 300ml of solvent was blown off from a filtered soln of benzoyl disulphide (25g) in acetone (350ml). The remaining acetone was decanted from the solid which was recrystd first from 300ml of 1:1 (v/v) EtOH/ethyl acetate, then from 300ml of EtOH, and finally from 240ml of 1:1 (v/v) EtOH/ethyl acetate. Yield about 40% [Pryor and Pickering *JACS* **84** 2705 *1962*]. *Handle in a fume cupboard because of* **TOXICITY** *and obnoxious odour.*

N-Benzoyl-o tolylhydroxylamine *[1143-94-4]* M 227.3, m 104°. Recrystd from aq EtOH.

3,4-Benzpyrene *[50-32-8]* M 252.3, m 177.5-178°. A soln of 250mg in 100ml of benzene was diluted with an equal vol of hexane, then passed through a column of alumina, Ca(OH)$_2$ and Celite (3:1:1). The adsorbed material was developed with a 2:3 benzene/hexane mixture. (It showed as an intensely fluorescent zone.) The main zone was eluted with 3:1 acetone/EtOH, and was transferred into 1:1 benzene-hexane by adding water. The soln was washed, dried with Na$_2$SO$_4$, evapd and crystd from benzene by the addn of MeOH [Lijinsky and Zechmeister *JACS* **75** 5495 *1953*]. **CARCINOGENIC.**

4'-Benzylacetophenone *[782-92-3]* M 210.3, m 73°. Crystd from EtOH (*ca* 1ml/g).

Benzyl alcohol *[100-51-6]* M 107.2, f.p. -15.3°, b 205.5°, 93°/10mm, n 1.54033, d 0.981. Usually purified by careful fractional distn at reduced pressure in the absence of air. Benzaldehyde, if present, can be detected by UV absorption at 283nm. Also purified by shaking with aq KOH and extracting with peroxide-free ethyl ether. After washing with water, the extract was treated with satd NaHS sol, filtered, washed and dried with CaO and distd under reduced pressure [Mathews *JACS* **48** 562 *1926*]. Peroxy compounds can be removed by shaking with a soln of Fe(II) followed by washing the alcohol layer with distd water and fractionally distd.

Benzylamine *[100-46-9]* M 107.2, b 178°/742mm, 185°/768mm, n 1.5392, d 0.981. Dried with NaOH or KOH, then distd from Na, under N$_2$, through a column packed with glass helices, taking the middle fraction. Has also been distd from zinc dust under reduced pressure.

Benzylamine hydrochloride *[3287-99-8]* M 143.6, m 248° (rapid heating). Crystd from water.

N-Benzylaniline *[103-32-2]* M 183.4, m 36°, b 306-307°, d 1.061. Crystd from pet ether (b 60-80°) (*ca* 0.5ml/g).

Benzyl bromide *[100-39-0]* M 171.0, m -4°, b 85°/12mm, 192°/760mm, n 1.575, d 1.438. Washed with conc H$_2$SO$_4$, water, 10% Na$_2$CO$_3$ or NaHCO$_3$ soln, and again with water. Dried with CaCl$_2$, Na$_2$CO$_3$ or MgSO$_4$, and fractionally distd in the dark, under reduced pressure. *Handle in a fume cupboard, extremely* **LACHRYMATORY.**

Benzyl chloride *[100-44-7]* M 126.6, m 139°, b 63°/8mm, n 1.538, d 1.100. Dried with MgSO$_4$ or CaSO$_4$, or refluxed with fresh Ca turnings, then fractionally distd under reduced pressure, collecting the middle fraction and storing with CaH$_2$ or P$_2$O$_5$. Has also been purified by passage through a column of alumina. *Strongly* **LACHRYMATORY.**

Benzyl chloroformate see **benzyloxycarbonyl chloride.**

N-Benzyl-β-chloropropionamide *[24752-66-7]* M 197.7, m 94°. Crystd from MeOH.

Benzyl cyanide *[140-29-4]* M 117.1, b 100°/8mm, 233.5°/760mm, n 1.52327, d 1.015. Benzyl isocyanide can be removed by shaking vigorously with an equal vol of 50% H$_2$SO$_4$ at 60°, washing with satd aq NaHCO$_3$, then half-satd NaCl soln, drying and fractionally distilling under reduced pressure. Distn from CaH$_2$ causes some decomposition of this compound: it is better to use P$_2$O$_5$. Other purification procedures include passage through a column of highly activated alumina, and distn from Raney nickel. *Precautions should be taken because of possible formation of free* **TOXIC** *cyanide; use a fume cupboard.*

Benzyldimethyloctadecylammonium chloride *[122-19-0]* **M 442.2, m 63°.** Crystd from acetone.

Benzyl disulphide see **dibenzyl disulphide.**

Benzyl ether *[103-50-4]* **M 198.3, b 298°, 158-160°/0.1mm, n 1.54057, d 1.043.** Refluxed over sodium, then distd under reduced pressure. Also purified by fractional freezing.

Benzyl ethyl ether *[539-30-0]* **M 136.2, b 186°, n 14955.** Dried with CaCl$_2$ or NaOH, then fractionally distd.

O-**Benzylhydroxylamine hydrochloride** *[2687-43-6]* **M 159.6, m 234-238°(sublimes).** Crystd from water or EtOH.

Benzylideneacetophenone see **benzalacetophenone.**

N-**Benzylideneaniline** *[538-51-2]* **M 181.2, m 48° (54°), b 300°/760mm.** Steam volatile and crystd from benzene or 85% EtOH.

S-**Benzyl-isothiouronium chloride** *[538-28-3]* **M 202.7, two forms, m 150° and 175°.** Crystd from 0.2M HCl (2ml/g) or EtOH and dried in air.

Benzylmalonic acid *[616-75-1]* **M 194.2, m 121°.** Crystd from benzene.

Benzylidene malononitrile *[2700-22-3]* **M 154.2, m 83-84°.** Recrystd from EtOH [Bernasconi *et al, JACS* **107** 3612 *1985*].

Benzyl mercaptan *[100-53-8]* **M 124.2, b 70.5-70.7°/9.5mm, n 1.5761, d 1.058.** Purified *via* the mercury salt [see Kern *JACS* **75** 1865 *1953*], which was crystd from benzene as needles (m 121°), and then dissolved in CHCl$_3$. Passage of H$_2$S gas regenerated the mercaptan. The HgS ppte was filtered off, and washed thoroughly with CHCl$_3$. The filtrate and washings were evapd to remove CHCl$_3$, then residue was fractionally distd under reduced pressure [Mackle and McClean, *TFS* **58** 895 *1962*].

Benzyl methyl ketone see **phenylacetone.**

Benzyl Orange *[589-02-6]* **M 405.5.** Crystd from water.

Benzyloxycarbonyl chloride *[501-53-1]* **M 170.6, b 103°/20mm, n 1.5190, d 1.195.** Commercial material is better than 95% pure and may contain some toluene, benzyl alcohol, benzyl chloride and HCl. After long storage (e.g. two years at 4°)' Greenstein and Winitz [*The Chemistry of the Amino Acids* Vol 2, p 890, J Wiley and Sons NY, *1961*] recommended that the liquid should be flushed with a stream of dry air, filtered and stored over sodium sulphate to remove CO$_2$ and HCl which are formed by decomposition. It may further be distilled from an oil bath at a temperature below 85° because Thiel and Dent [*Annalen* **301** 257 *1898*] stated that benzyloxycarbonyl chloride decarboxylates to benzyl chloride slowly at 100° and vigorously at 155°. Redistillation at higher vac below 85° yields material which shows no other peaks than those of benzyloxycarbonyl chloride by NMR spectroscopy.

N-**Benzyloxycarbonylglycyl-L-alaninamide** *[17331-79-2]* **M 279.3.** Recrystd from EtOH/ethyl ether.

N-**Benzyloxycarbonyl-*N'*-methyl-L-alaninamide** *[33628-84-1]* **M 236.3.** Recrystd from ethyl acetate.

p-**(Benzyloxy)phenol** *[103-16-2]* **M 200.2, m 122.5°.** Crystd from EtOH or water, and dried over P$_2$O$_5$ under vac. [Walter *et al, JACS* **108** 5210 *1986*.]

2-Benzylphenol *[28994-41-4]* **M 184.2, m 54.5°, b 312°/760mm, 175°/18mm.** Crystd from EtOH, stable form has **m 52°** and unstable form has **m 21°**.

4-Benzylphenol *[101-53-1]* **M 184.2, m 84°.** Crystd from water.

2-Benzylpyridine *[101-82-6]* **M 169.2, b 98.5°/4mm, n²⁶ 1.5771, d 1.054,**
4-Benzylpyridine *[2116-65-6]* **M 169.2, b 110.0°/6mm, n²⁶ 1.5814, d 1.065.** Dried with NaOH for several days, then distd from CaO under reduced pressure, redistilling the middle fraction.

4-*N*-Benzylsulphanilamide *[1709-54-2]* **M 262.3, m 175°.** Crystd from dioxane/water.

Benzyl sulphide *[538-74-9]* **M 214.3, m 50°.** Crystd from EtOH, then chromatographed on alumina using pentane as eluent, and finally recrystd from EtOH [Kice and Bowers *JACS* **84** 2390 *1962*].

Benzylthiocyanate *[3012-37-1]* **M 149.2, m 43°, b 256°(dec).** Crystd from EtOH or aq EtOH.

S-**Benzylthiuronium chloride** see *S*-benzylisothiuronium chloride.

Benzyl toluene-*p*-sulphonate *[1024-41-5]* **M 162.3, m 58°.** Crystd from pet ether (b 40-60°).

Benzyltrimethylammonium chloride *[56-93-9]* **M 185.7, m 238-239°(dec).** A 60% aq soln was evapd to dryness under vac on a steam bath, and then left in a vac desiccator containing a suitable dehydrating agent. The solid residue was dissolved in a small amount of boiling abs EtOH and pptd by adding an equal vol of ethyl ether and cooling. After washing, the ppte was dried under vac [Karusch *JACS* **73** 1246 *1951*].

Benzyltrimethylammonium hydroxide (Triton B) *[100-85-6]* **M 167.3, d 0.91.** A 38% soln (as supplied) was decolorized (charcoal), then evapd under reduced pressure to a syrup, with final drying at 75° and 1mm pressure. Prepared anhydrous by prol;onged drying over P_2O_5 in a vac desiccator.

Berbamine *[478-61-5]* **M 608.7, m 197-210°.** Crystd from pet ether.

Berberine *[2086-83-1]* **M 608.7, m 145°.** Crystd from pet ether.

Berberine hydrochloride (2H₂O) *[633-65-8]* **M 371.8, m 204-206°(dec).** Crystd from water.

Betaine *[107-43-7]* **M 117.1, m 301-305°(dec) (anhydrous).** Crystd from aq EtOH.

Biacetyl (Butan-2,3-dione) *[431-03-8]* **M 86.1, b 88°, n¹⁸·⁵1.3933, d 0.981.** Dried with anhydrous $CaSO_4$, $CaCl_2$ or $MgSO_4$, then vac distd under nitrogen, taking the middle fraction and storing it at Dry-ice temperature in the dark (to prevent polymerization).

Bibenzyl *[103-29-7]* **M 182.3, m 52.5-53.5°.** Crystd from hexane, MeOH, or 95% EtOH. Has also been sublimed under vac, and further purified by percolation through columns of silica gel and activated alumina.

Bicuculline *[485-49-4]* **M 367.4, m 215° (196°, 177°), [α]₅₄₆ +159° (c 1, CHCl₃).** Crystd by dissolving in CHCl₃ and adding MeOH or EtOH.

Bicyclohexyl *[92-51-3]* **M 166.3, b 238° (*cis-cis*), 217-219° (*trans-trans*).** Shaken repeatedly with aq $KMnO_4$ and with conc H_2SO_4, washed with water, dried , first from $CaCl_2$ then from sodium, and distd. [Mackenzie *JACS* **77** 2214 *1955*].

Bicyclo(3.2.1.)octane *[6221-55-2]* **M 110.2, m 141°.** Purified by zone melting.

Biguanide *[56-03-1]* M 101.1, m 130°. Crystd from EtOH.

Bilirubin *[635-65-4]* M 584.7, ε_{450nm} 55,600 in $CHCl_3$. Meso-type impurities eliminated by successive Soxhlet extraction with ethyl ether and MeOH. Then crystd from $CHCl_3$, and dried to constant weight at 80° under vac. [Gray *et al JCS* 2264 *1961*.]

Biliverdin *[114-25-0]* M 582.6, m >300°. Crystd from MeOH.

2,2'-Binaphthyl *[61-78-2]* M 254.3, m 188°. Crystd from benzene.

Biopterin *[36183-24-1]* M 237.2, m >300°(dec), $[\alpha]_{546}$ -63° (c , 2.0M HCl). Biopterin is best recrystd (90% recovery) by dissolving in 1% aq ammonia (*ca* 100 parts), and adding this soln dropwise to an equal vol of M aq formic acid at 100° and allowing to cool at 4° overnight. It is dried at 20° to 50°/01mm in the presence of P_2O_5 [Armarego, Waring and Paal *Aust J Chem* 35 785 *1982*]. Also crystd from *ca* 50 parts of water or 100 parts of hot 3M aq HCl by adding hot 3M aq NH_3 and cooling.

D-Biotin *[58-85-5]* M 244.3, m 230.2°(dec), $[\alpha]_{546}$ +108°(c 1, 0.1N NaOH). Crystd from hot water.

Biphenyl *[92-52-4]* M 154.2, m 70-71°, b 255°, d 0.992. Crystd from EtOH, MeOH aq MeOH, pet ether (b 40-60°) or glacial acetic acid. Freed from polar impurities by passage of its soln in benzene through an alumina column, followed by evapn of the benzene. Its soln in CCl_4 has been purified by distn under vac and by zone refining. Purified by treatment with maleic anhydride to remove anthracene-like impurities. Recrystd from EtOH followed by repeated vac sublimation and passage through a zone refiner. [Taliani and Bree *JPC* 88 2351 *1984*.]

p-**Biphenylamine** *[CAS:92-67-1]* M 169.2, m 53°, b 191°/15mm. Crystd from water. CARCINOGEN.

Biphenyl-2-carboxylic acid *[947-84-2]* M 198.2, m 114°, b 343-344°,
Biphenyl-4-carboxylic acid *[92-92-2]* M 198.2, m 228°. Crystd from benzene/pet ether or aq EtOH.

2,4'-Biphenyldiamine *[492-17-1]* M 184.2, m 45°, b 363°/760mm. Crystd from aq EtOH.

Biphenylene *[259-79-0]* M 152.2, m 152°. Recrystd from cyclohexane then sublimed in vac.

α-(4-Biphenylyl)butyric acid *[959-10-4]* M 240.3, m 175-177°,
γ-(4-Biphenyl)butyric acid *[6057-60-9]* M 240.3, m 118°. Crystd from MeOH.

2-Biphenylyl diphenyl phosphate *[132-29-6]* M 302.4, n^{25}1.5925. Vac distd, then percolated through an alumina column. Passed through a packed column maintained at 150° to remove residual traces of volatile materials by a countercurrent stream of nitrogen at reduced pressure. [Dobry and Keller *JPC* 61 1448 *1957*.]

2,2'-Bipyridyl *[366-18-7]* M 156.2, m 70.5°, b 273°. Crystd from hexane, or EtOH, or (after charcoal treament of a $CHCl_3$ soln) from pet ether. Also pptd from a conc soln in EtOH by addition of water. Dried in vac over P_2O_5. Further purification by chromatography on alumina or by sublimation. [Airoldi *et al, JCSDT* 1913 *1986*.]

4,4'-Bipyridyl *[553-26-4]* M 156.2, m 73°(hydrate), 114° (171-171°)(anhydrous), b 305°/760mm, 293°/743mm. Crystd from water, benzene/pet ether, ethyl acetate and sublimed *in vacuo* at 70°. Also purified by dissolving in 0.1M H_2SO_4 and twice pptd by addition of 1M NaOH to *p*H 8. Recrystd from EtOH. [Man *et al, JCSFT 1* 82 869 *1986*; Collman *et al, JACS* 109 4606 *1987*.]

2,2'-Bipyridylamine *[1202-34-2]* M 171.2, m 95.1°. Crystd from acetone.

2,2'-Biquinolyl *[119-91-5]* **M 256.3, m 196°.** Decolorized in CHCl₃ soln (charcoal), then crystd to constant melting point from EtOH or pet ether [Cumper, Ginman and Vogel *JCS* 1188 *1962*].

Bis-acrylamide (*N,N'*-methylene bisacrylamide) *[110-26-9]* **M 154.2.** Recrystd from MeOH (100g dissolved in 500ml boiling MeOH and filtered without suction in a warmed funnel. Allowed to stand at room temperature and then at -15°C overnight. Crystals collected with suction in a cooled funnel and washed with cold MeOH). Crystals air-dried in a warm oven. The **toxicity** of bis-acrylamide is similar to acrylamide.

Bis-(4-aminophenylmethane *[101-77-9]* **M 198.3, m 92-93°, b 232°/9mm.** Crystd from 95% EtOH.

2,5-Bis(2-benzothiazolyl)hydroquinone *[33450-09-8]* **M 440.3.** Purified by repeated crystn from dimethylformamide followed by sublimation in vac [Erusting *et al*, *JPC* **91** *1404 1987*].

Bis-(*p*-bromophenyl)ether *[53563-56-7]* **M 328.0, m 60.1-61.7°.** Crystd twice from EtOH, once from benzene and dried under vac [Purcell and Smith *JACS* **83** 1063 *1961*].

Bis-*N*-*tert*-butyloxycarbonyl-*l*-cystine, m 144.5-145°, [α]_D -133.2° (c 1.2, MeOH). Recrystd by dissolving in ethyl acetate and adding hexane [Ferraro *Biochem Preps* **13** 39 *1971*].

Bis-(*p*-*tert*-butylphenyl)phenyl phosphate *[115-87-7]* **M 438.5, b 281°/5mm, n²⁵ 1.5412.** Same as for 2-biphenylyl diphenyl phosphate.

Bis-(β-chloroethyl)amine hydrochloride *[821-48-7]* **M 178.5, m 214-215°.** Crystd from acetone.

Bis-(β-chloroethyl) ether *[111-44-4]* **M 143.0, b 178.8°, n 1.45750, d 1.220.** Washed with conc H₂SO₄, then Na₂CO₃ soln, drying with anhydrous Na₂CO₃, and finally passing through a 50cm column of activated alumina. Alternatively, washed with 10% ferrous sulphate soln, then water, and dried with CaSO₄. Distd at 94°/33mm.

N,N-Bis-(2-chloroethyl)2-naphthylamine *[494-03-1]* **M 268.3, m 54-56°, b 210°/5mm.** Crystd from pet ether. **CARCINOGENIC.**

Bis-(chloromethyl)durene *[3022-16-0]* **M 231.2, m 197-198°.** Crystd three times from benzene, then dried under vac in an Abderhalden pistol.

3,3'-Bis-(chloromethyl)oxacyclobutane *[78-71-1]* **M 155.0, m 18.9°.** Shaken with aq NaHCO₃ or FeSO₄ to remove peroxides. Septd, dried with anhydrous Na₂SO₄, then distd under reduced pressure from a little CaH₂ [Dainton, Ivin and Walmsley *TFS* **65** 17884 *1960*].

Bis-(2-chlorophenyl) phenyl phosphate *[597-80-8]* **M 395, b 254°/4mm, n²⁵ 1.5767.** Same as for 2-biphenylyl diphenyl phosphate.

2,2-Bis-(4-chlorophenyl)-1,1,1-trichloroethane (DDT) *[50-29-7]* **M 354.5, m 108°.** Crystd from *n*-propyl alcohol (5ml/g), then dried in air or in an air oven at 50-60°.

2,2'-Bis-[di-(carboxymethyl)-amino]diethyl   **ether,**   (HOOCCH₂)₂NCH₂CH₂OCH₂CH₂N-(CH₂COOH)₂ *[923-73-9]* **M 336.3,**
1,2-Bis-[2-di-(carboxymethyl)-aminoethoxy]ethane, (HOOCCH₂)₂NCH₂CH₂OCH₂CH₂OCH₂CH₂N-(CH₂COOH)₂ *[67-42-5]* **M 380.4.** Crystd from EtOH.

2,2-Bis-(4-chlorophenyl)-1,1-dichloroethane (*p,p*-DDD) *[72-54-8]* **M 320.1, m 111-112°.** Crystd from EtOH, and the purity checked by tlc.

4,4'-Bis-(dimethylamino)benzophenone *[90-93-7]* M 268.4, m 175°. Crystd from EtOH (25ml/g) and dried under vac.

Bis-(4-dimethylaminobenzylidene)benzidine *[6001-51-0]* M 454.5, m 318°. Crystd from nitrobenzene.

1,8-Bis-(dimethylamino)naphthalene (Proton sponge) *[20734-58-1]* M 214.3, m 47-48°. Crystd from EtOH and dried in a vac oven. Stored in the dark.

Bis-(dimethylthiocarbamyl)disulphide (tetramethylthiuram disulphide) *[137-26-8]* M 240.4, m 155-156°. Crystd from $CHCl_3$, by addn of EtOH.

1,2-Bis-(diphenylphosphine)ethane see **ethylenebis-(diphenylphosphine)**.

Bis-(2-ethoxyethyl) ether see **diethylene glycol diethyl ether**.

Bis-(2-ethylhexyl) 2-ethylhexyl phosphonate *[25103-23-5]* M 434.6, n^{25} 1.4473. Purified by stirring an 0.4M soln in benzene with an equal vol of 6M HCl at *ca* 60° for 8hr. The benzene layer was then shaken successively with equal vols of water (twice), aq 5% Na_2CO_3 (three times), and water (eight times), followed by evapn of the benzene and dissolved water under reduced pressure at room temperature (using a rotating evacuated flask). Stored in dry, dark conditions [Peppard *et al*, *J Inorg Nuclear Chem* **24** 1387 *1962*]. Vac distd, then percolated through an alumina column before finally passed through a packed column maintained at 150° where residual traces of volatile materials were removed by a counter-current stream of N_2 at reduced pressure [Dobry and Keller *JPC* **61** 1448 *1957*].

Bis-(2-ethylhexyl) phosphoric acid *[298-07-7]* M 322.4, b 209°/10mm, d 0.965. See Peppard, Ferraro and Mason [*J Inorg Nuclear Chem* **7** 231 *1958*] or Stewart and Crandall [*JACS* **73** 1377 *1951*].

N,N-**Bis-(2-hydroxyethyl)-2-aminoethanesulphonic acid** (BES) *[10191-18-1]* M 213.3, m 150-155°. Crystd from aq EtOH.

N,N-**Bis-(2-hydroxyethyl)glycine** (Bicine) *[150-25-4]* M 163.2, m 191-194°(dec). Crystd from 80% MeOH.

Bis-(2-hydroxyethyl)amino-tris-(hydroxymethyl)methane (Bis-Tris) *[6976-37-0]* M 209.2, m 89°. Crystd from hot 1-butanol. Dried in vac at 25°.

3,4-Bis-(4-hydroxyphenyl)hexane *[5635-50-7]* M 270.4, m 187°. Purified from diethylstilboestrol by zone refining.

Bis-(2-methoxyethyl) ether see **diglyme**.

1,4-Bismethylaminoanthraquinone (Disperse Blue 14) *[2475-44-7]* M 266.3, kmax 640 (594)nm. Purified by thin-layer chromatography on silica gel plates, using toluene/acetone (3:1) as eluent. The main band was scraped off and extracted with MeOH. The solvent was evapd and the dye was dried in a drying pistol [Land, McAlpine, Sinclair and Truscott *JCSFT 1* **72** 2091 *1976*].

Bis-(1-naphthylmethyl)amine *[5798-49-2]* M 329.4, m 62°. Crystd from pet ether.

N,N'-Bis-(nicotinic acid) hydrazide *[840-78-8]* M 227-228°. Crystd from water.

Bis-(4-nitrophenyl) ether *[101-63-3]* M 260.2, m 142-143°,
Bis-(4-nitrophenyl) methane *[1817-74-9]* M 258.2, m 183°. Crystd twice from benzene, and dried under vac.

Bisnorcholanic acid *[57761-00-9]* M 332.5, m 214° (α-form), 242° (β-form), 210-211° (γ-form), 184° (δ-form), 181° (ε-form). Crystd from EtOH (α-form), or acetic acid (all forms).

3,3'-Bis-(phenoxymethyl)oxacyclobutane *[1224-69-7]* M 270.3, m 67.5-68°. Crystd from MeOH.

1,4-Bis-(2-pyridyl-2-vinyl)benzene *[20218-87-5]* M 284.3. Recrystd from xylene, then chromatogrphy (in the dark) on basic silica gel (60-80-mesh), using CH_2Cl_2 as eluent. Vac sublimed in the dark to a cold surface at 10^{-3} torr.

Biuret *[108-19-0]* M 103.1, sinters at 218° and chars at 270°. Crystd from EtOH.

Bixin *[6983-79-5]* M 394.5, m 198°. Crystd from acetone (violet prisms)

Blue Tetrazolium *[1871-22-3]* M 727.7, m 254-255°(dec). Crystd from 95% EtOH/anhydrous ethyl ether, to constant absorbance at 254nm.

R-2-*endo*-Borneol *[464-43-7]* M 154.3, m 208° $[\alpha]_D$ +15.8° (in EtOH). Crystd from boiling EtOH (charcoal).

(±)-Borneol *[6627-72-1]* M 154.3, m 130°(dec). Crystd to constant melting point from pet ether (b 60-80°).

Brazilin *[474-07-7]* M 269.3, m 130°(dec). Crystd from EtOH.

Brilliant Cresyl Blue *[4712-70-3]* M 332.8. Crystd from pet ether.

Brilliant Green *[633-03-4]* M 482.7, m 209-211°(dec). Purified by pptn as the perchlorate from aq soln (0.3%) after filtering, heating to 75° and adjustment to *p*H 1-2. Recrystd from EtOH/water (1:4) [Kerr and Gregory *Analyst* 94 1036 *1969*].

N-**Bromosuccinimide** *[79-15-2]* M 138.0, m 107-109°. Crystd from $CHCl_3$/hexane or water and dried over $CaCl_2$. [Chow and Zhao *JOC* 52 1931 *1987*.]

4-Bromoacetanilide *[103-88-8]* M 214.1, m 167°. Crystd from aq MeOH or EtOH. Purified by zone refining.

Bromoacetic acid *[79-08-3]* M 138.9, m 50°, b 118°/15mm, 208°/760mm. Crystd from pet ether (b 40-60°). Ethyl ether soln passed through an alumina column, and the ether evapd at room temperature under vac.

Bromoacetone *[598-31-2]* M 137.0, b 31.5°/8mm. Stood with anhydrous $CaCO_3$, distd under low vac, and stored with $CaCO_3$ in the dark at 0°.

4-Bromoacetophenone *[99-90-1]* M 199.1, m 54°,
ω-Bromoacetophenone *[70-11-1]* M 199.1, m 57-58°. Crystd from EtOH, MeOH or from pet ether (b 80-100°). [Tanner *JOC* 52 2142 *1987*.]

4-Bromoaniline *[106-40-7]* M 172.0, m 66°. Crystd (with appreciable loss) from aq EtOH.

2-Bromoanisole *[578-57-4]* M 187.0, f.p. 2.5°, b 124°/40mm, n^{25} 1.5717, d 1.513,
4-Bromoanisole *[104-92-7]* M 187.0, f.p. 13.4°, b 124°/40mm, n^{25} 1.5617, d 1.495. Crystd by partial freezing (repeatedly), then distd at reduced pressure.

9-Bromoanthracene *[1564-64-3]* M 98-100°. Crystd from MeOH or EtOH followed by sublimation in vac. [Masnori *et al, JACS* 108 126 *1986*.]

4-Bromobenzal diacetate *[55605-27-1]* **M 287.1, m 95°.** Crystd from hot EtOH (3ml/g).

Bromobenzene *[108-86-1]* **M 157.0, b 155.9°, n 1.5588, n^{15} 1.56252, d 1.495.** Washed vigorously with conc H$_2$SO$_4$, then 10% NaOH or NaHCO$_3$ solns, and water. Dried with CaCl$_2$ or Na$_2$SO$_4$, or passed through activated alumina, before refluxing with, and distilling from, Ca turnings or sodium, using a glass helix-packed column.

4-Bromobenzenesulphonyl chloride *[98-58-8]* **M 255.5, m 74.3-75.1°.** Crystd from pet ether, or from ethyl ether cooled in powdered Dry-ice after the ether soln had been washed with 10% NaOH until colourless, then dried with anhydrous Na$_2$SO$_4$.

o-**Bromobenzoic acid** *[88-65-3]* **M 201.0, m 148.9°.** Crystd from benzene or MeOH.

m-**Bromobenzoic acid** *[585-76-2]* **M 201.0, m 155°.** Crystd from acetone/water, MeOH or acetic acid.

p-**Bromobenzoic acid** *[586-76-5]* **M 201.0, m 251-252°.** Crystd from MeOH, or MeOH/water mixture.

p-**Bromobenzophenone** *[90-90-4]* **M 261.1, m 81°.** Crystd from EtOH.

p-**Bromobenzyl bromide** *[589-15-1]* **M 249.9, m 60-61°,**
p-**Bromobenzyl chloride** *[589-17-3]* **M 123.5, m 40-41°, b 105-115°/12mm.** Crystd from EtOH. **LACHRYMATORY.**

p-**Bromobiphenyl** *[92-66-0]* **M 233.1, m 88.8-89.2°.** Crystd from abs EtOH and dried under vac.

1-Bromobutane see *n*-**butyl bromide.**

2-Bromobutane *[78-76-2]* **M 137.0, b 91.2°. n 1.4367, n^{25} 1.4341, d 1.255.** Washed with conc HCl, water, 10% aq NaHSO$_3$, and then water. Dried with CaCl$_2$, Na$_2$SO$_4$ or anhydrous K$_2$CO$_3$, and fractionally distd through a 1m column packed with glass helices.

(+)-3-Bromocamphor-8-sulphonic acid *[+: 14671-04-6],[endo: 21633-53-4]* **M 311.2, m 195-196°(anhydrous), [α]$_D$ +88.3° (in H$_2$O).** Crystd from water.

3-Bromocamphor-10-sulphonic acid *[24262-38-2]* **M 311.2, m 47.5°, [α]$_D$ +98.3° (in H$_2$O),**
(+)-3-Bromocamphor-8-sulphonic acid ammonium salt *[14575-84-9]* **M 328.2, m 270°(dec) [α]$_D$ +84.8° (in H$_2$O).** Crystd from water.

4-Bromo-4'-chlorobenzophenone *[27428-57-5]* **M 295.6.** Purified by zone refining [Lin and Hanson *JPC* **91** 2279 *1987*].

Bromocresol Green *[76-60-8]* **M 698.0, m 218-219°(dec).** Crystd from glacial acetic acid or dissolved in aq 5% NaHCO$_3$ soln and pptd from hot soln by dropwise addn of aq HCl. Repeated until the extinction did not increase (λ$_{max}$ 423nm).

Bromocresol Purple *[115-40-2]* **M 540.2, m 241-242°(dec).** Dissolved in aq 5% NaHCO$_3$ soln and pptd from hot soln by dropwise addn of aq HCl. Repeated until the extinction did not increase (λ$_{max}$ 419nm). Can also be crystd from benzene.

1-Bromodecane see *n*-**decyl bromide.**

p-**Bromo-*N,N*-dimethylaniline** *[586-77-6]* **M 200.1, m 55°, b 264°.** Refluxed for 3hr with two equivalents of acetic anhydride, then fractionally distd under reduced pressure

1-Bromo-2,4-dinitrobenzene *[584-48-5]* **M 247.0, m 75°.** Crystd from ethyl ether, isopropyl ether, 80% EtOH or abs EtOH.

Bromoethane see **ethyl bromide.**

Bromoform *[75-25-2]* **M 252.8, f.p. 8.1°, 55-56°/35mm, 149.6°/760mm. d^{15} 1.60053, n 1.5988, d^{15} 2.9038, d^{30} 2.86460.** Storage and stability of bromoform and chloroform are similar. Ethanol, added as a stabilizer, is removed by washing with water or with satd CaCl$_2$ soln, and the CHBr$_3$, after drying with CaCl$_2$ or K$_2$CO$_3$, is fractionally distd. Prior to distn, CHBr$_3$ has also been washed with conc H$_2$SO$_4$ until the acid layer no longer became coloured, then dil NaOH or NaHCO$_3$, and water. A further purification step is fractional crystn by partial freezing.

1-Bromoheptane see *n*-**heptyl bromide.**

2-Bromohexadecanoic acid (2-bromopalmitic acid) *[18263-25-7]* **M 335.3, m 53°.** Crystd from pet ether (60-80°).

1-Bromohexane see *n*-**hexyl bromide.**

4-Bromo-1-isopropylaminopentane hydrobromide M 208.1, m 167-167.5°. Crystd from acetone/ethyl ether.

Bromomethane see **methyl bromide.**

1-Bromo-2-methylpropane see **isopropyl bromide.**

2-Bromo-2-methylpropane *[507-19-7]* **M 137.0, b 71-73°, n 1.429, d 1.218.** Neutralised with K$_2$CO$_3$, distd, and dehydrated using molecular sieves (5A), then vac distd and degassed by freeze-pump-thaw technique. Sealed under vac.

2-Bromo-3-methylindole (2-bromoskatole) *[1484-28-2]* **M 210.1, m 102-104°.** Purified by chromatography on silica gel in CHCl$_3$/pet ether (1:2) followed by crystn from aq EtOH. [Phillips and Cohen *JACS* **108** 2023 *1986*.]

1-Bromonaphthalene *[90-11-9]* **M 207.1, b 118°/6mm, d 1.489.** Purified by passage through activated alumina, and three vac distns.

2-Bromonaphthalene *[580-13-2]* **M 207.1, m 59°.** Purified by fractional elution from a chromatographic column. Crystd from EtOH.

1-Bromo-2-naphthol *[573-97-7]* **M 223.1, m 76-78°,**
6-Bromo-2-naphthol *[15231-91-1]* **M 223.1, m 122-126°.** Crystd from EtOH.

x-Bromo-4-nitroacetophenone *[99-81-0]* **M 244.1, m 98°.** Crystd from benzene-pet ether.

o-**Bromonitrobenzene** *[577-19-5]* **M 202.1, m 43°,**
m-**Bromonitrobenzene** *[585-79-5]* **M 202.1, m 55-56°,**
p-**Bromonitrobenzene** *[586-78-7]* **M 202.1, m 127°.** Crystd twice from pet ether, using charcoal before the first crystn.

α-**Bromo-*p*-nitrotoluene** see *p*-**nitrobenzyl bromide.**

1-Bromooctane see *n*-**octyl bromide.**

1-Bromopentane see *n*-**amyl bromide.**

p-Bromophenacyl bromide *[99-73-0]* M 277.9, m 110-111°. Crystd from EtOH (*ca* 8ml/g).

o-Bromophenol *[95-56-7]* M 173.0, b 194°, d 1.490. Purified by at least two passes through a chromatographic column.

p-Bromophenol *[106-41-2]* M 173.0, m 64°. Crystd from CHCl$_3$, CCl$_4$, pet ether (b 40-60°), or water. and dried at 70° under vac for 2hr.

Bromophenol Blue *[115-39-9]* M 670.0, m 270-271°(dec). Crystd from benzene or acetone/glacial acetic acid, and air dried.

(4-Bromophenoxy)acetic acid *[1878-91-7]* M 231.1, m 158°,
β-(4-Bromophenoxy)propionic acid *[93670-18-9]* M 247.1, m 146°. Crystd from EtOH.

2-Bromo-4'-phenylacetophenone see *p*-phenylphenacyl bromide.

4-Bromophenylhydrazine *[589-21-9]* M 187.1, m 108-109°. Crystd from water.

4-Bromophenyl isocyanate *[2492-02-9]* M 189.0, m 41-42°. Crystd from pet ether (b 30-40°).

Bromopicrin *[464-10-8]* M 297.8, m 10.2-10.3°, b 85-87°/16mm, n 1.5790, d 2.7880. Steam distd, dried with anhydrous Na$_2$SO$_4$ and vac distd.

1-Bromopropane see *n*-propyl bromide.

2-Bromopropane see isopropyl bromide.

3-Bromopropene see allyl bromide.

β-Bromopropionic acid *[590-92-1]* M 153.0, m 62.5°. Crystd from CCl$_4$.

(3-Bromopropyl)benzene see 3-phenylpropyl bromide.

2-Bromopyridine *[109-04-6]* M 158.0, b 49.0°/2.7mm, n 1.5713, d 1.660. Dried over KOH for several days, then distd from CaO under reduced pressure, taking the middle fraction.

Bromopyrogallol Red *[16574-43-9]* M 576.2, m 300°. Crystd from 50% EtOH.

Bromopyruvic acid *[1113-59-3]* M 167.0, m 79-82°. Dried by azeotropic distn (toluene), and then recrystd from dry chloroform. Dried for 48hr at 20° (0.5 Torr) over P$_2$O$_5$. Stored at 0°. [Labandiniere *et al, JOC* **52** 157 *1987*.]

5-Bromosalicyl hydroxamic acid *[5798-94-7]* M 210.1, m 232°(dec). Crystd from EtOH.

N-Bromosuccinimide *[128-08-5]* M 178.0, m 183-184°(dec). N-Bromosuccinimide (30g) was dissolved rapidly in 300ml of boiling water and filtered through a fluted filter paper into a flask immersed in an ice bath, and left for 2hr. The crystals were filtered, washed thoroughly with *ca* 100ml of ice-cold water and drained on a Büchner funnel before drying under vac over P$_2$O$_5$ or CaCl$_2$ [Dauben and McCoy *JACS* **81** 4863 *1959*]. Has also been crystd from acetic acid or water (10 parts, washed in water and dried *in vacuo*. [Wilcox *et al, JACS* **108** 7693 *1986; Shell et al, JACS* **108** 121 *1986*; Phillips and Cohen *JACS* **108** 2013 *1986*.]

Bromotetronic acid *[21151-51-9]* **M 179.0, m 183°(dec)**. Decolorized, and free bromine was removed by charcoal treatment of an ethyl acetate soln, then recrystd from ethyl acetate [Schuler, Bhatia and Schuler *JPC* **78** 1063 *1974*].

Bromothymol Blue *[76-59-5]* **M 624.4, m 201-203°**. Dissolved in aq 5% NaHCO$_3$ soln and pptd from the hot soln by dropwise addn of aq HCl. Repeated until the extinction did not increase (λ_{max} 420nm).

α-Bromotoluene   see **benzyl bromide**.

p-**Bromotoluene** *[106-38-7]* **M 171.0, m 28°, b 184°, d 1.390**. Crystd from EtOH [Taylor and Stewart *JACS* **108** 6977 *1986*].

α-Bromo-4-toluic acid *[6232-88-8]* **M 215.1, m 229-230°**. Crystd from acetone.

Bromotrichloromethane *[75-62-7]* **M 198.3, f.p. -5.6°, b 104.1°, n 1.5061, d 2.01**. Washed with aq NaOH soln or dil Na$_2$CO$_3$, then with water, and dried with CaCl$_2$, BaO or MgSO$_4$ before distilling in diffuse light and storing in the dark. Has also been purified by treatment with charcoal and by fractional crystn by partial freezing. Purified by vigorous stirring with portions of conc H$_2$SO$_4$ until the acid did not discolour during several hours stirring. Washed with Na$_2$CO$_3$ and water, dried with CaCl$_2$ and then P$_2$O$_5$. Illuminated with a 1000W projection lamp at 6-in for 10hr, after making 0.01M in bromine. Passed through a 30 x 1.5cm column of activated alumina before fractionally distilling through a 12-in Vigreux column. The middle fraction was passed through a fresh activated alumina column [Firestone and Willard *JACS* **83** 3511 *1961*].

Bromotrifluoromethane (Freon) *[75-63-8]* **M 148.9, b -59°, d 1.590**. Passed through a tube containing P$_2$O$_5$ on glass wool into a vac system where it was frozen out in a quartz sample tube and degassed by a series of cycles of freezing, evacuating and thawing.

5-Bromovaleric acid *[2067-33-6]* **M 181.0, m 40°**. Crystd from pet ether.

α-Bromo-*p*-xylene *[104-81-4]* **M 185.1, m 35°, b 218-220°/740mm**. Crystd from EtOH.

Bromural *[496-67-3]* **M 223.1, m 154-155°**. Crystd from toluene, and air dried.

Brucine (H$_2$O) *[5892-11-5]* **M 430.5, m 178-179°, [α]$_{546}$ -149.9°** (anhydrous; c 1, in CHCl$_3$). Crystd once from water, as tetrahydrate, then suspended in CHCl$_3$ and shaken with anhydrous Na$_2$SO$_4$ (to dehydrate the brucine, which then dissolves). Pptd by pouring the soln into a large bulk of dry pet ether (b 40-60°), filtered and heated to 120° in high vac [Turner *JCS* 842 *1951*].

[α]-Brucine sulphate (hydrate) *[4845-99-2]* **M 887.0, m 180°(dec)**. Crystd from water.

Bufotenine hydrogen oxalate *[2963-79-3]* **M 294.3, m 96.5°**. Crystd from ether.

1,3-Butadiene *[106-99-0]* **M 54.1, b -2.6°**. Dried by condensing with a soln of aluminium triethyl in decahydronaphthalene; then flash distd. Also dried by passage over anhydrous CaCl$_2$ or distd from NaBH$_4$. Also purified by passage through a column packed with miolecular sieves (4Å), followed by cooling in Dry-ice/MeOH bath overnight, filtering off the ice and drying over CaH$_2$ at -78° and distd in a vacuum line.

n-**Butane** *[106-97-8]* **M 58.1, m -135°, b -0.5°**. Dried by passage over anhydrous Mg(ClO$_4$)$_2$ and molecular sieves type 4A. Air was removed by prolonged and frequent degassing at -107°.

1,4-Butanediol *[110-63-4]* **M 90.1, f.p. 20.4°, b 107-108°/4mm, 127°/20mm, n 1.4467, d 1.02.** Distd and stored over Linde type 4A molecular sieves, or crystd twice from anhydrous ethyl ether/acetone, and redistd. Also purified by recrystn from the melt and doubly distd in vac in the presence of Na_2SO_4.

meso-**2,3-Butanediol** *[513-85-9]* **M 90.1, m 25°.** Crystd from isopropyl ether.

2,3-Butanedione see **biacetyl**.

1-Butanethiol *[109-79-5]* **M 90.2, b 98.4°, n 1.44298, n^{25} 1.44034, d^{25} 0.837.** Dried with $CaSO_4$ or Na_2SO_4, then refluxed from magnesium; or dried with, and distd from CaO, under nitrogen [Roberts and Friend *JACS* **108** 7204 *1986*]. Has been separated from hydrocarbons by extractive distn with aniline.
Dissolved in 20% NaOH, extracted with a small amount of benzene, then steam distd, until clear. The soln was then cooled and acidified slightly with 15% H_2SO_4. The thiol was distd out, dried with $CaSO_4$ or $CaCl_2$, and fractionally distd under N_2 [Mathias and Filho *JPC* **62** 1427 *1958*]. Also purified by pptn as lead mercaptide from alcoholic soln, with regeneration by adding dil HCl to the residue after steam distn. *All operations should be carried out in a fume cupboard due to the toxicity and obnoxious odour of the thiol.*

2-Butanethiol *[513-53-1]* **M 90.2, b 37.4°/134mm, d^{25} 1.43385, d^{25} 0.8456.** Purified as for 1-butanethiol.

n-**Butanol** *[71-36-3]* **M 74.1, b 117.7°, n 1.39922, n^{15} 1.40118, d^{25} 0.80572.** Dried with $MgSO_4$, CaO, K_2CO_3, Ca or solid NaOH, followed by refluxing with, and distn from, calcium, magnesium activated with iodine, aluminium amalgam or sodium. Can also dry with molecular sieves, or by refluxing with *n*-butyl phthalate or succinate. (For method, see *Ethanol*.) *n*-Butanol can also be dried by efficient fractional distn, water passing over in the first fractn as a binary azeotrope (contains about 37% water). An ultraviolet-transparent distillate has been obtained by drying with magnesium amd distilling from sulphanilic acid. To remove bases, aldehydes and ketones, the alcohol has been washed with dil H_2SO_4, then $NaHSO_4$ soln; esters were removed by boiling for 1.5hr with 10% NaOH.
Also purified by adding 2g $NaBH_4$ to 1.5L butanol, gently bubbling with argon and refluxing for 1 day at 50°. Then added 2g of freshly cut sodium (washed with butanol) and refluxed for 1 day. Distd and the middle fraction collected [Jou and Freeman *JPC* **81** 909 *1977*].

2-Butanol see *sec*-butyl alcohol.

tert-**Butanol** see **2-methyl-2-propanol**.

2-Butanone *[78-93-0]* **M 72.1, b 79.6°, n 1.37850, n^{25} 1.37612, d 0.853.** In general, purification methods are the same as for acetone. Aldehydes can be removed by refluxing with $KMnO_4$ + CaO, until the Schiff aldehyde test is negative, prior to distn. Shaking with satd K_2CO_3, or passage through a small column of activated alumina, removes cyclic impurities. The ketone can be dried by careful distn (an azeotrope containing 11% water boils at 73.4°), or by $CaSO_4$, P_2O_5, Na_2SO_4, or K_2CO_3, followed by fractional distn. Purification as the bisulphite addition compound is achieved by shaking with excess satd Na_2SO_3, cooled to 0°, filtering off the ppte, washing with a little ethyl ether and drying in air; this is followed by decomposition with a slight excess of Na_2CO_3 soln and steam distn, the distillate being satd with K_2CO_3 so that the ketone can be separated, dried with K_2CO_3, filtered, and distd. Purification as the NaI addition compound (m 73-74°) is more convenient. (For details, see *Acetone*.) Small quantities of 2-butanone can be purified by conversion to the semicarbazone, recrystn to constant melting point, drying under vac over $CaCl_2$ and paraffin wax, refluxing for 30min with excess oxalic acid, followed by steam distn, salting out, drying and distn [Cowan, Jeffery and Vogel *JCS* 171 *1940*].

cis-**2-Butene** *[590-18-1]* **M 56.1, b 2.95-3.05°/746mm,**
trans-**2-Butene** *[624-64-9]* **M 56.1, b 0.3-0.4°/744mm.** Dried with CaH_2. Purified by gas chromatography.

2-Butene-1,4-dicarboxylic acid (*trans*-β-hydromuconic acid) *[4436-74-2]* **M 144.1, m 194-197°.** Crystd from boiling water, then dried at 50-60° in a vac oven.

But-3-en-2-one (methyl vinyl ketone) *[78-94-4]* **M 70.1, b 79-80°/760mm, d 0.842.** Dried with K_2CO_3, then Na_2SO_4, and fractionally distd.

2-Butoxyethanol (butyl cellosolve) *[111-76-2]* **M 118.2, b 171°/745mm, n 1.4191, d 0.903.** Peroxides can be removed by refluxing with anhydrous $SnCl_2$ or by passage under slight pressure through a column of activated alumina. Dried with anhydrous K_2CO_3 and $CaSO_4$, filtered and distd, or refluxed with, and distd from NaOH.

2-(2-Butoxyethoxy)ethanol see **diethylene glycol mono-*n*-butyl ether.**

n-**Butyl acetate** *[123-86-4]* **M 116.2, b 126.1°, n 1.39406, d 0.882.** Distd, refluxed with successive small portions of $KMnO_4$ until the colour persisted, dried with anhydrous $CaSO_4$, filtered and redistd.

tert-**Butyl acetate** [540-88-5 *[540-88-5]* **M 116.2, b 97-98°, d 0.72.** Washed with 5% Na_2CO_3 soln, then satd aq $CaCl_2$, dried with $CaSO_4$ and distd.

Butyl acrylate *[141-32-2]* **M 128.2, b 59°/25mm, n^{12} 1.4254, d 0.894.** Washed repeatedly with aq NaOH to remove inhibitors such as hydroquinone, then with distd water. Dried with $CaCl_2$. Fractionally distd under reduced pressure in an all-glass apparatus. The middle fraction was sealed under nitrogen and stored at 0° in the dark until used [Mallik and Das *JACS* **82** 4269 *1960*].

n-**Butyl alcohol** see *n*-**butanol.**

(±)-*sec*-Butyl alcohol *[15892-23-6]* **M 74.1, b 99.4°, d 0.808.** Purification methods are the same as for *n*-Butanol. These include drying with K_2CO_3 or $CaSO_4$, followed by filtration and fractional distn, refluxing with CaO, distn, then refluxing with magnesium and redistn; and refluxing with, then distn from CaH_2. Calcium carbide has also been used as a drying agent. Anhydrous alcohol is obtained by refluxing with *sec*-butyl phthalate or succinate. (For method see *Ethanol*.) Small amounts of alcohol can be purified by conversion to the alkyl hydrogen phthalate and recrystn [Hargreaves, *JCS* 3679 *1956*]. For purification of optical isomers, see Timmermans and Martin [*JCP* **25** 411 *1928*].

tert-**Butyl alcohol** *[75-65-0]* **M 74.1, m 25.7°, b 82.5°, n$^{25.5}$ 1.38516.** Synthesised commercially by the hydration of 2-methylpropene in dil H_2SO_4. Dried with CaO, K_2CO_3, $CaSO_4$ or $MgSO_4$, filtered and fractionally distd. Dried further by refluxing with, and distilling from, either magnesium activated with iodine, or small amounts of calcium, sodium or potassium, under nitrogen. Passage through a column of type 4A molecular sieve is another effective method of drying. So, also, refluxing with *tert*-butyl phthalate or succinate. (For method see *Ethanol*.) Other methods include refluxing with excess aluminium *tert*-butylate, or standing with CaH_2, distilling as needed. Further purification is achieved by fractional crystn by partial freezing, taking care to exclude moisture. tert-Butyl alcohol samples containing much water can be dried by adding benzene, so that the water distils off as a tertiary azeotrope, **b 67.3°.** Traces of isobutylene have been removed from dry tert-butyl alcohol by bubbling dry pre-purified nitrogen through for several hours at 40-50° before using.

n-**Butylamine** *[109-73-9]* **M 73.1, b 77.8°, n 1.4009, n^{25} 1.3992, d 0.740.** Dried with solid KOH, K_2CO_3, $LiAlH_4$, CaH_2 or $MgSO_4$, then refluxed with, and fractionally distd from P_2O_5, CaH_2, CaO or BaO. Further purified by pptn as the hydrochloride, m 213-213.5°, from ether soln by bubbling HCl gas into it. Repptd three times from EtOH by adding ether, followed by liberation of the free amine using excess strong base. The amine was extracted into ether, which was separated, dried with solid KOH, the ether removed by evapn and then the amine was distd. It was stored in a desiccator over solid NaOH [Bunnett and Davis *JACS* **82** 665 *1960*].

tert-**Butylamine** *[75-64-9]* M 73.1, b 42°, d 0.696. Dried with KOH or LiAlH$_4$. Distd from CaH$_2$ or BaO.

n-**Butyl *p*-aminobenzoate** *[94-25-7]* M 193.2, m 57-59°. Crystd from EtOH.

tert-**Butylammonium bromide** *[60469-70-7]* M 154.1. Recrystd several times from abs EtOH and thoroughly dried at 105°.

2-*tert*-**Butylanthracene** *[13719-97-6]* M 234.3, m 148-149°. Recrystd from EtOH and finally purified by tlc.

n-**Butylbenzene** *[104-51-8]* M 134.2, b 183.3°, n 1.48979, n^{25} 1.48742, d 0.860. Distd from sodium. Washed with small portions of conc H$_2$SO$_4$ until the acid was no longer coloured, then with water and aq Na$_2$CO$_3$. Dried with anhydrous MgSO$_4$, and distd twice from Na, collecting the middle fraction [Vogel *JCS* 607 *1948*].

tert-**Butylbenzene** *[98-06-6]* M 134.2, b 169.1°, n 1.49266, n^{25} 1.49024, d 0.867. Washed with cold conc H$_2$SO$_4$ until a fresh portion of acid was no longer coloured, then with 10% aq NaOH, followed by distd water until neutral. Dried with CaSO$_4$ and distd in a glass helices-packed column, taking the middle fraction.

n-**Butyl bromide** *[109-65-9]* M 137.0, b 101-102°, n 1.4399, n^{25} 1.4374, d^{25} 1.2678. Washed with conc H$_2$SO$_4$, water, 10% Na$_2$CO$_3$ and again with water. Dried with CaCl$_2$, CaSO$_4$ or K$_2$CO$_3$, and distd. Redistd after drying with P$_2$O$_5$, or passed through two columns containing 5:1 silica gel/Celite mixture and stored with freshly activated alumina.

sec-**Butyl bromide** see **2-bromobutane**.

Butyl carbitol see **diethylene glycol mono-*n*-butyl ether**.

4-*tert*-**Butylcatechol** *[98-29-3]* M 166.2, m 55-56°. Crystd from pet ether.

Butyl cellosolve see **2-butoxyethanol**.

n-**Butyl chloride** *[109-69-3]* M 92.6, b 78°, n 1.4021, d 0.886. Shaken repeatedly with conc H$_2$SO$_4$ (until no further colour developed in the acid), then washed with water, aq NaHCO$_3$ or Na$_2$CO$_3$, and more water. Dried with CaCl$_2$, or MgSO$_4$ (then with P$_2$O$_5$ if desired), decanted and fractionally distd. Alternatively, a stream of oxygen continuing *ca* three times as long as was necessary to obtain the first coloration of starch iodide paper by the exit gas. After washing with NaHCO$_3$ soln to hydrolyze ozonides and to remove the resulting organic acids, the liquid was dried and distd [Chien and Willard *JACS* **75** 6160 *1953*].

sec-**Butyl chloride** see **2-chlorobutane**.

tert-**Butyl chloride** *[507-20-0]* M 92.6, f.p. -24.6°, b 50.4°, n 1.38564, d 0.851. Purification methods commonly used for other alkyl halides lead to decomposition. Some impurities can be removed by photochlorination with a small amount of chlorine prior to use. The liquid can be washed with ice water, dried with CaCl$_2$ or CaCl$_2$ + CaO and fractionally distd. It has been further purified by repeated fractional crystn by partial freezing.

6-*tert*-**Butyl-1-chloro-2-naphthol** *[525-27-9]* M 232.7, m 76°, b 185°/15mm. Crystd from pet ether.

tert-**Butyl cyanide** *[630-18-2]* M 83.1, m 16-18°, b 104-106°, d 0.765. Purified by a two stage vac distn and degassed by freeze-pump-thaw technique. Stored under vac at 0°.

4-*tert*-Butyl-1-cyclohexanone *[98-53-3]* M 154.3, m 49-50°. Crystd from pentane.

***n*-Butyl disulphide** *[629-45-8]* M 178.4, b 110-113°/15mm, n^{22} 1.494, d 0.938. Shaken with lead peroxide, filtered and distd in vac under N_2.

***n*-Butyl ether** *[142-96-1]* M 130.2, b 52-53°/26mm, 142.0°/760mm, n 1.39925, n^{25} 1.39685, d 0.764. Peroxides (detected by the liberation of iodine from weakly acid (HCl) solns of 2% KI) can be removed by shaking 1L of ether with 5-10ml of a soln comprising 6.0g of ferrous sulphate and 6ml conc H_2SO_4 and 110ml of water, with aq Na_2SO_3, or with acidified NaI, water, then $Na_2S_2O_3$. After washing with dil NaOH, KOH, or Na_2CO_3, then water, the ether is dried with $CaCl_2$ and distd. It can be further dried by distn from CaH_2 or Na (after drying with P_2O_5), and stored in the dark with Na or NaH. The ether can also be purified by treating with CS_2 and NaOH, expelling the excess sulphide by heating. The ether is then washed with water, dried with NaOH and distd [Kusama and Koike *JCS Jap Pure Chem Sect* **72** 229 *1951*]. Other purification procedures include passage through an activated alumina column to remove peroxides, or through a column of silica gel, and distn after adding about 3% (v/v) of a 1M soln of MeMgI in *n*-butyl ether.

***n*-Butyl ethyl ether** *[628-81-9]* M 102.2, b 92.7°, n 1.38175, n^{25} 1.3800, d 0.751. Purified by drying with $CaSO_4$, by passage through a column of activated alumina (to remove peroxides), followed by prolonged refluxing with Na and then fractional distn.

***tert*-Butyl ethyl ether** *[637-92-3]* M 102.2, b 71-72°, d 0.741. Dried with $CaSO_4$, passed through an alumina column, and fractionally distd.

***n*-Butyl formate** *[592-84-7]* M 102.1, b 106.6°, n 1.3890, d 0.891. Washed with satd $NaHCO_3$ soln in the presence of satd NaCl, until no further reaction occurred, then with satd NaCl soln, dried ($MgSO_4$) and fractionally distd.

***tert*-Butyl hydroperoxide** *[75-91-2]* M 90.1, f.p. 5.4°, b 38°/18mm, n 1.4013, d 0.900. Alcoholic and volatile impurities can be removed by prolonged refluxing at 40° under reduced pressure, or by steam distn. For example, Bartlett, Benzing and Pincock [*JACS* **82** 1762 *1960*] refluxed at 30mm pressure in an azeotropic separation apparatus until rwo phases no longer separated, and then distilling at 41°/23mm. Pure material is stored under N_2, in the dark at 0°. Crude commercial material has been added to 25% NaOH below 30°, and the crystals of the sodium salt have been collected, washed twice with benzene and dissolved in distd water. After adjusting the pH of the soln to 7.5 by adding solid CO_2, the peroxide was extracted into pet ether, from which, after drying with K_2CO_3, it was recovered by distilling off the solvent under reduced pressure at room temperature [O'Brien, Beringer and Mesrobian *JACS* **79** 6238 *1957*]. Similarly, a soln in pet ether has been extracted with cold aq NaOH, and the hydroperoxide has been regenerated by adding at 0°, $KHSO_4$ at a pH not higher than 4.5, then extracted into ethyl ether, dried with $MgSO_4$, filtered and the ether evapd in a rotary evaporator under reduced pressure [Milac and Djokic *JACS* **84** 3098 *1962*]. *Care should be taken when handling this peroxide because of the possiblility of* **EXPLOSION.**

***n*-Butyl iodide** *[542-69-8]* M 184.0, b 130.4°, n^{25} 1.44967, d 1.616. Dried with $MgSO_4$ or P_2O_5, fractionally distd through a column packed with glass helices, taking the middle fraction and storing with calcium or mercury in the dark. Also purified by prior passage through activated alumina or by shaking with conc H_2SO_4 then washing with Na_2SO_3 soln. It has also been treated carefully with sodium to remove free HI and water, before distilling in a column containing copper turnings at the top. Another purification consisted of treatment with bromine, followed by extraction of free halogen with $Na_2S_2O_3$, washing with water, drying and fractional distn.

***sec*-Butyl iodide** see **2-iodobutane.**

***tert*-Butyl iodide** *[558-17-8]* M 184.0, b 100°(dec), d 1.544. Vac distn has been used to obtain a distillate which remained colourless for several weeks at -5°. More extensive treatment has been used by Boggs, Thompson and Crain [*JPC* **61** 625 *1957*] who washed with aq NaHSO$_3$ soln to remove free iodine, dried for 1hr with Na$_2$SO$_3$ at 0°, and purified by four or five successive partial freezings of the liquid to obtain colourless material which was stored at -78°.

n-**Butyl mercaptan** see **1-butanethiol.**

sec-**Butyl mercaptan** see **2-butanethiol.**

tert-**Butyl mercaptan** see **2-methyl-2-propanethiol.**

n-**Butyl methacrylate** *[97-88-1]* **M 142.2, b 49-52°/0.1mm,**
tert-**Butyl methacrylate** *[585-07-9]* **M 142.2.** Purified as for butyl acrylate.

2-*tert*-Butyl-4-methoxyphenol *[121-00-6]* **M 180.3, m 64.1°.** Fractionally distd in vac, then passed as a soln in CHCl$_3$ through alumina, and the solvent evapd from the eluate.

n-**Butyl methyl ether** *[628-28-4]* **M 88.2, b 70°, d 0.744.** Dried with CaSO$_4$, passed through an alumina column to remove peroxides, and fractionally distd.

dl-sec-**Butyl methyl ether** *[1634-04-4]* **M 88.2, b 54°, n 1.369.** Same as for *n*-butyl methyl ether.

tert-**Butyl methyl ketone** *[75-97-8]* **M 100.2, b 105°/746mm, 106°/760mm, n 1.401, d 0.814.** Refluxed with a little KMnO$_4$. Dried with CaSO$_4$ and distd.

8-*sec*-Butylmetrazole *[25717-83-3]* **M 194.3, m 70°.** Crystd from pet ether, and dried for 2 days under vac over P$_2$O$_5$.

p-tert-**Butylnitrobenzene** *[3282-56-2]* **M 179.2, m 28.4°.** Fractionally crystd three times by partially freezing a mixture of the mono-nitro isomers, then recryst from MeOH twice and dried under vac [Brown *JACS* **81** 3232 *1959*].

N-(*n*-Butyl)-5-nitro-2-furamide *[14121-89-2]* **M 212.2, m 89-90°.** Recrystd twice from EtOH/water mixture.

tert-**Butyl peracetate** *[107-71-1]* **M 132.2, b 23-24°/0.5mm, n^{25} 1.4030.** Washed with NaHCO$_3$ from a benzene soln, then redistd to remove benzene [Kochi *JACS* **84** 774 *1962*]. Handle with adequate protection due to possible **EXPLOSIVE** nature.

tert-**Butylperoxy isobutyrate** *[109-13-7]* **M 160.2, f.p. -45.6°.** After diluting 90ml of the material with 120ml of pet ether, the mixture was cooled to 5° and shaken twice with 90ml portions of 5% NaOH soln (also at 5°). The non-aqueous layer, after washing once with cold water, was dried at 0° with a mixture of anhydrous MgSO$_4$ and MgCO$_3$ containing *ca* 40% MgO. After filtering, this material was passed, twice, through a column of silica gel at 0° (to remove *tert*-butyl hydroperoxide). The soln was evapd at 0°/0.5-1mm to remove the solvent, and the residue was recrystd several times from pet ether at -60°, then subjected to high vac to remove traces of solvent [Milos and Golubovic *JACS* **80** 5994 *1958*]. *Handle with adequate protection due to possible* **EXPLOSIVE** *nature.*

tert-**Butylperphthalic acid** *[15042-77-0]* **M 238.2.** Crystd from ethyl ether and dried over H$_2$SO$_4$. **POSSIBLE EXPLOSIVE.**

p-tert-**Butylphenol** *[98-54-4]* **M 150.2, m 99°.** Crystd to constant melting point from pet ether (b 60-80°). Also purified *via* its benzoate, as for phenol.

p-tert-**Butylphenoxyacetic acid** *[1798-04-5]* **M 208.3, m 88-89°.** Crystd from pet ether/benzene mixture.

n-**Butylphenyl** *n*-**butylphosphonate** *[36411-99-1]* **M 270.3.** Crystd three times from hexane as its compound with uranyl nitrate. See *tributyl phosphate*.

p-tert-**Butylphenyl diphenylphosphate** *[981-40-8]* **M 382.4, b 261°/6mm, n^{25} 1.5522.** Purified by vac distn, and percolation through an alumina column, followed by passage through a packed column maintained at 150° to remove residual traces of volatile materials in a counter-current stream of N$_2$ at reduced pressure [Dobry and Keller *JPC* **61** 1448 *1957*].

n-**Butyl phenyl ether** *[1126-79-0]* **M 150.2, b 210.5°, d 0.935.** Dissolved in ethyl ether, washed first with 10% aq NaOH to remove traces of phenol, then repeatedly with distd water, followed by evapn of the solvent and distn under reduced pressure [Arnett and Wu *JACS* **82** 5660 *1960*].

N-tert-**Butyl α-phenyl nitrone** *[3376-24-7]* **M 177.2, m 73-74°.** Crystd from hexane.

Butyl phosphate *[126-73-8]* **M 266.3, b 289°/760mm, n 1.424, d 0.979.** Washed with water, then with 1% NaOH or 5% Na$_2$CO$_3$ for several hours, then finally with water. Dried under reduced pressure and vac distd.

Butyl phthalate *[84-74-2]* **M 278.4, f.p. -35°, b 340°/760mm, d 1.043.** Freed from alcohol by washing with water, or from acids and butyl hydrogen phthalate by washing with dil NaOH. Distd at 10 torr or less.

Butyl stearate *[123-95-5]* **M 340.6, m 26.3°, d 0.861.** Acidic impurities removed by shaking with 0.05M NaOH or a 2% NaHCO$_3$ soln, followed by several water washes, then purified by fractional freezing of the melt and fractional crystn from solvents with boiling points below 100°.

p-tert-**Butyltoluene** **[98-51-1] M 148.3, f.p. -53.2°, b 91°/28mm, n 1.4920, d 0.854.** A sample containing 5% of the *meta*-isomer was purified by selective mercuration. Fractional distn of the solid arylmercuric acetate, after removal from the residual hydrocarbon, gave pure *p-tert*-butyltoluene [Stock and Brown *JACS* **81** 5615 *1959*].

tert-**Butyl 2,4,6-trichlorophenyl carbonate** *[19065-08-5]* **M 297.6, m 64-66°.** Crystd from a mixture of MeOH (90ml) and water (6ml) using charcoal [Broadbent *et al, JCS(C)* 2632 *1967*].

n-**Butyl vinyl ether** *[111-34-2]* **M 100.2, b 93.3°, d 0.775.** After five washings with equal vols of water to remove alcohols (made slightly alkaline with KOH), the ether was dried with sodium and distd under vac, taking the middle fraction [Coombes and Eley *JCS* 3700 *1957*]. Stored over KOH.

2-Butyne *[503-17-3]* **M 54.1, b 0°/253mm, d 0.693.** Stood with sodium for 24hr, then fractionally distd at reduced pressure.

2-Butyne-1,4-diol *[110-65-6]* **M 86.1, m 54-57°.** Crystd from ethyl acetate.

n-**Butyraldehyde** *[123-72-8]* **M 72.1, b 74.8°, n^{15} 1.38164, n 1.37911, d 0.810.** Dried with CaCl$_2$ or CaSO$_4$, then fractionally distd under N$_2$. Lin and Day [*JACS* **74** 5133 *1952*] shook with batches of CaSO$_4$ for 10min intervals until a 5ml sample, on mixing with 2,5ml of CCl$_4$ containing 0.5g of aluminium isopropoxide, gave no ppte and caused the soln to boil within 2min. Water can be removed from *n*-butyraldehyde by careful distn as an azeotrope distilling at 68°. The aldehyde has also been purified through its bisulphite compound which, after decomposing with excess NaHCO$_3$ soln, was steam distd, extracted under N$_2$ into ether and, after drying, the extract was fractionally distd [Kyte, Jeffery and Vogel *JCS* 4454 *1960*].

Butyramide *[514-35-5]* **M 87.1, m 115°, b 230°.** Crystd from acetone, benzene, CCl$_4$-pet ether, 20% EtOH or water. Dried under vac over P$_2$O$_5$, CaCl$_2$ or 99% H$_2$SO$_4$.

n-**Butyric acid** *[107-92-6]* **M 88.1, f.p. -5.3°, b 163.3°, n 1.39796, n^{25} 1.39581, d 0.961.** Distd, mixed with KMnO$_4$ (20g/L), and fractionally redistd, discarding the first third [Vogel *JCS* 1814 *1948*].

n-**Butyric anhydride** *[106-31-0]* **M 158.2, b 198°, d 0.968.** Dried by shaking with P$_2$O$_5$, then distilling.

γ-**Butyrolactone** *[96-48-0]* **M 86.1, b 83.8°/12mm, d 1.124.** Dried with anhydrous CaSO$_4$, then fractionally distd. *Handle in a fume cupboard due to* **TOXICITY.**

Butyronitrile *[109-74-0]* **M 69.1, b 117.9°, n 1.3846, n ^{30} 1.37954, d 0.793.** Treated with conc HCl until the smell of the isonitrile had gone, then dried with K$_2$CO$_3$ and fractionally distd [Turner *JCS* 1681 *1956*]. Alternatively it was twice heated at 75° and stirred for several hours with a mixture of 7.7g Na$_2$CO$_3$ and 11.5g KMnO$_4$ per L of butyronitrile. The mixture was cooled, then distd. The middle fraction was dried over activated alumina. [Schoeller and Wiemann *JACS* **108** 22 *1986*.]

Caffeic acid *[331-39-5]* M 180.2, m 195°. Crystd from water.

Caffeine *[58-08-2]* M 194.2, m 237°. Crystd from water or absolute EtOH.

Calcein *[1461-15-0]* M 622.5. Free acid crystd from 50% aq MeOH, or 300mg sample in minimum amount of 0.1M NaOH, add 50ml 10-20% aq MeOH and filter. To the filtrate add 1M HCl to adjust to pH 2.5. Hold in refrigerator overnight and filter on a No 4 glass filter. Wash well with MeOH. [Wallach *et al, AC* **31** 456 *1959*.]

Calcon carboxylic acid *[3737-95-9]* M 428.4, m 300°. Purified through its *p*-toluidinium salt. The dye was dissolved in warm 20% aq MeOH and treated with *p*-toluidine to ppte the salt after cooling. Finally recrystd from hot water. [Itoh and Ueno *Analyst* **95** 583 *1970*.]

Calmagite *[3147-14-6]* M 358.4, m 300°. Crude sample was extracted with anhydrous ethyl ether [Lindstrom and Diehl *AC* **32** 1123 *1960*].

RS-Camphene *[565-00-4]* M 136.2, m 51-52°. Crystd twice from EtOH, then repeatedly melted and frozen at 30mm pressure. [Williams and Smyth *JACS* **84** 1808 *1962*.]

R-Camphor *[464-49-3]* M 136.2, m 178.8°, b 204°, $[\alpha]_{546}^{35}$ +59.6° (in EtOH). Crystd from EtOH, 50% EtOH/water, or pet ether or from glacial acetic acid by addition of water. It can be sublimed and also fractionally crystd from its own melt.

(1R,2S)(+)-Camphoric acid *[124-83-4]* M 200.2, m 186-188°, $[\alpha]_{546}$ +57° (c 1, EtOH). Crystd from water.

(±)-Camphoric anhydride *[595-30-2]* M 182.2, **transition temp 135°, m 223.5°.** Crystd from EtOH.

(1S) (+)Camphor-10-sulphonic acid *[3144-16-9]* M 232.3, m 193°(dec), $[\alpha]_{546}$ +27.5° (c 10, H_2O). Crystd from ethyl acetate and dried under vac.

S-Canavanine *[543-38-4]* M 176.2, m 184°, $[\alpha]_D^{17}$ +19.4° (c 2, H_2O),
S-Canavanine sulphate *[2219-31-0]* M 274.3, m 172°(dec), $[\alpha]_D$ +17.3° (c 3.2, H_2O). Crystd from aq EtOH.

Cannabinol *[521-35-7]* M 310.4, m 76-77°, b 185°/0.05mm. Crystd from pet ether. Sublimed.

Canthaxanthin (*trans*) *[514-78-3]* M 564.9, m 211-212°, $\varepsilon_{1cm}^{1\%}$ 2200 (470nm) in cyclohexane. Purified by chromatography on a column of deactivated alumina or magnesia, or on a thin layer of silica gel G (Merck), using dichloromethane/ethyl ether (9:1) to develop the chromatogram. Stored in the dark and in an inert atmosphere at -20°.

Capric acid (decanoic acid) *[334-48-5]* M 172.3, m 31.5°, b 148°/11mm, n^{25} 1.4239, d 0.8858. Purified by conversion to its methyl ester, b 114.0°/15mm (using excess MeOH, in the presence of H_2SO_4). After removal of the H_2SO_4 and excess MeOH, the ester was distd under vac through a 3ft column packed with glass helices. The acid was then obtained from the ester by saponification. [Trachtman and Miller *JACS* **84** 4828 *1962*.]

n-**Caproamide** *[628-02-4]* M 115.2, m 100°. Crystd from hot water.

Caproic acid *[142-62-1]* M 116.2, b 205.4°, n 1.4168, d^{20} 0.925. Dried with $MgSO_4$ and fractionally distilled from $CaSO_4$.

ε-Caprolactam *[105-60-2]* **M 113.2, m 70°.** Distd at reduced pressure, crystd from acetone or pet ether and redistd. Purified by zone melting. Very hygroscopic. Discolours in contact with air unless small amounts (0.2g/L) of NaOH, Na_2CO_3 or $NaBO_2$ are present. Crystd from a mixture of pet ether (185ml of b 70°) and 2-methyl-2-propanol (30ml), from acetone, or pet ether. Distd under reduced pressure and stored under nitrogen.

Capronitrile *[124-12-9]* **M 125.2, b 163.7°, n 1.4069, n^{25} 1.4048.** Washed twice with half-volumes of conc HCl, then with satd aq $NaHCO_3$, dried with $MgSO_4$, and distilled..

Caprylic acid see *n*-octanoic acid.

Capsorubin *[470-38-2]* **M 604.9, m 218°, λ_{max} 443, 468, 503 nm, in hexane.** Possible impurities: zeaxanthin and capsanthin. Purified by chromatography on a column of $CaCO_3$ or MgO. Crystd from benzene/pet ether or CS_2.

Captan *[133-06-2]* **M 300.5, m 172-173°.** Crystd from CCl_4.

p-(Carbamoylmethoxy)acetanilide *[14260-41-4]* **M 208.2, m 208°.** Crystd from water.

3-Carbamoyl-1-methylpyridinium chloride *[1005-24-9]* **M 172.6.** Crystd from MeOH.

Carbanilide *[102-07-8]* **M 212.3, m 242°.** Crystd from EtOH or a large vol (40ml/g) of hot water.

Carbazole *[86-74-8]* **M 167.2, m 240-243°.** Dissolved (60g) in conc H_2SO_4 (300ml), extracted with three 200ml portions of benzene, then stirred into 1600ml of an ice-water mixture. The ppte was filtered off, washed with a little water, dried, crystd from benzene and then from pyridine/benzene. [Feldman, Pantages and Orchin *JACS* **73** 4341 *1951*.] Has also been crystd from EtOH or toluene, sublimed in vac, zone-refined, and purified by tlc.

9-Carbazolacetic acid *[524-80-1]* **M 225.2, m 215°.** Crystd from ethyl acetate.

1-Carbethoxy-4-methylpiperazine hydrochloride *[532-78-5]* **M 204.7, m 168.5-169°.** Crystd from absolute EtOH.

Carbitol see **diethylene glycol monoethyl ether.**

Carbobenzoxy chloride see **benzyloxycarbonyl chloride.**

γ-Carboline *[244-63-3]* **M 168.2, m 225°.** Crystd from water.

Carbon Black Leached for 24hr with 1:1 HCl to remove oil contamination, then washed repeatedly with distd water. Dried in air, and eluted for one day each with benzene and acetone. Again dried in air at room temp, then heated in a vac for 24hr at 600° to remove adsorbed gases. [Tamamushi and Tamaki *TFS* **55** 1007 *1959*.]

Carbon disulphide *[75-15-0]* **M 76.1, b 46.3°, n 1.627, d 1.264.** Shaken for 3hr with three portions of $KMnO_4$ soln (5g/L), twice for 6hr with mercury (to remove sulphide impurities) until no further darkening of the interface occurred, and finally with a soln of $HgSO_4$ (2.5g/L) or cold, satd $HgCl_2$. Dried with $CaCl_2$, $MgSO_4$, or CaH_2 (with further drying by refluxing with P_2O_5), followed by fractional distn in diffuse light. **Alkali metals cannot be used as drying agents.** Has also been purified by standing with bromine (0.5ml/L) for 3-4hr, shaking with KOH soln, then copper turnings (to remove unreacted bromine), and drying with $CaCl_2$.
Small quantities of carbon disulphide have been purified (including removal of hydrocarbons) by mechanical agitation of a 45-50g sample with a soln of 130g of sodium sulphide in 150ml of water for 24hr at 35-40°. The aq sodium thiocarbonate soln was separated from unreacted CS_2, then pptd with

140g of copper sulphate in 350g of water, with cooling. After filtering off the copper thiocarbonate, it was decomposed by passing steam into it. The distillate was separated from water and distd from P_2O_5. [Ruff and Golla *Z anorg Chem* **138** 17 *1924*.]

Carbon tetrabromide *[558-13-4]* **M 331.7, m 92.5°.** Reactive bromide was removed by refluxing with dil aq Na_2CO_3, then steam distd, crystd from EtOH, and dried in the dark under vac. [Sharpe and Walker *JCS* 157 *1962*]. Can be sublimed at 70° at low pressure.

Carbon tetrachloride *[56-23-5]* **M 153.8, b 76.8°, d²⁵ 1.5842.** For many purposes, careful fractional distn gives adequate purification. Carbon disulphide can be removed by shaking vigorously for several hours with satd KOH, separating, and washing with water: this treatment is repeated. The CCl_4 is shaken with conc H_2SO_4 until there is no further coloration, then washed with water, dried with $CaCl_2$ or $MgSO_4$ and distd (from P_2O_5 if desired). **It must not be dried with sodium.** An initial refluxing with mercury for 2hr removes sulphides. Other purification steps include passage of dry CCl_4 through activated alumina, and distn from $KMnO_4$. Carbonyl containing impurities can be removed by percolation through a Celite column impregnated with 2,4-dinitrophenylhydrazine (DNPH), H_3PO_4 and water. (Prepared by dissolving 0.5g DNPH in 6ml of 85% H_3PO_4 by grinding together, then mixing with 4ml of distd water and 10g Celite.) [Schwartz and Parks *AC* **33** 1396 *1961*]. Photochlorination of CCl_4 has also been used: CCl_4 to which a small amount of chlorine has been added is illuminated in a glass bottle (e.g. for 24hr with a 200W tungsten lamp near it), and, after washing out the excess chlorine with 0.02M Na_2SO_3, the CCl_4 is washed with distd water and distd from P_2O_5. It can be dried by passing through 4A molecular sieves and distd. Another purification procedure is to wash CCl_4 with aq NaOH, then repeatedly with water and N_2 gas bubbled through the liquid for several hours. After drying over $CaCl_2$ it is percolated through silica gel and distd under dry N_2 before use [Klassen and Ross *JPC* **91** 3664 *1987*].

Carbon tetrafluoride *[75-73-0]* **M 88.0, b -15°.** Purified by repeated passage over activated charcoal at solid-CO_2 temperatures. Traces of air were removed by pumping while alternately freezing and melting. Alternatively, liquefied by cooling in liquid air and then fractionally distilled under vac. (The chief impurity originally present was probably CF_3Cl.)

Carbon tetraiodide *[507-25-5]* **M 519.6, m 168°(dec).** Sublimed *in vacuo*.

N,N'-**Carbonyldiimidazole** *[530-62-1]* **M 162.2, m 115.5-116°.** Crystd from benzene or tetrahydrofuran, in a dry-box.

Carbonyl sulphide *[463-58-1]* **M 60.1, b -47.5°.** Passed through traps containing satd aq lead acetate and then through a column of anhydrous $CaSO_4$.

Carbostyril see **2-hydroxyquinoline.**

(Carboxymethyl)trimethylammonium chloride hydrazide see **Girard reagent T.**

o-**Carboxyphenylacetonitrile** *[6627-91-4]* **M 161.2, m 114-115°.** Crystd (with considerable loss) from benzene or glacial acetic acid.

S-Carnosine *[305-84-0]* **M 226.2, m 246-250°(dec), $[\alpha]_D^{25}$ +20.5° (water).** Likely impurities: histidine, β-alanine. Crystd from water by adding EtOH in excess.

ξ-**Carotene** *[38894-81-4]* **M 536.9, m 38-42°, λ_{max} 378, 400, 425nm, $\varepsilon_{1cm}^{1\%}$ 2270 (400nm), in pet ether.** Purified by chromatography on 50% magnesia-HyfloSupercel, developing with hexane and eluting with 10% EtOH in hexane. It was crystd from toluene/MeOH. [Gorman *et al, JACS* **107** 4404 *1985*.] Stored in the dark under inert atmosphere at -20°.

α-Carotene *[74488-99-5]* **M 536.9, m 184-188°,** λ_{max} **422, 446, 474 nm, in hexane,** $\varepsilon_{1cm}^{1\%}$ **2725 (at 446nm), 2490 (at 474nm).** Purified by chromatography on columns of calcium hydroxide, alumina or magnesia. Crystd from CS_2/MeOH, toluene/MeOH, ethyl ether/pet ether, or acetone/pet ether. Stored in the dark, under inert atmosphere at -20°.

β-Carotene *[7235-40-7]* **M 536.9, m 178-180°,** $\varepsilon_{1cm}^{1\%}$ **2590 (450nm), 2280 (478nm), in hexane.** Purified by chromatography on magnesia column, thin layer of kieselguhr or magnesia. Crystd from CS_2/MeOH, ethyl ether/pet ether, acetone/pet ether or toluene/MeOH. Stored in the dark, under inert atmosphere, at -20°. Recrystd from 1:1 EtOH/$CHCl_3$ [Bobrowski and Das *JPC* **89** 5079 *1985*; Johnston and Scaiano *JACS* **108** 2349 *1986*].

γ-Carotene *[10593-83-6]* **M 536.9,** $\varepsilon_{1cm}^{1\%}$ **2055 (437nm), 3100 (462nm), 2720 (494nm).** Purified by chromatography on alumina or magnesia columns. Crystd from benzene/MeOH (2:1). Stored in the dark, under inert atmosphere, at 0°.

λ–Carrageenan *[9064-57-7]*. Pptd from a soln of 4g in 600ml cf water containing 12g of potassium acetate by addition of EtOH. The fraction taken pptd between 30 and 45% (v/v) EtOH. [Pal and Schubert *JACS* **84** 4384 *1962*.]

Catechin *[7295-85-4]* **M 272.3, m 177° (anhyd).** Crystd from hot water. Dried at 100°.

Catechol *[120-80-9]* **M 110.1, m 105°.** Crystd from benzene or toluene. Sublimed under vac. [Rozo *et al, AC* **58** 2988 *1986*.]

Cation exchange resin. Conditioned before use by successive washing with water, EtOH and water, and taken through two H^+-Na^+-H^+ cycles by successive treatment with M NaOH, water and M HCl then washed with water until neutral. (Ion exchange resins, BDH Handbook, 5th edn 1971).

β-Cellobiose *[528-50-7]* **M 342.3, m 228-229°(dec),** $[\alpha]_D^{25}$ **+33.3° (c 2, water).** Crystd from 75% aq EtOH.

Cellosolve see **2-ethoxyethanol.**

Cellulose triacetate *[9012-09-3]* **M 72,000-74,000.** Extracted with cold EtOH, dried in air, washed with hot distd water, again dried in air, then dried at 50° for 30min. [Madorsky, Hart and Straus *J Res Nat Bur Stand* **60** 343 *1958*.]

Cetane see *n*-hexadecane.

Cetyl acetate *[629-70-9]* **M 284.5, m 18.3°.** Vac distd twice, then crystd several times from ethyl ether/MeOH.

Cetyl alcohol (1-hexadecanol) *[36653-82-4]* **M 242.5, m 49.3°.** Crystd from aq EtOH or from cyclohexane. Purified by zone refining. Purity checked by gas chromatography.

Cetylamide *[629-54-9]* **M 255.4,**
Cetylamine (1-hexadecylamine) *[143-27-1]* **M 241.5, m 78°.** Crystd from thiophen-free benzene and dried under vac over P_2O_5.

Cetylammonium chloride *[1602-97-7]* **M 278.0.** Crystd from MeOH.

Cetyl bromide (1-bromohexadecane) *[112-82-3]* **M 305.4, m 15°, b 193-196°/14mm.** Shaken with H_2SO_4, washed with water, dried with K_2CO_3 and fractionally distd.

Cetyl ether *[4113-12-6]* **M 466.9, m 54°.** Vac distd then crystd several times from MeOH/benzene.

Cetylpyridinium chloride (H$_2$O) *[6004-24-6]* **M 358.0, m 80-83°.** Crystd from MeOH or EtOH/ethyl ether and dried *in vacuo*. [Moss *et al, JACS* **108** 788 *1986*; Lennox and McClelland *JACS* **108** 3771 *1986*.]

Cetyltrimethylammonium bromide (cetrimonium bromide) *[124-03-8]* **M 364.5, m 227-235°(dec).** Crystd from EtOH, EtOH/benzene or from wet acetone after extracting twice with pet ether. Shaken with anhydrous ethyl ether, filtered and dissolved in a little hot MeOH. After cooling in the refrigerator, the ppte was filtered at room temp and redissolved in MeOH. Anhydrous ether was added and, after warming to obtain a clear soln, it was cooled and crystalline material was filtered. [Duynstee and Grunwald *JACS* **81** 4540 *1959*; Hakemi *et al, JACS* **91** 120 *1987*.]

Cetyltrimethylammonium chloride *[112-02-7]* **M 320.0.** Crystd from acetone/ether mixture, EtOH/ether, or from MeOH. [Moss *et al, JACS* **109** 4363 *1987*.]

Chalcone see **benzalacetophenone.**

Charcoal. Charcoal (50g) was added to 1L of 6M HCl and boiled for 45min. The supernatant was discarded, and the charcoal was boiled with two more lots of HCl, then with distilled water until the supernatant no longer gave a test for chloride ion. The charcoal (which was now phosphate-free) was filtered on a sintered-glass funnel and air dried at 120° for 24hr. [Lippin, Talbert and Cohn *JACS* **76** 2871 *1954*.] The purification can be carried out using a Soxhlet extractor (without cartridge), allowing longer extraction times. Treatment with conc H$_2$SO$_4$ instead of HCl has been used to remove reducing substances.

Chaulmoogric acid *[502-30-7]* **M 280.4, m 68.5°, b 247-248°/20mm.** Crystd from pet ether or EtOH.

Chelerythrine *[2870-15-7]* **M 389.4, m 207°.** Crystd from CHCl$_3$ by addition of MeOH.

Chelex 100 *[11139-85-8]*. Washed successively with 2M ammonia, water, 2M nitric acid and water. Chelex 100 may develop an odour on long standing. This can be removed by heating to 80° for 2hr in 3M ammonia, then washing with water. [Ashbrook *J Chromat* **105** 151 *1975*.]

Chelidamic acid see **4-hydroxypyridine-2,6-dicarboxylic acid.**

Chelidonic acid *[6003-94-7]* **M 184.1, m 262°.** Crystd from aq EtOH.

Chenodesoxycholic acid *[474-25-9]* **M 392.6, m 143°,** $[\alpha]_{546}$ **+14° (c 2, EtOH).** Crystd from ethyl acetate.

Chimyl alcohol *[6145-69-3]* **M 316.5, m 64°.** Crystd from hexane.

Chloral *[75-87-6]* **M 147.4, b 98°.** Distd, then dried by distilling through a heated column of CaSO$_4$.

Chloralacetone chloroform *[512-47-0]* **M 324.9, m 65°.** Crystd from benzene.

2-Chloroacetophenone *[532-27-4]* **M 154.6, m 54-56°.** Crystd from MeOH [Tanner *JOC* **52** 2142 *1987*].

α-Chloralose *[15879-93-3]* **M 309.5, m 187°.** Crystd from EtOH or ethyl ether.

Chlorambucil *[305-03-3]* **M 304.2, m 64-66°.** Crystd from pet ether.

Chloramphenicol *[56-75-7]* **M 323.1, m 150.5-151.5°,** $[\alpha]_{546}$ **+25° (c 5, EtOH).** Crystd from water (sol 2.5mg/ml at 25°) or ethylene dichloride. Sublimed under high vac.

Chloroamphenicol palmitate *[530-43-8]* M 561.5, m 90°, [α]$_D^{26}$ +24.6° (c 5, EtOH). Crystd from benzene.

p-Chloranil *[118-75-2]* M 245.9, m 294.2-294.6° (sealed tube). Crystd from acetone, benzene, EtOH or toluene, drying under vac over P$_2$O$_5$, or from acetic acid, drying over NaOH in a vac desiccator. It can be sublimed under vac. Sample may contain significant amounts of the o-chloranil isomer as impurity. Purified by triple sublimation under vacuum. Recrystd before use.

Chloranilic acid *[87-88-7]* M 209.0, m 283-284°. A soln of 8g in 1L of boiling water was filtered while hot, then extracted twice at about 50° with 200ml portions of benzene. The aq phase was cooled in ice-water. The crystals were filtered off, washed with three 10ml portions of water, and dried at 115°. [Thamer and Voight *JPC* 56 225 *1952*.]

Chlorazol Sky Blue FF *[2610-05-1]* M 996.9. Freed from other electrolytes by adding aq sodium acetate to a boiling soln of the dye in distd water. After standing, the salted-out dye was filtered on a Büchner funnel, the process being repeated several times. Finally, the pptd dye was boiled several times with absolute EtOH to wash out any sodium acetate, then dried (as the sodium salt) at 105°. [McGregor, Peters and Petropolous *TFS* 58 1045 *1962*.]

α-Chloroacetamide *[79-07-2]* M 93.5, m 121°, b 224-225°/743mm. Crystd from acetone and dried under vac over P$_2$O$_5$.

p-Chloroacetanilide *[539-03-7]* M 169.6, m 179°. Crystd from EtOH or aq EtOH.

Chloroacetic acid *[79-11-8]* M 94.5, m 62.8°, b 189°. Crystd from CHCl$_3$, CCl$_4$, benzene or water. Dried over P$_2$O$_5$ or conc H$_2$SO$_4$ in a vac desiccator. Further purification by distn from MgSO$_4$, and by fractional crystn from the melt. Stored under vac or under dry N$_2$. [Bernasconi *et al, JACS* 107 3621 *1985*.]

Chloroacetic anhydride *[541-88-8]* M 171.0, m 46°, d 1.5494. Crystd from benzene.

Chloroacetone *[78-95-5]* M 92.5, b 119°/763mm, d 1.15. Dissolved in water and shaken repeatedly with small amounts of ethyl ether which extracts, preferentially, 1,1-dichloroacetone present as an impurity. The chloroacetone was then extracted from the aq phase using a large amount of ethyl ether, and distd at slightly reduced pressure. It was dried with CaCl$_2$ and stored at Dry-ice temperature. Alternatively, it was stood with CaSO$_4$, distd and stored over CaSO$_4$.

Chloroacetonitrile *[107-14-2]* M 75.5, b 125°. Refluxed with P$_2$O$_5$ for one day, then distd through a helices-packed column. Also purified by gas chromatography.

o-Chloroaniline *[95-51-2]* M 127.6, m -1.9°, b 208.8°, n 1.58807, d 1.213. Freed from small amounts of the p-isomer by dissolving in one equivalent of H$_2$SO$_4$ and steam distilling. The p-isomer remains behind as the sulphate. [Sidgwick and Rubie *JCS* 1013 *1921*.] An alternative method is to dissolve in warm 10% HCl (11ml/g of amine) and on cooling, the hydrochloride of o-chloroaniline separates out. The latter can be recrystd until the acetyl derivative has a constant melting point. (In this way, yields are better than for the recrystn of the picrate from EtOH or of the acetyl derivative from pet ether.) [King and Orton *JCS* 1377 *1911*.]

p-Chloroaniline *[106-47-8]* M 127.6, m 70-71°. Crystd from MeOH, pet ether (b 30-60°), or 50% aq EtOH, then benzene/pet ether (b 60-70°), then dried in a vac desiccator. Can be distd under vac (b 75-77°/33mm).

p-Chloroanisole *[623-12-1]* M 142.6, b 79°/11.5mm, 196.6°/760mm, n$^{25.5}$ 1.5326, d 1.164. Washed with 10% (vol) aq H$_2$SO$_4$ (three times), 10% aq KOH (three times), and then with water until neutral.

Dried with MgSO$_4$ and fractionally distd from CaH$_2$ through a glass helices-packed column under reduced pressure.

9-Chloroanthracene *[716-53-0]* **M 212.9, m 105-107°.** Crystd from EtOH. [Masnori *JACS* **108** 1126 *1986*.]

10-Chloro-9-anthraldehyde *[10527-16-9]* **M 240.7, m 217-219°.** Crystd from EtOH.

o-**Chlorobenzaldehyde** *[89-98-5]* **M 140.6. m 11°, b 213-214°, n 1.566, d 1.248.** Washed with 10% Na$_2$CO$_3$ soln, then fractionally distd in the presence of a small amount of catechol.

m-**Chlorobenzaldehyde** *[587-04-2]* **M 140.6, m 18°, b 213-214°, n 1.564, d 1.241.** Purified by low temp crystn from pet ether (b 40-60°).

p-**Chlorobenzaldehyde** *[104-88-1]* **M 140.6, m 47°.** Crystd from EtOH/water (3:1), then sublimed twice at 2mm pressure at a temperature slightly above the melting point.

Chlorobenzene *[108-90-7]* **M 112.6, b 131.7°, n 1.52480, d 1.107.** The main impurities are likely to be chlorinated impurities originally present in the benzene used in the synthesis of chlorobenzene, and also unchlorinated hydrocarbons. A common purification procedure is to wash several times with conc H$_2$SO$_4$ then with aq NaHCO$_3$ or Na$_2$CO$_3$, and water, followed by drying with CaCl$_2$, K$_2$CO$_3$ or CaSO$_4$, then with P$_2$O$_5$, and distn. It can also be dried with Linde 4A molecular sieve. Passage through, and storage over, activated alumina has also been used to obtain low conductance material. [Flaherty and Stern *JACS* **80** 1034 *1958*.]

p-**Chlorobenzenesulphonyl chloride** *[98-60-2]* **M 211.1, m 53°, b 141°/15mm.** Crystd from ether in powdered Dry-ice, after soln had been washed with 10% NaOH until colourless and dried with Na$_2$SO$_4$.

p-**Chlorobenzhydrazide** *[536-40-3]* **M 170.6, m 164°.** Crystd from water.

o-**Chlorobenzoic acid** *[118-91-2]* **M 156.6, m 139-140°.** Crystd successively from glacial acetic acid, aq EtOH, and pet ether (b 60-80°). Other solvents include hot water or toluene (*ca* 4ml/g). Crude material can be given an initial purification by dissolving 30g in 100ml of hot water containing 10g of Na$_2$CO$_3$, boiling with 5g of charcoal for 15min, then filtering and adding 31ml of 1:1 aq HCl: the ppte is washed with a little water and dried at 100°.

m-**Chlorobenzoic acid** *[535-80-8]* **M 156.6, m 158°.** Crystd successively from glacial acetic acid, aq EtOH and pet ether (b 60-80°).

p-**Chlorobenzoic acid** *[74-11-3]* **M 156.6, m 238-239°.** Same as for *m*-chlorobenzoic acid. Has also been crystd from hot water, and from EtOH.

o-**Chlorobenzonitrile** *[873-32-5]* **M 137.6, m 45-46°.** Crystd to constant melting point from benzene/pet ether (b 40-60°).

4-Chlorobenzophenone *[134-85-0]* **M 216.7, m 75-76°.** Recrystd for EtOH. [Wagner *et al, JACS* **108** 7727 *1986*.]

o-**Chlorobenzotrifluoride** *[88-16-4]* **M 180.6, b 152.3°,**
m-**Chlorobenzotrifluoride** *[98-15-7]* **M 180.6, b 137.6°,**
p-**Chlorobenzotrifluoride** *[98-56-6]* **M 180.6, b 138.6°.** Dried with CaSO$_4$, and distd at high reflux ratio through a silvered vacuum-jacketed glass column packed with one-eight inch glass helices [Potter and Saylor *JACS* **73** 90 *1951*].

p-**Chlorobenzyl chloride** *[104-83-6]* **M 161.0, m 28-29°, b 96°/15mm.** Dried with CaSO$_4$, then fractionally distd under reduced pressure. Crystd from heptane or dry ethyl ether. **LACHRYMATORY.**

p-**Chlorobenzylisothiuronium chloride** *[544-47-8]* **M 237.1, m 197°.** Crystd from conc HCl by addn of water.

1-Chlorobutane see *n*-**butyl chloride.**

2-Chlorobutane *[78-86-4]* **M 92.6, b 68.5°, n^{25} 1.3945, d 0.873.** Purified in the same way as *n*-butyl chloride.

Chlorocresol see **chloromethylphenol.**

Chlorocyclohexane *[542-18-7]* **M 118.6, b 142.5°, n^{25} 1.46265, d 1.00.** Washed several times with dil NaHCO$_3$, then repeatedly with distd water. Dried with CaCl$_2$ and fractionally distd.

2-Chloro-1,4-dihydroxybenzene see **chloroquinol.**

4-Chloro-3,5-dimethylphenol *[88-04-0]* **M 156.6, m 115.5°.** Crystd from benzene or toluene.

1-Chloro-2,4-dinitrobenzene *[97-00-7]* **M 202.6, m 51°.** Usually crystd from EtOH or MeOH. Has also been crystd from ethyl ether, benzene or isopropyl alcohol. A preliminary purification step has been to pass its soln in benzene through an alumina column. Also purified by zone refining.

4-Chloro-3,5-dinitrobenzoic acid *[118-97-8]* **M 246.6, m 159-161°.** Crystd from EtOH/water, EtOH or benzene.

Chloroethane see **ethyl chloride.**

2-Chloroethanol *[107-07-3]* **M 80.5, b 51.0°/31mm, 128.6°/760mm, n^{15} 1.44380, d 1.201.** Dried with, then distd from, CaSO$_4$ in the presence of a little Na$_2$CO$_3$ to remove traces of acid.

2-Chloroethyl bromide *[107-04-0]* **M 143.4, b 106-108°.** Washed with conc H$_2$SO$_4$, water, 10% Na$_2$CO$_3$ soln, and again with water, then dried with CaCl$_2$ and fractionally distd before use.

2-Chloroethyl vinyl ether *[110-75-8]* **M 106.6, b 109°/760mm, n 1.437, d 1.048.** Washed repeatedly with equal vols of water made slightly alkaline with KOH, dried with sodium, and distd under vac.

Chloroform *[67-66-3]* **M 119.4, b 61.2°, n^{15} 1.44858, d^{15} 1.49845, d^{10} 1.47060.** Reacts slowly with oxygen or oxidising agents, when exposed to air and light, giving, mainly, phosgene, Cl$_2$ and HCl. Commercial CHCl$_3$ is usually stabilized by addn of up to 1% EtOH or of dimethylaminoazobenzene. Simplest purifications involve washing with water to remove the EtOH, drying with K$_2$CO$_3$ or CaCl$_2$, refluxing with P$_2$O$_5$, CaCl$_2$, CaSO$_4$ or Na$_2$SO$_4$, and distilling. **It must not be dried with sodium.** The distd CHCl$_3$ should be stored in the dark to avoid photochemical formation of phosgene. As an alternative purification, CHCl$_3$ can be shaken with several small portions of conc H$_2$SO$_4$, washed thoroughly with water, and dried with CaCl$_2$ or K$_2$CO$_3$ before filtering and distilling. EtOH can be removed from CHCl$_3$ by passage through a column of activated alumina, or through a column of silica gel 4-ft long by 1.75-in diameter at a flow rate of 3ml/min. (The column, which can hold about 8% of its weight of EtOH, is regenerated by air drying and then heating at 600° for 6hr. It is pre-purified by washing with CHCl$_3$, then EtOH, leaving in conc H$_2$SO$_4$ for about 8hr, washing with water until the washings are neutral, then air drying, followed by activation at 600° for 6hr. Just before use it is reheated for 2hr to 154°.) [McLaughlin, Kaniecki and Gray *AC* **30** 1517 *1958*.]
Carbonyl-containing impurities can be removed from CHCl$_3$ by percolation through a Celite column impregnated with 2,4-dinitrophenylhydrazine, phosphoric acid and water. (Prepared by dissolving 0.5g

DNPH in 6ml of 85% H_3PO_4 by grinding together, then mixing with 4ml of distd water and 10g of Celite.) [Schwartz and Parks *AC* 33 1396 *1961*.] Chloroform can be dried by distn from powdered type 4A Linde molecular sieves. For use as a solvent in IR spectroscopy, chloroform is washed with water (to remove EtOH), then dried for several hours over anhydrous $CaCl_2$ and fractionally distd. This treatment removes material absorbing near 1600 cm^{-1}. (Percolation through activated alumina increases this absorbing impurity.) [Goodspeed and Millson *Chem Ind* 1594 *1967*.]

Chlorogenic acid *[327-97-9]* **M 354.3, m 208°, $[\alpha]_D^{25}$ -36° (c 1, H_2O).** Crystd from water. Dried at 110°.

5-Chloro-8-hydroxy-7-iodoquinoline *[130-26-7]* **M 305.5, m 178-179°.** Crystd from abs EtOH.

p-**Chloroiodobenzene** *[637-87-6]* **M 238.5, m 53-54°.** Crystd from EtOH.

2,3-Chloromaleic anhydride *[1122-17-4]* **M 166.9, m 112-115°.** Purified by sublimation in vacuum [Katakis *et al, JCSDT 1491 1986*].

2-Chloro-3-methylindole (2-chloroskatole) *[51206-73-6]* **M 165.6, m 114.5-115.5°.** Purified by chromatography on silica gel in CH_2Cl_2/pet ether (1:2), followed by recrystn from aq EtOH or aq acetic acid. [Phillips and Cohen *JACS* 108 2023 *1986*.]

4-Chloro-2-methylphenol *[1570-64-5]* **M 142.6, m 49°.** Purified by zone melting.

4-Chloro-3-methylphenol *[59-50-76]* **M 142.6, m 66°.** Crystd from pet ether.

1-Chloro-2-methylpropane see **isobutyl chloride.**

2-Chloro-2-methylpropane see *tert*-**butyl chloride.**

Chloromycetin see **chloramphenicol.**

Chloromycetin palmitate see **chloramphenicol palmitate.**

1-Chloronaphthalene *[90-13-1]* **M 162.6, f.p. -2.3°, b 136-136.5°/20mm, 259.3°/760mm, n 1.6326, d 1.194.** Washed with dil $NaHCO_3$, then dried with Na_2SO_4 and fractionally distd under reduced pressure. Alternatively, before distn, it was passed through a column of activated alumina, or dried with $CaCl_2$, then distd from sodium. It can be further purified by fractional crystn by partial freezing or by crystn of its picrate to constant melting point (132-133°) from EtOH, and recovering from the picrate.

2-Chloronaphthalene *[91-58-7]* **M 162.6, m 61°, b 264-266°.** Crystd from 25% EtOH/water and dried under vac.

1-Chloro-2 naphthol *[633-99-8]* **M 178.6,**
2-Chloro-1-naphthol *[606-40-6]* **M 178.6,**
4-Chloro-1-naphthol *[604-44-4]* **M 178.6, m 120-121°.** Crystd from EtOH.

4-Chloro-2-nitroaniline *[89-63-4]* **M 172.6, m 116-116.5°.** Crystd from hot water or EtOH/water and dried for 10hr at 60° under vac.

2-Chloro-4-nitrobenzamide *[3011-89-0]* **M 200.6, m 172°.** Crystd from EtOH.

o-**Chloronitrobenzene** *[88-73-3]* **M 157.6, m 32.8-33.2°.** Crystd from EtOH, MeOH or pentane (charcoal).

m-Chloronitrobenzene   *[121-73-3]* **M 157.6, m 45.3-45.8°.** Crystd from MeOH or 95% EtOH (charcoal), then pentane.

p-Chloronitrobenzene   *[100-00-5]* **M 157.6, m 83.5-84°.** Crystd from 95% EtOH (charcoal).

1-Chloronitroethane   *[625-47-8]* **M 109.5, b 37-38°/20mm, n 1.4224, n^{25} 1.4235.** Dissolved in alkali, extracted with ether (discarded), then the aq phase was acidified with hydroxylamine hydrochloride, and the nitro compound fractionally distd under reduced pressure. [Pearson and Dillon *JACS* **75** 2439 *1953*.]

2-Chloro-5-nitropyridine   *[4548-45-2]* **M 158.5, m 108°.** Crystd from benzene or benzene/pet ether.

α-**Chloro-***m*-**nitrotoluene** see *m*-**nitrobenzyl chloride.**

1-Chloropentane see *n*-**amyl chloride.**

m-Chloroperbenzoic acid   *[937-14-4]* **M 172.6, m 92-94°(dec).** Recrystd from CH$_2$Cl$_2$ [Traylor and Mikztal *JACS* **109** 2770 *1987*]. The peracid was freed from *m*-chlorobenzoic acid by dissolving 50g/L of benzene and washing with an aq soln buffered at pH 7.4 (NaH$_2$PO$_4$/NaOH) (5 x 100ml). The organic layer was dried over MgSO$_4$ and carefully evaporated under vac. *Necessary care should be taken in case of* **EXPLOSION.** The soln was crystd twice from CH$_2$Cl$_2$/Et$_2$O and stored at 0°. [Bortolini *et al*, *JOC* **52** 5093 *1987*.]

o-Chlorophenol   *[95-57-8]* **M 128.6, m 8.8°, b 176°.** Passed at least twice through a gas chromatograph column.

m-Chlorophenol   *[108-43-0]* **M 128.6, m 33°, b 214°.** Could not be obtained solid by crystn from pet ether. Purified by distn under reduced pressure.

p-Chlorophenol   *[106-48-9]* **M 128.6, m 43°.** Distd, then crystd from pet ether (b 40-60°) or hexane, and dried under vac over P$_2$O$_5$ at room temp. [Bernasconi and Paschalis *JACS* **108** 2969 *1986*.]

Chlorophenol Red   *[4430-20-0]* **M 423.3, λ$_{max}$573nm.** Crystd from glacial acetic acid.

p-Chlorophenoxyacetic acid   *[122-88-3]* **M 186.6, m 157°,**
α-*p*-Chlorophenoxypropionic acid   *[3307-39-9]* **M 200.6, m 116°,**
β-*p*-Chlorophenoxypropionic acid   *[3284-79-5]* **M 200.6, m 138°.** Crystd from EtOH.

m-Chlorophenylacetic acid   *[1878-65-5]* **M 170.6, m 74°,**
p-Chlorophenylacetic acid   *[1878-66-6]* **M 170.6, m 106°.** Crystd from EtOH/water.

3-(*p*-**Chlorophenyl)-1,1-dimethylurea**   *[150-68-5]* **M 198.7, m 171°.** Crystd from MeOH.

o-Chlorophenyl diphenyl phosphate   *[115-85-5]* **M 360.7, b 236°/4mm, n^{25} 1.5707.** Purified by vac distn, percolated through a column of alumina, then passed through a packed column maintained by a countercurrent stream of nitrogen at reduced pressure [Dobry and Keller *JPC* **61** 1448 *1957*].

4-Chloro-1,2-phenylenediamine   *[95-83-0]* **M 142.6, m 69-70°.** Recrystd from pet. ether.

p-Chlorophenyl isocyanate   *[104-12-1]* **M 153.6, m 29-30°.** Crystd from pet ether (b 30-40°).

p-**Chlorophenyl** *o*-**nitrobenzyl ether, M 263.7, m 69°,**
p-**Chlorophenyl** *p*-**nitrobenzyl ether**   *[5442-44-4]* **M 263.7, m 102°.** Crystd from EtOH.

Chloropicrin *[76-06-2]* **M 164.5, b 112°**. Dried with $MgSO_4$ and fractionally distd. **EXTREMELY NEUROTOXIC, use appropriate precautions.**

1-Chloropropane see *n*-**propyl chloride.**

2-Chloropropane see **isopropyl chloride.**

Chloro-2-propanone see **chloroacetone.**

3-Chloropropene see **allyl chloride.**

α-Chloropropionic acid *[598-78-7]* **M 108.5, b 98°/3mm, n 1.4535, d 1.182.** Dried with P_2O_5 and fractionally distd under vac.

β-Chloropropionic acid *[107-94-8]* **M 108.5, m 41°.** Crystd from pet ether or benzene.

γ-Chloropropyl bromide *[109-70-6]* **M 157.5, b 142-145°, n^{25} 1.4732.** Washed with conc H_2SO_4, water, 10% Na_2CO_3 soln, water again and then dried with $CaCl_2$ and fractionally distd just before use [Akagi, Oae and Murakami *JACS* **78** 4034 *1956*].

6-Chloropurine *[87-42-3]* **M 154.6, m 179°(dec).** Crystd from water.

2-Chloropyridine *[109-09-1]* **M 113.6, b 49.0°/7mm, n 1.5322, d 1.20.** Dried with NaOH for several days, then distd from CaO under reduced pressure.

3-Chloropyridine *[626-60-8]* **M 113.6, b 148°, n 1.5304, d 1.194.** Distd from KOH pellets.

4-Chloropyridine *[626-61-9]* **M 113.6, b 85-86°/100mm, 147-148°/760mm.** Dissolved in distd water and excess of 6M NaOH was added to give *p*H 12. The organic phase was separated and extracted with four volumes of ethyl ether. The combined extracts were filtered through paper to remove water and the solvent evapd. The dark brown residual liquid was kept under high vacuum [Vaidya and Mathias *JACS* **108** 5514 *1986*]. It can be distd but readily darkens thus it is best kept as the hydrochloride *[7379-35-3]* **M 150.1, m 163-165°(dec).**

Chloroquinol *[615-67-8]* **M 144.5, m 106°.** Crystd from $CHCl_3$ or toluene.

2-Chloroquinoline *[612-62-4]* **M 163.6, m 34°, b 147-148°/15mm, n^{25} 1.62923, d^{35} 1.2351.** Purified by crystn of its picrate to constant melting point (123-124°) from benzene, regenerating the base and distilling under vac [Cumper, Redford and Vogel *JCS* 1183 *1962*]. 2-Chloroquinoline can be crystd from EtOH.

8-Chloroquinoline *[611-33-6]* **M 163.6, b 171-171.5°/24mm, n 1.64403, d 1.2780.** Purified by crystn of its $ZnCl_2$ complex (m 228°) from aq EtOH.

4-Chlororesorcinol *[95-88-5]* **M 144.6, m 105°.** Crystd from boiling CCl_4 (10g/L), after charcoal, and air dried.

5-Chlorosalicaldehyde *[635-93-8]* **M 156.6, m 98.5-99°.** Steam distd, then crystd from aq EtOH.

N-Chlorosuccinimide *[128-09-6]* **M 133.5, m 149-150°.** Rapidly crystd from benzene, or glacial acetic acid and washed well with water then dried *in vacuo*. [Phillips and Cohen *JACS* **108** 2023 *1986*.].

8-Chlorotheophylline *[85-18-7]* **M 214.6, m 311°(dec).** Crystd from water.

p-Chlorothiophenol *[106-54-7]* **M 144.6, m 51-52°.** Recrystd from aq EtOH [D'Sousa *et al, JOC* **52** 1720 *1987*].

α-Chlorotoluene see **benzyl chloride.**

o-Chlorotoluene *[95-49-8]* **M 126.6, b 159°, n 1.5255, d 1.083.** Dried for several days with $CaCl_2$, then distd from Na using a glass helices-packed column.

p-Chlorotoluene *[106-43-4]* **M 126.6, f.p. 7.2°, b 162.4°, n 1.5208, d 1.07.** Dried with BaO, fractionally distd, then fractionally crystd by partial freezing.

2-Chlorotriethylamine hydrochloride *[869-24-9]* **M 172.1, m 208-210°.** Crystd from abs MeOH (to remove highly coloured impurities).

Chlorotrifluoroethylene *[79-38-9]* **M 116.5, b -26 to -24°.** Scrubbed with 10% KOH soln, then 10% H_2SO_4 soln to remove inhibitors, and dried. Passed through silica gel.

Chlorotrifluoromethane *[75-72-9]* **M 104.5, m -180°, b -81.5°.** Main impurities were CO_2, O_2, and N_2. The CO_2 was removed by passage through satd aq KOH, followed by conc H_2SO_4. The O_2 was removed using a tower packed with activated copper on kieselguhr at 200°, and the gas dried over P_2O_5.

Chlorotriphenylmethane *[76-83-5]* **M 278.8, m 112-113°.** Crystd from benzene soln (100ml) containing a little acetyl chloride, by addn of 200ml of pet ether and cooling. Alternatively, a soln in ethyl ether was satd with dry HCl (by dripping conc HCl into conc H_2SO_4 and passing the gas through P_2O_5 towers) at 0°, then cooled in a Dry-ice/acetone bath. The crystals so obtained were recrystd from pet ether (b 30-60°) using Dry-ice/acetone baths [Thomas and Rochow *JACS* **79** 1843 *1957*].

4-Chloro-3,5-xylenol see **4-Chloro-3,5-dimethylphenol.**

Cholamine chloride hydrochloride see **(2-aminoethyl)trimethylammonium chloride hydrochloride.**

5-β-Cholanic acid *[546-18-9]* **M 360.6, m 164-165°, $[\alpha]^{14}$ +21.7° ($CHCl_3$).** Crystd from EtOH.

Cholanthrene *[479-23-2]* **M 254.3, m 173°.** Crystd from benzene/ethyl ether.

Cholestane *[481-21-0]* **M 372.7, m 80°, $[\alpha]_{546}$ +29.5° (c 2, $CHCl_3$).** Crystd from ethyl ether/EtOH.

5α-Cholestan-3β-ol *[80-97-7]* **M 388.7, m 142-143°(monohydrate), $[\alpha]_{546}$ +28° (c 1, $CHCl_3$), $[\alpha]_D$ +27.4° (in $CHCl_3$).** Crystd from EtOH or slightly aq EtOH. [Mizutani and Whitten *JACS* **107** 3621 *1985*.]

2-Cholestene *[102850-21-5]* **M 370.6, m 75-76°, $[\alpha]_D^{24}$ +64°.** Recrystd from MeOH or ethyl ether/acetone. [Berzbrester and Chandran *JACS* **109** 174 *1987*.]

Cholesterol *[57-88-5]* **M 386.7, m 148.9-149.4°, $[\alpha]_D^{25}$ -35° (hexane).** Crystd from ethyl acetate, EtOH or isopropyl ether/MeOH. [Hiromitsu and Kevan *JACS* **109** 4501 *1987*.] For extensive details of purification through the dibromide, see Fieser [*JACS* **75** 5421 *1953*] and Schwenk and Werthessen [*Arch Biochem Biophys* **40** 334 *1952*], and by repeated crystn from acetic acid, see Fieser [*JACS* **75** 4395 *1953*].

Cholesteryl acetate *[604-35-3]* **M 428.7, m 112-115°, $[\alpha]_{546}$ -51° (c 5, $CHCl_3$).** Crystd from *n*-pentanol.

Cholesteryl myristate *[1989-52-2]* **M 597.0.** Crystd from *n*-pentanol. Purified by column chromatography with MeOH and evapd to dryness. Dissolved in water and pptd with HCl (spot 1) or passed through a cation-exchange column (spot 2). Finally, dried in vac over P_2O_5. [Malanik and Malat *Anal Chim Acta* **76** 464 *1975*.]

Cholesteryl oleate *[303-43-5]* M 651.1, m 48.8-49.4°. Purified by chromatography on silica gel.

Cholic acid *[81-25-4]* M 408.6, m 198-200°, $[\alpha]_{546}$ +41° (c 0.6, EtOH). Crystd from EtOH. Dried under vac at 94°.

Choline chloride *[67-48-1]* M 139.6. *Extremely deliquescent*. Purity checked by $AgNO_3$ titration or by titration of free base after passage through an anion-exchange column. Crystd from absolute EtOH, or EtOH-ethyl ether, dried under vac and stored in a vac desiccator over P_2O_5 or $Mg(ClO_4)_2$.

Chromazurol S *[1667-99-8]* M 539.3, λ_{max} 540nm, ϵ 7.80 x 10^4 (10M HCl). Crude material (40g) is dissolved in water (250ml) and filtered. Then added conc HCl (50ml) to filtrate, with stirring. Ppte filtered off, washed with HCl (2M) and dried. Redissolved in water (250ml) and pptn repeated twice more in water bath at 70°. Then dried under vac over solid KOH (first) and P_2O_5 [Martynov *et al*, *Zh Analit Khim* **32** 519 *1977*].

Chromotropic acid *[148-25-4]* M 120.3. Crystd from water by addition of EtOH.

Chrysene *[218-01-9]* M 228.3, m 255-256°. Purified by chromatography on alumina from pet ether in a darkened room. Its soln in benzene was passed through the column of decolorizing charcoal, then crystd by concentration of the eluate. Also purified by crystn from benzene or benzene-pet ether , and by zone refining. [Gorman *et al*, *JACS* **107** 4404 *1985*.] It was freed from 5*H*-benzo[*b*]carbazole by dissolving in *N,N*-dimethylformamide and successively adding small portions of alkali and iodomethane until the fluorescent colour of the carbazole anion no longer appeared when alkali was added. The chrysene (and alkylated 5*H*-benzo[*b*]carbazole) separated on addition of water. Final purification was by crystn from ethylcyclohexane and from 2-methoxyethanol [Bender, Sawicki and Wilson *AC* **36** 1011 *1964*]. It can be sublimed in a vac.

γ-Chymotrypsin (*EC* 3.4.21.1) *[9004-07-3]*. Crystd twice from four-tenths satd ammonium sulphate soln, then dissolved in 1mM HCl and dialysed against 1mM HCl at 2-4°. The soln was stored at 2° [Lang, Frieden and Grunwald *JACS* **80** 4923 *1958*].

Cinchonidine *[485-71-2]* M 294.4, m 210.5°, $[\alpha]_{546}$ -127.5° (c 0.5, EtOH). Crystd from aq EtOH.

Cinchonine *[118-10-5]* M 294.4, m 265°, $[\alpha]_{546}$ +268° (c 0.5, EtOH). Crystd from EtOH or ethyl ether.

Cincophen see **2-phenyl-4-quinolinecarboxylic acid.**

1,8-Cineole *[478-82-6]* M 154.2, f.p. 1.3°, b 176.0°, d 0.9251. Purified by dilution with an equal vol of pet ether, then satn with dry HBr. The ppte was filtered off, washed with small portions of pet ether, and then cineole was regenerated by stirring the crystals with water. It can also be purified through its *o*-cresol or resorcinol addition compounds. Stored with sodium until required.

trans-**Cinnamic acid** *[140-10-3]* M 148.2, m 134.5-135°. Crystd from benzene, CCl_4, hot water, water/EtOH (3:1), or 20% aq EtOH. Dried at 60° under vac.

Cinnamic anhydride *[538-56-7]* M 278.4, m 136°. Crystd from benzene or toluene/pet ether (b 60-80°).

N-**Cinnamoyl-*N*-phenylhydroxylamine** *[7369-44-0]* M 239.3, m 158-163°. Recrystd from EtOH.

Cinnamyl alcohol *[104-54-1]* M 134.2, m 33°, b 143.5°/14mm, λ_{max} 251nm (ε 18,180). Crystd from ethyl ether/pentane.

Cinnoline *[253-66-7]* M 130.2, m 38°. Crystd from pet ether. Kept under N_2 in sealed tubes in the dark at 0°.

Citraconic acid *[498-23-7]* M 130.1, m 91°. Steam distd and crystd from EtOH/ligroin.

Citranaxanthin *[3604-90-8]* M 456.7, m 155-156°, $\varepsilon_{1cm}^{1\%}$ 410 (349nm), 275 (466nm) in hexane. Purified by chromatography on a column of 1:1 magnesia-HyfloSupercel. Crystd from pet ether. Stored in the dark, under inert atmosphere, at 0°.

Citric acid (H_2O) *[5949-29-1]* M 210.1, m 156-157°, 153° (anhyd). Crystd from water.

S-Citrulline (2-amino-5-ureidopentanoic acid) *[2436-90-0]* M 175.2, m 222°, $[\alpha]_D$ +24.2° (in 5M HCl). Likely impurities are arginine, and ornithine. Crystd from water by adding 5 vols of EtOH. Also crystd from water by addn of MeOH.

β-Cocaine *[50-36-2]* M 303.4, m 98°, $[\alpha]_D$ -15.8° ($CHCl_3$). Crystd from EtOH.

Cocarboxylase *[532-40-1]* M 416.8, m 195°(dec). Crystd from EtOH slightly acidified with HCl.

Cofazimine *[2030-63-9]* M 473.5. Recrystd from acetone.

Codeine *[76-57-3]* M 299.4, m 154-156°, $[\alpha]_D$ -138° (in EtOH). Crystd from water or aq EtOH. Dried at 80°.

Colchicine *[64-86-8]* M 399.5, m 155-157°(dec), $[\alpha]_{546}$ -570° (c 1, H_2O). Commercial material contains up to 4% desmethylcolchicine. Purified by chromatography on alumina, eluting with $CHCl_3$ [Ashley and Harris *JCS* 677 *1944*]. Alternatively, an acetone soln on alkali-free alumina has been used, and eluting with acetone [Nicholls and Tarbell *JACS* **75** 1104 *1953*].

Colchicoside *[477-29-2]* M 547.5, m 216-218°. Crystd from EtOH.

2,4,6-Collidine see **2,4,6-trimethylpyridine**.

Conessine *[546-06-5]* M 356.6, m 125° $[\alpha]_D$ -1.9° (in $CHCl_3$), +25.3° (in EtOH). Crystd from acetone.

Congo Red *[573-58-0]* M 696.7, λ_{max} 497nm. Crystd from aq EtOH (1:3). Dried in air.

Convallatoxin *[508-75-8]* M 550.6, m 238-239°. Crystd from ethyl acetate.

Coproporphyrin I *[531-14-6]* M 654.7, λmax 591, 548, 401nm in 10% HCl. Crystd from pyridine/glacial acetic acid.

Coprostane (5α-cholestane) *[481-20-0]* M 372.7, m 72°. Crystd from EtOH.

4,5-Coprosten-3-ol (cholest-4-ene-3β-ol) *[517-10-2]* M 386.7, m 132°. Crystd from MeOH/ethyl ether.

Coprosterol (cholestan-3β-ol) *[80-97-7]* M 388.7, m 101°. Crystd from MeOH.

Coronene *[191-07-1]* M 300.4, m 438-440°, λ_{max} 345nm (log ε 4.07). Crystd from benzene or toluene, then sublimed in vac.

Cortisol *[50-23-7]* M 362.4, m 217-220°, λ_{max} 242nm (log ε 4.20). Crystd from EtOH.

Corticosterone *[50-22-6]* M 346.5, m 180-182°, λ_{max} 240nm. Crystd from acetone or EtOH.

Cortisone *[53-06-5]* M 360.5, m 230-231°, $[\alpha]_{546}$ +225° (c 1, in EtOH). Crystd from 95% EtOH or acetone.

Cortisone-21-acetate *[50-04-4]* M 402.5, m 242-243°, $[\alpha]_{546}$ +227° (c 1, in $CHCl_3$). Crystd from acetone.

Coumalic acid *[500-05-0]* M 140.1, m 205-210°(dec). Crystd from MeOH.

4-Coumaric acid see *p*-hydroxycinnamic acid.

Coumarilic acid see **benzofuran-2-carboxylic acid.**

Coumarin *[91-64-5]* M 146.2, m 68-69°, b 298°. Crystd from ethanol or water and sublimed *in vacuo* at 43° [Srinivasan and deLevie *JPC* **91** 2904 *1987*].

Coumarin-3-carboxylic acid *[531-81-7]* M 190.2, m 188°(dec). Crystd from water.

Creatine (H_2O) *[6020-87-7]* M 131.1, m 303°. Likely impurities are creatinine and other guanidino compounds. Crystd from water as monohydrate. Dried under vac over P_2O_5 to give anhydrous material.

Creatinine *[60-27-5]* M 113.1, m 260°(dec). Likely impurities are creatine and ammonium chloride. Dissolved in dil HCl, then neutralised by adding ammonia. Recrystd from water by adding excess of acetone.

o-**Cresol** *[95-48-7]* M 108.1, m 30.9°, b 191°, n^{41} 1.53610, n^{46} 1.53362. Can be freed from *m*- and *p*-isomers by repeated fractional distn, Crystd from benzene by addn of pet ether. Fractional crystd by partial freezing of its melt.

m-**Cresol** *[108-39-4]* M 108.1, f.p. 12.0°, b 202.7°, n 1.5438, d 1.034. Separation of the *m* and *p*-cresols requires chemical methods, such as conversion to their sulphonates [Brüchner *AC* **75** 289 *1928*]. An equal volume of H_2SO_4 is added to *m*-cresol, stirred with a glass rod until soln is complete. Heat for 3hr at 103-105°. Dilute carefully with 1-1.5 vols of water, heat to boiling point and steam distil until all unsulphonated cresol has been removed. Cool and extract residue with ether. Evaporate the soln until the boiling point reaches 134° and steam distil off the *m*-cresol. Another purification involves distn, fractional crystn from the melt, then redistn. Freed from *p*-cresol by soln in glacial acetic acid and bromination by about half of an equivalent amount of bromine in glacial acetic acid. The acetic acid was distd off, then fractional distn of the residue under vac gave bromocresols from which 4-bromo-*m*-cresol was obtained by crystn from hexane. Addn of the bromocresol in glacial acetic acid slowly to a reaction mixture of HI and red phosphorus or (more smoothly) of HI and hypophosphorus acid, in glacial acetic acid, at reflux, removed the bromine. After an hour, the soln was distd at atmospheric pressure until layers were formed. Then it was cooled and diluted with water. The cresol was extracted with ether, washed with water, $NaHCO_3$ soln and again with water, dried with a little $CaCl_2$ and distd [Baltzly, Ide and Phillips *JACS* **77** 2522 *1955*].

p-**Cresol** *[106-44-5]* M 108.1, m 34.8°, b 201.9°, n^{41} 1.53115, n^{46} 1.52870. Can be separated from *m*-cresol by fractional crystn of its melt. Purified by distn, by pptn from benzene soln with pet ether, and *via* its benzoate, as for phenol. Dried under vac over P_2O_5. Has also been crystd from pet ether (b 40-60°) and by conversion to sodium *p*-cresoxyacetate which, after crystn from water was decomposed by heating with HCl in an autoclave [Savard *Ann Chim (Paris)* **11** 287 *1929*].

o-Cresol Red *[1733-12-6]* **M 382.4, m 290°(dec).** Crystd from glacial acetic acid. Air dried. Dissolved in aq 5% $NaHCO_3$ soln and pptd from hot soln by dropwise addn of aq HCl. Repeated until extinction coefficients did not increase.

o-Cresotic acid (methylsalicylic acid) *[83-40-9]* **M 152.2, m 163-164°,**
m-Cresotic acid *[50-85-1]* **M 152.2, m 177°,**
p-Cresotic acid *[89-56-5]* **M 152.2, m 151°.** Crystd from water.

Crocetin diethyl ester *[5056-14-4]* **M 384.5, m 218-219°,** $\varepsilon_{1cm}^{1\%}$ **2340 (400nm), 3820 (422nm), 3850 (450nm) in pet ether.** Purified by chromatography on a column of silica gel G. Crystd from benzene. Stored in the dark, under an inert atmosphere, at 0°.

Crotonaldehyde *[127-73-9]* **M 70.1, b 104-105°.** Fractionally distd under N_2, through a short Vigreux column. Stored in sealed ampoules.

trans-**Crotonic acid** *[3724-65-0]* **M 86.1, m 72-72.5°.** Distd under reduced pressure. Crystd from pet ether (b 60-80°) or water, or by partial freezing of the melt.

(E)- and (Z)-Crotonitrile (mixture) *[4786-20-3]* **M 67.1, b 120-121°, n 1.4595, d 1.091.** Separated by preparative glc on a column using 5% FFAP on Chromosorb G. [Lewis *et al, JACS* **108** 2818 *1986*.]

Crotyl bromide *[29576-14-5]* **M 135.0, b 103-105°/740mm, n^{25} 1.4792.** Dried with $MgSO_4$, $CaCO_3$ mixture. Fractionally distd through an all-glass Todd column.

15-Crown-5 *[33100-27-5]* **M 220.3, b 93-96°/0.1mm, n 1.465, d 1.113.** Dried over 3A molecular sieves.

18-Crown-6 *[17455-13-9]* **M 264.3, m 37-39°.** Recrystd from acetonitrile and vac dried. Purified by pptn of 18-crown-6/nitromethane 1:2 complex with Et_2O/nitromethane (10:1 mixture). The complex is decomposed in vacuum and distilled under reduced pressure. Also recrytstd from acetonitrile and vac dried.

Cryptopine *[482-74-6]* **M 369.4, m 220-221°.** Crystd from benzene.

Cryptoxanthin *[472-70-8]* **M 552.9,** $\varepsilon_{1cm}^{1\%}$ **2370 (452nm), 2080 (480nm) in pet ether.** Purified by chromatography on MgO, $CaCO_3$ or deactivated alumina, using EtOH or ethyl ether to develop the column. Crystd from $CHCl_3$/EtOH. Stored in the dark, under inert atmosphere, at -20°.

Crystal Violet Chloride *[548-62-9]* **M 408.0.** Crystd from water (20ml/g), the crystals being separated from the chilled soln by centrifugation, then washed with chilled EtOH and ethyl ether and dried under vac. The carbinol was pptd from an aq soln of the dye, using excess NaOH, then dissolved in HCl and recrystd from water as the chloride [Turgeon and La Mer *JACS* **74** 5988 *1952*].

Cumene *[98-82-8]* **M 120.2, b 69-70°/41mm, 152.4°/760mm, n 1.49146, n^{25} 1.48892, d 0.864.** Usual purification is by washing with several small portions of conc H_2SO_4 (until the acid layer is no longer coloured), then with water, 10% aq Na_2CO_3, again with water, and drying with $MgSO_4$, $MgCO_3$ or Na_2SO_4, followed by fractional distn. It can then be dried with, and distd from, Na, NaH or CaH_2. Passage through columns of alumina or silica gel removes oxidation products. Has also been steam distd from 3% NaOH, and azeotropically distd with 2-ethoxyethanol (which was subsequently removed by washing out with water).

Cumene hydroperoxide *[80-15-9]* **M 152.2, b 60°/0.2mm, n^{24} 1.5232, d 1.028.** Purified by adding 100ml of 70% material slowly and with agitation to 300ml of 25% NaOH in water, keeping the temperature below 30°. The resulting crystals of the sodium salt were filtered off, washed twice with 25 ml portions of benzene, then stirred with 100ml of benzene for 20min. After filtering off the crystals and repeating the washing, they were suspended in 100ml of distd water and the *p*H was adjusted to 7.5

by addn of 4M HCl. The free hydroperoxide was extracted into two 20ml portions of *n*-hexane, and the solvent was evapd under vac at room temperature, the last traces being removed at 40-50° and 1mm [Fordham and Williams *Canad J Res* 27B 943 *1949*]. Petroleum ether, **but not ethyl ether**, can be used instead of benzene, and powdered solid CO_2 can replace the 4M HCl. *The material is potentially explosive.*

Cupferron *[135-20-6]* M 155.2, m 162.5-163.5°. Crystd from EtOH (charcoal), washed with ethyl ether and air dried.

Cuprein *[524-63-0]* M 310.4, m 202°, $[\alpha]_D^7$ -176° (in MeOH). Crystd from EtOH.

Cuproin see **2,2'-biquinolyl.**

Curcumin *[458-37-7]* M 368.4, m 183°. Crystd from EtOH or acetic acid.

Cyanamide *[420-04-2]* M 42.0, m 41°. Crystd from ethyl ether, then vac distd at 80° (370mm). Hygroscopic.

Cyanoacetamide *[107-91-5]* M 84.1, m 119.4°. Crystd from MeOH/dioxane (6:4), then water. Dried over P_2O_5 under vac.

Cyanoacetic acid *[372-09-8]* M 85.1, m 70.9-71.1°. Crystd to constant melting point from benzene/acetone (2:3), and dried over silica gel.

Cyanoacetic acid hydrazide *[140-87-4]* M 99.1, m 114.5-115°. Crystd from EtOH.

4-Cyanoacetophenone see **4-acetylbenzonitrile.**

p-**Cyanoaniline** *[873-74-5]* M 118.1, m 85-87°. Crystd from water, or EtOH, and dried in a vac for 6hr at 40°. [Edidin *et al, JACS* 109 3945 *1987.*]

9-Cyanoanthracene *[1210-12-4]* M 203.2, m 134-137°. Purified by crystn from EtOH or toluene, and vac sublimed in the dark and in an inert atmosphere [Ebied *et al JCSFT 1* 76 2170 *1980*; Kikuchi *et al, JPC* 91 574 *1987*].

9-Cyanoanthracene photodimer *[33998-38-8]* M 406.4. Purified by dissolving in the minimum amount of $CHCl_3$ followed by addn of EtOH [Ebied *et al JCSFT 1* 75 1111 *1979*; 76 2170 *1980*].

p-**Cyanobenzoic acid** *[619-65-8]* M 147.1, m 219°. Crystd from water.

Cyanoguanidine *[461-58-5]* M 84.1, m 209.5°. Crystd from water or EtOH.

p-**Cyanophenol** *[767-00-0]* M 119.1, m 113°. Crystd from pet ether, benzene or water and kept under vac over P_2O_5. [Bernasconi and Paschelis *JACS* 108 2969 *1986*.]

3-Cyanopyridine *[100-54-9]* M 104.1, m 50°. Crystd to constant melting point from *o*-xylene/hexane.

4-Cyanopyridine *[100-48-1]* M 104.1, m 76-79°. Crystd from dichloromethane/ethyl ether mixture.

Cyanuric acid *[108-80-5]* M 120.1, m >300°. Crystd from water. Dried at room temperature in a desiccator under vac.

Cyanuric chloride *[108-77-0]* M 184.4, m 154°. Crystd from CCl_4 or pet ether (b 90-100°), and dried under vac. Recrystd twice from anhydrous benzene immediately before use [Abuchowski *et al, JBC* 252 3582 *1977*].

Cyclobutane-1,1-dicarboxylic acid *[5445-51-2]* **M 144.1, m 158°.** Crystd from ethyl acetate.

trans-**Cyclobutane-1,2-dicarboxylic acid** *[1124-13-6]* **M 144.1, m 131°.** Crystd from benzene.

Cyclobutanone *[1191-95-3]* **M 70.1, b 96-97°, n^{25} 1.4189, d 0.931.** Treated with dil aq KMnO$_4$, dried with molecular sieves and fractionally distd. Purified *via* the semicarbazone, then regenerated, dried with CaSO$_4$, and distd in a spinning-band column. Alternatively, purified by preparative gas chromatography using a Carbowax 20-M column at 80°. (This treatment removes acetone.)

Cyclodecanone *[1502-06-3]* **M 154.2, m 21-24°, b 100-102°/12mm.** Purified by sublimation.

α-Cyclodextrin (H$_2$O) *[10016-20-3]* **M 972.9, m >280°(dec), [α]$_{546}$ +175° (c 10, H$_2$O).** Recrystd from 60% aq EtOH, then twice from water, and dried for 12hr in vac at 80°. Also purified by pptn from water with 1,1,2-trichloroethylene. The ppte was isolated, washed and resuspended in water. This was boiled to steam distil the trichloroethylene. The soln was freeze-dried to recover the cyclodextrin. [Armstrong *et al*, *JACS* **108** 1418 *1986*.]

β-Cyclodextrin (H$_2$O) *[7585-39-9]* **M 1135.0, m >300°(dec), [α]$_{546}$ +170° (c 10, H$_2$O).** Recrystd from water and dried for 12hr in vac at 110°. The purity was assessed by tlc on cellulose with a fluorescent indicator. [Taguchi, *JACS* **108** 2705 *1986*; Tabushi *et al*, *JACS* **108** 4514 *1986*.]

Cycloheptane *[291-64-5]* **M 98.2, b 114.4°, n 1.4588, d 0.812.** Distd from sodium, under nitrogen.

Cycloheptanone *[502-42-1]* **M 112.2, b 105°/80mm, 172.5°/760, n^{24} 1.4607, d 0.952.** Shaken with aq KMnO$_4$ to remove material absorbing around 230-240nm, then dried with Linde type 13X molecular sieves and fractionally distd.

Cycloheptatriene *[544-25-2]* **M 92.1. b 114-115°, n 1.522, d 0.895.** Washed with alkali, then fractionally distd.

Cycloheptimidazole see **1,3-diaza-azulene.**

1,3-Cyclohexadiene *[592-57-4]* **M 80.1, b 83-84°, n 1.4707, d 0.840.** Distd from NaBH$_4$.

Cyclohexane *[110-82-7]* **M 84.2, f.p. 6.6°, b 80.7°, n 1.42623, n^{25} 1.42354, d^{24} 0.77410.** Commonly, washed with conc H$_2$SO$_4$ until the washings are colourless, followed by water, aq Na$_2$CO$_3$ or 5% NaOH, and again with water until neutral. It is next dried with P$_2$O$_5$, Linde type 4A molecular sieves, CaCl$_2$, or MgSO$_4$ then Na and distd. Cyclohexane has been refluxed with, and distd from Na, CaH$_2$, LiAlH$_4$ (which also removes peroxides), sodium/potassium alloy, or P$_2$O$_5$. Traces of benzene can be removed by passage through a column of silica gel that has been freshly heated: this gives material suitable for ultraviolet and infrared spectroscopy. If there is much benzene in the cyclohexane, most of it can be removed by a preliminary treatment with nitrating acid (a cold mixture of 30ml conc HNO$_3$ and 70ml of conc H$_2$SO$_4$) which converts benzene into nitrobenzene. The impure cyclohexane and the nitrating acid are placed in an ice bath and stirred vigorously for 15min, after which the mixture is allowed to warm to 25° during 1hr. The cyclohexane is decanted, washed several times with 25% NaOH, then water dried with CaCl$_2$, and distd from sodium. Carbonyl-containing impurities can be removed as described for chloroform. Other purification procedures include passage through columns of activated alumina and repeated crystn by partial freezing. Small quantities may be purified by chromatography on a Dowex 710-Chromosorb W gas-liquid chromatographic column.

Cyclohexane-1,2-diaminetetraacetic acid (H$_2$O; CDTA) *[13291-61-7]* **M 364.4.** Dissolved in aq NaOH as its disodium salt, then pptd by adding HCl. The free acid was filtered off and boiled with distd water to remove traces of HCl [Bond and Jones *TFS* **55** 1310 *1959*]. Recrystd from water and dried under vac.

trans-1,2-Cyclohexanediol *[1460-57-7]* M 116.2, m 104°. Crystd from acetone and dried at 50° for several days.

cis-1,3-Cyclohexanediol *[931-17-9]* M 116.2, m 86°. Crystd from ethyl acetate and acetone.

trans-1,3-Cyclohexanediol *[5515-64-0]* M 116.2, m 117°. Crystd from ethyl acetate.

cis-1,4-Cyclohexanediol *[556-58-9]* M 116.2, m 102.5°. Crystd from acetone (charcoal), then dried and sublimed under vac.

Cyclohexane-1,3-dione *[504-02-9]* M 112.1, m 107-108°. Crystd from benzene.

Cyclohexane-1,4-dione *[637-88-7]* M 112.1, m78°. Crystd from water, then benzene.

Cyclohexane-1,2-dione dioxime (Nioxime) *[492-99-9]* M 142.2, m 189-190°. Crystd from alcohol/water and dried in vac at 40°.

Cyclohexanol *[108-93-0]* M 100.2, m 25.2°, b 161.1°, n 1.466, n^{25} 1.4365, n^{30} 1.4629, d 0.9459. Refluxed with freshly ignited CaO, or dried with Na_2CO_3, then fractionally distd. Redistd from Na. Further purified by fractional crystn from the melt in dry air. Peroxides and aldehydes can be removed by prior washing with ferrous sulphate and water, followed by distillation under nitrogen from 2,4-dinitrophenylhydrazine, using a short fractionating column: water distils as the azeotrope. Dry cyclohexanol is *very hygroscopic*.

Cyclohexanone *[108-94-1]* M 98.2, f.p. -16.4°, b 155.7°, n^{15} 1.45203. n 1.45097, d 0.947. Dried with $MgSO_4$, $CaSO_4$, Na_2SO_4 or Linde type 13X molecular sieves, then distd. Cyclohexanol and other oxidisable impurities can be removed by treatment with chromic acid or dil $KMnO_4$. More thorough purification is possible by conversion to the bisulphite addition compound, or the semicarbazone, followed by decompn with Na_2CO_3 and steam distn. [For example, equal weights of the bisulphite adduct (crystd from water) and Na_2CO_3 are dissolved in hot water and, after steam distn, the distillate is saturated with NaCl and extracted with benzene which is then dried and the solvent evapd prior to further distn.]

Cyclohexanone oxime *[100-64-1]* M 113.2, m 90°. Crystd from water or pet ether (b 60-80°).

Cyclohexanone phenylhydrazone *[946-82-7]* M 173.3, m 77°. Crystd from EtOH.

Cyclohexene *[110-83-8]* M 82.2, b 83°, n 1.4464, n^{25} 1.4437, d 0.810. Freed from peroxides by washing with successive portions of dil acidified ferrous sulphate, or with $NaHSO_3$ soln then with distd water, dried with $CaCl_2$ or $CaSO_4$, and distd under N_2. Alternative methods of removing peroxides include passage through a column of alumina, refluxing with sodium wire or cupric stearate (then distg from sodium). Diene is removed by refluxing with maleic anhydride before distg under vac. Treatment with 0.1moles of MeMgI in 40ml of ethyl ether removes traces of oxygenated impurities. Other purification procedures include washing with aq NaOH, drying and distg under N_2 through a spinning band column; redistg from CaH_2; storage with sodium wire; and passage through a column of alumina, under N_2, immediately before use. Stored in a refrigerator under argon. [Woon *et al*, *JACS* 108 7990 1986; Wong *et al*, *JACS* 109 3428 1987.]

Cycloheximide *[68-81-9]* M 281.4, m 119.5-121°. Crystd from water/MeOH (4:1), amyl acetate, isopropyl acetate/isopropyl ether or water.

Cyclohexylamine *[108-91-8]* M 99.2, b 134.5°, n 1.45926, n^{25} 1.4565, d 0.866, d^{25} 0.8625. Dried with $CaCl_2$ or $LiAlH_4$, then distd from BaO, KOH or Na, under N_2. Also purified by conversion to the hydrochloride, several crystns from water, then liberation of the amine with alkali and fractional distn under N_2.

Cyclohexylbenzene *[827-52-1]* **M 160.3, f.p. 6.8°, b 237-239°, n 1.5258, d 0.950.** Purified by fractional distn, and fractional freezing.

Cyclohexyl bromide *[108-85-0]* **M 156.3, b 72°/29mm, n^{25} 1.4935, d 0.902.** Shaken with 60% aq HBr to remove the free alcohol. After separation from the excess HBr, the sample was dried and fractionally distd.

Cyclohexyl chloride *[542-18-7]* **M 118.6, b 142-142.5°, n 1.462, d 1.000.** Dried with $CaCl_2$ and distd.

Cyclohexylidene fulvene *[3141-04-6]* **M 134.2.** Purified by column chromatography and eluted with *n*-hexane [Abboud *et al, JACS* **109** 1334 *1987*].

Cyclohexyl methacrylate *[101-43-9]* **168.2, b 81-86°/0.1mm, n 1.458, d 0.964.** Purification as for methyl methacrylate.

1-Cyclohexyl-5-methyltetrazole *[7707-57-5]* **M 166.2, m 124-124.5°.** Crystd from abs EtOH, then sublimed at 115°/3mm.

Cycloleucine see **1-amino-1-cyclopentanecarboxylic acid.**

Cyclononanone *[3350-30-9]* **M 140.2, m 142.0-142.8°, b 220-222°.** Repeatedly sublimed at 0.05-0.1mm pressure.

Cyclooctadiene *[29965-97-7]* **M 108.2; [1,3-cyclooctadiene,** *1700-10-3,* **M 108.2, b 142-144°/760mm, n 1.494, d 0.873],[cis-cis-1,5-cyclooctadiene,** *1552-12-1,* **M 108.2, n 1.494, d 0.880.].** Purified by glc.

Cyclooctanone *[502-49-8]* **M 126.2, m 42°.** Purified by sublimation after drying with Linde type 13X molecular sieves.

1,3,5,7-Cyclooctatetraene *[629-20-9]* **M 104.2, b 141-141.5°, n^{25} 1.5350. d 1.537.** Purified by shaking 3ml with 20ml of 10% aq $AgNO_3$ for 15min, then filtering off the silver nitrate complex as a ppte. The ppte was dissolved in water and added to cold conc ammonia to regenerate the cyclooctatetraene which was fractionally distd under vac onto molecular sieves and stored at 0°. It was passed through a dry alumina column before use [Broadley *et al, JSCDT* 373 *1986*.]

*cis***-Cyclooctene** *[931-87-3]* **M 110.2, b 32-34°/12mm, n 1.470, d 0.848.** The *cis*-isomer was freed from the *trans*-isomer by fractional distn through a spinning-band column, followed by preparative gas chromatography on a Dowex 710-Chromosorb W glc column. It was passed through a short alumina column immediately before use [Collman *et al, JACS* **108** 2588 *1986*.] It has also been distd in a dry nitrogen glove box from powdered fused NaOH through a Vigreux column and then passed through activated neutral alumina before use [Wong *et al, JACS* **109** 4328 *1987*].

Cyclopentadecanone *[502-72-7]* **M 224.4, m 63°.** Sublimation is better than crystn from aq EtOH.

Cyclopentadiene *[542-92-7]* **M 66.1, b 41-42°.** Dried with $Mg(ClO_4)_2$ and distd.

Cyclopentane *[287-92-3]* **M 70.1, b 49.3°, n 1.40645, n^{25} 1.4340, d 0.745.** Freed from cyclopentene by two passages through a column of carefully dried and degassed activated silica gel.

Cyclopentane-1,1-dicarboxylic acid *[5802-65-3]* **M 158.1, m 184°.** Recrystd from water.

Cyclopentanone *[120-92-3]* **M 84.1, b 130-130.5°, n 1.4370, n²⁵ 1.4340, d 0.947.** Shaken with aq KMnO₄ to remove materials absorbing around 230 to 240nm. Dried with Linde type 13X molecular sieves and fractionally distd. Has also been purified by conversion to the NaHSO₃ adduct which, after crystallising four times from EtOH/water (4:1), was decomposed by adding to an equal weight of Na₂CO₃ in hot water. The free cyclopentanone was steam distd from the soln. The distillate was satd with NaCl and extracted with benzene which was then dried and evapd; the residue was distd [Allen, Ellington and Meakins *JCS* 1909 *1960*].

Cyclopentene *[142-29-0]* **M 68.1, b 45-46°, n 1.4228, d 0.772.** Freed from hydroperoxide by refluxing with cupric stearate. Fractionally distd from Na. Chromatographed on a Dowex 710-Chromosorb W glc column. Methods for **cyclohexene** should be applicable here. Also washed with 1M NaOH soln followed by water. It was dried over anhydrous Na₂SO₄, distd over powdered NaOH under nitrogen, and passed through neutral alumina before use [Woon *et al*, *JACS* **108** 7990 *1986*.] It was distd in a dry nitrogen atmosphere from powdered fused NaOH through a Vigreux column, and then passed through activated neutral alumina before use [Wong *et al*, *JACS* **109** 3428 *1987*].

Cyclopropane *[75-19-4]* **M 42.1, b -34°.** Washed with a soln of HgSO₄, and dried with CaCl₂, then Mg(ClO₄)₂.

Cyclopropane-1,1-dicarboxylic acid *[598-10-7]* **M 130.1, m 140°.** Recrystd from CHCl₃.

Cyclopropyldiphenylcarbinol *[5785-66-0]* **M 224.3, m 86-87°.** Crystd from *n*-heptane.

Cyclopropyl methyl ketone *[765-43-5]* **M 84.1, b 111.6-111.8°/752mm, n 1.4242, d 0.850.** Stored with anhydrous CaSO₄, distd under nitrogen. Redistd under vac.

R-Cycloserine see **R-4-amino-3-isoxazolidone.**

Cyclotetradecane *[295-17-0]* **M 192.3, m 56°.** Recrystd twice from aq EtOH then sublimed *in vacuo* [Dretloff *et al*, *JACS* **109** 7797 *1987*.]

Cyclotetradecanone *832-10-0]* **M 206.3, m 25°, b 145°/10mm, n 1.480, d 0.926.** It was converted to the semicarbazone which was recrystd from EtOH and reconverted to the free cyclotetradecanone by hydrolysis [Dretloff *et al*, *JACS* **109** 7797 *1987*.]

Cyclotrimethylenetrinitramine (RDX) *[121-82-4]* **M 222.2, m 203.8°(dec).** Crystd from acetone. **EXPLOSIVE.**

p-**Cymene** *[99-87-6]* **M 134.2, b 177.1°, n 1.4909, n²⁵ 1.4885, d 0.8569.** Washed with cold, conc H₂SO₄ until there is no further colour change, then repeatedly with water, 10% aq Na₂CO₃ and water again. Dried with Na₂SO₄, CaCl₂ or MgSO₄, and distd. Further purification steps include steam distn from 3% NaOH, percolation through silica gel or activated alumina, and a preliminary refluxing for several days over powdered sulphur.

(±)-**Cysteic acid** *[3024-83-7]* **M 169.2, m 260°(dec),**
S-Cysteic acid (H₂O) *[498-40-8]* **M 187.2, m 289°, [α]_D +8.7° (water).** Likely impurities are cystine and oxides of cysteine. Crystd from water by adding 2 vols of EtOH.

L-Cysteine hydrochloride (H₂O) *[52-89-1]* **M 157.6, m 175-178°(dec), [α]_D²⁵ +6.53° (5M HCl).** Likely impurities are cystine and tyrosine. Crystd from MeOH by adding ethyl ether, or from hot 20% HCl. Dried under vac over P₂O₅. Hygroscopic.

(±)-**Cysteine hydrochloride** *[10318-18-0]* **M 157.6.** Crystd from hot 20% HCl; dried under vac over P₂O₅.

L-Cystine *[56-89-3]* M 240.3, $[\alpha]_D^{18.5}$ -229° (c 0.92 in M HCl). Cystine disulphoxide was removed by treating an aq suspension with H_2S. The cystine was filtered off, washed with distd water and dried at 100° under vac over P_2O_5. Crystd by dissolving in 1.5M HCl, then adjusting to neutral with ammonia. Likely impurities are D-cystine, *meso*-cystine and tyrosine.

Cytidine *[65-46-3]* M 243.2, m 230°(dec), $[\alpha]_{546}$ +37° (c 9, H_2O). Crystd from 90% aq EtOH.

Cytisine *[485-35-8]* M 190.3, m 152-153°. Crystd from acetone.

Cytosine *[71-30-7]* M 111.1, m 320-325°(dec). Crystd from water.

DDT see **1,1,1-trichloro-2,2-bis(p-chlorophenyl)ethane.**

Decahydronaphthalene (mixed isomers) *[91-17-8]* M 138.2, b 191.7°, n 1.476, d 0.886. Stirred with conc H_2SO_4 for some hours. Then the organic phase was separated, washed with water, satd aq Na_2CO_3, again with water, dried with $CaSO_4$ or CaH_2 (and perhaps dried further with Na), filtered and distd under reduced pressure (b 63-70°/10mm). Also purified by repeated passage through long columns of silica gel previously activated at 200-250°, followed by distn from $LiAlH_4$ and storage under N_2. Type 4A molecular sieves can be used as a drying agent. Storage over silica gel removes water and other polar substances.

cis-**Decahydronaphthalene** *[493-01-6]* M 138.2, f.p. -43.2°, b 195.7°, n 1.48113, d 0.897,
trans-**Decahydronaphthalene** *[493-02-7]* M 138.2, f.p. -30.6°, b 187.3°, n 1.46968, d 0.870. Purification methods described for the mixed isomers are applicable. The individual isomers can be separated by very efficient fractional distn, followed by fractional crystn by partial freezing. The *cis*-isomer reacts preferentially with $AlCl_3$ and can be removed from the *trans*-isomer by stirring the mixture with a limited amount of $AlCl_3$ for 48hr at room temperature, filtering and distilling.

Decalin see **decahydronaphthalene.**

Decamethylene glycol see **decane-1,10-diol.**

n-**Decane** *[124-18-5]* M 142.3, b 174.1°, n 1.41189, n^{25} 1.40967, d 0.770. It can be purified by shaking with conc H_2SO_4, washing with water, aq $NaHCO_3$, and more water, then drying with $MgSO_4$, refluxing with Na and distilling. Passed through a column of silica gel or alumina. It can also be purified by azeotropic distn with 2-butoxyethanol, the alcohol being washed out of the distillate, using water; the decane is next dried and redistilled. It can be stored with NaH. Further purification can be achieved by preparative gas chromatography on a column packed with 30% SE-30 (General Electric methyl-silicone rubber) on 42/60 Chromosorb P at 150° and 40psig, using helium [Chu *JCP* **41** 226 *1964*]. Also purified by zone refining.

Decan-1,10-diol *[112-47-0]* M 174.3, m 72.5-74°. Crystd from dry ethylene dichloride.

Decane-1,10-dicarboxylic acid see **1,10-dodecanedioic acid.**

Decanoic acid see **capric acid.**

n-**Decanol** *[112-30-1]* M 158.3, f.p. 6.0°, b 110-119°/0.1mm, n 1.434, d 0.823. Fractionally distd in an all-glass unit at 10mm pressure (b 110°), then fractionally crystd by partial freezing. Also purified by preparative glc, and by passage through alumina before use.

Decyl alcohol see **decanol.**

n-**Decyl bromide** *[112-29-8]* M 221.2, b 117-118°/15.5mm, d 1.066. Shaken with H_2SO_4, washed with water, dried with K_2CO_3, and fractionally distd.

Decyltrimethylammonium bromide *[2082-84-0]* M 280.3. Crystd from 50% (v/v) EtOH/ethyl ether, or from acetone and washed with ether. Dried under vac at 60°. Also recrystd from EtOH and dried over silica gel. [Dearden and Wooley *JPC* **91** 2404 *1987*.]

Dehydro-L(+)-ascorbic acid *[490-83-5]* M 174.1, m 196°(dec), $[\alpha]_{546}$ +42.5° (c 1,in H_2O),
7-Dehydrocholesterol *[434-16-2]* M 384.7, m 142-143°. Crystd from MeOH.

Dehydrocholic acid *[81-23-2]* M 402.5, m 237°, $[\alpha]_{546}$ -159° (c 1, in $CHCl_3$). Crystd from acetone.

Dehydroepiandrosterone *[54-43-0]* **M 288.4, m 140-141° and 152-153° (dimorphic).** Crystd from MeOH and sublimed in vac.

Delphinine *[561-07-9]* **M 559.7, m 197-199°.** Crystd from EtOH.

3-Deoxy-D-allose *[6605-21-6]* **M 164.2, [α]$_D$ +8° (c 0.25 in H$_2$O).** Obtained from ethyl ether as a colourless syrup.

Deoxybenzoin *[451-40-1]* **M 196.3, m 60°, b 177°/12mm, 320°/760mm.** Crystd from EtOH.

Deoxycholic acid *[83-44-3]* **M 392.6, m 176°, [α]$_{546}$ +64° (c 1, EtOH).** Refluxed with CCl$_4$ (50ml/g), filtered, evapd under vac at 25°, recrystd from acetone and dried under vac at 155° [Trenner *et al JACS* **76** 1196 *1954*]. A soln of (cholic acid-free) material (100ml) in 500ml of hot EtOH was filtered, evapd to less than 500ml on a hot plate, and poured into 1500ml of cold ethyl ether. The ppte, filtered by suction, was crystd twice from 1-2 parts of abs EtOH, to give an alcoholate, **m 118-120°**, which was dissolved in EtOH (100ml for 60g) and poured into boiling water. After boiling for several hours the ppte was filtered off, dried, ground and dried to constant weight [Sobotka and Goldberg *BJ* **26** 555 *1932*]. Deoxycholic acid was also freed from fatty acids and cholic acid by silica gel chromatography by elution with 0.5% acetic acid in ethyl acetate [Tang *et al JACS* **107** 4058 *1985*].

11-Deoxycorticosterone *[64-85-7]* **M 330.5, m 141-142°, [α]$_D$ +178° and [α]$_{546}$ +223° (c 1, in EtOH).** Crystd from ethyl ether.

2-Deoxy-β-galactose *[1949-89-9]* **M 164.2, m 126-128°.** Crystd from ethyl ether.

2-Deoxy-α-D-glucose *[154-17-6]* **M 164.2, m 146°.** Crystd from MeOH/acetone.

2-Deoxy-β-L-ribose M 134.1, m 77°,
2-Deoxy-β-D-ribose *[533-67-5]* **M 134.1, m 86-87°.** Crystd from ethyl ether.

Desoxycholic acid see **deoxycholic acid.**

Desthiobiotin *[533-48-2]* **M 214.3, m 156-158°,**
Desyl bromide *[484-50-0]* **M 275.2, m 57.1-57.5°.** Crystd from 95% EtOH.

Dextrose see **D-glucose.**

Diacetamide *[625-77-4]* **M 101.1, m 75.5-76.5°, b 222-223°.** Purified by crystn from MeOH [Arnett and Harrelson *JACS* **109** 809 *1987*].

Diacetyl see **biacetyl.**

p-**Diacetylbenzene** *[1009-61-6]* **M 162.2, m 113-5-114.2°.** Crystd from benzene and vac dried over CaCl$_2$. Also dissolved in acetone, treated with Norit, evapd and recrystd from MeOH [Wagner *et al, JACS* **108** 7727 *1986*.]

Diadzein see **4',7'-dihydroxyisoflavone.**

Diamantane *[2292-79-7]* **M 188.3, m 234-235°.** Purified by repeated crystn from MeOH or pentane. Also dissolved in methylene dichloride, washed with 5% aq NaOH and water, and dried (MgSO$_4$). The soln was concentrated to a small vol, an equal weight of alumina was added, and the solvent evapd. The residue was placed on an activated alumina column (*ca* 4 x weight of diamantane) and eluted with pet ether (b 40-60°). Eight sublimations and twenty zone refining experiments gave material **m 251°** of 99.99% purity by differential analysis [*Tetrahedron lett* 3877 *1970*; *JCS(C)* 2691 *1972*].

3,6-Diaminoacridine hydrochloride *[952-23-8]* **M 245.7, m 270°(dec), ε$_{456}$ 4.3 x 10^4.** First purified by pptn of the free base by adding aq NH$_3$ soln to an aq soln of the hydrochloride or hydrogen sulphate, drying the ppte and subliming at 0.01mm Hg [Müller and Crothers *Eur J Biochem*, **54** 267 *1975*].

3,6-Diaminoacridine sulphate (proflavin sulphate) *[1811-28-5]* **M 516.6, λ$_{max}$ 456nm.** An aq soln, after treatment with charcoal, was concentrated, chilled overnight, filtered and the ppte, was rinsed with a little ethyl ether. The ppte was dried in air, then overnight in a vac oven at 70°.

1.3-Diaminoadamantane *[702-79-4]* **M 164.3, m 52°.** Purified by zone refining.

1,4-Diaminoanthraquinone *[128-95-0]* **M 238.3, m 268°.** Purified by thin-layer chromatography on silica gel using toluene/acetone (9:1) as eluent. The main band was scraped off and extracted with MeOH. The solvent was evapd and the quinone was dried in a drying pistol [Land, McAlpine, Sinclair and Truscott *JCSFT 1* **72** 2091 *1976*]. Crystd from EtOH in dark violet crystals.

1,5-Diaminoanthraquinone *[129-44-2]* **M 238.3, m 319°.** Recrystd from EtOH or acetic acid [Flom and Barbara *JPC* **89** 4481 *1985*].

2,6-Diaminoanthraquinone *[131-14-6]* **M 238.3, m 310-320°.** Crystd from pyridine. Column-chromatographed to remove a fluorescent impurity, then crystd from EtOH.

3,4-Diaminobenzoic acid *[619-05-6]* **M 152.2, m 213°(dec),**
3,5-Diaminobenzoic acid *[535-87-5]* **M 152.2, m 235-240°(dec).** Crystd from water.

3,3'-Diaminobenzidine tetrahydrochloride (2H$_2$O) *[7411-49-6]* **M 396.1, m >300°(dec).** Dissolved in water and pptd by adding conc HCl, then dried over solid NaOH.

4,4'-Diaminobiphenyl see **benzidine.**

1,4-Diaminobutane dihydrochloride (putrescine hydrochloride) *[333-93-7]* **M 161.1, m >290°.** Crystd from EtOH/water.

1,2-Diaminocyclohexanetetraacetic acid see **cyclohexane-1,2-diaminetetraacetic acid.**

1,2-Diamino-4,5-dichlorobenzene *[5348-42-5]* **177.0, m 163°.** Refluxed with activated charcoal in CH$_2$Cl$_2$, followed by recrystn from ethyl ether/pet ether or pet ether [Koolar and Kochi *JOC* **52** 4545 *1987*].

2,2'-Diaminodiethylamine (diethylenetriamine) *[111-40-0]* **M 103.2, b 208°, n 1.483, d 0.95.** Dried with Na and distd, preferably under reduced pressure, or in a stream of N$_2$.

4,4'-Diamino-3,3'-dinitrobiphenyl *[6271-79-0]* **M 274.2, m 275°.** Crystd from aq EtOH.

4,4'-Diaminodiphenylamine *[537-65-5]* **M 199.3, m 158°.** Crystd from water.

4,4'-Diaminodiphenylmethane *[101-77-9]* **M 198.3, m 91.6-92°.** Crystd from water or benzene.

3,3'-Diaminodipropylamine *[56-18-8]* **M 131.2, b 152°/50mm, n 1.481, d 0.938.** Dried with Na and distd under vac.

6,9-Diamino-2-ethoxyacridine *[442-16-0]* **M 257.3, m 226°.** Crystd from 50% EtOH.

2,7-diaminofluorene *[524-64-4]* **M 196.3, m 165°,**
2,4-Diamino-6-hydroxypyrimidine *[56-06-4]* **M 126.1, m 260-270°(dec),**

1,5-Diaminonaphthalene *[2243-62-1]* M 158.2, m 190°. Crystd from water.

1,8-Diaminonaphthalene *[479-27-6]* M 158.2, m 66.5°. Crystd from water or aq EtOH, and sublimed in a vac.

2,3-Diaminonaphthalene *[771-97-1]* M 158.2, m 199°. Crystd from water, or dissolved in 0.1M HCl, heated to 50°. After cooling, the soln was extracted with decalin to remove fluorescent impurities and centrifuged.

2,4-Diamino-5-phenylthiazole *[490-55-1]* M 191.3, m 163-164°(dec). Crystd from aq EtOH or water. Stored in the dark under N_2.

d,l-**2,6-Diaminopimelic acid** *[2577-62-0]* M 190.2, m 313-315°(dec). Crystd from water.

1,3-Diaminopropane dihydrochloride *[10517-44-9]* M 147.1, m 243°. Crystd from EtOH/water.

1,3-Diaminopropan-2-ol *[616-29-5]* M 90.1, m 38-40°. Dissolved in an equal amount of water, shaken with charcoal and vac distd at 68°/0.1mm. It is too viscous to be distd through a packed column.

2,3-Diaminopyridine *[452-58-4]* M 109.1, m 116°,
2,6-Diaminopyridine *[141-86-6]* M 109.1, m 121.5°. Crystd from benzene and sublimed in a vac.

3,4-Diaminopyridine *[54-96-6]* M 109.1, m 218-219°. Crystd from benzene and stored under H_2 because it is deliquescent and absorbs CO_2.

meso-**2,3-Diaminosuccinic acid** *[50817-04-4]* M 148.1, m 305-306°(dec, and sublimes). Crystd from water.

Diaminotoluene see **toluenediamine.**

3,5-Diamino-1,2,4-triazole *[1455-77-2]* M 99.1, m 206°. Crystd from water or EtOH.

Di-*n*-amyl *n*-amylphosphonate *[6418-56-0]* M 292.4. Purified by three crystns of its compound with uranyl nitrate from hexane. For method see *tributyl phosphate.*

2,5-Di-*tert*-amylhydroquinone *[79-74-3]* M 250.4, m 185.8-186.5°. Crystd under N_2 from boiling glacial acetic acid (7ml/g) plus boiling water (2.5ml/g) [Stolow and Bonaventura *JACS* **85** 3636 *1963*].

Di-n-amyl phthalate *[131-18-0]* M 306.4, b 204-206°/11mm. n 1.4885, d^{25} 1.0230. Washed with aq Na_2CO_3, then distd water. Dried with $CaCl_2$ and distd under reduced pressure. Stored in a vac desiccator over P_2O_5.

1,3-Diaza-azulene *[275-94-5]* M 130.1, m 120°. Recrystd repeatedly from deaerated cyclohexane in the dark.

1,4-Diazabicyclo[2.2.2]octane (Dabco, TED) see **triethylenediamine.**

1,8-Diazabiphenylene *[259-84-7]* M 154.2,
2,7-Diazabiphenylene *[31857-42-8]* M 154.2. Recrystd from cyclohexane, then sublimed in a vac.

Diazoaminobenzene *[27195-22-8]* M 197.2, m 99°. Crystd from pet ether (b 60-80°), 60% MeOH/water or 50% aq EtOH (charcoal) containing a small amount of KOH. Also purified by chromatography on alumina. Stored in the dark.

6-Diazo-5-oxo-L-norleucine *[157-03-9]* M 171.2, m 145-155°(dec). Crystd from EtOH.

Dibenzalacetone *[538-58-9]* **M 234.3, m 112°.** Crystd from hot ethyl acetate (2.5ml/g) or EtOH.

Dibenz[a,h]anthracene *[53-70-3]* **M 278.4, m 266-267°.** The yellow-green colour (due to other pentacyclic impurities) has been removed by crystn from benzene or by selective oxidation with lead tetraacetate in acetic acid [Moriconi *et al JACS* **82** 3441 *1960*].

Dibenzo-18-crown-6 *[14187-32-7]* **M 360.4, m 163-164°.** Crystd from benzene, *n*-heptane or toluene and dried under vacuum at room temperature for several days. [Szezygiel *JPC* **91** 1252 *1987*.]

Dibenzo-18-crown-8 *[14174-09-5]* **M 448.5, m 103-106°.** Recrystd from EtOH, and vac dried at 60° over P_2O_5 for 16hr. [Delville *et al, JACS* **109** 7293 *1987*.]

Dibenzofuran *[132-64-9]* **M 168.2, m 82.4°.** Dissolved in ethyl ether, then shaken with two portions of aq NaOH (2M), washed with water, separated and dried ($MgSO_4$). After evaporating the ether, dibenzofuran was crystd from aq 80% EtOH and dried under vac. [Cass *et al JCS* 1406 *1958*.] High purity material was obtained by zone refining.

Dibenzopyran (xanthene) *[92-83-1]* **M 182.2,m 100.5°, b 310-312°.** Crystd from 95% EtOH.

Dibenzothiophen *[132-65-0]* **M 184.3, m 99°.** Purified by chromatography on alumina with pet ether, in a darkened room. Crystd from water or EtOH.

Dibenzoylmethane *[120-46-7]* **M 224.3, m 80°.** Crystd from pet ether or MeOH.

Dibenzoyl peroxide see **benzoyl peroxide.**

Di-*O*-benzoyl-R-tartaric acid (H_2O) *[17026-42-5]* **M 376.3, $[\alpha]_{546}$+136° (c 2 in EtOH),**
Di-*O*-benzoyl-S-tartaric acid (H_2O) *[2743-38-6]* **M 376.3, $[\alpha]_{546}$-136° (c 2 in EtOH),**
Crystd from water as monohydrate, **m 88-89°,** and crystd from xylene as anhydrous acid, **m 173°.**

2,3,6,7-Dibenzphenanthrene *[222-93-5]* **M 276.3, m 257°.** Crystd from xylene.

Dibenzyl disulphide *[150-60-7]* **M 246.4, m 71-72°.** Crystd from EtOH.

Dibenzyl ketone *[102-04-5]* **M 210.3, m 34.0°.** Fractionally crystd from its melt, then crystd from pet ether. Stored in the dark.

Dibenzyl sulphide *[528-74-9]* **M 214.3, m 48.5°.** Crystd from EtOH/water (10:1), or repeatedly from purified hot ethyl ether. Vac dried at 30° over P_2O_5, fused under nitrogen and redried.

2,4'-Dibromoacetophenone see *p*-bromophenacyl bromide.

2,4-Dibromoaniline *[615-57-6]* **M 250.9, m 79-80°.** Crystd from aq EtOH.

9,10-Dibromoanthracene *[523-27-3]* **M 336.0, m 226°.** Recrystd from xylene and vac sublimed [Johnston *et al, JACS* **109** 1291 *1987*].

***p*-Dibromobenzene** *[106-37-6]* **M 235.9, m 87.8°.** Steam distd, crystd from EtOH or MeOH and dried in the dark under vac. Purified by zone melting.

2,5-Dibromobenzoic acid *[610-71-9]* **M 279.9, m 157°.** Crystd from water or EtOH.

4,4' Dibromobiphenyl *[92-86-4]* **M 312.0, m 164°, b 355-360°/760mm.** Crystd from MeOH.

trans-1,4-Dibromobut-2-ene *[821-06-7]* M 213.9, m 54°, b 85°/10mm. Crystd from ligroin.

Dibromodeoxybenzoin *[15023-99-1]* M 354.0, m 111.8-112.7°. Crystd from acetic acid.

Dibromodichloromethane *[594-18-3]* M 242.7, m 22°. Crystd repeatedly from its melt, after washing with aq $Na_2S_2O_3$ and drying with BaO.

1,2-Dibromoethane *[106-93-4]* M 187.9, f 10.0°, b 29.1°/10mm, 131.7°/760mm, n^{15} 1.54160, d 2.179. Washed with conc HCl or H_2SO_4, then water, aq $NaHCO_3$ or Na_2CO_3, more water, and dried with $CaCl_2$. Fractionally distd. Alternatively, kept in daylight with excess bromine for 2hr, then extracted with aq Na_2SO_3, washed with water, dried with $CaCl_2$, filtered and distd. It can also be purified by fractional crystn by partial freezing. Stored in the dark.

4',5'-Dibromofluorescein *[596-03-2]* M 490.1, m 285°. Crystd from aq 30% EtOH.

5,7-Dibromo-8-hydroxyquinoline *[521-74-4]* M 303.0, m 196°. Crystd from acetone/EtOH. It can be sublimed.

2,5-Dibromonitrobenzene *[3460-18-4]* M 280.9, m 84°. Crystd from acetone.

2,6-Dibromo-4-nitrophenol *[99-28-5]* M 280.9, m 143-144°. Crystd from aq EtOH.

2,4-Dibromophenol *[615-58-7]* M 251.9, m 37°. Crystd from $CHCl_3$ at -40°.

2,6-Dibromophenol *[608-33-3]* M 251.9, m 56-57°. Vac distd (at 18mm), then crystd from cold $CHCl_3$ or from EtOH/water.

1,3-Dibromopropane *[109-64-8]* M 201.9, f -34.4°, b 165°, n 1.522, d 1.977. Washed with dil aq Na_2CO_3, then water. Dried and fractionally distd under reduced pressure.

5,7-Dibromo-8-quinolinol see 5,7-dibromo-8-hydroxyquinoline.

meso 2,3-Dibromosuccinic acid *[526-78-3]* M 275.9, m 288-290° (sealed tube, dec). Crystd from distd water, keeping the temperature below 70°.

1,2-Dibromotetrafluoroethane *[124-73-2]* M 259.8, b 47.3°/760mm. Washed with water, then with weak alkali. Dried with $CaCl_2$ or H_2SO_4 and distd. [Locke *et al JACS* 56 1726 *1934*.] Also purified by gas chromatography on a silicone DC-200 column.

α-α'-Dibromo-*o*-xylene *[91-13-4]* M 264 0, m 95°, b 129-130°/4.5mm. Crystd from $CHCl_3$

α-α'-Dibromo-*m*-xylene *[626-15-3]* M 264.0, m 77°, b 156-160°/12mm. Crystd from acetone.

α-α'-Dibromo-*p*-xylene *[623-24-5]* M 264.0, m 145-147°, b 155-158°/12-15mm, 245°/760mm. Crystd from benzene or chloroform.

Di-*n*-butylamine *[111-92-2]* M 129.3, b 159°, n 1.41766, d 0.761. Dried with $LiAlH_4$, CaH_2 or KOH pellets, filtered and distd from BaO or CaH_2.

α-Dibutylamino-α-(*p*-methoxyphenyl)acetamide *[519-88-0]* M 292.4, m 134°. Crystd from EtOH containing 10% ethyl ether.

p-Di-*tert*-butylbenzene *[1571-86-4]* M 190.3, m 80°. Crystd from ethyl ether, EtOH and dried under vac over P_2O_5 at 55°. [Tanner *et al, JOC* 52 2142 *1987*.]

Di-*tert*-butyl peroxide *[110-05-4]* **M 146.2, n 1.398, d 0.794.** Washed with aq AgNO$_3$ to remove olefinic impurities. Then washed with water, dried over MgSO$_4$, and finally passed through a column of alumina to remove hydroperoxide impurities. [Jackson *et al, JACS* **107** 208 *1985*.]

Di-*n*-butyl *n*-butylphosphonate *[78-46-6]* **M 250.3, b 150-151°/10mm, 160-162°/20mm, n$_{25}$1.4302.** Purified by three crystns of its compound with uranyl nitrate, from hexane. For method, see *tributyl phosphate*.

Dibutylcarbitol *[112-73-2]* **M 218.3, b 125-130°/0.1mm, n 1.424, d 0.883.** Freed from peroxides by slow passage through a column of activated alumina. The eluate was shaken with Na$_2$CO$_3$ (to remove any remaining acidic impurities), washed with water, and stored with CaCl$_2$ in a dark bottle [Tuck *JCS* 3202 *1957*].

3,5-Di-*tert*-butyl catechol *[1020-31-1]* **M 222.3, m 96-99°.** Recrystd three times from pentane [Funabiki *et al, JACS* **108** 2921 *1986*.]

2,6-Di-*tert*-butyl-*p*-cresol (BHT) *[128-37-0]* **M 230.4, m 71.5°.** Dissolved in *n*-hexane at room temperature, then cooled with rapid stirring, to -60°. The ppte was separated, redissolved in hexane, and the process was repeated until the mother liquor was no longer coloured. The final product was stored under N$_2$ at 0° [Blanchard *JACS* **82** 2014 *1960*]. Also crystd from EtOH, MeOH, benzene, *n*-hexane, methylcyclohexane or pet ether (b 60-80°), and dried under vac.

Di-*n*-butyl cyclohexylphosphonate *[1085-92-3]* **M 245.4.** The compound with uranyl nitrate was crystd three times from hexane. For method see *tributyl phosphate*.

2,6-Di-*tert*-butyl-4-dimethylaminomethylphenol *[88-27-7]* **M 263.4.** Crystd from *n*-hexane.

Di-*tert*-butyldiperphthalate *[2155-71-7]* **M 310.3.** Crystd from ethyl ether. Dried over H$_2$SO$_4$.

Di-*n*-butyl ether see *n*-butyl ether.

2,6-Di-*tert*-butyl-4-ethylphenol *[4130-42-1]* **M 234.4, m 42-44°.** Cryst from aq EtOH or *n*-hexane.

2,5-Di-*tert*-butylhydroquinone *[88-58-4]* **M 222.3, m 222-223°.** Crystd from benzene or glacial acetic acid.

2,4-Di-*tert*-butyl-4-isopropylphenol *[5427-03-2]* **M 248.4, m 39-41°.** Crystd from *n*-hexane or aq EtOH.

2,6-Di-*tert*-butyl-4-methylphenol see **2,6-di-*tert*-butyl-*p*-cresol.**

Di-*tert*-butylperoxide (*tert*-butyl peroxide) *[110-05-4]* **M 146.2, n 1.3889, d 0.794.** Washed with aq AgNO$_3$ to remove olefinic impurities, water and dried. Freed from *tert*-butyl hydroperoxide by passage through an alumina column, and two high vac distns from room temp to a liquid-air trap [Offenbach and Tobolsky *JACS* **79** 278 *1957*]. *The necessary protection from* **EXPLOSION** *should be used.*

2,6-Di-*tert*-butylphenol *[128-39-2]* **M 206.3, m 37-38°.** Crystd from aq EtOH or *n*-hexane.

Dibutyl phthalate *[84-74-2]* **M 278.4, b 206°/20mm, 340°/760mm, d 1.4929, n^{25} 1.4901, d^{25} 1.0426.** Washed with dil NaOH (to remove any butyl hydrogen phthalate), aq NaHCO$_3$ (charcoal), then distd water. Dried with CaCl$_2$, distd under vac, and stored in a desiccator over P$_2$O$_5$.

2,6-Di-*tert*-butylpyridine, *[585-48-8]* **M 191.3, b 100-101°/23mm, n 1.474, d 0.852.** Redistd from KOH pellets.

Di-n-butyl sulphide *[544-40-1]* **M 146.3, α-form b 182°, β-form 190-230°(dec).** Washed with aq 5% NaOH, then water. Dried with $CaCl_2$ and distd from sodium.

Di-n-butyl sulphone *[598-04-9]* **M 162.3, m 43.5°.** Purified by zone melting.

3,5-Dicarbethoxy-1,4-dihydrocollidine *[632-93-9]* **M 267.3, m 131-132°.** Crystd from hot EtOH/water.

Dichloramine-T *[473-34-7]* **M 240.1, m 83°.** Crystd from pet ether (b 60-80°) or $CHCl_3$/pet ether. Dried in air.

Dichloroacetic acid *[79-43-6]* **M 128.9, m 13.5°, b 95.0-95.5°/17-18mm, n 1.4658, d 1.5634.** Crystd from benzene or pet ether. Dried with $MgSO_4$ and fractionally distd. [Bernasconi *et al, JACS* **107** 3612 *1985*.]

sym-**Dichloroacetone** *[534-07-6]* **M 127.0, m 45°, b 173°, d 1.383.** Crystd from benzene. Distd under vac.

Dichloroacetonitrile *[3018-12-0]* **M 110.0.** Purified by gas chromatography.

2,4-Dichloroaniline *[554-00-7]* **M 162.0, m 63°.** Crystd from EtOH/water. Also crystd from EtOH and dried *in vacuo* for 6hr at 40° [Moore *et al, JACS* **108** 2257 *1986*; Edidin *et al, JACS* **109** 3945 *1987*].

3,4-Dichloroaniline *[95-76-1]* **M 162.0, m 71.5°.** Crystd from MeOH.

9,10-Dichloroanthracene *[605-49-1]* **M 247.1, m 214-215°.** Purified by crystn from MeOH or EtOH, followed by sublimation under reduced pressure. [Masnori and Kochi *JACS* **107** 7880 *1985*.]

2,4-Dichlorobenzaldehyde *[874-42-0]* **M 175.0, m 72°.** Crystd from EtOH or ligroin.

2,6-Dichlorobenzaldehyde *[83-38-5]* **M 175.0, m 70.5-71.5°.** Crystd from EtOH/water or pet ether (b 30-60°).

o-**Dichlorobenzene** *[95-50-1]* **M 147.0, b 81-82°/31-32mm, 180.5°/760mm, n 1.55145, n^{25} 1.54911, d 1.306.** Contaminants may include the *p*-isomer and trichlorobenzene [Suslick *et al JACS* **106** 4522 *1984*]. It was shaken with conc or fuming H_2SO_4, washed with water, dried with $CaCl_2$, and distd from CaH_2 or sodium in a glass-packed column. Low conductivity material (*ca* 10^{-10} mhos) has been obtained by refluxing with P_2O_5, fractionally distilled and passed through a column packed with silica gel or activated alumina: it was stored in a dry-box under N_2 or with activated alumina.

m-**Dichlorobenzene** *[541-77-1]* **M 147.0, b 173.0°, n 1.54586, n^{25} 1.54337, d 1.289.** Washed with aq 10% NaOH, then with water until neutral, dried and distd. Conductivity material (*ca* 10^{-10} mhos) has been prepared by refluxing over P_2O_5 for 8hr, then fractionally distilling, and storing with activated alumina. *m*-Dichlorobenzene dissolves rubber stoppers.

p-**Dichlorobenzene** *[106-46-7]* **M 147.0, m 53.0°, b 174.1°, n^{60} 1.52849, d 1.241.** *o*-Dichlorobenzene is a common impurity. Has been purified by steam distn, crystn from EtOH or boiling MeOH, air-dried and dried in the dark under vac. Also purified by zone refining.

2,2'-Dichlorobenzidine *[84-68-4]* **M 253.1, m 165°.** Crystd from EtOH.

3,3'-Dichlorobenzidine *[91-94-1]* **M 253.1, m 132-133°.** Crystd from EtOH or benzene. **CARCINOGEN.**

2,4-Dichlorobenzoic acid *[50-84-0]* **M 191.0, m 163-164°.** Crystd from aq EtOH (charcoal), then benzene (charcoal). It can also be recrystd from water.

2,5-Dichlorobenzoic acid *[50-79-3]* **M 191.0, m 154°, b 301°/760mm.** Crystd from water.

2,6-Dichlorobenzoic acid *[50-30-6]* **M 191.0, m 141-142°.** Crystd from EtOH and sublimed in a vac.

3,4-Dichlorobenzoic acid *[51-44-5]* **M 191.0, m 206-207°.** Crystd from aq EtOH (charcoal) or acetic acid.

3,5-Dichlorobenzoic acid *[51-36-5]* **M 191.0, m 188°.** Crystd from EtOH and sublimed in a vac.

2,6-Dichlorobenzonitrile *[1194-65-6]* **M 172.0, m 145°.** Crystd from acetone.

4,4'-Dichlorobenzophenone *[90-98-2]* **M 251.1, m 145-146°.** Recrystd from EtOH [Wagner *et al, JACS* **108** 7727 *1986*].

2,5-Dichloro-1,4-benzoquinone *[615-93-0]* **M 177.0, m 161-162°.** Recrystd twice from 95% EtOH as yellow needles [Beck *et al, JACS* **108** 4018 *1986*].

2,6-Dichloro-1,4-benzoquinone *[697-91-6]* **M 177.0.** Recrystd from pet ether (b 60-70°) [Carlson and Miller *JACS* **107** 479 *1985*].

3,4-Dichlorobenzyl alcohol *[1805-32-9]* **M 177.0, m 38-39°.** Crystd from water.

2,3-Dichloro-1:3-butadiene *[1653-19-6]* **M 123.0, b 41-43°/85mm, 98°/760mm.** Crystd from pentane to constant melting point about -40°. A mixture of *meso* and *d,l* forms was separated by gas chromatography on an 8m stainless steel column (8mm i.d.) with 20% DEGS on Chromosorb W (60-80 mesh) at 60° and 80ml He/min. [Su and Ache *JPC* **80** 659 *1976*.]

1,1-Dichloro-2,2-bis-(*p*-chlorophenyl)ethane *[72-54-8]* **M 320.1, m 109-111°.** Purity checked by TLC.

4,6-Dichloro-*o*-cresol see **2,4-dichloro-6-methylphenol**.

2,3-Dichloro-5,6-dicyano-*p*-benzoquinone (DDQ) *[84-58-2]* **M 227.0, m 203°** (dec). Crystd from $CHCl_3$, $CHCl_3$/benzene (4:1), or benzene and stored at 0°. [Pataki and Harvey *JOC* **52** 2226 *1987*].

β,β'-Dichlorodiethyl ether *[111-44-4]* **M 143.0, b 79-80°/20mm, 176-177.0°/743mm, n 1.457, d 1.219.** Peroxide formation occurs rapidly, especially if distn is attempted at atmospheric pressure. After drying with NaOH pellets for 2 days, the ether was distd under N_2 at reduced pressure. The distillate was made 10^{-6}M in catechol to diminish peroxide formation, and was redistd immediately before use.

1,2-Dichloro-1,2-difluoroethane *[431-08-7]* **M 134.9.** For purification of diastereoisomeric mixture, with resolution into *meso* and *rac* forms, see Machulla and Stocklin [*JPC* **78** 658 *1974*].

Dichlorodifluoromethane *[75-71-8]* **M 120.9, b -25°.** Passage through satd aq KOH then conc H_2SO_4, and a tower packed with activated copper on kielselguhr at 200° removed CO_2 and O_2. A trap cooled to -29° removed a trace of high boiling material.

2,5-Dichloro-3,6-dihydroxy-*p*-benzoquinone see **chloranilic acid**.

1,3-Dichloro-5,5'-dimethylhydantoin *[118-52-5]* **M 197.0, m 136°.** Crystd from $CHCl_3$.

Dichlorodimethylsilane *[75-78-5]* **M 129.1, m -76°, n 1.404, d 1.064.** Purified by zone melting.

1,1-Dichloroethane *[75-34-3]* **M 99.0, b 57.3°, $n^{15}1.41975$, d^{15} 1.18350, d 1.177.** Shaken with conc H_2SO_4 or aq $KMnO_4$, then washed with water, satd aq $NaHCO_3$, again with water, dried with K_2CO_3 and distd from CaH_2 or $CaSO_4$. Stored over silica gel.

1,2-Dichloroethane *[107-06-2]* **M 99.0, b 83.4°, $n^{15}1.44759$, d 1.256.** Usually prepared by chlorinating ethylene, so that likely impurities include higher chloro derivatives and other chloro compounds depending on the impurities originally present in the ethylene. It forms azeotropes with water, MeOH, EtOH, trichloroethylene, CCl_4 and isopropanol. Its azeotrope with water (containing 8.9% water, and b 77°) can be used to remove gross amounts of water prior to final drying. As a preliminary purification step, it can be steam distd.
Shaken with conc H_2SO_4 (to remove alcohol added as an oxidation inhibitor), washed with water, then dil KOH or aq Na_2CO_3 and again with water. After an initial drying with $CaCl_2$, $MgSO_4$ or by distn, it is refluxed with P_2O_5, $CaSO_4$ or CaH_2 and fractionally distd. Carbonyl-containing impurities can be removed as described for chloroform.

1,2-Dichloroethylene **M 96.9, b 60° (*cis*), d 1.284, b 48° (*trans*), d 1.257.** Shaken successively with conc H_2SO_4, water, aq $NaHCO_3$ and water. Dried with $MgSO_4$ and distn. separated the *cis*- and *trans*-isomers.

cis-**1,2-Dichloroethylene** *[156-59-2]* **b 60.4°, $n^{15}1.44903$, n 1.4495, d 1.2830.** Purified by careful fractional distn., followed by passage through neutral activated alumina. Also by shaking with mercury, drying with K_2CO_3 and distn. from $CaSO_4$.

trans-**1,2-Dichloroethylene** *[156-60-5]* **b 47.7°, $n^{15}1.45189$, n 1.4462, d 1.2551.** Dried with $MgSO_4$, and fractionally distd. under CO_2. Fractional crystn at low temperatures has also been used.

5,7-Dichloro-8-hydroxyquinoline *[773-76-2]* **M 214.1, m 180-181°.** Crystd from acetone/EtOH.

Dichloromaleic anhydride *[1122-17-4]* **M 167.0, m 120°.** Purified by sublimation *in vacuo* [Katakis *et al, JCSDT* 1491 *1986*].

Dichloromethane *[75-09-2]* **M 84.9, b 40.0°, n 1.42456, $n^{25}1.4201$, d 1.325.** Shaken with portions of conc H_2SO_4 until the acid layer remained colourless, then washed with water, aq 5% Na_2CO_3, $NaHCO_3$ or NaOH, then water again. Pre-dried with $CaCl_2$, and distd from $CaSO_4$, CaH_2 or P_2O_5. Stored away from bright light in a brown bottle with Linde type 4A molecular sieves, in an atmosphere of dry N_2. Other purification steps include washing with aq. $Na_2S_2O_3$, passage through a column of silica gel, and removal of carbonyl-containing impurities as described under **Chloroform**. It has also been purified by treatment with basic alumina, distd, and stored over molecular sieves under nitrogen [Puchot *et al, JACS* 108 2353 *1986*].

3,9-Dichloro-7-methoxyacridine *[86-38-4]* **M 278.1, m 160-161°.** Crystd from benzene.

5,7-Dichloro-2-methyl-8-hydroxyquinoline *[72-80-0]* **M 228.1, m 114-115°.** Crystd from EtOH.

2,4-Dichloro-6-methylphenol *[1570-65-6]* **M 177.0, m 55°, b 129-132°/40mm.** Crystd from water.

2,4-Dichloro-1-naphthol *[2050-76-2]* **M 213.1, m 106-107°.** Crystd from MeOH.

2,3-Dichloro-1,4-naphthoquinone *[117-80-6]* **M 227.1, m 193°.** Crystd from EtOH.

2,5-Dichloro-4-nitroaniline *[6627-34-5]* **M 207.0, m 157-158°.** Crystd from EtOH, then sublimed.

2,6-Dichloro-4-nitroaniline *[99-30-9]* **M 207.0, m 193°.** Crystd from aq EtOH or benzene/EtOH.

2,5-Dichloro-1-nitrobenzene *[89-61-2]* **M 192.0, m 56°,**

3,4-Dichloro-1-nitrobenzene *[99-54-7]* **M 192.0, m 43°.** Crystd from absolute EtOH.

2,4-Dichloro-6-nitrophenol *[609-89-2]* **M 208.0, m 122-123°.** Crystd from acetic acid.

2,6-Dichloro-4-nitrophenol *[618-00-4]* **M 208.0, m 125°.** Crystd from EtOH and dried *in vacuo* over anhydrous MgSO$_4$.

Dichlorophen [2,2'-methylenebis(4-chlorophenol)] *[97-23-4]* **M 269.1, b 177-178°.** Crystd from toluene.

2,3-Dichlorophenol *[576-24-9]* **M 163.0, m 57°.** Crystd from ether.

2,4-Dichlorophenol *[120-83-2]* **M 163.0, m 42-43°.** Crystd from pet ether (b 30-40°). Purified by repeated zone melting, using a P$_2$O$_5$ guard tube to exclude moisture. Very hygroscopic when dry.

2,5-Dichlorophenol *[583-78-8]* **M 163.0, m 58°, b 211°/744mm.** Crystd from ligroin and sublimed.

3,4-Dichlorophenol *[95-77-2]* **M 163.0, m 68°, b 253.5°/767mm,**
3,5-Dichlorophenol *[591-35-5]* **M 163.0, m 68°, b 122-124°/8mm, 233-234°/760mm.** Crystd from pet ether/benzene mixture.

2,6-Dichlorophenol-indophenol sodium salt (2H$_2$O) *[620-45-1]* **M 326.1, ε = 2.1 x 10^4 at 600nm and pH 8.** Dissolved in 0.001M phosphate buffer, *p*H 7.5 (alternatively, about 2g of the dye was dissolved in 80ml of M HCl), and extracted into ethyl ether. The extract was washed with water, extracted with aq 2% NaHCO$_3$, and the sodium salt of the dye was pptd by adding NaCl (30g/100ml of NaHCO$_3$ soln), then filtered off, washed with dil NaCl soln and dried.

2,4-Dichlorophenoxyacetic acid (2,4-D) *[94-75-7]* **M 221.0, m 146°,**
α-(2,4-Dichlorophenoxy)propionic acid (2,4-DP) *[120-36-5]* **m 117°,** Crystd from MeOH.

2,4-Dichlorophenylacetic acid *[19719-28-9]* **M 205.0, m 131°,**
2,6-Dichlorophenylacetic acid *[6575-24-2]* **M 205.0, m 157-158°.** Crystd from aq EtOH.

3-(3,4-Dichlorophenyl)-1,1-dimethyl urea (Diuron) *[330-54-1]* **M 233.1.** Crystd four times from 95% EtOH [Beck *et al, JACS* **108** 4018 *1986*].

4,5-Dichloro-*o*-phenylenediamine *[5348-42-5]* **M 177.1.** Dried over Na$_2$SO$_4$. Recrystd from hexane.

4,5-Dichlorophthalic acid *[56962-08-4]* **M 235.0, m 200° (dec to anhydride).** Crystd from water.

1,2-Dichloropropane *[78-87-5]* **M 113°, b 95.9-96.2°, n 1.439, d 1.158.** Distd from CaH$_2$.

2,2-Dichloropropane *[594-20-7]* **M 113.0, b 69.3°, n 1.415, d 1.090.** Washed with aq Na$_2$CO$_3$ soln, then distd water, dried over CaCl$_2$ and fractionally distd.

1,3-Dichloro-2-propanone see *sym*-dichloroacetone.

2,6-Dichloropyridine *[2402-78-0]* **M 148.0, m 87-88°,**
3,5-Dichloropyridine *[2457-47-8]* **M 148.0, m 64-65°.** Crystd from EtOH.

4,7-Dichloroquinoline *[86-98-6]* **M 198.1, m 86.4-87.4°, b 148°/10mm.** Crystd from MeOH or 95% EtOH.

5,7-Dichloro-8-quinolinol see **5,7-dichloro-8-hydroxyquinoline.**

2,6-Dichlorostyrene *[28469-92-3]* **M 173.0, b 72-73°/2mm, n 1.5798, d 1.4045.** Purified by fractional crystn from the melt and by distn.

p-α-**Dichlorotoluene** see *p*-**chlorobenzyl chloride.**

2,4-Dichlorotoluene *[95-73-8]* **M 161.1, m -13.5°, b 61-62°/3mm, n 1.5513, d 1.250,**
2,6-Dichlorotoluene *[118-69-4]* **M 161.1, b 199-200°/760mm, n 1.548, d 1.254,**
3,4-Dichlorotoluene *[95-75-0]* **M 161.1, m -16°, b 205°/760mm, n 1.549, d 1.2541.** Recrystd from EtOH at low temp or fractionally distd.

α,α'-Dichloro-*p*-xylene *[623-25-6]* **M 175.1, m 100°.** Crystd from benzene and dried under vac.

Dicinnamalacetone *[622-21-9]* **M 314.4, m 146°.** Crystd from benzene/isooctane (1:1).

Dicumyl peroxide *[80-43-3]* **M 270.4, m 39-40°.** Crystd from 95% EtOH (charcoal). Stored at 0°. *Potentially* **EXPLOSIVE.**

1,2-Dicyanobenzene *[91-15-6]* **M 128.1, m 141°.** Recrystd from hot toluene.

9,10-Dicyanoanthracene *[1217-45-4]* **M 228.2.** Recrystd twice from pyridine [Mattes and Farid *JACS* **108** 7356 *1986*].

1,4-Dicyanobenzene *[623-26-7]* **M 128.1, m 222°.** Crystd from EtOH.

Dicyanodiamide see **cyanoguanidine.**

1,4-Dicyanonaphthalene *[3029-30-9]* **M 178.2, m 206°.** Purified by crystn and sublimed in a vac.

Dicyclohexyl-18-crown-6 *[16069-36-6]* **M 372.5.** Purified by chromatography on neutral alumina and eluting with an ether/hexane mixture [see *Inorg Chem* **14** 3132 *1975*]. Dissolved in ether at *ca* 40°, and spectroscopic grade MeCN was added to the soln which was then chilled. The crown ether pptd and was filtered off. It was dried *in vacuo* at room temp [Wallace *JPC* **89** 1357 *1985*].

Di-*n*-decylamine *[1120-49-6]* **M 297.6, m 34°. b 153°/1mm, 359°/760mm.** Dissolved in benzene and pptd as its bisulphate by shaking with 4M H_2SO_4. Filtered. Washed with benzene, separating by centrifugation, then the free base was liberated by treating with aq NaOH [McDowell and Allen *JPC* **65** 1358 *1961*].

Didodecylamine *[3007-31-6]* **M 353.7, m 51.8°.** Crystd from EtOH/benzene under N_2.

Didodecyldimethylammonium bromide *[3282-73-3]* **M 463.6.** Recrystd from acetone, acetone/ether mixture, then from ethyl acetate, washed with ether and dried in a vac oven at 60° [Chen *et al JPC* **88** 1631 *1984*; Rupert *et al, JACS* **107** 2628 *1985*; Halpern *et al, JACS* **108** 3920 *1986*; Allen *et al, JPC* **91** 2320 *1987*].

Dienosterol m 227-228°. Crystd from EtOH.

Diethanolamine *[111-42-2]* **M 105.1, m 28°, b 154-155°/10mm, 270°/760mm.** Fractionally distd twice, then fractionally crystd from its melt.

1,2-Diethoxyethane see **ethylene glycol diethyl ether.**

N,N-**Diethylacetamide** *[2235-46-3]* **M 157.2, b 86-88°, n 1.474, d 0.994.** Dissolved in cyclohexane, shaken with anhydrous BaO and then filtered. The procedure was repeated three times, and the cyclohexane was distd off at 1 atmosphere pressure. The crude amide was also fractionally distd three times from anhydrous BaO.

Diethyl acetamidomalonate *[1068-90-2]* M 217.2, m 96°. Crystd from benzene/pet ether.

Diethylamine *[109-89-7]* M 73.1, b 55.5°, n 1.38637, d 0.707. Dried with LiAlH$_4$ or KOH pellets. Refluxed with, and distd from, BaO or KOH. Converted to the *p*-toluenesulphonamide and crystd to constant melting point from dry pet ether (b 90-120°), then hydrolysed with HCl, excess NaOH was added, and the amine distd through a tower of activated alumina, redistd and dried with activated alumina before use [Swift *JACS* 64 115 *1942*].

Diethylamine hydrochloride *[660-68-4]* M 109.6, m 223.5°. Crystd from abs EtOH. Also crystd from dichloroethane/MeOH.

trans-**4-(Diethylamino)azobenzene** *[3588-91-8]* M 320.5, m 171°. Purified by column chromatography [Flamigni and Monti *JPC* 89 3702 *1985*].

N,N-**Diethylaniline** *[91-66-7]* M 149.2, b 216.5°, n 1.5409, d 0.938. Refluxed for 4hr with half its weight of acetic anhydride, then fractionally distd under reduced pressure (b 92°/10mm).

5,5-Diethylbarbituric acid *[57-44-3]* M 184.2, m 188-192°. Crystd from water or EtOH. Dried in a vac over P$_2$O$_5$.

N,N'-**Diethylcarbanilide** *[611-92-7]* M 240.3, m 79°. Crystd from EtOH.

Diethyl carbitol see **Diethylene glycol diethyl ether.**

Diethyl carbonate *[105-58-8]* M 118.1, b 126.8°, n^{15} 1.38654, n^{25} 1.38287, d 0.975. Washed (100ml) with an aq 10% Na$_2$CO$_3$ (20ml), satd CaCl$_2$ (20ml), then water (30ml). Dried by standing with solid CaCl$_2$ for 1hr. (Prolonged contact should be avoided because slow combination with CaCl$_2$ occurs.) Fractionally distd.

1,1'-Diethyl-2,2'-cyanine chloride Crystd from EtOH and dried in a vac oven at 80° for 4hr.

N,N-**Diethylcyclohexylamine** *[91-65-6]* M 155.3, b 193°/760mm, m 1.4562, d 0.850. Dried with BaO and fractionally distd.

O,O-**Diethyl-*S*-2-diethylaminoethyl phosphorothiolate** *[78-53-5]* M 269.3, m 98-99°. Crystd from isopropanol/ethyl ether.

sym-**Diethyldiphenylurea** see *N,N*-diethylcarbanilide.

Diethyl sulphide *[110-81-6]* M 122.3, b 154-155°, n 1.506, d 0.993. Dried with silica gel or MgSO$_4$ and distd under reduced pressure (optionally from CaCl$_2$).

Diethylene glycol *[111-46-6]* M 106.1, f.p. -10.5°, b 244.3°, n^{15} 1.4490, n 1.4475, d 1.118. Fractionally distd under reduced pressure (b 133°/14mm), then fractionally crystd by partial freezing.

Diethylene glycol diethyl ether *[112-36-7]* M 162.2, b 85-86°/10mm, 188.2-188.3°/751mm, d 0.909. Dried with MgSO$_4$, then CaH$_2$ or LiAlH$_4$, under N$_2$. If sodium is used the ether should be redistd, alone to remove any products which may be formed by the action of sodium on the ether. As a preliminary purification, the crude ether (2L) can be refluxed for 12hr with 25ml of conc HCl in 200ml of water, under reduced pressure, with slow passage of N$_2$ to remove aldehydes and other volatile substances. After cooling, addn of sufficient solid KOH pellets (slowly and with shaking until no more dissolve) gives two liquid phases. The upper of these is decanted, dried with fresh KOH pellets, decanted, then refluxed over, and distd from, sodium.

Diethylene glycol dimethyl ether see **diglyme.**

Diethylene glycol mono-*n*-butyl ether *[112-34-5]* **M 162.2, b 69-70°/0.3mm, 230.5°/760, n 1.4286, d 0.967.** Dried with anhydrous K_2CO_3 or $CaSO_4$, filtered and fractionally distd. Peroxides can be removed by refluxing with stannous chloride or a mixture of $FeSO_4$ and $KHSO_4$ (or, less completely, by filtration under slight pressure through a column of activated alumina).

Diethylene glycol monoethyl ether *[111-90-0]* **M 134.2, b 201.9°, m 1.4273, n^{25} 1.4254, d 0.999.** Ethylene glycol can be removed by extracting 250g in 750ml of benzene with 5ml portions of water, allowing for phase separation, until successive aqueous portions show the same vol increase. Dried, and freed from peroxides, as described for diethylene glycol mono-*n*-butyl ether.

Diethylene glycol monomethyl ether *[111-77-3]* **M 120.2, b 194°, n 1.423, d 1.010.** Purified as for diethylene glycol mono-*n*-butyl ether.

Diethylenetriamine see **2,2'-diaminodiethylamine.**

Diethylenetriaminepenta-acetic acid *[67-43-6]* **M 393.4, m 219-220°.** Crystd from water. Dried under vac or at 110°. [Bielski and Thomas *JACS* **109** 7761 *1987*.]

Diethyl ether see **ethyl ether.**

Diethyl fumarate *[623-91-6]* **M 172.2, b 218°, n 1.441, d 1.052.** Washed with aq 5% Na_2CO_3, then with satd $CaCl_2$ soln, dried with $CaCl_2$ and distd.

Di-(2-ethylhexyl)phosphoric acid *[27215-10-7]* **M 322.4.**
Contaminants of commercial samples include the monoester, polyphosphates, pyrophosphate, 2-ethylhexanol and metal impurities. Dissolved in *n*-hexane to give an 0.8M soln. Washed with an equal vol of M HNO_3, then with satd $(NH_4)_2CO_3$ soln, with 3M HNO_3, and twice with water [Petrow and Allen *AC* **33** 1303 *1961*]. Similarly, the impure sodium salt, after scrubbing with pet ether, has been acidified with HCl and the free organic acid has been extracted into pet ether and purified as above. For purification *via* the copper salt see McDowell *et al* [*J Inorg Nuclear Chem* **38** 2127 *1976*].

Di-(2-ethylhexyl)phthalate *[117-81-7]* **M 390.6, b 384°, n 1.4863, d 0.9803.** Washed with Na_2CO_3 soln, then shaken with water. After the resulting emulsion had been broken by adding ether, the ethereal soln was washed twice with water, dried ($CaCl_2$), and evapd. The residual liquid was distd several times under reduced pressure, then stored in a vac desiccator over P_2O_5 [French and Singer *JCS* 1424 *1956*]

Diethyl ketone (3-pentanone) *[96-22-0]* **M 86.1, b 102.1°, n 1.392, d 0.8099.** Dried with anhydrous $CaSO_4$ or $CuSO_4$, and distd from P_2O_5 under N_2 or under reduced pressure. Further purification by conversion to the semicarbazone (recrystd to constant **m** 139°, from EtOH) which, after drying under vac over $CaCl_2$ and paraffin wax, was refluxed for 30min with excess oxalic acid, then steam distd and salted out with K_2CO_3. Dried with Na_2SO_4 and distd [Cowan, Jeffrey and Vogel *JCS* 171 *1940*].

Diethyl phthalate *[84-66-2]* **M 222.2. b 172°/12mm, b 295°/760mm, n 1.5022, d^{25} 1.1160.** Washed with aq Na_2CO_3, then distd. water, dried ($CaCl_2$), and distd under reduced pressure. Stored in a vac desiccator over P_2O_5.

2,2-Diethyl-1,3-propanediol *[115-76-4]* **M 132.2, m 61.4-61.8°.** Crystd from pet ether (b 65-70°).

Diethylstilboesterol *[56-25-1]* **M 268.4, m 169-172°.** Crystd from benzene.

Diethyl succinate *[123-25-1]* M 174.2, b 105°/15mm, n 1.4199, d 1.047. Dried with MgSO$_4$, and distd at 15mm pressure.

Diethyl sulphate *[64-67-5]* M 154.2, b 96°/15mm, 118°/40mm, n 1.399, d 1.177. Washed with aq 3% Na$_2$CO$_3$ (to remove acidic material), then distd water, dried (CaCl$_2$), filtered and distd. *Causes blisters to the skin.*

Diethyl sulphide *[352-93-2]* M 90.2, m 0°/15mm, 90.1°/760mm, n 1.443, d 0.837. Washed with aq 5% NaOH, then water, dried with CaCl$_2$ and distd from sodium. Can also be dried with MgSO$_4$ or silica gel. Alternative purification is *via* the Hg(II) chloride complex [(Et)$_2$S.2HgCl$_2$] (see dimethyl sulphide).

Diethyl terephthalate *[636-09-0]* M 222.2, m 44°, 142°/2mm, 302°/760mm. Crystd from toluene and distd under reduced pressure.

sym-**Diethylthiourea** *[105-55-5]* M 132.2, m 76-77°. Crystd from benzene.

Digitonin *[11024-24-1]* M 1229.3, m >270°(dec), [α]$_{546}$ -63° (c 3, MeOH). Crystd from aq 85% EtOH or MeOH/ethyl ether.

Digitoxigenin *[142-62-4]* M 374.5, m 253°, [α]$_{546}$ +21° (c 1, MeOH). Crystd from aq 40% EtOH.

D(+)-Digitoxose *[527-52-6]* M 148.2, m 112°, [α]$_{546}$ +57° (c 1, H$_2$O). Crystd from MeOH/ethyl ether, or ethyl acetate.

Diglycolic acid *[110-99-6]* M 134.1, m 148° (monohydrate). Crystd from water.

Diglycyl glycine *[556-33-2]* M 189.2, m 246°(dec). Crystd from water or water/EtOH and dried at 110°.

Diglyme *[111-46-6]* M 134.2, b 62°/17mm, 75°/35mm, 160°/760mm. n 1.4087, d 0.917. Dried with NaOH pellets or CaH$_2$, then refluxed with, and distd (under reduced pressure) from Na, CaH$_2$, LiAlH$_4$, NaBH$_4$ or NaH. These operations were carried out under N$_2$. The amine-like odour of diglyme has been removed by shaking with a weakly acidic ion-exchange resin (Amberlite IR-120) before drying and distn. Addn of 0.01% NaBH$_4$ to the distillate inhibits peroxidation. Purification as for dioxane. Also passed through a 12-in column of molecular sieves to remove water and peroxides.

Digoxin *[20830-75-5]* M 781.0, m 265°(dec), [α]$_{546}$ +14.0° (c 10, pyridine). Crystd from aq EtOH or aq pyridine.

4,4'-Di-*n*-heptyloxyazoxybenzene *[2635-26-9]* M 426.6. Crystd from hexane and dried by heating under vac.

Dihexadecyl phosphate *[2197-63-9]* M 546.9, m 75°. Crystd from MeOH [Lukac *JACS* 106 4387 *1984*].

9,10-Dihydroanthracene *[613-31-0]* M 180.3, m 110-110.5°,
Dihydrochloranil *[1198-5-6]* M 247.9. Crystd from EtOH [Rabideau *et al*, *JACS* 108 8130 *1986*].

Dihydrocholesterol see **cholestanol.**

Dihydrocinnamic acid *[501-52-0]* M 150.2, m 48-49°. Crystd from pet ether (b 60-80°).

Dihydropyran *[110-87-2]* **M 84.1, b 84.4°/742mm, n 1.441, d 0.922.** Partially dried with Na_2CO_3, then fractionally distd. The fraction **b 84-85°**, was refluxed with Na until hydrogen was no longer evolved when fresh Na was added. It was then dried, and distd again.

Dihydrotachysterol *[67-96-9]* **M 398.7, m 125-127°.** Crystd from 90% MeOH.

1,2-Dihydroxyanthraquinone see **alizarin.**

1,4-Dihydroxyanthraquinone see **quinizarin.**

1,5-Dihydroxyanthraquinone see **anthrarufin.**

1,8-Dihydroxyanthraquinone *[117-10-2]* **M 240.1, m 193-197°.** Crystd from EtOH and sublimed in a vac.

2,4-Dihydroxyazobenzene *[2051-85-6]* **M 214.2, m 228°.** Crystd from hot EtOH (charcoal).

2,3-Dihydroxybenzaldehyde *[24677-78-9]* **M 138.1, m 135-136°,**
2,4-Dihydroxybenzoic acid *[95-01-2]* **M 154.1, m 226-227°(dec).** Crystd from water.

2,5-Dihydroxybenzoic acid *[490-79-9]* **M 154.1, m >200°(dec).** Crystd from benzene/acetone. Dried in a vac desiccator over silica gel.

2,6-Dihydroxybenzoic acid *[303-07-1]* **M 154.1, m 167°(dec).** Dissolved in aq $NaHCO_3$ and the soln was washed with ether to remove non-acidic material. The acid was pptd by adding H_2SO_4, and recrystd from water. Dried under vac and stored in the dark [Lowe and Smith *JCSFT 1* **69** 1934 *1973*].

2,4-Dihydroxybenzophenone *[131-56-6]* **M 214.2, m 145.5-147°.** Recrystd from MeOH.

2,5-Dihydroxybenzyl alcohol *[495-08-9]* **M 140.1, m 100°.** Crystd from $CHCl_3$. Sublimed.

2,2'-Dihydroxybiphenyl *[1806-29-7]* **M 186.2, m 108.5-109.5°.** Repeatedly crystd from toluene, then sublimed at $60°/10^{-4}$mm.

3α,7α-Dihydroxycholanic acid *[474-25-9]* **M 239.6, m 143°, $[α]_{546}$ +14° (c 2, EtOH).** Crystd from ethyl acetate.

6,7-Dihydroxycoumarin (esculetin) *[305-01-1]* **M 178.2, m 268-270°(dec).** Crystd from glacial acetic acid.

2,2'-Dihydroxy-6,6'-dinaphthyl disulphide, m 220-223°. Recryst from hot glacial acetic acid. [Barnett and Seligman *Science* **116** 323 *1952*.]

7,8-Dihydroxycoumarin *[486-35-1]* **M 178.2, m 256°(dec).** Crystd from aq EtOH. Sublimed.

(*N,N*-Dihydroxyethyl)glycine *[150-25-4]* **M 163.2, m 193°(dec).** Dissolved in a small vol of hot water and pptd with EtOH, twice. Repeated once more but with charcoal treatment of the aq soln, and filtered before addn of EtOH.

3,4-Dihydroxyisoflavone *[578-86-9]* **M 256.3, m 234-236°.** Crystd from aq 50% EtOH.

Dihydroxymaleic acid (dihydroxyfumaric acid hydrate) *[133-38-0]* **M 148.1, m 155°(dec).** Crystd from water.

5,7-Dihydroxy-4'-methoxyflavone *[491-80-5]* **M 284.3, m 261°.** Crystd from 95% EtOH.

1,8-Dihydroxy-3-methylanthraquinone (chrysophanic acid) *[481-74-3]* **M 245.3, m 196°.** Crystd from EtOH or benzene and sublimed in a vac.

1,5-Dihydroxynaphthalene *[83-56-7]* **M 160.2, m 165°(dec).** Crystd from nitromethane.

1,6-Dihydroxynaphthalene *[575-44-0]* **M 160.2, m 138-139° (with previous softening).** Crystd from benzene/EtOH after treatment with charcoal.

2,5-Dihydroxyphenylacetic acid (homogentisic acid) *[451-13-8]* **M 168.2, m 152°.** Crystd from EtOH/CHCl$_3$.

S-β-(3,4-Dihydroxyphenyl)alanine (DOPA) *[59-92-7]* **M 197.2, m 285.5°(dec), [α]$_D$ -12.0° (1M HCl).** Likely impurities are vanillin, hippuric acid, 3-methoxytyrosine and 3-aminotyrosine. Crystd by dissolving in dil HCl and adding dil ammonia to give *p*H 5, under N$_2$. Alternatively, crystd from aq EtOH. Unstable in aq alkali.

2,3-Dihydroxytoluene *[452-86-8]* **M 124.1, m 65-66°.** Crystd from benzene and purity checked by tlc.

***p*-Diiodobenzene** *[624-38-4]* **M 329.9, m 132-133°.** Crystd from EtOH or boiling MeOH, then air dried.

1,2-Diiodoethane *[624-73-7]* **M 281.9, m 81-84°, d 2.134.** Dissolved in ether, washed with satd aq Na$_2$S$_2$O$_3$, drying it over MgSO$_4$ and evaporating the ether *in vacuo* [Molander *et al, JACS* 109 453 *1987*].

5,7-Diiodo-8-hydroxyquinoline *[83-73-8]* **M 397.0, m 214-215°(dec).** Crystd from xylene and dried at 70° in a vac.

Diiodomethane *[75-11-6]* **M 267.8, m 6.1°, b 66-70°/11-12mm, d 3.325.** Fractionally distd under reduced pressure, then fractionally crystd by partial freezing, and stabilized with silver wool if necessary. It has also been purified by drying over CaCl$_2$ and fractionally distd from Cu powder.

5,7-Diiodo-8-quinolinol see **5,7-diiodo-8-hydroxyquinoline.**

S-3,5-Diiodotyrosine (iodogorgoic acid) *[300-39-0]* **M 469.0, m 204°(dec), [α]$_D$ +1.5° (in 1M HCl).** Likely impurities are tyrosine, 3-iodotyrosine and iodide ions. Crystd from cold dil ammonia by adding acetic acid to give *p*H 6. It can also be crystd from aq 70% EtOH.

Diisooctyl phenylphosphonate *[49637-59-4]* **M 378.5, n^{25} 1.4780.** Vac distd, percolated through a column of alumina, then passed through a packed column maintained at 150° to remove residual traces of volatile materials in a countercurrent stream of N$_2$ at reduced pressure [Dobry and Keller *JPC* 61 1448 *1957*].

Diisopropanolamine *[110-97-4]* **M 133.2, m 41-44°, d 1.004.** Repeatedly crystd from dry ethyl ether.

Diisopropylamine *[108-18-9]* **M 101.2, b 83.5°/760mm, n 1.39236, d 0.720.** Distd from NaOH, or refluxed over Na wire or NaH for three minutes and distd into a dry receiver under N$_2$.

Diisopropyl ether see **isopropyl ether.**

Diisopropylethylamine *7087-68-5]* **M 129.3, b 127°.** Distd from ninhydrin, then from KOH [Dryland and Sheppard, *JCSFT 1* 125 *1986*].

Diisopropyl ketone *[565-80-0]* M 114.2, b 123-125°, n 1.400, d 0.801. Dried with $CaSO_4$, shaken with chromatographic alumina and fractionally distd.

Diketene *[674-82-8]* M 84.1, m -7°, b 66-68°/90mm, n 1.4376, n^{25} 1.4348, d 1.440. Diketene polymerizes violently in the presence of alkali. Distd at reduced pressure, then fractionally crystd by partial freezing (using as a cooling bath a 1:1 soln of $Na_2S_2O_3$ in water, cooled with Dry-ice until slushy, and stored in a Dewar flask). Freezing proceeds slowly, and takes about a day for half completion. The crystals are separated and stored in a refrigerator under N_2.

2,2'-Diketospirilloxanthin *[24009-17-4]* M 624.9, m 225-227°, ϵ_{1cm}1% 550(349nm), 820(422nm), 2125(488nm), 2725(516nm), 2130(551nm) in hexane. Purified by chromatography on a column of partially deactivated alumina. Crystd from acetone/pet ether. Stored in the dark, in an inert atmosphere at 0°.

Dilauroyl peroxide *[105-74-8]* M 398.6, m 53-55°. Crystd from *n*-hexane. *Potentially* **EXPLOSIVE.**

Dilituric acid see **5-*N*-nitrobarbituric acid.**

Dimedone *[126-81-8]* M 140.2, m 148-149°. Crystd from acetone (*ca* 8ml/g), water or aq EtOH. Dried in air.

1,2-Dimercapto-3-propanol *[59-52-9]* M 124.2, b 82-84°/0.8mm, n 1.5732, d 1.239,
1,3-Dimercapto-2-propanol *[584-04-3]* M 124.2, b 82°/1.5mm. Pptd as the mercury mercaptide (see Björberg *Ber* **75** 13 *1942*], regenerated with H_2S, and distd at 2.7mm [Rosenblatt and Jean *AC* 951 *1955*].

4,4'-Dimethoxyazobenzene *[2396-60-3]* M 242.3, m 162.7-164.7°. Chromatographed on basic alumina, eluted with benzene. Crystd from 2:2:1 (v/v) methanol/ethanol/benzene.

4,4'-Dimethoxyazoxybenzene *[1562-94-3]* M 258.3, m 165°. Crystd from hot 95% EtOH, dried, then sublimed in a vac onto a cold finger.

3,4-Dimethoxybenzaldehyde see **veratraldehyde.**

o-**Dimethoxybenzene** (veratrole) *[91-16-7]* M 137.2, m 23°, b 208.5-208.7°, n^{25} 1.53232, d 1.085. Steam distd. Fractionally distd from BaO, CaH_2 or Na. Crystd from benzene or low-boiling pet ether at 0°. Fractionally crystd from its melt. Stored over anhydrous Na_2SO_4.

m-**Dimethoxybenzene** *[151-10-0]* M 137.2, b 212-213°, n 1.5215, d 1.056. Extracted with aq NaOH, and water, then dried. Fractionally distd from BaO or Na.

p-**Dimethoxybenzene** *[150-78-7]* M 137.2, m 57.2-57.8°. Steam distd. Crystd from benzene, MeOH or EtOH. Dried under vac. Also sublimes under vac.

2,4-Dimethoxybenzoic acid *[91-52-1]* M 182.2, m 109°,
2,6-Dimethoxybenzoic acid *[1466-76-8]* M 182.2, m 186-187°. Crystd from water.

3,4-Dimethoxybenzoic acid *[93-07-2]* M 182.2, m 181-182°. Crystd from water or aq acetic acid.

3,5-Dimethoxybenzoic acid *[1132-21-4]* M 182.2, m 185-186°. Crystd from water, EtOH or aq acetic acid.

p,p'-**Dimethoxybenzophenone** *[90-96-0]* M 242.3, m 144.5°. Crystd from abs EtOH.

2,6-Dimethoxybenzoquinone *[530-55-2]* M 168.1, m 256°. Crystd from acetic acid. Sublimes in a vac.

1,1-Dimethoxyethane (acetaldehyde dimethyl acetal) *[534-15-6]* **M 90.1, b 212°/760mm, n 1.4140, d 0.828.** Purified by glc.

1,2-Dimethoxyethane (glyme) *[110-71-4]* **M 90.1, b 84°, n 1.380, d 0.867.** Traces of water and acidic materials have been removed by refluxing with Na, K or CaH$_2$, decanting and distilling from Na, K, CaH$_2$ or LiAlH$_4$. Reaction has been speeded up by using vigorous high-speed stirring and molten potassium. For virtually complete elimination of water, 1,2-dimethoxyethane has been dried with Na-K alloy until a characteristic blue colour was formed in the solvent at Dry-ice/cellosolve temperatures: the solvent was kept with the alloy until distd for use [Ward *JACS* **83** 1296 *1961*]. Alternatively, glyme, refluxed with benzophenone and Na-K, was dry enough if, on distn, it gave a blue colour of the ketyl immediately on addition to benzophenone and sodium [Ayscough and Wilson *JCS* 5412 *1963*]. Also purified by distn under N$_2$ from sodium benzophenone ketyl (see above).

3,5-Dimethoxy-4-hydroxycinnamic acid see 3-(4-Hydroxy-3,5-dimethoxy)acrylic acid.

5,6-Dimethoxy-1-indanone *[2107-69-9]* **M 192.2, m 118-120°.** Crystd from MeOH, then sublimed in a vac.

Dimethoxymethane (methylal) *[109-87-5]* **M 76.1, b 42.3°, n^{15} 1.35626, n 1.35298, d 0.860.** The main impurity is MeOH, which can be removed by treatment with sodium wire, followed by fractional distn from sodium. The solvent is kept dry by storing in contact with molecular sieves. Alternately, technical dimethoxymethane was stood with paraformaldehyde and a few drops of H$_2$SO$_4$ for 24hr, then distd. It could also be purified by shaking with an equal volume of 20% NaOH, leaving for 30min, and distilling. Methods of purification used for acetal are probably applicable to methylal.

1,4-Dimethoxynaphthalene *[10075-62-4]* **M 188.2, m 87-88°,**
1,5-Dimethoxynaphthalene *[10075-63-5]* **M 188.2, m 183-184°.** Crystd from EtOH.

2,6-Dimethoxyphenol *[91-10-1]* **M 154.2, m 54-56°.** Purified by zone melting or sublimation in a vac.

3,4-Dimethoxyphenyl acetic acid (homoveratric acid) *[93-40-3]* **M 196.2, m 97-99°.** Crystd from water or benzene/ligroin.

3,5-Dimethoxyphenylacetonitrile *[13388-75-5]* **M 177.1.** Crystd from MeOH. [Sankaraman *et al,* *JACS* **109** 5235 *1987*.]

4,4'-Dimethoxythiobenzophenone *[958-80-5]* **M 258.3, m 120°.** Recrystd from a mixture of cyclohexane/dichloromethane (4:1).

2,6-Dimethoxytoluene *[5673-07-4]* **M 152.2, m 39-41°.** Sublimed *in vacuo* [Sankaraman *et al, JACS* **109** 5235 *1987*].

Dimethyl acetal see 1,1-dimethoxyethane.

N,N-**Dimethylacetamide** *[127-19-5]* **M 87.1, b 58.0-58.5°/11.4mm, n 1.437, d 0.940.** Shaken with BaO for several days, refluxed with BaO for 1hr, then fractionally distd under reduced pressure, and stored over molecular sieves.

β,β-**Dimethylacrylic acid** (senecioic acid) *[541-47-9]* **M 100.1, m 68°.** Crystd from hot water or pet ether (b 60-80°).

Dimethylamine *[124-40-4]* **M 45.1, fp -92.2°, b 0°/563mm, 6.9°/760mm.** Dried by passage through a KOH-filled tower, or using sodium at 0° during 18hr.

Dimethylamine hydrochloride *[506-59-2]* **M 81.6, m 171°.** Crystd from hot $CHCl_3$ or abs EtOH. Also recrystd from MeOH/ether soln. Dried in a vac desiccator over H_2SO_4, then P_2O_5.

p-Dimethylaminoazobenzene (Methyl Yellow) *[60-11-7]* **M 225.3, m 118-119°(dec).** Crystd from acetic acid or isooctane, or from 95% EtOH by adding hot water and cooling. Dried over KOH under vac at 50°. **CARCINOGEN.**

p-Dimethylaminobenzaldehyde (Ehrlich's Reagent) **M 149.2, m 74-75°.** Crystd from water, hexane, or from EtOH (2ml/g), after charcoal treatment, by adding excess of water. Also dissolved in aq acetic acid, filtered, and pptd with ammonia. Finally recrystd from EtOH.

p-Dimethylaminobenzoic acid *[619-84-1]* **M 165.2, m 242.5-243.5°(dec).** Crystd from EtOH/water.

p-Dimethylaminobenzophenone *[530-44-9]* **M 225.3, m 92-93°.** Crystd from EtOH.

dl-4-Dimethylamino-2,2-diphenylvaleramide *[5985-87-5]* **M 296.4, m 183-184°.** Crystd from aq EtOH.

(-)-L-4-Dimethylamino-2,2-diphenylvaleramide *[6078-64-4]* **M 296.4, 136.5-137.5°.** Crystd from pet ether or EtOH.

2-Dimethylaminoethanol *[108-01-0]* **M 89.1, b 134.5-135.5°, n 1.4362, d 1.430.** Dried with anhydrous K_2CO_3 or KOH, and fractionally distd.

4-Dimethylaminopyridine *[1122-58-3]* **M 122.2, m 108-109°, b 191°.** Recrystd from toluene [Sadownik *et al, JACS* **108** 7789 *1986*].

N,N-Dimethylaniline *[121-69-7]* **M 121.2, f.p.2°, b 84°/15mm, 193°/760mm, n^{25} 1.5556, d 0.956.** Primary and secondary amines (including aniline and monomethylaniline) can be removed by refluxing for some hours with excess acetic anhydride, and then fractionally distilling. Crocker and Jones (*JCS* 1808 *1959*) used four volumes of acetic anhydride, then distd off the greater part of it, and took up the residue in ice-cold dil HCl. Non-basic materials were removed by ether extraction, then the dimethylaniline was liberated with ammonia, extracted with ether, dried, and distd under reduced pressure. Metzler and Tobolsky (*JACS* **76** 5178 *1954*) refluxed with only 10% (w/w) of acetic anhydride, then cooled and poured into excess 20% HCl, which, after cooling, was extracted with ethyl ether. (The amine hydrochloride, remains in the aqueous phase.) The HCl soln was cautiously made alkaline to phenolphthalein, and the amine layer was drawn off, dried over KOH and fractionally distd under reduced pressure, under nitrogen. Suitable drying agents for dimethylaniline include NaOH, BaO, $CaSO_4$, and CaH_2.
Other purification procedures include the formation of the picrate, prepared in benzene soln and crystd to constant melting point, then decomposed with warm 10% NaOH and extracted into ether: the extract was washed with water, and distd under reduced pressure. The oxalate has also been used. The base has been fractionally crystd by partial freezing and also from aq 80% EtOH then from abs EtOH. It has been distd from zinc dust, under nitrogen.

2,6-Dimethylaniline *[87-62-7]* **M 121.2, f.p. 11°, b 210-211°/736mm, n 1.5604, d 0.974.** Converted to its hydrochloride which, after recrystn, was decomposed with alkali to give the free base. Dried over KOH and fractionally distd.

3,4-Dimethylaniline *[95-64-7]* **M 121.2, m 51°, b 116-118°/25mm, b 226°/760mm.** Crystd from ligroin.

9,10-Dimethylanthracene *[781-43-1]* **M 206.3, m 180-181°.** Crystd from EtOH, and by recrystn from the melt.

1,3-Dimethylbarbituric acid *[769-42-6]* **M 156.1, m 123°.** Crystd from water and sublimed in a vac. Also purified by dissolving 10g in 100ml of boiling $CCl_4/CHCl_3$ (8:2) (1g charcoal), filtered and cooled to 25°. Dried *in vacuo* [Kohn *et al, AC* **58** 3184 *1986*].

7,12-Dimethylbenz[a]anthracene *[57-97-6]* **M 256.4, m 122-123°.** Purified by chromatography on alumina. Crystd from acetone/EtOH.

5,6-Dimethylbenzimidazole *[582-60-5]* **M 146.2, m 205-206°.** Crystd from ethyl ether. Sublimed at 140°/3mm.

2,3-Dimethylbenzoic acid *[603-79-21]* **M 150.2, m 146°.** Crystd from EtOH and·is volatile in steam.

2,4-Dimethylbenzoic acid *[611-01-8]* **M 150.2, m 126-127°, b 267°/727mm.** Crystd from EtOH, and sublimed in a vac.

2,5-Dimethylbenzoic acid *[610-72-0]* **M 150.2, m 134°, b 268°/760mm,**
2,6-Dimethylbenzoic acid *[632-46-2]* **M 150.2, m 117°.** Steam distd, and crystd from EtOH.

3,4-Dimethylbenzoic acid *[619-04-5]* **M 150.2, m 166°.** Crystd from EtOH and sublimed in a vac.

3,5-Dimethylbenzoic acid *[419-06-9]* **M 150.2, m 170°.** Distd in steam, crystd from water or EtOH and sublimed in a vac.

4,4'-Dimethylbenzophenone *[54323-31-8]* **M 210.3, m 95°, b 333-334°/725mm.** Purified by zone refining.

2,5-Dimethyl-1,4-benzoquinone *[137-18-8]* **M 136.1, m 124-125°.** Crystd from EtOH.

2,6-Dimethyl-1,4-benzoquinone *[527-61-7]* **M 136.1, m 72° (sealed tube).** Crystd from water/EtOH (8:1).

2,3-Dimethylbenzothiophene *[31317-17-6]* **M 212.3, b 123-124°/10mm, n^{19} 1.6171.** Fractionated through a 90cm Monel spiral column.

N,N-**Dimethylbenzylamine** *[103-83-3]* **M 135.2, b 66-67°/15mm, 181°/760mm, n 1.501, d 0.900.** Refluxed with acetic anhydride for 24hr, then fractionally distd. The middle fraction was dried with KOH, distd under reduced pressure, and stored under vac. Distn of the amine with zinc dust, at reduced pressure, under nitrogen, has also been used.

4,4'-Dimethyl-2,2'-bipyridine *[1134-35-6]* **M 184.2, m 175-176°.** Crystd from ethyl acetate. [Elliott *et al, JACS* **107** 4647 *1985*.]

1,1'-Dimethyl-4,4'-bipyridylium dichloride (3H_2O; Methyl Viologen) *[1910-42-5]* **M 311.2, m >300°(dec).** Recrystd from MeOH/acetone mixture. Also crystd three times from abs EtOH [Bancroft *et al AC* **53** 1390 *1981*]. Dried at 80° in a vac.

N,N-**Dimethylbiuret** *[7710-35-2]* **M 131.1.** Purified by repeated crystn from the melt.

2,3-Dimethyl-1,3-butadiene *[513-81-5]* **M 82.2, m -69-70°, b 68-69°/760mm, n 1.4385, d 0.727.** Distd from $NaBH_4$, and purified by zone melting.

1,3-Dimethylbutadiene sulphone *[10033-92-8]* **M 145.2, m 40.4-41.0°.** Crystd from ethyl ether.

2,2-Dimethylbutane *[75-83-2]* **M 86.2, b 49.7°, n 1.36876, n^{25} 1.36595, d 0.649.** Distd azeotropically with MeOH, then washed with water, dried and distd.

2,3-Dimethylbutane *[79-29-8]* **M 86.2, b 58.0°, n 1.37495, n^{25} 1.37231, d 1.375.** Distd from sodium, passed through a column of silica gel (activated by heating in nitrogen to 350° before use) to remove unsaturated impurities, and again distd from sodium. Also distilled azeotropically with MeOH, then washed with water, dried and redistd.

2,3-Dimethylbut-2-ene *[563-79-1]* **M 84.2, b 72-73°/760mm, n 1.41153, d 0.708.** Purified by glc on a column of 20% squalene on chromosorb P at 50° [Flowers and Rabinovitch *JPC* **89** 563 *1985*]. Also washed with 1M NaOH soln followed by H$_2$O. Dried over Na$_2$SO$_4$, distd over powdered KOH under nitrogen and passed through activated alumina before use. [Woon *et al, JACS* **108** 7990 *1986*; Wong *et al, JACS* **109** 3428 *1987*.]

Dimethylcarbamoyl chloride *[79-44-7]* **M 107.5, m -33°, b 34°/0.1mm, n 1.4511, d 1.172.** Must dist under high vac to avoid decomposition.

3,3'-Dimethylcarbanilide *[620-50-8]* **M 240.3, m 225°.** Crystd from ethyl acetate.

Dimethyl carbonate *[616-38-5]* **M 90.1, m 4.65°, b 90-91°, n 1.369, d 1.070.** Contains small amounts of water and alcohol which form azeotropes. Stood for several days in contact with Linde type 4A molecular sieves, then fractionally distd. The middle fraction was frozen slowly at 2°, several times, retaining 80% of the solvent at each cycle.

N,N-**Dimethyl-*p*-chlorobenzene** *[698-69-1]* **M 155.6.** Purified by vacuum sublimation [Guarr *et al, JACS* **107** 5104 *1985*].

cis-and trans-**1,4-Dimethylcyclohexane** *[589-90-2]* **M 112.2, b 120°, n 1.427, d 0.788.** Freed from olefines by shaking with conc H$_2$SO$_4$, washing with water, drying and fractionally distilling.

5,5-Dimethyl-1,3-cyclohexanedione see **dimedone**.

1,2-Dimethylcyclohexene *[1674-10-8]* **M 110.2, b 135-136°/760mm, n 1.4591, d 0.826.** Passed through a column of basic alumina and distd.

1,5-Dimethyl-1,5-diazaundecamethylene polymethobromide (Hexadimethrene, polybrene) *[28728-55-4]*. Purified by chromatography on Dowex 50 and/or by filtration through alumina before use [Frank *Hoppe-Seyler's Z Physiol Chemie* **360** 997 *1979*].

Dimethyldihydroresorcinol see **dimedone**.

2,9-Dimethyl-4,7-diphenyl-1,10-phenanthroline *[4733-39-5]* **M 360.5, m >280°.** Purified by recrystn from benzene.

Dimethyl disulphide *[624-92-0]* **M 94.2, f.p. -98°, b 40°/12mm, 110°/760mm, n 1.5260, d 1.0605.** Passed through neutral alumina before use.

Dimethyl ether see **methyl ether**.

2,2-Dimethylethyleneimine *[2658-24-4]* **M 71.1, b 70.5-71.0°.** Freshly distd from sodium before use.

N,N-**Dimethylformamide** (DMF) *[68-12-2]* **M 73.1, b 76°/39mm, 153°/760mm, n 1.4297, n^{25} 1.4269, d 0.948.** Decomposes slightly at its normal boiling point to give small amounts of dimethylamine and carbon monoxide. The decomposition is catalysed by acidic or basic materials, so that even at room temperature DMF is appreciably decomposed if allowed to stand for several hours with solid KOH, NaOH or CaH$_2$. If these reagents are used as dehydrating agents, therefore, they should not be refluxed with the DMF. Use of CaSO$_4$, MgSO$_4$, silica gel or Linde type 4A molecular sieves is preferable, followed by distn under reduced pressure. This procedure is adequate for most laboratory purposes. Larger amounts of water can be removed by azeotropic distn with benzene (10% v/v, previously dried over CaH_2), at atmospheric pressure: water and benzene distil below 80°. The liquid remaining in the distn flask is further dried by adding $MgSO_4$ (previously ignited overnight at 300-400°) to give 25g/L. After shaking for one day, a further quantity of $MgSO_4$ is added, and the DMF distd at 15-20mm pressure through a 3-ft vacuum-jacketed column packed with steel helices. However, $MgSO_4$ is an inefficient drying agent, leaving about 0.01M water in the final DMF. More efficient drying (to around 0.001-0.007M water) is achieved by standing with powdered BaO, followed by decanting before distn, with alumina powder (50g/L; previously heated overnight to 500-600°), and distilling from more of the alumina; or by refluxing at 120-140° for 24hr with triphenylchlorosilane (5-10g/L), then distilling at *ca* 5mm pressure [Thomas and Rochow JACS **79** 1843 *1957*]. Free amine in DMF can be detected by colour reaction with 1-fluoro-2,4-dinitrobenzene.

For efficiency of desiccants in drying dimethyl formamide see Burfield and Smithers [*JOC* **43** 3966 *1978*].

[For review on purification, tests from purity and physical properties, see Juillard *PAC* **49** 885 *1977*.]

d,l-**2,4-Dimethylglutaric acid**   *[2121-67-7]* M 160.2, m 144-145°. Distd in steam and crystd from ether/pet ether.

3,3-Dimethylglutaric acid   *[4839-46-7]* M 160.2, m 103-104°, b 89-90°/2mm, 126-127°/4.5mm. Crystd from water, benzene or ether/pet ether. Dried in a vac.

3,3-Dimethylglutarimide   *[1123-40-6]* M 141.2, m 144-146°. Recrystd from EtOH [Arnett and Harrelson *JACS* **109** 809 *1987*].

N,N-**Dimethylglycinehydrazide hydrochloride**   *[539-64-0]* M 153.6, m 181°. Crystd by adding EtOH to a conc aq soln.

Dimethylglyoxime   *[95-45-4]* M 116.1, m 240°. Crystd from EtOH (10ml/g) or aq EtOH.

2,5-Dimethyl-2,4-hexadiene   *[764-13-6]* M 110.2, f.p. 14.5°, b 132-134°, n 1.4796, d 0.773. Distd, then repeatedly fractionally crystd by partial freezing. Immediately before use, the material was passed through a column containing Woelm silica gel (activity I) and Woelm alumina (neutral) in separate layers.

2,2-Dimethylhexane   *[590-73-8]* M 114.2, m -121.2°, b 107°, d 0.695,
2,5-Dimethylhexane   *[592-13-2]* M 114.2, m -91.2°, b 109°, d 0.694. Dried over type 4A molecular sieves and distd.

2,5-Dimethylhexane-2,5-diol   *[110-03-2]* M 146.2, m 88-90°. Purified by fractional crystn. Then the diol was dissolved in hot acetone, treated with activated charcoal, and filtered while hot. The soln was cooled and the diol was filtered off and washed well with cold acetone. The crystn process was repeated several times and the crystals were dried under a vac in a freeze-drying apparatus [Goates *et al JCSFT 1* **78** 3045 *1982*].

5,5-Dimethylhydantoin   *[77-71-4]* M 128.1, m 177-178°. Crystd from EtOH and sublimed in a vac.

1,1-Dimethylhydrazine   *[57-14-7]* M 60.1, b 60.1°/702mm, n 1.408, d 0.790. Fractionally distd through a 4-ft column packed with glass helecies. Pptd as its oxalate from ethyl ether soln. After crystn from 95% EtOH, the salt was decomposed with aq satd NaOH, and the free base was distd, dried over BaO and redistd [McBride and Kruse *JACS* **79** 572 *1957*]. Distn and storage should be under nitrogen.

4,6-Dimethyl-2-hydroxypyrimidine   *[108-79-2]* M 124.1, m 198-199°. Crystd from abs EtOH (charcoal).

1,2-Dimethylimidazole *[1739-84-0]* **M 96.1, b 206°/760mm, d 1.084.** Crystd from benzene and stored at 0-4°. [Gorun *et al*, *JACS* **109** 4244 *1987*.]

1,1-Dimethylindene *[18636-55-0]* **M 144.2.** Purified by gas chromatography.

Dimethyl itaconate *[617-52-7]* **M 158.2, m 38°, b 208°, d 1.124.** Crystd from MeOH by cooling to -78°.

Dimethylmaleic anhydride *[766-39-2]* **M 126.1, m 96°, b 225°/760mm.** Distd from benzene/ligroin and sublimed in a vac.

Dimethylmalonic acid *[595-46-0]* **M 132.1, m 192-193°.** Crystd from benzene/pet ether and sublimed in a vac with slight decomposition.

1,5-Dimethylnaphthalene *[571-61-9]* **M 156.2, m 81-82°, b 265-266°.** Crystd from 85% aq EtOH.

2,3-Dimethylnaphthalene *[581-40-8]* **M 156.2, m 104-104.5°,**
2,6-Dimethylnaphthalene *[581-42-0]* **M 156.2, m 110-111°, b 122.5-123.5°/10mm, 261-262°/760mm.** Distd in steam and crystd from EtOH.

3,3'-Dimethylnaphthidine (4,4'-diamino-3,3'-dimethyl-1,1'-binaphthyl) *[13138-48-2]* **M 312.4, m 213°.** Recrystd from EtOH or pet ether (b 60-80°).

N,N-Dimethyl-*m*-nitroaniline *[619-31-8]* **M 166.1, m 60°.** Crystd from EtOH.

N,N-Dimethyl-*p*-nitroaniline *[100-23-2]* **M 166.1, m 164.5-165.2°.** Crystd from EtOH or aq EtOH. Dried under vac.

N,N-Dimethyl-*p*-nitrosoaniline *[138-89-6]* **M 150.2, m 85.8-86.0°.** Crystd from pet ether or $CHCl_3/CCl_4$. Dried in air.

N,N-Dimethyl-*p*-nitrosoaniline hydrochloride *[42344-05-8]* **M 186.7, m 177°.** Crystd from hot water in the presence of a little HCl.

2,6-Dimethyl-2,4,6-octatriene *[7216-56-0]* **M 136.2, b 80-82°/15mm, ε_{278nm} 42,870.** Repeated distn at 15mm through a long column of glass helices, the final distn being from sodium under nitrogen.

Dimethylolurea *[140-95-4]* **M 120.1, m 123°.** Crystd from aq 75% EtOH.

Dimethyl oxalate *[553-90-2]* **M 118.1, m 54°, b 163-165°, d 1.148.** Crystd repeatedly from EtOH. Degassed under nitrogen high vac and distd.

3,3-Dimethyloxetane *[6921-35-3]* **M 86.1, b 79.2-80.3°/760mm.** Purified by gas chromatography using a 2m silicone oil column.

2,3-Dimethylpentane *[565-59-3]* **M 100.2, b 89.8°, n 1.39197, n^{25} 1.38946, d 0.695.** Purified by azeotropic distn with EtOH, followed by washing out the EtOH with water, drying and distn [Streiff *et al*, *J Res Nat Bur Stand* **37** 331 *1946*].

2,4-Dimethylpentane *[108-08-7]* **M 100.2, 80.5°, n 1.3814, n^{25} 1.37882, d 0.763.** Extracted repeatedly with conc H_2SO_4, washed with water, dried and distd. Percolated through silica gel (previously heated in nitrogen to 350°). Purified by azeotropic distn with EtOH, followed by washing out the EtOH with water, drying and distn.

4,4-Dimethyl-1-pentene *[762-62-9]* **M 98.2, b 72.5°/760mm, n 1.3918, d 0.6827.** Purified by passage through alumina before use [Traylor *et al*, *JACS* **109** 3625 *1987*].

Dimethyl peroxide *[690-02-8]* **M 62.1, b 13.5°/760mm, n 1.3503, d 0.8677.** Purified by repeated trap-to-trap fractionation until no impurities could be detected by gas IR spectroscopy [Haas and Oberhammer *JACS* **106** 6146 *1984*]. *All necessary precautions should be taken in case of* **EXPLOSION.**

2,9-Dimethyl-1,10-phenanthroline *[484-11-7]* **M 208.3, m 162-164°.** Purified as hemihydrate from water, and as anhydrous from benzene.

2,3-Dimethylphenol *[526-75-0]* **M 122.2, m 75°, b 120°/20mm, 218°/760mm.** Crystd from aq EtOH.

2,5-Dimethylphenol *[95-87-1]* **M 122.2, m 73°, b 211.5°/762mm.** Crystd from EtOH/ether.

2,6-Dimethylphenol *[576-26-1]* **M 122.2, m 49°, b 203°/760mm.** Fractionally distd under nitrogen, crystd from benzene or hexane, and sublimed at 38°/10mm.

3,4-Dimethylphenol *[95-65-8]* **M 122.2, m 65°, b 225°/757mm,**
3,5-Dimethylphenol *[108-68-9]* **M 122.2, m 68°, b 219°.** Heated with an equal weight of conc H_2SO_4 at 103-105° for 2-3hr, then diluted with four volumes of water, refluxed for 1hr, and either steam distd or extracted repeatedly with ethyl ether after cooling to room temperature. The steam distillate was also extracted and evaporated to dryness. (The purification process depends on the much slower sulphonation of 3,5-dimethylphenol than most of its likely contaminants.) [Kester *IEC* **24** 770 *1932*.] It can also be crystd from water, hexane or pet ether, and vac sublimed. [Bernasconi and Paschalis *JACS* **108** *1986*.]

N,N-**Dimethyl-2-(α-phenyl-*o*-tolyloxy)ethylamine hydrochloride** see **phenyltoloxamine hydrochloride.**

N,N-**Dimethyl-*p*-phenylazoaniline** see *p*-**dimethylaminoazobenzene.**

1,2-Dimethyl-3-phenyl-5-pyrazolone see **antipyrine.**

Dimethyl phthalate *[131-11-3]* **M 194.2, b 282°, n 1.5149, d 1.190, d^{25} 1.1865.** Washed with aq Na_2CO_3, then distd water, dried ($CaCl_2$) and distd under reduced pressure (b 151-152°/0.1mm).

2,2-Dimethyl-1,3-propanediol *[126-30-7]* **M 104.2, m 128.4-129.4°, b 208°/760mm.** Crystd from benzene or acetone/water (1:1).

2,2-Dimethyl-1-propanol (neopentyl alcohol) *[75-84-3]* **M 88.2, m 52°, b 113.1°/760mm.** Difficult to distil because it is a solid at ambient temperatures. Purified by fractional crystallisation and sublimation.

N,N-**Dimethylpropionamide** *[758-96-3]* **M 101.2, b 175-178°, n 1.440, d 0.920.** Shaken over BaO for 1-2 days, then distd at reduced pressure.

2,5-Dimethylpyrazine *[123-32-0]* **M 108.1, b 156°, n 1.502, d 0.990.** Purified *via* its picrate (m 150°)[Wiggins and Wise *JCS* 4780 *1956*].

3,5-Dimethylpyrazole *[67-51-6]* **M 96.1, m 107-108°.** Crystd from cyclohexane or water. [Barszez *et al*, *JCSDT* 2025 *1986*.]

Dimethylpyridine see **lutidine.**

2,3-Dimethylquinoxaline *[2379-55-7]* **M 158.2, m 106°.** Crystd from distd water.

2,4-Dimethylresorcinol *[634-65-1]* **M 138.1, m 149-150°.** Crystd from pet ether (b 60-80°).

meso-α,β-**Dimethylsuccinic acid**   *[608-40-2]* **M 146.1, m 211°.**   Crystd from EtOH/ether or EtOH/chloroform.

2,2-Dimethylsuccinic acid   *[597-43-3]* **M 146.1, m 141°.**   Crystd from EtOH/ether or EtOH/chloroform.

(±)-2,3-Dimethylsuccinic acid   *[13545-04-5]* **M 146.1, m 129°.**   Crystd from water.

Dimethyl sulphide   *[75-18-3]* **M 62.1, f.p. -98.27°, b 0°/172mm, 37.5-38°/760mm, n^{25} 1.4319, d^{21} 0.8458.** Purified *via* the Hg(II) chloride complex by dissolving 1 mole of Hg(II)Cl$_2$ in 1250ml of EtOH and slowly adding the boiling alcoholic soln of dimethyl sulphide to give the right ratio for 2(CH$_3$)$_2$S.3HgCl$_2$. After recrystn of the complex to constant melting point, 500g of complex is heated with 250ml conc HCl in 750ml of water. The sulphide is separated, washed with water, and dried with CaCl$_2$ and CaSO$_4$. Finally, it is distd under reduced pressure from sodium.

2,4-Dimethylsulpholane   *[1003-78-7]* **M 148.2, b 128°/77mm, d^{25} 1.1314.**   Vac distd.

Dimethyl sulphone   *[67-71-0]* **M 94.1, m 109°.**   Crystd from water. Dried over P$_2$O$_5$.

Dimethyl sulphoxide (DMSO)   *[67-68-5]* **M 78.1, m 18.0-18.5°, b 75.6-75.8°/12mm, 190°/760mm, n 1.479, d 1.100.** Colourless, odourless, very hygroscopic liquid, synthesised from dimethyl sulphide. The main impurity is water, with a trace of dimethyl sulphone. The Karl-Fischer test is applicable. It is dried with Linde types 4A or 13X molecular sieves, by prolonged contact and passage through a column of the material, then distd under reduced pressure. Other drying agents include CaH$_2$, CaO, BaO and CaSO$_4$. It can also be fractionally crystd by partial freezing. More extensive purification is achieved by standing overnight with freshly heated and cooled chromatographic grade alumina. It is then refluxed for 4hr over CaO, dried over CaH$_2$, and then fractionally distd at low pressure. For efficiency of desiccants in drying dimethyl sulphoxide see Burfield and Smithers [*JOC* **43** 3966 *1978*; Sato *et al*, *JCSDT* 1949 *1986*].

Dimethyl terephthalate   *[120-61-6]* **M 194.2, m 150°.**   Purified by zone melting.

N,N-**Dimethylthiocarbamoyl chloride**   *[16420-13-6]* **M 123.6, m 42-43°, b 64-65°/0.1mm.**   Crystd twice from pentane.

N,N-**Dimethyl-*o*-toluidide**   *[609-72-3]* **M 135.2, b 68°/10mm, 211-211.5°/760mm, n 1.53664, d 0.937.** Isomers and other bases have been removed by heating in a water bath for 100hr with two equivalents of 20% HCl and two and a half volumes of 40% aq formaldehyde, then making the soln alkaline and separating the free base. After washing well with water it was distd at 10mm pressure and redistd at ambient pressure [von Braun and Aust *Ber* **47** 260 *1914*]. Other procedures include drying with NaOH, distilling from zinc in an atmosphere of nitrogen under reduced pressure, and refluxing with excess of acetic anhydride in the presence of conc H$_2$SO$_4$ as catalyst, followed by fractional distn under vac.

N,N-**Dimethyl-*m*-toluidide**   *[121-72-2]* **M 135.2, b 211.5-212.5°, d 0.93,**
N,N-**Dimethyl-*p*-toluidide**   *[99-97-8]* **M 135.2, b 93-94°/11mm, b 211°, n 1.5469, d 0.937.**   Methods described for *N,N*-dimethylaniline are applicable. Also dried over BaO, distd and stored over KOH.

1,3-Dimethyluracil   *[874-14-6]* **M 140.1, m 121-122°.**   Crystd from EtOH/ether.

sym-**Dimethylurea**   *[96-31-1]* **M 88.1, m 106°.**   Crystd from acetone/ethyl ether by cooling in an ice bath. Also crystd from EtOH and dried at 50° and 5mm for 24hr [Bloemendahl and Somsen *JACS* **107** 3426 *1985*.]

Di-β-naphthol   *[41024-90-21]* **M 286.3, m 218°.**   Crystd from toluene or benzene (10ml/g).

β,β'-Dinaphthylamine *[532-18-3]* M 269.3,m 170.5°. Crystd from benzene.

2,4-Dinitroaniline *[97-02-9]* M 183.1, m 180°, ε$_{348nm}$ 12,300 in dil aq HClO$_4$. Crystd from boiling EtOH by adding one-third volume of water and cooling slowly. Dried in a steam oven.

2,6-Dinitroaniline *[606-22-4]* M 183.1, m 139-140°. Purified by chromatography on alumina, then crystd from benzene or EtOH.

2,4-Dinitroanisole *[119-27-7]* M 198.1, m 94-95°. Crystd from aq EtOH.

3,5-Dinitroanisole *[5327-44-6]* M 198.1, m 105-106°. Purified by repeated crystn from water.

1,2-Dinitrobenzene *[528-29-0]* M 168.1, m 116.5°. Crystd from EtOH.

1,3-Dinitrobenzene *[99-65-0]* M 168.1, m 90.5-91°. Crystd from alkaline EtOH soln (20g in 750ml 95% EtOH at 40°, plus 100ml of 2M NaOH) by cooling and adding 2.5L of water. The ppt, after filtering off, washing with water and sucking dry, was crystd from 120ml, then 80ml of abs EtOH [Callow, Callow and Emmens *BJ* 32 1312 *1938*]. Has also been crystd from MeOH, CCl$_4$ and ethyl acetate. Can be sublimed in vac. [Tanner *JOC* 52 2142 *1987*.]

1,4-Dinitrobenzene *[100-25-4]* M 168.1, m 173°. Crystd from EtOH or ethyl acetate. Dried under vac over P$_2$O$_5$. Can be sublimed in vac.

2,4-Dinitrobenzenesulphenyl chloride *[528-76-7]* M 234.6, m 96°. Crystd from CCl$_4$.

2,4-Dinitrobenzenesulphonyl chloride *[1656-44-6]* M 266.6, m 102°. Crystd from benzene or benzene/pet ether.

3,3'-Dinitrobenzidine see 4,4'-Diamino-3,3'-dinitrobiphenyl.

2,4-Dinitrobenzoic acid *[610-30-3]* M 212.1, m 183°. Crystd from aq 20% EtOH (10ml/g), dried at 100°.

2,5-Dinitrobenzoic acid *[610-28-6]* M 212.1, m 179.5-180°. Crystd from distd water. Dried in a vac desiccator.

2,6-Dinitrobenzoic acid *[603-12-3]* M 212.1, m 202-203°. Crystd from water.

3,4-Dinitrobenzoic acid *[528-45-0]* M 212.1, m 166°. Crystd from EtOH by addn of water.

3,5-Dinitrobenzoic acid *[99-34-3]* M 212.1, m 205°. Crystd from distd water or 50% EtOH (4ml/g). Dried in a vac desiccator or at 70° over BaO under vac for 6hr.

4,4'-Dinitrobenzoic anhydride *[902-47-6]* M 406.2, m 189-190°. Crystd from acetone.

3,5-Dinitrobenzoyl chloride *[99-34-3]* M 230.6, m 69.5°. Crystd from CCl$_4$ or pet ether (b 40-60°). It reacts readily with water, and should be kept in sealed tubes or under pet ether.

2,2'-Dinitrobiphenyl *[2436-96-6]* M 244.2, m 123-124°,
2,4'-Dinitrobiphenyl *[606-81-5]* M 244.2, m 92.7-93.7°. Crystd from EtOH.

4,4'-Dinitrobiphenyl *[1528-74-1]* M 244.2, m 240.9-241.8°. Crystd from benzene, EtOH (charcoal) or acetone. Dried under vac over P$_2$O$_5$.

2,4-Dinitrochlorobenzene *[97-00-7]* **M 202.6, m 51° (stable form), 43° (unstable form), b 315°/760mm.** Crystd from EtOH (stable form), or from ether (unstable form).

4,6-Dinitro-*o*-cresol *[534-52-1]* **M 198.1, m 85-86°,**
2,4-Dinitrodiphenylamine *[961-68-2]* **M 259.2, m 157°.** Crystd from aq EtOH.

4,4'-Dinitrodiphenylurea *[587-90-6]* **M 302.2, m 312°(dec).** Crystd from EtOH. Sublimes in vac.

2,4-Dinitrofluorobenzene (Sanger's reagent) *[70-34-8]* **M 186.1, m 25-27°, b 133°/2mm, 140-141°/5mm. d 1.483.** Crystd from ether or EtOH. Vac distd through a Todd Column. If it is to be purified by distn *in vacuo*, the distn unit must be allowed to cool before air is allowed into the apparatus otherwise the residue carbonizes spontaneously and an **EXPLOSION** may occur. The material is a **SKIN IRRITANT** and may cause serious dermatitis.

3,5-Dinitro-2-hydroxybenzoic acid see **3,5-dinitrosalicylic acid.**

3,4-Dinitro-2-methylbenzoic acid see **3,5-dinitro-*o*-toluic acid.**

1,8-Dinitronaphthalene *[602-38-0]* **M 218.2, m 170-171°.** Crystd from benzene.

2,4-Dinitro-1-naphthol (Martius Yellow) *[605-69-6]* **M 234.2, m 81-82°.** Crystd from benzene or aq EtOH.

2,4-Dinitrophenetole *[610-54-8]* **M 240.2, m 85-86°.** Crystd from aq EtOH.

2,4-Dinitrophenol *[51-28-5]* **M 184.1, m 114°.** Crystd from benzene, EtOH, EtOH/water or water acidified with dil HCl, then recrystd from CCl_4. Dried in an oven and stored in a vac desiccator over $CaSO_4$.

2,5-Dinitrophenol *[329-71-5]* **M 184.1, m 108°.** Crystd from water containing a little EtOH.

2,6-Dinitrophenol *[573-56-8]* **M 184.1, m 63.0-63.7°.** Crystd from benzene/cyclohexane, aq EtOH, water or benzene/pet ether (b 60-80°, 1:1).

3,4-Dinitrophenol *[577-71-9]* **M 184.1, m 138°.** Steam distd and crystd from water and air-dried. **CAUTION** - explosive when dry, store with 10% water.

3,5-Dinitrophenol *[586-11-8]* **M 184.1, m 126°.** Crystd from benzene or $CHCl_3$/pet ether. Should be stored with 10% water because it is **EXPLOSIVE** when dry.

2,4-Dinitrophenylacetic acid *[643-43-6]* **M 226.2, m 179°(dec).** Crystd from water.

2,4-Dinitrophenylhydrazine *[119-26-6]* **M 198.1, m 200°(dec).** Crystd from butan-1-ol, dioxane, EtOH or ethyl acetate.

2,2-Dinitropropane *[595-49-3]* **M 162.1, m 53.5°.** Crystd from EtOH or MeOH. Dried over $CaCl_2$ or under vac for 1hr just above the melting point.

2,4-Dinitroresorcinol *[519-44-8]* **M 200.1, m 160°.** Crystd from aq EtOH.

3,5-Dinitrosalicylic acid *[609-99-4]* **M 228.1, m 173-174°.** Crystd from water.

2,6-Dinitrothymol *[303-21-9]* **M 240.2, m 53-54°.** Crystd from aq EtOH.

2,3-Dinitrotoluene *[602-01-7]* **M 182.1, m 63°.** Distd in steam and crystd from water or benzene/pet ether. Stored with 10% water. *Could be* **EXPLOSIVE** *when dry.*

2,4-Dinitrotoluene *[121-14-2]* **M 182.1, m 70.5-71.0°.** Crystd from acetone, isopropanol or MeOH. Dried under vac over H_2SO_4. Purified by zone melting.

2,5-Dinitrotoluene *[619-15-8]* **M 182.1, m 51.2°.** Crystd from benzene.

2,6-Dinitrotoluene *[606-20-2]* **M 182.1, m 64.3°.** Crystd from acetone.

3,4-Dinitrotoluene *[610-39-9]* **M 182.1, m 61°.** Distil in steam and cryst from benzene/pet ether. Store with 10% of water to avoid **EXPLOSION**.

2,5-Dinitro-*o*-toluic acid *[28169-46-2]* **M 226.2, m 206°.** Crystd from water.

2,4-Dinitro-*m*-xylene *[603-02-1]* **M 196.2, m 83-84°.** Crystd from EtOH.

Dinonyl phthalate (mainly 3,5,5-trimethylhexyl isomer) *[28553-12-0]* **M 418.6, n 1.4825, d 0.9640.** Washed with aq Na_2CO_3, then shaken with water. Ether was added to break the emulsion, and the soln was washed twice with water, and dried ($CaCl_2$). After evaporating the ether, the residual liquid was distd three times under reduced pressure. It was stored in a vac desiccator over P_2O_5

Dioctadecyldimethylammonium bromide *[3700-67-2]* **M 570.5.** Crystd from acetone [Lukac *JACS* **106** 4387 *1984*].

Dioctyl phenylphosphonate *[1754-47-8]* **M 378.8, n^{25} 1.4780, d 1.485.** Purified as described under diisooctyl phenylphosphonate.

Diosgenin *[512-04-9]* **M 294.5, m 204-207°, $[\alpha]_D^{25}$ -129° (in Me_2CO).** Crystd from acetone.

1,3-Dioxane *[505-22-6]* **M 88.1, b 104.5°/751mm, n 1.417, d 1.040.** Dried with sodium and fractionally distd.

1,3-Dioxalane *[646-06-0]* **M 74.1, b 75.0-75.2°, n^{21} 1.3997, d 1.0600.** Dried with solid NaOH, KOH or $CaSO_4$, and distd from sodium or sodium amalgam. Barker *et al* [*JCS* 802 *1959*] heated 34ml of dioxalane under reflux with 3g of PbO_2 for 2hr, then cooled and filtered. After adding xylene (40ml) and PbO_2 (2g) to the filtrate, the mixture was fractionally distd. Addition of xylene (20ml) and sodium wire to the main fraction (b 70-71°) led to vigorous reaction, following which the mixture was again fractionally distd. Xylene and sodium additions were made to the main fraction (b 73-74°) before it was finally distd.

1,4-Dioxane *[123-91-1]* **M 88.1, f.p. 11.8°, b 101.3°, n^{15} 1.4236, n^{25} 1.42025, d^{25} 1.0292.** Prepared commercially either by dehydration of ethylene glycol with H_2SO_4 and heating ethylene oxide or bis(β-chloroethyl)ether with NaOH.
Usual impurities are acetaldehyde, ethylene acetal, acetic acid, water and peroxides. Peroxides can be removed (and the aldehyde content decreased) by percolation through a column of activated alumina (80g per 100-200ml solvent), by refluxing with $NaBH_4$ or anhydrous stannous chloride and distilling, or by acidification with conc HCl, shaking with ferrous sulphate and leaving in contact with it for 24hr before filtering and purifying further.
Hess and Frahm [*Ber* **71** 2627 *1938*] refluxed 2L of dioxane with 27ml conc HCl amd 200ml water for 12hr with slow passage of nitrogen to remove acetaldehyde. After cooling the soln KOH pellets were added slowly and with shaking until no more would dissolve and a second layer had separated. The dioxane was decanted, treated with fresh KOH pellets to remove any aq phase, then transferred to a clean flask where it was refluxed for 6-12hr with sodium, then distd from it. Alternatively, Kraus and Vingee [*JACS* **56** 511 *1934*] heated on a steam bath with solid KOH until fresh addition of KOH gave

no more resin (due to acetaldehyde). After filtering through paper, the dioxane was refluxed over sodium until the surface of the metal was not further discoloured during several hours. It was then distd from sodium.

The acetal (b 82.5°) is removed during fractional distn. Traces of benzene, if present, can be removed as the benzene/MeOH azeotrope by distn in the presence of MeOH. Distn from LiAlH$_4$ removes aldehydes, peroxides and water. Dioxane can be dried using Linde type 4X molecular sieves. Other purification procedures include distn from excess C$_2$H$_5$MgBr, refluxing with PbO$_2$ to remove peroxides, fractional crystn by partial freezing and the addition of KI to dioxane acidified with aq HCl. Dioxane should be stored out of contact with air, preferably under N$_2$.

A detailed purification procedure is as follows: Dioxane was stood over ferrous sulphate for at least 2 days, under nitrogen. Then water (100ml) and conc HCl (14ml) / litre of dioxane were added (giving a pale yellow colour). After refluxing for 8-12hr with vigorous N$_2$ bubbling, pellets of KOH were added to the warm soln to form two layers and to discharge the colour. The soln was cooled rapidly with more KOH pellets being added (magnetic stirring) until no more dissolved in the cooled soln. After 4-12hr, if the lower phase was not black, the upper phase was decanted rapidly into a clean flask containing sodium, and refluxed over sodium (until freshly added sodium remained bright) for 1hr. The middle fraction was collected (and checked for minimum absorbency below 250nm). The distillate was fractionally frozen three times by cooling in a refrigerator, with occasional shaking or stirring. This material was stored in a refrigerator. For use it was thawed, refluxed over sodium for 48hr, and distilled into a container for use. All joints were clad with Teflon tape.

Coetzee and Chang [*PAC* **57** 633 *1985*] dried the solvent by passing it slowly through a column (20g/L) of 3A molecular sieve activated by heating at 250° for 24hr. Impurities (including peroxides) were removed by passing the effluent slowly through a column packed with type NaX zeolite (pellets ground to 0.1mm size) activated by heating at 400° for 24hr, or chromatographic grade basic Al$_2$O$_3$ activated by heating at 250° for 24hr. After removal of peroxides the effluent was refluxed several hours over sodium wire, excluding moisture, distilled under nitrogen or argon and stored in the dark.

One of the best tests of purity of dioxane is the formation of the purple disodium benzophenone complex during reflux and its persistence on cooling. (Benzophenone is better than fluorenone for this purpose, and for the storing of the solvent.) [Carter, McClelland and Warhurst *TFS* **56** 343 *1960*.]

S-1,2-Dipalmitin   *[761-35-3]* M 568-9, m 68-69° [α]$_D$ -2.9° (c 8, CHCl$_3$), [α]$_{546}$ +1.0° (c 10, CHCl$_3$/MeOH, 9:1). Crystd from chloroform/pet ether.

d,l-βγ-**Dipalmitoylphosphatidyl choline**   *[2797-68-4]* M 734.1, m 230-233°. Recrystd from chloroform and dried for 48hr at 10^{-5} torr [O'Leary and Levine *JPC* **88** 1790 *1984*].

2,5-Di-*tert*-pentylhydroquinone see **2,5-Di-*tert*-amylhydroquinone**.

4,4'-Di-*n*-pentyloxyazoxybenzene   *[64242-26-8]* M 370.5. Crystd from acetone, and dried by heating under vac.

Diphenic acid   *[482-05-3]* M 242.2, m 228-229°. Crystd from water.

Diphenic anhydride   *[6050-13-1]* M 466.3, m 217°. After removing free acid by extraction with cold aq Na$_2$CO$_3$, the residue has been crystd from acetic anhydride and dried at 100°.

N,N-**Diphenylacetamidine**   *[621-09-0]* M 210.3, m 131°. Crystd from EtOH, then sublimed under vac at *ca* 96° onto a "finger" cooled in solid CO$_2$/MeOH, with continuous pumping to free it from occluded solvent.

Diphenylacetic acid   *[117-34-0]* M 212.3, m 147.4-148.4°. Crystd from benzene or aq 50% EtOH.

Diphenylacetonitrile   *[86-29-3]* M 193.3, m 73-75°. Crystd from EtOH or pet ether (b 90-100°).

Diphenylacetylene (tolan)   *[501-65-5]* M 178.2, m 62.5°, b 90-97°/0.3mm. Crystd from EtOH.

Diphenylamine *[122-39-4]* **M 169.2, m 62.0-62.5°**. Crystd from pet ether, MeOH, or EtOH/water. Dried under vac.

Diphenylamine-2-carboxylic acid *[91-40-7]* **M 213.2, m 184°**,
Diphenylamine-2,2'-dicarboxylic acid *[579-92-0]* **M 257.2.** Crystd from EtOH.

9,10-Diphenylanthracene *[1499-10-1]* **M 330.4, m 248-249°**. Crystd from acetic acid or xylene [Baumstark *et al, JOC* **52** 3308 *1987*].

N-**Diphenylanthranilic acid** see **diphenylamine-2-carboxylic acid.**

N,N'-**Diphenylbenzidine** *[531-91-9]* **M 336.4, m 245-247°**. Crystd from toluene or ethyl acetate. Stored in the dark.

trans-trans-**1,4-Diphenylbuta-1,3-diene** *[886-65-7]* **M 206.3, m 153-153.5°**. Its soln in pet ether (b 60-70°) was chromatographed on an alumina-Celite column (4:1) and the column was washed with the same solvent. The main zone was cut out, eluted with ethanol and transferred to pet ether,which was then dried and evaporated [Pinckard, Wille and Zechmesiter *JACS* **70** 1938 *1948*]. Recrystd from hexane.

sym-**Diphenylcarbazide** *[140-22-7]* **M 242.3, m 172°**. A common impurity is phenylsemicarbazide which can be removed by chromatography [Willems *et al, Anal Chim Acta* **51** 544 *1970*]. Crystd from EtOH or glacial acetic acid.

1,5-Diphenylcarbazone *[538-62-5]* **M 240.3, m 124-127°**. Crystd from EtOH (*ca* 5ml/g), and dried at 50°. A commercial sample, nominally *sym*-diphenylcarbazone but of **m** 154-156°, was a mixture of diphenylcarbazide and diphenylcarbazone. The former was removed by dissolving 5g of the crude material in 75ml of warm EtOH, then adding 25g Na_2CO_3 dissolved in 400ml of distd water. The alkaline soln was cooled and extracted six times with 50ml portions of ethyl ether (discarded). Diphenylcarbazone was then pptd by acidifying the alkaline soln with 3M HNO_3 or glacial acetic acid. It was filtered on a Büchner funnel, air dried, and stored in the dark [Gerlach and Frazier *AC* **30** 1142 *1958*]. Other impurities were phenylsemicarbazide and diphenylcarbodiazone. Impurities can be detected by chromatography [Willems *et al, Anal Chim Acta* **51** 544 *1970*].

Diphenylcarbinol see **benzhydrol.**

Diphenyl carbonate *[102-09-0]* **M 214.2, m 80°**. Purified by sublimation, and by preparative gas chromatography with 20% Apiezon on Embacel, and crystn from EtOH.

Diphenyl diselenide *[1666-13-3]* **M 312.1, m 62-64°**. Recrystd twice from hexane [Kice and Purkiss *JOC* **52** 3448 *1987*].

Diphenyl disulphide *[882-33-7]* **M 218.3, m 60.5°**. Crystd from MeOH. [Alberti *et al, JACS* **108** 3024 *1986*.] Crystd repeatedly from hot ethyl ether, then vac dried at 30° over P_2O_5, fused under nitrogen and redried, the whole procedure being repeated, with a final drying under vac for 24hr. Also recrystd from hexane/EtOH soln. [Burkey and Griller *JACS* **107** 246 *1985*.]

sym-**Diphenylethane** see **bibenzyl.**

1,1-Diphenylethanol *[599-67-7]* **M 198.3, m 80-81°**. Crystd from *n*-heptane. [Bromberg *et al, JACS* **107** 83 *1985*.]

Diphenyl ether see **Phenyl ether.**

1,1-Diphenylethylene *[530-48-3]* **M 180.3, b 268-270°, n 1.6088, d 1.024.** Distd under reduced pressure from KOH. Dried with CaH_2 and redistd.

N,N'-**Diphenylethylenediamine** (Wanzlick's reagent) *[150-61-8]* **M 212.3, m 67.5°.** Crystd from aq EtOH.

N,N'-**Diphenylformamide** *[622-15-1]* **M 197.2, m 142° (137°).** Crystd from abs EtOH, gives the hydrate with aq EtOH.

Diphenylglycollic acid see **benzilic acid.**

1,3-Diphenylguanidine *[102-06-7]* **M 211.3, m 148°.** Crystd from toluene, aq acetone or EtOH, and vac dried.

1,6-Diphenyl-1,3,5-hexatriene *[1720-32-7]* **M 232.3, m 200-203°.** Crystd from $CHCl_3$ or $EtOH/CHCl_3$ (1:1).

5,5-Diphenylhydantoin *[57-41-0]* **M 252.3, m 293-295°.** Crystd from EtOH.

1,1-Diphenylhydrazine *[530-50-7]* **M 184.2, m 126°.** Crystd from hot EtOH containing a little ammonium sulphide or H_2SO_3 (to prevent atmospheric oxidation), preferably under nitrogen. Dried in a vac desiccator.

Diphenyl hydrogen phosphate *[838-85-7]* **M 250.2, m 99.5°.** Crystd from $CHCl_3$/pet ether.

1,3-Diphenylisobenzofuran *[5471-63-6]* **M 270.3, m 129-130°.** Recrystd from EtOH or $EtOH/CHCl_3$ (1:1) under red light or from benzene in the dark.

Diphenylmethane *[101-81-5]* **M 168.2, m 25.4°.** Sublimed under vac, or distd at 72-75°/4mm. Crystd from EtOH. Purified by fractional crystn of the melt.

Diphenylmethanol see **benzhydrol.**

1,1-Diphenylmethylamine *[530-50-7]* **M 183.2, m 34°.** Crystd from water

Diphenylmethyl chloride *[90-99-3]* **M 202.7, m 17.0°, b 167°/17mm, n 1.5960.** Dried with Na_2SO_4 and fractionally distd under reduced pressure.

Diphenylnitrosamine see *N*-**nitrosodiphenylamine.**

4,7-Diphenyl-1,10-phenanthroline *[1662-01-7]* **M 332.4, m 218-220°.** Recrystd from toluene.

1,9-Diphenyl-1,3,6,8-nonatetraen-5-one see **dicinnamalacetone.**

all-trans-**1,8-Diphenyl-1,3,5,8-octatetraene** *[3029-40-1]* **M 258.4, m 235-237°.** Crystd from EtOH.

2,5-Diphenyl-1,3,4-oxadiazole (PPD) *[725-12-2]* **M 222.3, m 70° (hydrate), 139-140° (anhydrous), b 231°/13mm, 248°/16mm.** Crystd from EtOH and sublimed *in vacuo.*

2,5-Diphenyloxazole (PPO) *[92-71-7]* **M 221.3, m 74°, b 360°/760mm.** Distd in steam and crystd from ligroin.

N,N'-**Diphenyl-*p*-phenylenediamine** *[39529-22-1]* **M 260.3, m 148-149°.** Crystd from chlorobenzene/pet ether or benzene. Has also been crystd from aniline, then extracted three times with abs EtOH.

Diphenylphosphinic acid *[1707-03-5]* M 218.2, m 194-195°. Recrystd from 95% EtOH and dried under vac at room temperature.

1,1-Diphenyl-2-picrylhydrazyl *[1707-75-1]* M 394.3, m 178-179.5°. Crystd from $CHCl_3$, or benzene/pet ether (1:1), then degassed at 100° and $<10^{-5}$mm Hg for *ca* 50hr to decompose the 1:1 molar complex formed with benzene.

1,3-Diphenyl-1,3-propanedione see **dibenzoylmethane**.

1,3-Diphenyl-2-propanone see **dibenzyl ketone**.

2,2-Diphenylpropionic acid *[5558-66-7]* M 226.3, m 173-174°,
3,3-Diphenylpropionic acid *[606-83-7]* M 226.3, m 155°. Crystd from EtOH.

Diphenyl sulphide *[139-66-2]* M 186.3, b 145°/8mm, n 1.633, d 1.114. Washed with aq 5% NaOH, then water. Dried with $CaCl_2$, then with sodium. The sodium was filtered off and the diphenyl sulphide was distd under reduced pressure.

Diphenyl sulphone *[127-63-9]* M 218.3, m 125°, b 378°(dec). Crystd from ethyl ether. Purified by zone melting.

Diphenylthiocarbazone see **Dithizone**.

sym-**Diphenylthiourea** (thiocarbanilide) *[102-08-9]* M 228.3, m 154°. Crystd from boiling EtOH by adding hot water and allowing to cool.

1,3-Diphenyltriazene see **diazoaminobenzene**.

1,1-Diphenylurea *[603-54-3]* M 212.3, m 238-239°. Crystd from MeOH.

sym-**Diphenylurea** see **carbanilide**.

Diphosphopyridine nucleotide (NAD, DPN) *[53-84-9]* M 663.4. Purified by chromatography on Dowex-1 ion-exchange resin. The column was prepared by washing with 3M HCl until free of material absorbing at 260nm, then with water, 2M sodium formate until free of chloride ions and, finally, with water. NAD, as an 0.2% soln in water, adjusted with NaOH to *p*H 8, was adsorbed on the column, washed with water, and eluted with 0.1M formic acid. Fractions with strong absorption at 360nm were combined, acidified to *p*H 2.0 with 2M HCl, and cold acetone (*ca* 5L/g of NAD) was added slowly and with constant agitation. It was left overnight in the cold, then the ppte was collected in a centrifuge, washed with pure acetone and dried under vac over $CaCl_2$ and paraffin wax shavings [Kornberg in *Methods in Enzymology* Eds Colowick and Kaplan, Academic Press, New York **Vol 3** p876 *1957*]. The purity is checked by reduction to NADH (with EtOH and yeast alcohol dehydrogenase) which has ε_{340nm} 6220 $M^{-1}cm^{-1}$.

Dipicolinic acid (pyridine-2,6-dicarboxylic acid) *[499-83-2]* M 167.1, m 255°(dec), λ_{max} 270nm. Crystd from water, and sublimed in a vac.

N,N-Di-*n*-propylaniline *[2217-07-4]* M 177.3, b 127°/10mm, 238-241°/760mm. Refluxed for 3hr with acetic anhydride, then fractionally distd under reduced pressure.

Dipropylene glycol *[110-98-5]* M 134.2, b 109-110°/8mm, n 1.441,d 1.022. Fractionally distd below 15mm pressure, using packed column and taking precautions to avoid absorption of water.

Di-*n*-propyl ether see *n*-**propyl ether**

Di-*n*-propyl ketone *[123-19-3]* **M 114.2, b 143.5°, n 1.40732, d 0.8143.** Dried with CaSO₄, then distd from P₂O₅ under nitrogen.

Di-*n*-propyl sulphide *[111-47-7]* **M 118.2, b 141-142°, n 1.449, d 0.870.** Washed with aq 5% NaOH, then water. Dried with CaCl₂ and distd from sodium [Dunstan and Griffiths *JCS* 1344 *1962*].

Di-(4-pyridoyl)hydrazine *[4329-75-3]* **M 246.2, m 254-255°.** Crystd from water.

α,α'-Dipyridyl see α,α'-bipyridyl.

2,2'-Dipyridylamine *[1202-34-2]* **M 171.2, m 84° and remelts at 95° after solidifying, b 176-178°/13mm, 307-308°/760mm.** Crystd from benzene or toluene [Blakley and De Armond *JACS* 109 4895 *1987*].

1,2-Di-(4-pyridyl)-ethane *[4916-57-8]* **M 184.2.** Crystd from cyclohexane/benzene (5:1).

***trans*-1,2-Di-(4-pyridyl)-ethylene** *[1135-32-6]* **M 182.2, m 153-154°.** Crystd from water (1.6g/100ml at 100°).

1,3-Di-(4-pyridyl)-propane *[17252-51-6]* **M 198.3, m 60.5-61.5°.** Crystd from *n*-hexane/benzene (5:1).

α,α'-Diquinolyl see α,α'-biquinolyl.

S-1,2-Distearin *[1188-58-5]* **M 625.0, m 76-77°, [α]_D -2.8° (c 6.3, CHCl₃), [α]₅₄₆ +1.4° (c 10, CHCl₃/MeOH, 9:1).** Crystd from chloroform/pet ether.

2,5-Distyrylpyrazine *[14990-02-4]* **M 284.3, m 219°.** Recrystd from xylene; chromatographed on basic silica gel (60-80 mesh) using methylene chloride as eluent, then vac sublimed on to a cold surface at 10⁻³ torr [Ebied *JCSFT 1* 78 3213 *1982*]. Operations should be carried out in the dark.

1,3-Dithiane *[505-23-7]* **M 120.2, m 54°.** Crystd from 1.5 times its weight of MeOH at 0°, and sublimed at 40-50°/0.1mm.

2,2'-Dithiobis(benzothiazole) *[120-78-8]* **M 332.2, m 180°.** Crystd from benzene.

4,4'-Dithiodimorpholine *[103-34-3]* **M 236.2, m 124-125°.** Crystd from hot aq dimethylformamide.

1,4-Dithioerythritol (DTE, *erythro*-2,3-dihydroxy-1,4-dithiobutane) *[6892-68-8]* **M 154.3, m 82-84°.** Crystd from ether/hexane and stored in the dark at 0°.

Dithiooxamide (rubeanic acid) *[79-40-3]* **M 120.2, m >300°.** Crystd from EtOH and sublimed in a vac.

RS-1,4-Dithiothreitol (Cleland's reagent) *[27565-41-9]* **M 154.3, m 42-43°.** Crystd from ether and sublimed at 37°/0.005mm. Should be stored at 0°.

Dithizone *[60-10-6]* **M 256.3, ratio of D₆₂₀nm/D₄₅₀nm should be ≥1.65, ε₆₂₀nm 3.4 x 10⁴ (CHCl₃).** The crude material is dissolved in CCl₄ to give a concentrated soln. This is filtered through a sintered glass funnel and shaken with 0.8M aq ammonia to extract dithizonate ion. The aqueous layer is washed with several portions of CCl₄ to remove undesirable materials. The aq layer is acidified with dil H₂SO₄ to precipitate pure dithizone. It is dried in a vac. When only small amounts of dithizone are required, purification by paper chromatography is convenient. [Cooper and Hibbits *JACS* 75 5084 *1933*.] Instead of CCl₄, CHCl₃ can be used, and the final extract, after washing with water, can be evapd in air at 40-50° and dried in a desiccator.

Di-*p*-tolyl carbonate *[621-02-3]* **M 242.3, m 115°.** Purified by glc with 20% Apiezon on Embacel followed by sublimation *in vacuo*.

N,N'-**Di-*o*-tolylguanidine** *[97-39-2]* **M 239.3, m 179° (175-176°).** Crystd from aq EtOH.

Di-*p*-tolylphenylamine *[20440-95-3]* **M 273.4, m 108.5°.** Crystd from EtOH.

Di-*p*-tolyl phenylphosphonate *[94548-75-1]* **M 388.3, n²⁵ 1.5758.** Purified as described under diisooctyl phenylphosphonate.

Di-*p*-tolyl sulphone *[599-66-6]* **M 278.3, m 158-159°, b 405°.** Crystd repeatedly from ethyl ether. Purified by zone melting.

Di-*m*-tolylurea see **3,3-dimethylcarbanilide.**

Djenkolic acid *[498-59-9]* **M 254.1, m 300-350°(dec).** Crystd from a large vol of water.

Docosane *[629-97-0]* **M 310.6, m 47°, b 224°/15mm.** Crystd from EtOH or ether.

Docosanoic acid (behenic acid) *[112-85-6]* **M 340.6, m 81-82°.** Crystd from ligroin.

1-Docosanol *[661-19-8]* **M 182.3, m 70.8°.** Crystd from ether or chloroform/ether.

n-**Dodecane** *[112-40-3]* **M 170.3, b 97.5-99.5°/5mm, 216°/760mm, n 1.42156, d 0.748.** Passed through a column of Linde type 13X molecular sieves. Stored in contact with, and distd from, sodium. Passed through a column of activated silica gel. Has been crystd from ethyl ether at -60°. Unsaturated dry material which remained after passage through silica gel has been removed by catalytic hydrogenation (Pt₂O) at 45lb/in², followed by fractional distn under reduced pressure [Zook and Goldey *JACS* **75** 3975 *1953*]. Also purified by partial crystn from the melt.

Dodecane-1,12-dioic acid *[693-23-2]* **M 230.3, m 129°, b 245°/10mm.** Crystd from water, 75% or 95% EtOH, or glacial acetic acid.

1-Dodecanthiol *[112-55-0]* **M 202.4, b 111-112°/3mm, 153-155°/24mm, n 1.458, d 0.844.** Dried with CaO for several days, then distd from CaO.

Dodecanoic see **lauric acid.**

1-Dodecanol *[112-53-8]* **M 186.3, m 24°, d²⁴ 0.8309 (liquid).** Crystd from aq EtOH, and vac distd in a spinning-band column.

Dodecyl alcohol see **1-dodecanol.**

Dodecylammonium butyrate *[17615-97-3]* **M 273.4, m 39-40°.** Recrystd from *n*-hexane.

Dodecylammonium propionate *[17448-65-6]* **M 259.4, m 55-56°.** Recrystd from hexanol/pet ether (b 60-80°).

Dodecyldimethylamine oxide *[1643-20-5]* **M 229.4.** Crystd from acetone or ethyl acetate. [Bunton *et al, JOC* **52** 3832 *1987*.]

Dodecyl ether *[4542-57-8]* **M 354.6, m 33°.** Vac distd, then crystd from MeOH/benzene.

1-Dodecylpyridinium chloride *[104-71-5]* **M 301.9, m 68-70°.** Purified by repeated crystn from acetone (charcoal); twice recrystd from EtOH [Chu and Thomas *JACS* **108** 6270 *1986*].

Dodecyltrimethylammonium bromide *[1119-94-4]* **M 308.4.** Purified by repeated crystn from acetone. Washed with ethyl ether and dried in a vac oven at 60° [Dearden and Wooley *JPC* **91** 2404 *1987*].

Dodecyltrimethylammonium chloride *[112-00-5]* **M 263.9.** Dissolved in MeOH, treated with active charcoal, filtered and dried *in vacuo* [Waldenburg *JPC* **88** 1655 *1984*], or recrystd several times from 10% EtOH in acetone. Also repeatedly crystd from EtOH/ether or MeOH.

Dulcin see *p*-phenethylurea.

Dulcitol *[608-66-2]* **M 182.2, m 188-189°, b 276-280°/1.1mm.** Crystd from water by addition of EtOH.

Durene (1,2,4,5-tetramethylbenzene) *[95-93-2]* **M 134.2, m 79.5-80.5°.** Chromatographed on alumina, and recrystd from aq EtOH or benzene. Zone-refining removes duroaldehydes. Dried under vac. [Yamauchi *et al*, *JPC* **89** 4804 *1985*.] It has also been sublimed *in vacuo* [Johnston *et al*, *JACS* **109** 1291 *1987*].

Duroquinone (tetramethylbenzoquinone) *[527-17-3]* **M 164.2, m 110-111°.** Crystd from 95% EtOH. Dried under vac.

β-Ecdyson *[5289-74-7]* M 480.7, m 242°, $[\alpha]_D$ +64.7° (c 1, EtOH). Crystd from water or tetrahydrofuran/pet ether.

Echinenone *[432-68-8]* M 550.8, m 178-179°, $\varepsilon_{1cm}^{1\%}$ 2160 (458nm) in pet ether. Purified by chromatography on partially deactivated alumina or magnesia, or by using a thin layer of silica gel G with 4:1 cyclohexane/ethyl ether as the developing solvent. Stored in the dark at -20°.

Eicosane *[112-95-8]* M 282.6, m 36-37°, b 205°/15mm, n^{40} 1.43453, $d^{36.7}$ 0.7779. Crystd from EtOH.

Elaidic acid *[112-79-8]* M 282.5, m 44.5°. Crystd from acetic acid, then EtOH.

Ellagic acid (2H$_2$O) *[476-66-4]* M 302.2, m >360°. Crystd from pyridine.

Elymoclavine *[548-43-6]* M 254.3. Crystd from MeOH.

Enniatin A *[11113-62-5]* M 681.9, m 122-122.5°. Crystd from EtOH/water.

Eosin *[548-24-3]* M 624.1, λ_{max} 514nm. Freed from inorganic halides by repeated crystn from butan-1-ol.

1R,2S-(-)Ephedrine *[299-42-3]* M 165.2, m 40°, b 225°, $[\alpha]_{546}$ -47° and $[\alpha]_D$ -40° (c 5, 2.2M HCl). Crystd from aq 70% EtOH. Dehydrated by vac distn, the distillate being allowed to crystallise in a vac to prevent the uptake of CO$_2$ and water vapour. The anhydrous base was then recrystd from dry ether [Fleming and Saunders *JCS* 4150 *1955*].

(-)Ephedrine hydrochloride *[50-36-3]* M 201.7, m 218°, $[\alpha]_{546}$ -48° (c 5, 2M HCl). Crystd from water.

Epichlorohydrin *[106-89-8]* M 92.5, b 115.5°, n 1.438, d 1.180. Distd at atmospheric pressure, heated on a steam bath with one-quarter its weight of CaO, then decanted and fractionally distd.

R(-)Epinephrine (adrenalin) *[51-43-4]* M 183.2, m 215°(dec), $[\alpha]_{546}$ -61° (c 5, 0.5M HCl). Dissolved in dil aq acid, then pptd by addn of dil aq ammonia or alkali carbonates. (Epinephrine readily oxidises in neutral alkaline soln. This can be diminished if a little sulphite is added).

1,2-Epoxybutane *[106-88-7]* M 72.1, b 66.4-66.6°, n 1.3841, d 0.837. Dried with CaSO$_4$, and fractionally distd through a long (126cm) glass helices-packed column. The first fraction contains a water azeotrope.

(+)-Equilenine *[517-09-9]* M 266.3, m 258-259°, $[\alpha]_D^{16}$ +87° (c 7.1, H$_2$O). Crystd from EtOH.

Ergocornine *[564-36-3]* M 561.7, m 182-184°. Crystd with solvent of crystn from MeOH.

Ergocristine *[511-08-0]* M 573.7, m 165-170°. Crystd with 2 moles of solvent of crystn, from benzene.

Ergocryptine *[511-09-1]* M 575.7, m 212-214°. Crystd with solvent of crystn, from acetone, benzene or methanol.

Ergosterol *[57-87-4]* M 396.7, m 165-166°, $[\alpha]_{546}$ -171° (in CHCl$_3$). Crystd from ethyl acetate, then from ethylene dichloride.

Ergotamine *[113-15-5]* M 581.6, m 212-214°(dec). Crystd from benzene, then dried by prolonged heating in high vac. Very hygroscopic.

Ergotamine tartrate *[379-79-3]* M 657.1, m 203°(dec). Crystd from MeOH.

Eriochrome Black T *[1787-61-7]* M 416.4, $\varepsilon_{1cm}^{1\%}$ 656 (620nm) at pH 10, using the dimethylammonium salt. The sodium salt (200g) was converted to the free acid by stirring with 500ml of 1.5M HCl, and, after several minutes, the slurry was filtered on a sintered-glass funnel. The process was repeated and the material was air dried after washing with acid. It was extracted with benzene for 12hr in a Soxhlet extractor, then the benzene was evaptd and the residue was air dried. A further desalting with 1.5M HCl (1L) was followed by crystn from dimethylformamide (in which it is very soluble) by forming a satd soln at the boiling point, and allowing to cool slowly. The crystalline dimethylammonium salt so obtained was washed with benzene and treated repeatedly with dil HCl to give the insoluble free acid which, after air drying, was dissolved in alcohol, filtered and evapd. The final material was air dried, then dried in a vac desiccator over $Mg(ClO_4)_2$ [Diehl and Lindstrom, *AC* 31 414 *1959*].

Eriochrome Blue Black R *[2538-85-4]* M 416.4. Freed from metallic impurities by three pptns from aq soln by addn of HCl. The pptd dye was dried at 60° under vac.

Erucic acid *[112-86-7]* M 338.6, m 33.8°, b 358°/400mm. Crystd from MeOH.

meso-**Erythritol** *[149-32-6]* M 122.1, m 122°. Crystd from distd water and dried at 60° in a vac oven.

Erythrityl tetranitrate *[7297-25-8]* M 302.1, m 61°,
β-Erythroidine *[466-81-9]* M 273.3. Crystd from EtOH.

Erythrosin B *[568-63-8]* M 879.9, λ_{max} 525nm. Crystd from water.

1,3,5-Estratrien-3-ol-17-one (Estrone, Folliculin) *[50-27-1]* M 270.4, m 260-261°, polymorphic also m 254° and 256°, $[\alpha]_{546}$ +198° (c 1, dioxane). Crystd from EtOH.

1,3,5-Estratrien-3β,16α,17β-triol (Estriol) *[53-16-7]* M 288.4, m 283°, $[\alpha]_{546}$ +66° (c 1, dioxane). Crystd from EtOH/ethyl acetate.

Estriol see **1,3,5-Estratrien-3β,16α,17β-triol.**

Estrone see **1,3,5-Estratrien-3-ol-17-one.**

Ethane *[74-84-0]* M 30.1, f.p. -172°, b -88°, d_4^0 1.0493 (air = 1). Ethylene can be removed by passing the gas through a sintered-glass disc into fuming H_2SO_4 then slowly through a column of charcoal satd with bromine. Bromine and HBr were removed by passage through firebrick coated with *N,N*-dimethyl-*p*-toluidine. The ethane was also passed over KOH pellets (to remove CO_2) and dried with $Mg(ClO_4)_2$. Further purification was by several distns of liquified ethane, using a condensing temperature of -195°. Yang and Gant [*JPC* 65 1861 *1961*] treated ethane by standing it for 24hr at room temperature in a steel bomb containing activated charcoal treated with bromine. They then immersed the bomb in a Dry-ice/acetone bath and transferred the ethane to an activated charcoal trap cooled in liquid nitrogen. (The charcoal had previously been degassed by pumping for 24hr at 450°.) By allowing the trap to warm slowly, the ethane was distd, retaining only the middle third. Removal of methane was achieved using Linde type 13X molecular sieves (previously degassed by pumping for 24hr at 450°) in a trap which, after cooling in Dry-ice/acetone, was satd with ethane. After pumping for 10min, the ethane was recovered by warming the trap to room temperature. (The final gas contained less than 10^{-4} mole % of either ethylene or methane.)

Ethanethiol (ethyl mercaptan) *[540-63-6]* M 62.1, b 32.9°/704mm, d^{25} 0.83147. Dissolved in aq 20% NaOH, extracted with a small amount of benzene and then steam distd until clear. After cooling, the alkaline soln was acidified slightly with 15% H_2SO_4 and the thiol was distd off, dried with $CaSO_4$,

CaCl$_2$ or 4A molecular sieves, and fractionally distd under nitrogen [Ellis and Reid *JACS* **54** 1674 *1932*].

Ethanol *[64-17-5]* **M 46.1, b 78.3°, n 1.36139, d^{15} 0.79360, d^5 0.78506.** Usual impurities of fermentation alcohol are fusel oils (mainly higher alcohols, especially pentanols), aldehydes, esters, ketones and water. With synthetic alcohol, likely impurities are water, aldehydes, aliphatic esters, acetone and ethyl ether. Traces of benzene are present in ethanol that has been dehydrated by azeotropic distillation with benzene. Anhydrous ethanol is very hygroscopic. Water (down to 0.05%) can be detected by formation of a voluminous ppte when aluminium ethoxide in benzene is added to a test portion, Rectified spirit (95% ethanol) is converted to *absolute* (99.5%) ethanol by refluxing with freshly ignited CaO (250g/L) for 6hr, standing overnight and distilling with precautions to exclude moisture.

Numerous methods are available for further dehydration of *absolute* ethanol. Lund and Bjerrum [*Ber* **64** 210 *1931*] used reaction with magnesium ethoxide, prepared by placing 5g of clean dry magnesium turnings and 0,5g of iodine (or a few drops of CCl$_4$) in a 2L flask, followed by 50-75ml of *absolute* ethanol, and warming the mixture until a vigorous reaction occurs. When this subsides, heating is continued until all the magnesium is converted to magnesium ethoxide. Up to 1L of ethanol is added and, after an hour's reflux, it is distd off. The water content should be below 0.05%. Walden, Ulich and Laun [*Z physik Chem* **114** 275 *1925*] used amalgamated aluminium chips, prepared by degreasing aluminium chips, treating with alkali until hydrogen was vigorously evolved, washing with water until the washings were weakly alkaline and then stirring with 1% HgCl$_2$ soln. After 2min, the chips were washed quickly with water, then alcohol, then ether, and dried with filter paper. (The amalgam became warm.) These chips were added to the ethanol, which was then gently warmed for several hours until evolution of hydrogen ceased. The alcohol was distd and aspirated for some time with pure dry air. Smith [*JCS* 1288 *1927*] reacted 1L of *absolute* ethanol in a 2L flask with 7g of clean dry sodium, and added 25g of pure ethyl succinate (27g of pure ethyl phthalate was an alternative), and refluxed the mixture for 2hr in a system protected from moisture, and then distd the ethanol. A modification used 40g of ethyl formate, instead, so that sodium formate separated out and, during reflux, the excess of ethyl formate decomposed to CO and ethanol.

Dehydrating agents suitable for use with ethanol include Linde type 4A molecular sieves, calcium metal, and CaH$_2$. The calcium hydride (2g) was crushed to a powder and dissolved in 100ml *absolute* ethanol by gently boiling. About 70ml of the ethanol were distd off to remove any ammonia before the remainder was poured into 1L of *ca* 99.9% ethanol in a still, where it was boiled under reflux for 20hr, while a slow stream of pure, dry hydrogen was passed through. It was then distd [Rüber *Z Elektrochem* **29** 334 *1923*]. If calcium was used for drying, about ten times the theoretical amount should be taken, and traces of ammonia would be removed by passing dry air into the vapour during reflux.

Ethanol can be freed from traces of basic materials by distn from a little 2,4,6-trinitrobenzoic acid or sulphanilic acid. Benzene can be removed by fractional distn after adding a little water (the benzene/water/ethanol azeotrope distils at 64.9°); the alcohol is then redried using one of the methods described above. Alternatively, careful fractional distn can separate benzene as the benzene/ethanol azeotrope (b 68.2°). Aldehydes can be removed from ethanol by digesting with 8-10g of dissolved KOH and 5-10g of aluminium or zinc per L, followed by distn. Another method is to heat under reflux with KOH (20g/L) and AgNO$_3$ (10g/L) or to add 2.5-3g of lead acetate in 5ml of water to 1L of ethanol, followed (slowly and without stirring) by 5g of KOH in 25ml of ethanol: after 1hr the flask is shaken thoroughly, then set aside overnight before filtering and distilling. The residual water can be removed by standing the distillate over activated aluminium amalgam for 1 week, then filtering and distilling. Distn of ethanol from Raney nickel eliminates catalyst poisons.

Other purification procedures include pre-treatment with conc H$_2$SO$_4$ (3ml/L) to eliminate amines, and with KMnO$_4$ to oxidise aldehydes, followed by refluxing with KOH to resinify aldehydes, and distilling to remove traces of H$_3$PO$_4$ and other acidic impurities after passage through silica gel, and drying over CaSO$_4$. Water can be removed by azeotropic distn with dichloromethane (azeotrope boils at 38.1° and contains 1.8% water) or 2,2,4-trimethylpentane.

Ethanolamine see **1-aminoethanol.**

Ethidium bromide *[1239-45-8]* **M 384.3, m 260-262°.** Crystd from MeOH [Lamos *et al, JACS* **108** 4278 *1986*]. **POSSIBLE CARCINOGEN.**

S-Ethionine *[13073-35-3]* **M 163.2, m 282°(dec), [α]$_D^{25}$ +23.7° (in 5M HCl).** Likely impurities are *N*-acetyl-(R and S)-ethionine, S-methionine, and R-ethionine. Crystd from water by adding 4 volumes of EtOH.

3-Ethoxy-*N,N*-diethylaniline *[1846-92-2]* **M 193.3, b 141-142°/15mm.** Refluxed for 3hr with acetic anhydride, then fractionally distd under reduced pressure.

2-Ethoxyethanol *[110-80-5]* **M 90.1, b 134.8°, n 1.40751, d 0.931.** Dried with CaSO$_4$ or K$_2$CO$_3$, filtered and fractionally distd. Peroxides can be removed by refluxing with anhydrous SnCl$_2$ or by filtration under slight pressure through a column of activated alumina.

2-(2-Ethoxyethoxy)ethanol see **diethylene glycol monoethyl ether.**

2-Ethoxyethyl ether [*bis*-(2-ethoxyethyl) ether] *[112-36-7]* **M 162.2, b 76°/32mm, n 1.412, d 0.910.** Refluxed with LiAlH$_4$ for several hours, distd under reduced pressure and stored with CaH$_2$ under nitrogen. Also passed through (alkaline) alumina.

2-Ethoxyethyl methacrylate *[2370-63-0]* **M 158.2, b 91-93°/35mm, n 1.429, d 0.965.** Purified as described under methyl methacrylate.

1-Ethoxynaphthalene *[5328-01-8]* **M 172.2, b 136-138°/14mm, 282°/760mm, n 1.604, d 1.061.** Fractionally distd (twice) under a vac, then dried with, and distd under a vac from, sodium.

2-Ethoxynaphthalene *[2224-00-2]* **M 172.2, m 35.6-36.0°, b 142-143°/12mm.** Crystd from pet ether. Dried under vac.

Ethyl acetate *[141-78-6]* **M 88.1, b 77.1°, n 1.37239, n^{25} 1.36979, d 0.9003.** The commonest impurities are water, EtOH and acetic acid. These can be removed by washing with aq 5% Na$_2$CO$_3$, then with satd aq CaCl$_2$ or NaCl, and drying with K$_2$CO$_3$, CaSO$_4$ or MgSO$_4$. More efficient drying is achieved if the solvent is further dried with P$_2$O$_5$, CaH$_2$ or molecular sieves before distn. CaO has also been used. Alternatively, ethanol can be converted to ethyl acetate by refluxing with acetic anhydride (*ca* 1ml per 10ml of ester); the liquid is then fractionally distd, dried with K$_2$CO$_3$ and redistd.

Ethyl acetoacetate *[141-97-9]* **M 130.1, b 71°/12mm, 100°/80mm, n 1.419, d 1.026.** Shaken with small amounts of satd aq NaHCO$_3$ (until no further effervescence), then with water. Dried with MgSO$_4$ or CaCl$_2$. Distd under reduced pressure.

Ethyl acrylate *[140-88-5]* **M 100.1, b 99.5°, n 1.406, d 0.922.** Washed repeatedly with aq NaOH until free from inhibitors such as hydroquinone, then washed with satd aq CaCl$_2$ and distd under reduced pressure.

Ethyl alcohol see **ethanol.**

Ethylamine *[75-04-7]* **M 45.1, b 16.6°/760mm, d 1.3663.** Condensed in an all-glass apparatus cooled by circulating ice-water, and stored with KOH pellets below 0°.

Ethylamine hydrochloride *[557-66-4]* **M 81.5, m 109-110°.** Crystd from abs EtOH or MeOH/CHCl$_3$.

Ethyl *o*-aminobenzoate *[94-09-7]* **M 165.2, m 92°.** Crystd from EtOH/water and air dried.

p-**Ethylaniline** *[589-16-2]* **M 121.2, 88°/8mm, n 1.554, d 0.975.** Dissolved in benzene, then acetylated. The acetyl derivative was recrystallised from benzene/pet ether, and hydrolysed by refluxing 50g with

500ml of water and 115ml of conc H_2SO_4 until the soln becomes clear. The amine sulphate was isolated, suspended in water and solid KOH was added to regenerate the free base, which was separated, dried and distd from zinc dust under a vac {Berliner and Berliner *JACS* 76 6179 1954].

Ethylbenzene *[100-41-6]* **M 106.2, b 136.2°, n 1.49594, n^{25} 1.49330, d 0.867.** Shaken with cold conc H_2SO_4 until a fresh portion of acid remained colourless, then washed with aq 10% NaOH or NaHCO_3, followed by distd water until neutral. Dried with MgSO_4 or CaSO_4, then dried further with, and distd from, sodium, sodium hydride or CaH_2. Can also be dried by passing through silica gel. Sulphur-containing impurities have been removed by prolonged shaking with mercury. Also purified by fractional freezing.

Ethyl benzoate *[93-89-0]* **M 150.2, b 98°/19mm, 212.4°/760mm, n^{15} 1.5074, n^{25} 1.5043, d 1.046.** Washed with aq 5% Na_2CO_3, then satd CaCl_2, dried with CaSO_4 and distd under reduced pressure.

Ethyl bis-(2,4-dinitrophenyl)acetate *[5833-18-1]* **M 358.3, m 150-153°.** Crystd from toluene as pale yellow crystals.

Ethyl bixin *[6895-43-8]* **M 436.6, m 138°.** Crystd from EtOH.

Ethyl bromide *[74-96-4]* **M 109.0, b 0°/165mm, 38°/745mm, n 1.4241, d 1.460.** The main impurities are usually EtOH and water, with both of which it forms azeotropes. Ethanol and unsaturated compounds can be removed by washing with conc H_2SO_4 until no further coloration is produced. The ethyl bromide is then washed with water, aq Na_2CO_3, and water again, then dried with CaCl_2, MgSO_4 or CaH_2, and distd. from P_2O_5. Olefinic impurities can also be removed by storing the ethyl bromide in daylight with elementary bromine, later removing the free bromine by extraction with dil aq Na_2SO_3, drying the ethyl bromide with CaCl_2 and fractionally distilling. Alternatively, unsaturated compounds can be removed by bubbling oxygen containing *ca* 5% ozone through the liquid for an hour, then washing with aq Na_2SO_3 to hydrolyse ozonides and remove hydrolysis products, followed by drying and distn.

Ethyl bromoacetate *[105-36-2]* **M 167.0, b 158-158.5°/758mm, n 1.450, d 1.50,**
Ethyl α-bromopropionate *[535-11-5]* **M 181.0, b 69-70°/25mm, n 1.447, d 1.39.** Washed with satd aq Na_2CO_3 (three times), 50% aq CaCl_2 (three times) and satd aq NaCl (twice). Dried with MgSO_4, CaCl_2 or CaCO_3, and distd.

2-Ethyl-1-butanol *[97-95-0]* **M 102.2, b 146.3°, n^{15} 1.4243, n^{25} 1.4205.** Dried with CaSO_4 for several weeks, filtered and fractionally distd.

2-Ethylbut-1-ene *[760-21-4]* **M 84.1, b 66.6°, n 1.423, d 0.833.** Washed with satd aq NaOH, then water. Dried with CaCl_2, filtered and fractionally distd.

Ethyl *n*-butyrate *[105-54-4]* **M 116.2, b 49°/50mm, 119-120°/760mm, n 1.393, d 0.880.** Dried with anhydrous CuSO_4 and distd under dry nitrogen.

Ethyl carbamate (urethane) *[51-79-6]* **M 88.1, m 48.0-48.6°.** Crystd from benzene.

***N*-Ethylcarbazole** *[86-28-2]* **M 195.3, m 69-70°.** Recrystd from EtOH, EtOH/water or isopropanol and dried below 55°.

Ethyl carbonate *[105-58-8]* **M 118.1, b 124-125°, n 1.385, d 0.975.** Washed with aq 10% Na_2CO_3, then aq satd CaCl_2. Dried with MgSO_4 and distd.

Ethyl chloride *[75-00-3]* **M 64.5, b 12.4°, n 1.3676, d 0.8978.** Passed through absorption towers containing, successively, conc H_2SO_4, NaOH pellets, P_2O_5 on glass wool, or soda-lime, CaCl_2, P_2O_5.

Condensed into a flask containing CaH_2 and fractionally distd. Has also been purified by illumination in the presence of bromine at $0°$ using a 1000W lamp, followed by washing, drying and distn.

Ethyl chloroacetate *[105-39-5]* **M 122.6, b 143-143.2°, n^{25} 1.4192, d 1.150.** Shaken with satd aq Na_2CO_3 (three times), aq 50% $CaCl_2$ (three times) and satd aq NaCl (twice). Dried with Na_2SO_4 or $MgSO_4$ and distd. **LACHRYMATORY.**

Ethyl chloroformate *[541-41-3]* **M 108.5, m -81°, b 94-95°, n 1.3974, d 1.135.** Washed several times with water, redistd using an efficient fractionating column at atmospheric pressure and a $CaCl_2$ guard tube to keep free from moisture [Hamilton and Sly *JACS* **47** 435 *1925*; Saunders, Slocombe and Hardy, *JACS* **73** 3796 *1951*]. **LACHRYMATORY AND TOXIC.**

Ethyl cinnamate *[103-36-6]* **M 176.2, f.p. 6.7°, b 127°/6mm, 272.7°/768mm, n 1.55983, d 1.040.** Washed with aq 10% Na_2CO_3, then water, dried ($MgSO_4$), and distd. The purified ester was saponified with aq KOH, and, after acidifying the soln, cinnamic acid was isolated, washed and dried. The ester was reformed by refluxing for 15hr the cinnamic acid (25g) with abs EtOH (23g), conc H_2SO_4 (4g) and dry benzene (100ml), after which it was isolated, washed, dried and distd under reduced pressure [Jeffery and Vogel *JCS* 658 *1958*].

Ethyl *trans*-crotonate *[623-70-1]* **M 114.2, b 137°, n 1.425, d 0.917.** Washed with aq 5% Na_2CO_3, washed with satd aq $CaCl_2$, dried with $CaCl_2$ and distd.

Ethyl cyanoacetate *[105-56-6]* **M 113.1, b 206.0°, n 1.41751, d 1.061.** Shaken several times with aq 10% Na_2CO_3, washed well with water, dried with Na_2SO_4 and fractionally distd.

Ethylcyclohexane *[1678-91-7]* **M 112.2, b 131.8°, n 1.43304, n^{25} 1.43073, d 0.789.** Purified by azeotropic distn with 2-ethoxyethanol, then the alcohol was washed out with water and, after drying, the ethylcyclohexane was redistd.

Ethyl cyclohexanecarboxylate *[3289-28-9]* **M 156.2, b 76-77°/10mm, 92-93°/34mm, n 1.420, d 0.960.** Washed with M sodium hydroxide soln, then water, dried with Na_2SO_4 and distd.

Ethyl dibromoacetate *[105-36-2]* **M 245.9, b 81-82°/14.5mm, n_D^{22} 1.4973.** Washed briefly with conc aq $NaHCO_3$, then with aq $CaCl_2$. Dried with $CaSO_4$ and distd under reduced pressure.

Ethyl α,β-dibromo-β-phenylpropionate *[5464-70-0]* *[erythro: 30983-70-1]* **M 336.0, m 75°.** Crystd from pet ether (b 60-80°).

Ethyl dichloroacetate *[535-15-9]* **M 157.0, b 131.0-131.5°/401mm, n 1.438, d 1.28.** Shaken with aq 3% $NaHCO_3$ to remove free acid, washed with distd water, dried for 3 days with $CaSO_4$ and distd under reduced pressure.

Ethylene *[74-85-1]* **M 28.0, m -169.4°, b -102°/700mm.** Purified by passage through a series of towers containing molecular sieves or anhydrous $CaSO_4$ or a cuprous ammonia soln, then conc H_2SO_4, followed by KOH pellets. Alternatively, ethylene has been condensed in liquid nitrogen, with melting, freezing and pumping to remove air before passage through an activated charcoal trap, followed by a further condensation in liquid air. A sputtered sodium trap has also been used, to remove oxygen.

Ethylene bis(diphenylphosphine) (DIPHOS) *[1663-45-2]* **M 398.4, m 139-140°.** Crystd from EtOH [Backvell *et al, JOC* **52** 5430 *1987*].

Ethylenebis[(*o*-hydroxyphenyl)glycine] *[1170-02-1]* **M 360.4, m 249°(dec).** Purified by extensive Soxhlet extraction with acetone [Bonadies and Carrano *JACS* **108** 4088 *1986*].

[Ethylene bis(oxyethylenenitrilo)]tetraacetic acid   see ethylene glycol bis(β-aminoethylether)-tetraacetic acid.

Ethylene carbonate (1,3-dioxalan-2-one) *[96-49-1]* M 88.1, m 37°, n^{40} 1.4199, d 1.32. Dried over P_2O_5 then fractionally distd at 10mm pressure. Crystd from dry ethyl ether.

Ethylene chlorohydrin see 2-chloroethanol.

Ethylenediamine *[107-15-3]* M 60.1, f.p. 11.0°, b 117.0°, n 1.45677, n^{30} 1.4513, d 0.897. Forms a constant-boiling (b 118.5°) mixture with water (15%) [hygroscopic and miscible with water]. Recommended purification procedure [Asthana and Mukherjee in Coetzee, 1982 cf p 53]: to 1L of ethylenediamine was added 70g of type 5A Linde molecular sieves and shaken for 12hr. The liquid was decanted and shaken for a further 12hr with a mixture of CaO (50g) and KOH (15g). The supernatant was fractionally distd (at 20:1 reflux ratio) in contact with freshly activated molecular sieves. The fraction distilling at 117.2° at 760mm was collected. Finally it was fractionally distilled from sodium metal. All distns and storage of ethylenediamine should be carried out under nitrogen to prevent reaction with CO_2 and water. Material containing 30% water was dried with solid NaOH (600g/L), heated on a water bath for 10hr. Above 60°, separation into two phases took place. The hot ethylenediamine layer was decanted off, refluxed with 40g of sodium for 2hr and distd [Putnam and Kobe *Trans Electrochem Soc* **74** 609 *1938*]. Ethylenediamine is usually distd under nitrogen. Type 5A Linde molecular sieves (70g/L), then a mixture of 50g of CaO and 15g of KOH/L, with further dehydration of the supernatant with molecular sieves has also been used for drying this diamine, followed by distn from molecular sieves and, finally, from sodium metal. A spectroscopically improved material was obtained by shaking with freshly baked alumina (20g/L) before distn.

N,N'-Ethylenediaminediacetic acid *[5657-17-0]* M 176.2, m 222-224°(dec),
Ethylenediamine dihydrochloride *[333-18-6]* M 133.0. Crystd from water.

Ethylenediaminetetraacetic acid (EDTA) *[60-00-4]* M 292.3, m 253°(dec). Dissolved in aq KOH or ammonium hydroxide, and ppted with dil HCl or HNO_3, twice. Boiled twice with distd water to remove mineral acid, then recrystd from water or dimethylformamide. Dried at 110°.

Ethylene dibromide see 1,2-dibromoethane.

Ethylene dichloride see 1,2-dichloroethane.

Ethylene dimethacrylate *[97-90-5]* M 198.2, b 98-100°/5mm, n 1.456, d 1.053. Distd through a short Vigreux column at about 1mm pressure, in the presence of 3% (w/w) of phenyl-β-naphthylamine.

Ethylene dimyristate *[627-84-9]* M 482.8, m 61.7°. Crystd from benzene-MeOH or ethyl ether-MeOH, and dried in a vac desiccator.

(Ethylenedinitrilo)tetraacetic acid see ethylenediaminetetraacetic acid.

Ethylene dipalmitate *[624-03-3]* M 538.9, m 69.1°,
Ethylene distearate *[627-83-8]* M 595.0, m 75.3°. Crystd from benzene-MeOH or ethyl ether-MeOH and dried in a vac desiccator.

Ethylene glycol *[107-21-1]* M 62.1, b 68°/4mm, 197.9°/760mm. n^{15} 1.43312. n^{25} 1.43056, d 1.0986. Very hygroscopic, and also likely to contain higher diols. Dried with CaO, $CaSO_4$, $MgSO_4$ or NaOH and distd under vac. Further dried by reaction with sodium under nitrogen, refluxed for several hours and distd. The distillate was then passed through a column of Linde type 4A molecular sieves and finally distd under nitrogen, from more molecular sieves. Fractionally distd.

Ethylene glycol bis(β-aminoethylether)-N,N'-tetraacetic acid (EGTA) *[67-42-5]* M 380.4, m >245°(dec). Dissolved in aq NaOH, pptd by addn of aq HCl, washed with water and dried at 100° in a vac.

Ethylene glycol diacetate *[111-55-7]* M 146.2, b 190.2°, n 1.4150, d^{25} 1.4188, d 1.106. Dried with CaCl$_2$, filtered (taking precautions to exclude moisture) and fractionally distd under reduced pressure.

Ethylene glycol dibutyl ether *[112-48-1]* M 174.3, b 78-80°/0.2mm, n 1.42, d 1.105. Shaken with aq 5% Na$_2$CO$_3$, dried with MgSO$_4$ and stored with chromatographic alumina to prevent peroxide formation.

Ethylene glycol diethyl ether *[629-14-1]* M 118.2, b 121.5°, n 1.392, d 0.842. After refluxing for 12hr, a mixture of the ether (2L), conc HCl (27ml) and water (200ml), with slow passage of nitrogen, the soln was cooled, and KOH pellets were added slowly and with shaking until no more dissolved. The organic layer was decanted, treated with some KOH pellets and again decanted. It was refluxed with, and distd from sodium immediately before use. Alternatively, after removal of peroxides by treatment with activated alumina, the ether has been refluxed in the presence of the blue ketyl formed by sodium-potassium alloy with benzophenone, then distd.

Ethylene glycol dimethyl ether (monoglyme) *[110-71-4]* M 90.1, b 85°, n 1.379, d 0.866. Purified by distn from LiAlH$_4$ or sodium.

Ethylene glycol monobutyl ether see **2-butoxyethanol.**

Ethylene glycol monoethyl ether see **2-ethoxyethanol.**

Ethylene glycol monomethyl ether see **2-methoxyethanol.**

Ethylene oxide *[75-21-8]* M 44.0, b 13.5°/746mm, n^7 1.3597, d^{10} 0.882. Dried with CaSO$_4$, then distd from crushed NaOH. Has also been purified by its passage, as a gas, through towers containing solid NaOH.

Ethylene thiourea *[96-45-7]* M 102.2, m 203-204°. Crystd from EtOH or amyl alcohol.

Ethylene urea *[120-93-4]* M 86.1, m 131°. Crystd from MeOH (charcoal).

Ethylenimine (aziridine) *[151-56-4]* M 43.1, b 55.5°/760mm, d 0.8321. Dried with BaO, and distd from sodium under nitrogen. **TOXIC.**

Ethyl ether *[60-29-7]* M 74.1, b 34.6°/760mm, n^{15} 1.3555, n 1.35272, d 0.714. Usual impurities are water, EtOH, diethyl peroxide (which is explosive when concentrated), and aldehydes. Peroxides [detected by liberation of iodine from weakly acid (HCl) solutions of KI, or by the blue colour in the ether layer when 1mg of Na$_2$Cr$_2$O$_7$ and 1 drop of dil H$_2$SO$_4$ in 1ml of water is shaken with 10ml of ether] can be removed in several different ways. The simplest method is to pass dry ether through a column of activated alumina (80g Al$_2$O$_3$/700ml of ether). More commonly, 1L of ether is shaken repeatedly with 5-10ml of a soln comprising 6.0g of ferrous sulphate and 6ml of conc H$_2$SO$_4$ in 110ml of water. Aqueous 10% Na$_2$SO$_3$ or stannous chloride can also be used. The ether is then washed with water, dried for 24hr with CaCl$_2$, filtered and dried further by adding sodium wire until it remains bright. The ether is stored in a dark cool place, until distd from sodium before use. Peroxides can also be removed by wetting the ether with a little water, then adding excess LiAlH$_4$ or CaH$_2$ and leaving to stand for several hours. (This also dried the ether.)
Werner [*Analyst* **58** 335 *1933*] removed peroxides and aldehydes by adding 8g AgNO$_3$ in 60ml of water to 1L of ether, then 100ml of 4% NaOH and shaking for 6min. Fierz-David [*Chimia* **1** 246 *1947*] shook 1L of ether with 10g of a zinc-copper couple. (This reagent was prepared by suspending zinc dust in 50ml of hot water, adding 5ml of 2M HCl and decanting after 20sec, washing twice with water, covering with 50ml of water and 5ml of 5% cuprous sulphate with swirling. The liquid was decanted and

discarded, and the residue was washed three times with 20ml of ethanol and twice with 20ml of ethyl ether).

Aldehydes can be removed from ethyl ether by distn from hydrazine hydrogen sulphate, phenyl hydrazine or thiosemicarbazide. Peroxides and oxidisable impurities have also been removed by shaking with strongly alkaline satd $KMnO_4$ (with which the ether was left to stand in contact for 24hr), followed by washing with water, conc H_2SO_4, water again, then drying ($CaCl_2$) and distn from sodium, or sodium containing benzophenone to form the ketyl. Other purification procedures include distn from sodium triphenylmethide or butyl magnesium bromide, and drying with solid NaOH or P_2O_5.

2-Ethylethylenimine *[25449-67-9]* **M 71.1, b 88.5-89°.** Freshly distd from sodium before use. **TOXIC.**

Ethyl formate *[109-94-4]* **M 74.1, b 54.2°, n 1.35994, n^{25} 1.3565, d 0.921, d^{30} 0.909.** Free acid or alcohol is removed by standing with anhydrous K_2CO_3, with occasional shaking, then decanting and distilling from P_2O_5. Alternatively, the ester can be stood wih CaH_2 for several days, then distd from fresh CaH_2. Cannot be dried with $CaCl_2$ because it reacts rapidly with the ester to form a crystalline compound.

2-Ethyl-1-hexanol *[104-76-7]* **M 130.2, b 184.3°, n 1.431, d 0.833.** Dried with sodium, then fractionally distd.

2-Ethylhexyl vinyl ether *[37769-62-3]* **M 156.3.** Usually contains amines as polymerization inhibitors. These are removed by fractional distn.

Ethylidene dichloride see **1,1-dichloroethane.**

Ethyl iodide *[75-03-6]* **M 156.0, b 72.4°, n ^{15} 1.5682, n^{25} 1.5104, d 1.933.** Drying with P_2O_5 is unsatisfactory, and with $CaCl_2$ is incomplete. It is probably best to dry with sodium wire and distil [Hammond *et al JACS* **82** 704 *1960*]. Exposure of ethyl iodide to light leads to rapid decomposition, with the liberation of iodine. Free iodine can be removed by shaking with several portions of dil aq $Na_2S_2O_3$ (until the colour is discharged), followed by washing with water, drying (with $CaCl_2$, then sodium), and distn. The distd ethyl iodide is stored, over mercury, in a dark bottle away from direct sunlight. Other purification procedures include passage through a 60cm column of silica gel, followed by distn); and treatment with elemental bromine, extraction of free halogen with $Na_2S_2O_3$ soln, followed by washing with water, drying and distn. Free iodine and HI have also been removed by direct distn through a LeBel-Henninger column containing copper turnings. Purification by shaking with alkaline solns, and storage over silver, are reported to be unsatisfactory.

Ethyl isobutyrate *[623-48-3]* **M 116.2, b 110°, n 1.388. d 0.867.** Washed with aq 5% Na_2CO_3, then with satd aq $CaCl_2$. Dried with $CaSO_4$ and distd.

3-Ethylisothionicotinamide *[10605-12-6]* **M 166.2, m 164-166°(dec).** Crystd from EtOH.

Ethyl isovalerate *[108-64-5]* **M 130.2, b 134.7°, n 1.39621, n^{25} 1.3975, d 0.8664.** Washed with aq 5% Na_2CO_3, then satd aq $CaCl_2$. Dried with $CaSO_4$ and distd.

Ethyl malonate *[105-53-3]* **M 160.2, b 92°/22mm, 198-199°/760mm, n 1.413, d 1.056, d^{25} 1.0507.** The ester (250g) has been heated on a steam bath for 36hr with abs EtOH (125ml) and conc H_2SO_4 (75ml), then fractionally distd under reduced pressure.

Ethyl mercaptan see **ethanethiol.**

Ethyl methacrylate *[97-63-2]* **M 114.2, b 59°/100mm, n 1.515, d 0.915.** Washed successively with 5% aq $NaNO_2$, 5% $NaHSO_3$, 5% NaOH, then water. Dried with $MgSO_4$, added 0.2% (w/w) of phenyl-β-naphthylamine, and distd through a short Vigreux column [Schultz *JACS* **80** 1854 *1958*].

Ethyl methyl ether *[540-67-0]* **M 60.1, b 10.8°, d° 0.725.** Dried with $CaSO_4$, passed through an alumina column (to remove peroxides), then fractionally distd.

Ethyl methyl ketone (methyl ethyl ketone, MEK) see **2-butanone.**

3-Ethyl-2-methyl-2-pentene *[19780-67-7]* **M 112.2, b 114.5°/760mm.** Purified by preparative glc on a column of 20% squalene on Chromosorb P at 70°.

3-Ethyl-4-methylpyridine *[529-21-5]* **M 121.2, b 76°/12mm, 194.5°/750mm, n 1.510, d 0.947.** Dried with solid NaOH, and fractionally distd.

5-Ethyl-2-methylpyridine *[104-90-5]* **M 121.2, b 178.5°/765mm, n 1.497, d 0.919.** Purified by conversion to the picrate, crystn, and regeneration of the free base.

N-Ethylmorpholine *[100-74-3]* **M 115.2, b 138-139°/763mm, n 1.445, d 0.912.** Distd twice, then converted by HCl gas into the hydrochloride (extremely deliquescent) which was crystd from anhydrous EtOH-acetone (1:2) [Herries, Mathias and Rabin *BJ* 85 127 *1962*].

Ethyl p-nitrobenzoate *[99-77-4]* **M 195.2, m 56°.** Dissolved in ethyl ether and washed with aq alkali, then the ether was evapd and the solid recrystd from EtOH.

Ethyl orthoformate *[122-51-0]* **M 148.2, b 144°/760mm, n 1.391, d 0.892.** Shaken with aq 2% NaOH, dried with solid KOH andd distd from sodium through a 20cm Vigreux column.

Ethyl Orange *[13545-67-0]* **M 372.4.** Recrystd twice from water.

o-Ethylphenol *[90-00-6]* **M 122.2, f.p. 45.1°, b 210-212°, n 1.537, d 1.020,**
p-Ethylphenol *[123-07-9]* **M 122.2, m 47-48°, b 218.0°/762mm, n^{25} 1.5239.** Non-acidic impurities were removed by passing steam through a boiling soln containing 1 mole of the phenol and 1.75 moles of NaOH (as aq 10% soln). The residue was cooled and acidified with 30% (v/v) H_2SO_4, and the free phenol was extracted into ethyl ether. The extract was washed with water, dried with $CaSO_4$ and the ether was evapd. The phenol was distd at 100mm pressure through a Stedman gauze-packed column. It was further purified by fractional crystn by partial freezing, and by zone refining, under nitrogen [Biddiscombe *et al JCS* 5764 *1963*]. Alternative purification is *via* the benzoate, as for phenol.

Ethyl phenylacetate *[101-97-3]* **M 164.2, b 99-99.3°/14mm, n 1.499, d 1.030.** Shaken with satd aq Na_2CO_3 (three times), aq 50% $CaCl_2$ (twice) and satd aq NaCl (twice). Dried with $CaCl_2$ and distd under reduced pressure.

2-Ethyl-2-phenylglutarimide see **3-ethyl-3-phenyl-2,6-piperidinedione.**

3-Ethyl-5-phenylhydantoin *[86-35-1]* **M 204.2, m 94°.** Crystd from water.

N-Ethyl-5-phenylisoxazolinium-3'-sulphonate *[4156-16-5]* **M 253.3, m 220°(dec).** [Lamas *et al, JACS* 108 5543 *1986*.],
3-Ethyl-3-phenyl-2,6-piperidinedione *[77-21-4]* **M 217.3, m 84°.** Crystd from ethyl ether or ethyl acetate/pet ether.

Ethyl propionate *[105-37-3]* **M 102.1, b 99.1°, n^{15} 1.38643, n 1.38394, d 0.891.** Treated with anhydrous $CuSO_4$ and distd under nitrogen.

2-Ethylpyridine *[100-71-0]* **M 107.2, b 148.6°, d 0.942,**
4-Ethylpyridine *[536-75-4]* **M 107.2, b 168.2-168.3°, d 0.942.** Dried with BaO, and fractionally distd. Purified by conversion to the picrate, recrystn and regeneration of the free base followed by distn.

4-Ethylpyridine-1-oxide *[14906-55-9]* **M 123.1, m 109-110°.** Crystd from acetone/ether.

1-Ethylquinolinium iodide see **quinolinium ethiodide.**

Ethyl Red *[76058-33-8]* **M 197.4, m 150-152°.** Crystd from EtOH/ethyl ether.

Ethyl stearate *[111-61-5]* **M 312.5, m 33°, b 213-215°/15mm.** The solid portion was separated from the partially solid starting material, then crystd twice from EtOH, dried by azeotropic distn with benzene, and fractionally distd in a spinning-band column at low pressure [Welsh *TFS* **55** 52 *1959*].

Ethyl sulphide *[352-93-2]* **M 90.2, b 92.1°, n^{15} 1.44550, d 0.835.** Fractionally distd from sodium metal.

N-Ethyl thiourea *[625-53-6]* **M 104.2, m 110°.** Crystd from EtOH, MeOH or ether.

Ethyl trichloroacetate *[515-84-4]* **M 191.4, b 100-100.5°/30mm, d 1.383.** Shaken with satd aq Na_2CO_3 (three times), aq 50% $CaCl_2$ (three times), satd aq NaCl (twice), then distd with $CaCl_2$ and distd under reduced pressure.

Ethyl vinyl ether *[109-92-2]* **M 72.1, b 35.5°, d 0.755.** Contains polymerization inhibitors (usually amines, e.g. triethanolamine) which can be removed by fractional distn. Redistd from sodium. **LACHRYMATORY.**

Ethynylbenzene see **phenylacetylene.**

Etiocholane (5β-androsterone) *[438-23-3]* **M 260.5, m 78-80°.** Crystd from acetone.

Etiocholanic acid *[438-08-4]* **M 304.5, m 228-29°.** Crystd from glacial acetic acid.

Etioporphyrin I *[448-71-5]* **M 478.7.** Crystd from pyridine or $CHCl_3$-pet ether.

Farnesyl pyrophosphate *[13058-04-3]*, *[E,E: 372-97-4]* **M 382.3.** Purified by chromatography on Whatman No3 MM paper in a system of isopropanol-isobutanol-ammonia-water (40:20:1:30) (v/v). Stored as the Li or NH_4 salt at 0°.

Ferulic acid see **4-hydroxy-3-methoxycinnamic acid.**

Flavin adenine dinucleotide (diNa, $2H_2O$ salt), see Chapter 6.

Flavin mononucleotide (Na, $2H_2O$ salt), see Chapter 6.

Flavone *[525-82-6]* **M 222.3, m 99-100°.** Crystd from pet ether.

Fluoranthene *[206-44-0]* **M 202.3, m 110-111°.** Purified by chromatography of CCl_4 solns on alumina, with benzene as eluent. Crystd from EtOH, MeOH or benzene. Purified by zone melting. [Gorman *et al, JACS* **107** 4404 *1985*.]

2-Fluorenamine *[153-78-6]* **M 181.2, m 131-132°.** Crystd from EtOH.

9-Fluorenamine *[525-03-1]* **M 181.2, m 64-65°.** Crystd from hexane.

Fluorene *[86-73-7]* **M 166.2, m 114.7-115.1^o, b 160°/15mm.** Purified by chromatography of CCl_4 or pet ether (b 40-60°) soln on alumina, with benzene as eluent. Crystd from 95% EtOH, 90% acetic acid and again from EtOH. Crystn using glacial acetic acid retained an impurity which was removed by partial mercuration and pptn with LiBr [Brown, Dubeck and Goldman *JACS* **84** 1229 *1962*]. Has also been crystd from hexane, or benzene/EtOH, distd under vac and purified by zone refining. [Gorman *et al, JACS* **107** 4404 *1985*.]

9-Fluorenone *[486-25-9]* **M 180.2, m 82.5-83.0^o, b 341°/760mm.** Distd under high vac. Crystd from abs EtOH, MeOH or benzene/pentane. [Ikezawa *JACS* **108** 1589 *1986*.] Also twice recrystd from toluene and sublimed under reduced pressure [Saltiel *JACS* **108** 2674 *1986*].

N-2-Fluorenylacetamide *[53-96-3]* **M 223.3, m 194-195°.** Crystd from EtOH/water. **CARCINOGEN.**

Fluorescein *[2321-07-5]* **M 320.0, ε_{495nm} 7.84 x 10^4 (in 10^{-3}M NaOH).** Dissolved in dil aq NaOH, filtered and pptd by adding dil (1:1) HCl. The process was repeated twice more and the fluorescein was dried at 100°. Alternatively, it has been crystd from acetone by allowing the soln to evaporate at 37° in an open beaker. Also recrystd from EtOH and dried in a vac oven.

Fluoracetamide *[640-19-7]* **M 77.1, m 108°.** Crystd from chloroform.

Fluorobenzene *[462-06-6]* **M 96.1, b 84.8^o, n 1.46573, n^{30} 1.4610, d 1.025.** Dried for several days with P_2O_5, then fractionally distd.

o-Fluorobenzoic acid *[445-29-4]* **M 140.1, m 127°.** Crystd from 50% aq EtOH, then zone melted or vac sublimed at 130-140°.

m-Fluorobenzoic acid *[445-38-9]* **M 140.1, m 124°.** Crystd from 50% aq EtOH, then vac sublimed at 130-140°.

p-Fluorobenzoic acid *[456-22-4]* **M 140.1, m 182°.** Crystd from 50% aq EtOH, then zone melted or vac sublimed at 130-140°.

3-Fluoro-4-hydroxyphenylacetic acid *[458-09-3]* **M 170.1, m 33°.** Crystd from water.

1-Fluoro-4-nitrobenzene   *[350-46-9]* **M 141.1, m 27° (stable form), 21.5° (unstable form), b 205.3°/735mm, 95-97.5°/22mm, 86.6°/14mm.** Crystd from EtOH.

1-Fluoro-4-nitronaphthalene   *[341-92-4]* **M 191.2, m 80°.** Recrystd from EtOH as yellow needles [Bunce *et al, JOC* **52** 4214 *1987*].

o-**Fluorophenol**   *[367-12-4]* **M 112.1, m 16°, b 53°/14mm, n 1.514, d 1.257.** Passed at least twice through a gas chromatographic column for small quantities, or fractionally distd under reduced pressure.

p-**Fluorophenoxyacetic acid**   *[405-79-8]* **M 170.1, m 106°.** Crystd from EtOH.

p-**Fluorophenylacetic acid**   *[405-50-5]* **M 154.1, m 86°.** Crystd from heptane.

p-**Fluorophenyl-*o*-nitrophenyl ether**   *[448-37-3]* **M 247.2, m 62°.** Crystd from EtOH.

o-**Fluorotoluene**   *[95-52-3]* **M 110.1, b 114.4°, n 1.475, d 1.005,**
m-**Fluorotoluene**   *[352-70-5]* **M 110.1, b 116.5°, n^{27} 1.46524, d 1.00,**
p-**Fluorotoluene**   *[352-32-9]* **M 116.0°, n 1.46884, d 1.00.** Dried with P_2O_5 or $CaSO_4$ and fractionally distd through a silvered vacuum-jacketed glass column with 1/8th-in glass helices. A high reflux ratio is necessary because of the closeness of the boiling points of the three isomers [Potter and Saylor *JACS* **37** 90 *1951*].

Folic acid   *[75708-92-8]* **M 441.4, m >250°(dec) [α]$_D^{25}$ +23° (c 0.5, 0.1N NaOH).** Crystd from hot water after extraction with butanol see Blakley [*BJ* **65** 331 *1957*] and Kalifa, Furrer, Bieri and Viscontini [*Helv Chim Acta* **61** 2739 *1978*].

Formaldehyde   *[50-00-0]* **M 30.0, m 92°, b -79.6°/20mm, d^{-20} 0.815.** Commonly contains added MeOH. Addn of KOH soln (1 mole KOH: 100 moles HCHO) to 40% formaldehyde soln, or evapn to dryness, gives paraformaldehyde polymer which, after washing with water, is dried in a vac desiccator over P_2O_5 or H_2SO_4. Formaldehyde is regenerated by heating the paraformaldehyde to 120° under vac, or by decomposing it with barium peroxide. The monomer, a gas, is passed through a glass-wool filter cooled to -48° in $CaCl_2$/ice mixture to remove particles of polymer, then dried by passage over P_2O_5 and either condensed in a bulb immersed in liquid nitrogen or absorbed in ice-cold conductivity water.

Formamide   *[75-12-7]* **M 45.0, f.p. 2.6°, b 103°/9mm, 210.5°/760mm(dec), n 1.44754, n^{25} 1.44682, d 1.13.** Formamide is easily hydrolysed by acids and bases. It also reacts with peroxides, acid halides, acid anhydrides, esters and (on heating) alcohols; while strong dehydrating agents convert it to a nitrile. It is very hygroscopic. Commercial material often contains acids and ammonium formate. Vorhoek [*JACS* **58** 2577 *1956*] added some bromothymol blue to formamide and then neutralised it with NaOH before heating to 80-90° under reduced pressure to distil off ammonia and water. The amide was again neutralised and the process was repeated until the liquid remained neutral on heating. Sodium formate was added, and the formamide was reduced under reduced pressure at 80-90°. The distillate was again neutralised and redistd. It was then fractionally crystd in the absence of CO_2 and water by partial freezing.
Formamide (specific conductance 2 x 10^{-7} ohm^{-1} cm^{-1}) of low water content was dried by passage through a column of 3A molecular sieves, then deionized by treatment with a mixed-bed ion-exchange resin loaded with H$^+$ and HCONH$^-$ ions (using sodium formamide in formamide)[Notley and Spiro *JCS(B)* 362 *1966*].

Formamidine sulphinic acid   *[1758-73-2]* **M 108.1, m 124-126°(dec).** Dissolved in five parts of aq 1:1% $NaHSO_3$ at 60-63° (charcoal), then crystd slowly, with agitation, at 10°. Filtered. Dried immediately at 60° [Koniecki and Linch *AC* **30** 1134 *1958*].

Formanilide   *[103-70-8]* **M 121.1, m 50°, b 166°/14mm, 216°/120mm, d 1.14.**   Crystd from ligroin/xylene.

Formic acid *[64-18-6]* **M 46.0 (anhydr), f.p. 8.3°, b 25°/40mm, 100.7°/760mm, n 1.37140, n^{25} 1.36938, d 1.22.** Anhydrous formic acid can be obtained by direct fractional distillation under reduced pressure, the receiver being cooled in ice-water. The use of P$_2$O$_5$ or CaCl$_2$ as dehydrating agents is unsatisfactory. Reagent grade 88% formic acid can be satisfactorily dried by refluxing with phthalic anhydride for 6hr and then distilling. Alternatively, if it is left in contact with freshly prepared anhydrous CuSO$_4$ for several days about one half of the water is removed from 88% formic acid: distn removes the remainder. Boric anhydride (prepared by melting boric acid in an oven at a high temperature, cooling in a desiccator, and powdering) is a suitable dehydrating agent for 98% formic acid; after prolonged stirring with the anhydride the formic acid is distd under vac. Formic acid can be further purified by fractional crystn using partial freezing.

D(-)-Fructose *[57-48-7]* **M 180.2, m 103-106°, [α]$_{546}$ -190° (after 1hr, c 10, H$_2$O).** Dissolved in an equal weight of water (charcoal, previously washed with water to remove any soluble material), filtered and evapd under reduced pressure at 45-50° to give a syrup containing 90% of fructose. After cooling to 40°, the syrup was seeded and kept at this temperature for 20-30hr with occasional stirring. The crystals were removed by centrifugation, washed with a small quantity of water and dried to constant weight under a vac over conc H$_2$SO$_4$. For higher purity, this material was recrystd from 50% aq ethanol [Tsuzuki, Yamazaki and Kagami *JACS* **72** 1071 *1950*].

Fructose-1,6-diphosphate (trisodium salt) *[38099-82-0]* **M 406.1.** For purification *via* the acid strychnine salt, see Neuberg, Lustig and Rothenberg [*Arch Biochem* **3** 33 *1943*]. The calcium salt can be partially purified by soln in ice-cold M HCl (1g per 10ml) and repptn by dropwise addition of 2M NaOH: the ppte and supernatant are heated on a boiling water bath for a short time, then filtered and the ppte is washed with hot water. The magnesium salt can be pptd from cold aq soln by adding four volumes of EtOH.

Fructose-6-phosphate *[643-13-0]* **M 260.1.** Crystd as the barium salt from water by adding four volumes of EtOH. The barium can be removed by passage through the H$^+$ form of a cation exchange resin and the free acid collected by freeze-drying.

D(+)-Fucose *[3615-37-0]* **M 164.2, m 144°, [α]$_{546}$ +89° (after 24hr, c 10 in H$_2$O).** Crystd from EtOH.

Fumagillin *[101993-69-5]* **M 458.5, m 194-195°, [α]$_D^{25}$ -26.2° (in 95% EtOH).** Forty grams of a commercial sample containing 42% fumagillin, 45% sucrose, 10% antifoam agent and 3% of other impurities were digested with 150ml of CHCl$_3$. The insoluble sucrose was filtered off and washed with CHCl$_3$. The combined CHCl$_3$ extracts were evapd almost to dryness at room temperature under reduced pressure. The residue was triturated with 20ml of MeOH and the fumagillin was filtered off by suction. It was crystd twice from 500ml of hot MeOH by standing overnight in a refrigerator. (The long chain fatty ester used as antifoam agent was still present, but was then removed by repeated digestion, on a steam bath, with 100ml of ethyl ether.) For further purification, the fumagillin (10g) was dissolved in 150ml of 0.2M ammonia, and the insoluble residue was filtered off. The ammonia soln (cooled in running cold water) was then brought to *p*H 4 by careful addn of M HCl with constant shaking in the presence of 150ml of CHCl$_3$. (Fumagillin is acid-labile and must be removed rapidly from the aq acid soln.) The CHCl$_3$ extract was washed several times with distd water, dried (Na$_2$SO$_4$) and evapd under reduced pressure. The solid residue was washed with 20ml of MeOH. The fumagillin was filtered by suction, then crystd from 200ml of hot MeOH. [Tarbell *et al, JACS* **77** 5610 *1955*.] Alternatively, 10g of fumagillin in 100ml CHCl$_3$ was passed through a silica gel (5g) column to remove tarry material, and the CHCl$_3$ was evapd to leave an oil which gave fumagillin on crystn from amyl acetate. It recrystallises from MeOH (charcoal). The fumagillin was stored in dark bottles in the absence of oxygen and at low temperatures. [Schenk, Hargie and Isarasena *JACS* **77** 5606 *1955*.]

Fumaric acid *[110-17-8]* **M 116.1, m 289.5-291.5° (sealed tube).** Crystd from hot M HCl or water. Dried at 100°.

2-Furfuraldehyde see **furfural.**

Furan *[110-00-9]* **M 68.1, b 31.3°, n 1.4214, d 1.42.** Shaken with aq 5% KOH, dried with $CaSO_4$ or Na_2SO_4, then distd under nitrogen, from KOH or sodium, immediately before use. A trace of hydroquinone could be added as an inhibitor of oxidation.

2-Furanacrylic acid *[539-47-9]* **M 138.1, m 141°.** Crystd from water or pet ether (b 80-100°)(charcoal).

Furan-2-carboxylic acid *[88-14-2]* **M 112.1, m 133-134°, b 141-144°/20mm, 230-232°/760mm.** Crystd from hot water (charcoal), dried at 120° for 2hr, then recrystd from $CHCl_3$, and again dried at 120° for 2hr. For use as a standard in volumetric analysis, good quality commercial acid should be crystd from $CHCl_3$ and dried as above or sublimed at 130-140° at 50-60mm or less.

Furan-3-carboxylic acid *[488-93-7]* **M 112.1, m 122-123°,**
Furan-3,4-dicarboxylic acid *[3387-26-6]* **M 156.1, m 217-218°.** Crystd from water.

Furfural *[98-01-1]* **M 96.1, b 59-60°/15mm, 161°/760mm, n 1.52608, d 1.159.** Unstable to air, light and acids. Impurities include formic acid, β-formylacrylic acid and furan-2-carboxylic acid. Distd over an oil bath from 7% (w/w) Na_2CO_3 (added to neutralise acids, especially pyromucic acid). Redistd from 2% (w/w) Na_2CO_3, and then, finally fractionally distd under vac. It is stored in the dark. [Evans and Aylesworth *IECAE* **18** 24 *1926*.]
Impurities resulting from storage can be removed by passage through chromatographic grade alumina. Furfural can be separated from impurities other than carbonyl compounds by the bisulphite addition compound.

Furfuryl alcohol *[98-00-0]* **M 98.1, b 68-69°/20mm, 170.0°/750mm, n 1.4873, n^{30} 1.4801, d 1.132.** Distd under reduced pressure to remove tarry material, shaken with aq $NaHCO_3$, dried with Na_2SO_4 and fractionally distd under reduced pressure from Na_2CO_3. Further dried by shaking with Linde 5A molecular sieves.

Furfuryl amine *[617-89-0]* **M 97.1, b 142.5-143°/735mm, n 1.489, d 1.059.** Distd under nitrogen from KOH through a column packed with glass helices.

Furil *[492-94-4]* **M 190.2, m 165-166°.** Crystd from MeOH or benzene (charcoal).

2-Furoic acid see **furan-2-carboxylic acid.**

Furoin *[552-86-3]* **M 192.2, m 135-136°.** Crystd from MeOH (charcoal).

Furylacrylic acid see **2-furanacrylic acid.**

Galactaric Acid (mucic acid)   *[526-99-6]* M 210.1, m 212-213°(dec).   Dissolved in the minimum volume of dil aq NaOH, and pptd by adding dil HCl. The temperature should be kept below 25°.

D-Galactonic acid   *[576-36-3]* M 196.2, m 148°. Crystd from EtOH.

D(-)-Galactono-1,4-lactone   *[2782-07-2]* M 178.1, m 134-137°, $[\alpha]_D$ -78° (in H_2O). Crystd from EtOH.

D(+)-Galactosamine hydrochloride   *[1772-03-8]* M 215.6, m 181-185°, $[\alpha]_D^{25}$ +96.4° (after 24hr, c 3.2 in H_2O).   Dissolved in a small volume of water. Then added three volumes of EtOH, followed by acetone until faintly turbid and stood overnight in a refrigerator. [Roseman and Ludoweig *JACS* **76** 301 *1954*.]

α-D-Galactose   *[59-23-4]* M 180.2, m 167-168°, $[\alpha]_D$ +80.4° (after 24hr, c 4 in H_2O).   Crystd twice from aq 80% EtOH at -10°, then dried under vac over P_2O_5.

Gallic acid (H_2O)   *[149-91-7]* M 188.1, m 253°(dec). Crystd from water.

Genistein   *[446-72-0]* M 270.2, m 297-298°. Crystd from 60% aq EtOH or water.

Genistin   *[529-59-9]* M 432.4, m 256°. Crystd from 80% EtOH/water.

α-Gentiobiose   *[16750-26-8]* M 342.3, m 86°. Crystd from MeOH (retains solvent of crystn).

β-Gentiobiose   *[554-91-6]* M 342.3, m 190-195°. Crystd from EtOH.

Geraniol   *[106-24-1]* M 154.3, b 230°, n 1.4766, d 0.879.   Purified by ascending chromatography or by thin layer chromatography on plates of kieselguhr G with acetone/water/liquid paraffin (130:70:1) as solvent system. Hexane/ethyl acetate (1:4) is also suitable. Also purified by glc on a silicone-treated column of Carbowax 20M (10%) on Chromosorb W (60-80 mesh). [Porter *PAC* **20** 499 *1969*.] Stored in full, tightly sealed containers in the cool, protected from light.

Geranylgeranyl pyrophosphate   *[6699-20-3]* M 450.5. Purified by counter-current distribution between two phases of a butanol/isopropyl ether/ammonia /water mixture (15:5:1:19) (v/v), or by chromatography on DEAE-cellulose (linear gradient of 0.02M KCl in 1mM Tris buffer, pH 8.9). Stored as a powder at 0°.

Geranyl pyrophosphate   *[763-10-0]* M 314.2. Purified by paper chromatography on Whatman No 3 MM paper in a system of isopropyl alcohol/isobutyl alcohol/ammonia/water (40:20:1:39), R_F 0.77-0.82. Stored in the dark as the ammonium salt at 0°.

Gibberillic acid   *[77-06-5]* M 346.4, m 233-235°(dec), $[\alpha]_{546}$ +92° (c 1, MeOH). Crystd from ethyl acetate.

Girard Reagent T   *[123-46-6]* M 167.6, m 192°. Crystd from abs EtOH.

Glucamine   *[488-43-7]* M 181.2, m 127°. Crystd from MeOH.

D-Gluconamide   *[3118-85-2]* M 197.2, m 144°, $[\alpha]_D^{23}$ +31° (c 2, H_2O). Crystd from EtOH.

D-Glucono-δ-lactone   *[90-80-2]* M 178.1, m 152-153°, $[\alpha]_{546}$ +76° (c 4, H_2O). Crystd from ethylene glycol monomethyl ether and dried for 1hr at 110°.

Glucosamine   *[3416-24-8]*   **M 179.2, m 110°(dec).** Crystd from MeOH.

D-Glucosamine hydrochloride   *[66-84-2]*   **M 215.6, m >300°, $[\alpha]_D^{25}$ +71.8°   (after 20hr, c 4, H$_2$O).** Crystd from 3M HCl, water, and finally water/EtOH/acetone as for galactosamine hydrochloride.

α-D-Glucose   *[492-62-6]*   **M 180.2, m 146°, $[\alpha]_D$ +52.5°   (after 24hr, c 4, H$_2$O).** Recrysts slowly from aq 80% EtOH, then vac dried over P$_2$O$_5$. Alternatively, crystd from water at 55°, then dried for 6hr in a vac oven between 60-70° at 2mm.

β-D-Glucose   *[50-99-7]*   **M 180.2, m 148-150°.** Crystd from hot glacial acetic acid.

α-D-Glucose pentaacetate   *[604-68-2]*   **M 390.4, m 110-111°, $[\alpha]_{546}$ +119°   (c 5, CHCl$_3$),**
β-D-Glucose pentaacetate   *[604-69-3]*   **M 390.4, m 131-132°, $[\alpha]_{546}$ +5°   (c 5, CHCl$_3$).** Crystd from MeOH or EtOH.

D-Glucose phenylhydrazone   *[534-97-4]*   **M 358.4, m 208°.**   Crystd from aq EtOH.

Glucose-1-phosphate   *[59-56-3]*   **M 260.1.** Two litres of 5% aq soln was brought to pH 3.5 with glacial acetic acid (+ 3g of charcoal, and filtered). An equal volume of EtOH was added, the pH was adjusted to 8.0 (glass electrode) and the soln was stored at 3° overnight. The ppte was filtered off, dissolved in 1.2L of distd water, filtered and an equal vol of EtOH was added. After standing at 0° overnight, the crystals were collected at the centrifuge, and washed with 95% EtOH, then abs EtOH, ethanol/ethyl ether (1:1), and ethyl ether. [Sutherland and Wosilait, *JBC* **218** 459 *1956*.] Its barium salt can be crystd from water and EtOH. Heavy metal impurities can be removed by passage of an aq soln (*ca* 1%) through an Amberlite IR-120 column (in the appropriate H$^+$, Na$^+$ or K$^+$ forms).

Glucose-6-phosphate   *[sodium salt 54010-71-8]*   **M 260.1.** Can be freed from metal impurities as described for glucose-1-phosphate. Its barium salt can be purified by solution in dil HCl and pptn by neutralising the soln. The ppte is washed with small vols of cold water and dried in air.

D-Glucuronic acid   *[6556-12-3]*   **M 194.1, m 165°, $[\alpha]_D$ +36°   (c 3, H$_2$O).** Crystd from EtOH or ethyl acetate.

D-Glucuronolactone   *[32449-92-6]*   **M 176.1, m 175-177°, $[\alpha]_{546}$ +22°   (after 24hr, c 10, H$_2$O).** Crystd from water.

L-Glutamic acid   *[56-86-0]*   **M 147.1, m 224-225°(dec), $[\alpha]_D^{25}$ +31.4°   (c 5, 5M HCl).** Crystd from water acidified to pH 3.2 by adding 4 vols of EtOH, and dried at 110°. Likely impurities are aspartic acid and cysteine.

L-Glutamic acid-5-benzyl ester   *[1676-73-9]* **M 237.3, m 179-181° $[\alpha]_{589}$ 19.3° (c 1, HOAc.** Recryst from water and stored at 0°. [Estrin *Biochem Preps* **13** 25 *1971*.]

L-Glutamine   *[56-85-9]*   **M 146.2, m 184-185°, $[\alpha]_D^{25}$ +31.8°   (M HCl).** Likely impurities are glutamic acid, ammonium pyroglutamate, tyrosine, asparagine, isoglutamine, arginine. Crystd from water.

Glutaraldehyde   *[111-30-8]*   **M 100.1, b 71°/10mm, as 50% aq soln.** Likely impurities are oxidation products - acids, semialdehydes and polymers. It can be purified by repeated washing with activated charcoal (Norit XX) followed by vacuum filtration, using 15-20g charcoal/100ml of glutaraldehyde soln.
Vac distn at 60-65°/15mm, discarding the first 5-10%, was followed by dilution with an equal vol of freshly distd water at 70-75°, using magnetic stirring under nitrogen. The soln is stored at low temp (3-

4°), in a tightly stoppered container, and protected from light.   Standardised by titration with hydroxylamine. [Anderson *J Histochem Cytochem* 15 652 *1967*.]

Glutaric acid   *[110-94-1]*   **M 132.1, m 97.5-98°.** Crystd from benzene, $CHCl_3$, distd water or benzene containing 10% (w/w) of ethyl ether. Dried under vac.

Glutathione   *[70-18-8]*   **M 307.3, m 195°(dec), $[\alpha]_D^{27}$ -21.3°** (c 2, H_2O). Crystd from 50% aq EtOH.

dl−**Glyceraldehyde**   *[56-82-6]*   **M 90.1, m 145°.** Crystd from EtOH/ethyl ether.

Glycerol   *[56-81-5]*   **M 92.1, m 18.2°, b 182°/20mm, 290°/760mm, n^{25} 1.47352, d 1.261 .** Glycerol was dissolved in an equal vol of *n*-butanol (or *n*-propanol, amyl alcohol or liquid ammonia) in a water-tight container, was cooled and seeded while slowly revolving in an ice-water slurry. The crystals were collected by centrifugation, then washed with cold acetone or isopropyl ether. [Hass and Patterson *IEC* 33 615 *1941*.] Coloured impurities can be removed from substantially dry glycerol by extraction with 2,2,4-trimethylpentane.   Alternatively, glycerol can be decolorized and dried by treatment with activated charcoal and alumina, followed by filtering. Glycerol can be distd at 15mm in a stream of dry nitrogen, and stored in a desiccator over P_2O_5. Crude glycerol can be purified by digestion with conc H_2SO_4 and saponification with a lime paste, then re-acidified with H_2SO_4, filtered, treated with an anion exchange resin and fractionally distd under vac.

Glycinamide hydrochloride   *[1668-10-6]*   **M 110.5, m 186-189°** (207-208°). Crystd from EtOH.

Glycine see **aminoacetic acid.**

Glycine ethyl ester hydrochloride   *[623-33-6]* **M 136.9, m 145-146°,**
Glycine hydrochloride   *[6000-43-7]* **M 111.5, m 176-178°.**   Crystd from abs EtOH.

Glycine methyl ester hydrochloride   *[5680-79-5]*   **M 125.6, m 174°(dec).** Crystd from MeOH.

Glycine p-nitrophenyl ester hydrobromide. M 277.1,m 214° (dec). Recryst from MeOH by adding ethyl ether. [Alners *et al, Biochem Preps* 13 22 *1971*.]

Glycocholic acid   *[475-31-0]*   **M 465.6, m 154-155°, $[\alpha]_{546}$ +37°** (c 1, EtOH). Crystd from hot water. Dried at 100°.

D(+)-Glycogen   *[9005-79-2]*   **M 25,000-100,000, m 270-280°(dec), $[\alpha]_{546}$ +216°** (c 5, H_2O). A 5% aq soln (charcoal) was filtered and an equal vol of EtOH was added. After standing overnight at 3° the ppte was collected by centrifugation and washed with abs EtOH, then EtOH/ethyl ether (1:1), and ethyl ether. [Sutherland and Wosilait *JBC* 218 459 *1956*.]

Glycol dimethyl ether see **1,2-Dimethoxyethane.**

Glycollic acid   *[79-14-1]*   **M 76.1, m 81°.** Crystd from ethyl ether.

N-**Glycylaniline**   *[555-48-6]*   **M 150.2.** Crystd from water.

Glycylglycine   *[556-50-3]*   **M 132.1, m 260-262°(dec).** Crystd from aq 50% EtOH or water at 50-60° by addition of EtOH. Dried at 110°.

Glycylglycine hydrochloride   *[13059-60-4]*   **M 168.6.** Crystd from 95% EtOH.

Glycylglycylglycine see **diglycylglycine**

Glycyl-L-proline   *[704-15-4]*   M 172.2, m 185°.   Crystd from water at 50-60° by addition of EtOH.

dl−**Glycylserine**   *[687-38-7]*   M 162.2, m 207°(dec).   Crystd from water (charcoal) by addition of EtOH.

Glycyrrhizic acid ammonium salt (3H$_2$O)   *[53956-04-0]*   M 823.0, m 210°(dec).   Crystd from glacial acetic acid, then dissolved in ethanolic ammonia and evapd.

Glyoxal bis(2-hydroxyanil)   *[1149-16-2]*   M 240.3, m 210-213°, ε_{294nm} 9880.   Crystd from MeOH or EtOH.

Glyoxaline see **imidazole**.

Glyoxylic acid *[298-12-4]* M 74.0, m 98°(anhydr), 50-52°(monohydrate).   Crystd from water as the monohydrate.

Gramicidin S *[113-73-5]* M 1141.4, m 268-270°.   Crystd from EtOH.

Gramine *[87-52-5]* M 174.3, m 134°.   Crystd from ethyl ether, ethanol or acetone.

Griseofulvin *[126-07-8]* M 352.8, m 220°, $[\alpha]_D^{22}$ +365° (c 1, acetone).   Crystd from benzene.

Guaiacic acid *[500-40-3]* M 328.4, m 99-100.5°.   Crystd from EtOH.

Guaiacol *[90-05-1]* M 124.1, m 32°, b 106°/24mm, 205°/746mm.   Crystd from benzene/pet ether or distd.

Guaiacol carbonate *[553-17-3]* M 274.3, m 88.1°.   Crystd from EtOH.

Guanidine *[113-00-8]* M 59.1.   Crystd from water/EtOH under nitrogen.   Very deliquescent and absorbs CO$_2$ from the air readily.

Guanidine carbonate *[593-85-1]* M 180.2, m 197°.   Crystd from MeOH.

Guanidine hydrochloride *[50-01-1]* M 95.5, m 181-183°.   Crystd from hot methanol by chilling to about -10°, with vigorous stirring.   The fine crystals were filtered through fritted glass, washed with cold (-10°) methanol, dried at 50° under vac for 5hr.   (The product is more pure than that obtained by crystn at room temperature from methanol by adding large amounts of ethyl ether.)   [Kolthoff *et al*, *JACS* **79** 5102 *1957*.]

Guanosine (H$_2$O) *[118-00-3]* M 283.2, m 240-250°(dec), $[\alpha]_{546}$ -86° (c 1, 0.1M NaOH),
Guanylic acid *[85-32-5]* M 363.2, m 208°(dec).   Crystd from water.   Dried at 110°.

Haematin   *[15489-90-4]* **M 633.5, m 200°(dec).**   Crystd from pyridine. Dried at 40° under vac.

Haematoporphyrin dimethyl ester   *[33070-12-1]*   **M 626.7, m 212°.**   Crystd from $CHCl_3$/MeOH.

Haematoxylin   *[517-28-2]* **M 302.3, m 100-120°.**   Crystd from dil aq $NaHSO_3$ until colourless.

Haemin   *[16009-13-5]*   **M   652.0,   m   >300°(dec).**   Crystd   from   glacial   acetic   acid   or $CHCl_3$/pyridine/acetic acid.

Harmine   *[442-51-3]* **M 212.3, m 261°(dec).**   Crystd from MeOH.

Harmine hydrochloride (hydrate) *[343-27-1]* **M 248.7, m 280°(dec).**   Crystd from water.

Hecogenine acetate   *[915-35-5]* **M 472.7, m 265-268°, $[\alpha]_D^{23}$ -4.5° (c 1, $CHCl_3$).**   Crystd from MeOH.

Heparin (sodium salt)   *[9041-08-1]*.   Dissolved in 0.1M NaCl (1g/100ml) and pptd by addition of EtOH (150ml).

Heptadecanoic acid   *[506-12-7]* **M 270.5, m 60-61°, b 227°/100mm.**   Crystd from MeOH or pet ether.

1-Heptadecanol   *[1454-85-9]* **M 256.5, m 54°.**   Crystd from acetone.

Heptafluoro-2-iodopropane   *[677-69-0]* **M 295.9.**   Purified by gas chromatography on a triacetin column, followed by bulb-to-bulb distn at low temperature.

n-**Heptaldehyde**   *[111-71-7]* **M 114.2, b 40.5°/12mm, 152.8°/760mm, n^{25} 1.4130, d 0.819.**   Dried with $CaSO_4$ or Na_2SO_4 and fractionally distd under reduced pressure. More extensive purification by pptn as the bisulphite compound (formed by adding the aldehyde to satd aq $NaHSO_3$) which was filtered off and recrystd from hot water.   The crystals, after being filtered and washed well with water, were hydrolysed by adding 700ml of aq Na_2CO_3 (12.5% w/w of anhydrous Na_2CO_3) per 100g of aldehyde. The aldehyde was then steam distd, separated, dried with $CuSO_4$ and distd under reduced pressure in a slow stream of nitrogen. [McNesby and Davis *JACS* **76** 2148 *1954*.]

n-**Heptaldoxime**   *[629-31-2]* **M 129.2, m 53-55°.**   Crystd from 60% aq EtOH.

n-**Heptane**   *[142-18-5]* **M 100.2, b 98.4°, n 1.38765, n^{25} 1.38512, d 0.684.**   Passage through a silica gel column greatly reduces the ultraviolet absorption of *n*-heptane. (The silica gel is previously heated to 350° before use.)   For more extensive purification, heptane is shaken with successive small portions of conc H_2SO_4 until the lower (acid) layer remains colourless. The heptane is then washed successively with water, aq 10% Na_2CO_3, water (twice), and dried with $CaSO_4$, $MgSO_4$ or $CaCl_2$. It is distd from sodium. *n*-Heptane can be distd azeotropically with methanol, then the methanol can be washed out with water and, after drying, the heptane is redistd.   Other purification procedures include passage through activated basic alumina, drying with CaH_2, storage with sodium, and stirring with 0.5N $KMnO_4$ in 6N H_2SO_4 for 12hr after treatment with conc H_2SO_4.   Carbonyl-containing impurities have been removed by percolation through a column of impregnated Celite made by dissolving 0.5g of 2,4-dinitrophenylhydrazine in 6ml of 85% H_3PO_4 by grinding together, then adding 4ml of distd water and 10g Celite. [Schwartz and Parks *AC* **33** 1396 *1961*.]

4-Heptanone see **diisopropyl ketone.**

Hept-1-ene   *[592-76-7]* **M 98.2, b 93°/771mm, n 1.400, d 0.698.**   Distd from sodium, then carefully fractionally distd using an 18-in gauze-packed column. Can be purified by azeotropic distn with EtOH.

Contained the 2- and 3-isomers as impurities. These can be removed by gas chromatography using a Carbowax column at 70°.

n-Heptyl alcohol *[111-70-6]* **M 116.2, b 175.6°, n 1.425,d 0.825.** Shaken with successive lots of alkaline KMnO$_4$ until the colour persisted for 15min, then dried with K$_2$CO$_3$ or CaO, and fractionally distd.

n-Heptylamine *[111-68-2]* **M 115.2, b 155°, n 1.434, d 0.775.** Dried in contact with KOH pellets for 24hr, then decanted and fractionally distd.

n-Heptyl bromide *[629-04-9]* **M 179.1, b 70.6°/19mm, 180°/760mm, n 1.45, d 1.140.** Shaken with conc H$_2$SO$_4$, washed with water, dried with K$_2$CO$_3$, and fractionally distd.

Hesperetin *[520-33-2]* **M 302.3, m 227-228°.** Crystd from ethyl acetate.

Hesperidin *[520-26-3]* **M 610.6, m 258-262°, [α]$_{546}$ -82° (c 2, pyridine).** Dissolved in dil aq alkali and pptd by adjusting the pH to 6-7.

Hexachlorobenzene *[118-74-1]* **M 284.8, m 230.2-231.0°.** Crystd repeatedly from benzene. Dried under vac over P$_2$O$_5$.

Hexachloro-1,3-butadiene *[87-68-3]* **M 260.8, m 39°, b 283-284°(dec)/733mm, d 1.665.** Vac distd at less than 15mm pressure.

1,2,3,4,5,6-Hexachlorocyclohexane *[319-84-6]* **M 290.8, m 158° (α-), 312° (β-), 112.5° (γ-isomer).** Crystd from EtOH. Purified by zone melting.

Hexachlorocyclopentadiene *[77-47-4]* **M 272.8, b 80°/1mm, n^{25} 1.5628, d 1.702.** Dried with MgSO$_4$. Distd under vac in nitrogen.

Hexachloroethane *[67-72-1]* **M 236.7, m 187°.** Steam distd, then crystd from 95% EtOH. Dried in the dark under vac.

Hexachlorocyclotriphosphazene *[940-71-6]* **M 354.0, m 113-114°.** Purified by sublimation and twice crystd from hexane [Meirovitch *et al, JPC* **88** 1522, *1984*; Alcock *et al, JACS* **106** 5561 *1984*.]

Hexacosane *[630-01-3]* **M 366.7, m 56.4°, b 169°/0.05mm, 205°/1mm, 262°/15mm.** Distd under vac and crystd from ethyl ether.

Hexacosanoic acid *[506-46-7]* **M 396.7, m 88-89°.** Crystd from EtOH.

n-Hexadecane (Cetane) *[544-76-3]* **M 226.5, m 18.2°, b 105°/0.1mm, n 1.4345, n^{25} 1.4325, d 0.773.** Passed through a column of silica gel and distd under vac in a column packed with Pyrex helices. Stored over silica gel. Crystd from acetone, or fractionally crystd by partial freezing.

1,14-Hexadecandioic acid. *[505-54-4]* **M 286.4, m 126°.** Crystd from EtOH or ethyl acetate.

Hexadecanoic acid *[57-10-3]* **M 256.4, m 126°.** Purified by slow (overnight) recrystn from hexane. Some samples were also crystd from acetone, EtOH or ethyl acetate. Crystals were stood in air to lose solvent, or were pumped on a vacuum line. [Iwahashi *et al, JCSFT 1* **81** 973 *1985*.]

1-Hexadecyl- see **cetyl-.**

Hexadecyltrimethylammonium bromide (CTAB) *[57-09-0]* **M 364.4.** Recrystd once from acetone, acetone/water or acetone and <5% MeOH and dried under vac at 60°. Also crystd from abs EtOH. [Dearden and Wooley *JPC* **91** 2404 *1987*.]

1,5-Hexadiene *[592-42-7]* M 82.2, b 59.6°, n 1.4039, d 0.694. Distd from $NaBH_4$.

Hexadimethrine bromide see **1,5-dimethyl-1,5-diazaundecamethylene polymethobromide**

Hexaethylbenzene *[87-85-4]* M 246.3, m 128.7-129.5°. Crystd from benzene or benzene/EtOH.

Hexafluoroacetone *[684-16-2]* M 166.1, m -129°, (trihydrate m 18-21°), b -28°. Dehydrated by passage of the vapour over P_2O_5. Ethylene was removed by passing the dried vapour through a tube containing Pyrex glass wool moistened with conc H_2SO_4. Further purification was by low temperature distn using Warde-Le Roy stills. Stored in the dark at -78°. [Holmes and Kutschke *TFS* 58 333 *1962*.]

Hexafluorobenzene *[392-56-3]* M 186.1, m 5.1°, b 79-80°, n 1.378, d 1.61. Main impurities are incompletely fluorinated benzenes. Purified by standing in contact with oleum for 4hr at room temperature, repeating until the oleum does not become coloured. Washed several times with water, then dried with P_2O_5. Final purification was by repeated fractional crystn.

Hexafluoroethane *[76-16-4]* M 138.0, b -79°. Purified for pyrolysis studies by passage through a copper vessel containing CoF_3 at *ca* 270°, and held for 3hr in a bottle with a heated (1300°) platinum wire. It was then fractionally distd. [Steunenberg and Cady *JACS* 74 4165 *1962*.]

1,1,1,3,3,3-Hexafluoropropan-2-ol *[920-66-1]* M 168.1, b 57-58°/760mm, n^{22} 1.2750, d 1.4563. Distd from 3A molecular sieves, retaining the middle fraction.

Hexamethylbenzene *[87-85-4]* M 162.3, m 165-165.5°. Sublimed, then crystd from abs EtOH, benzene, EtOH/benzene or EtOH/cyclohexane. Also purified by zone melting. Dried under vac over P_2O_5.

Hexamethyl(Dewar)benzene *[7641-77-2]* M 162.3, m 7°, b 60°/20mm, n 1.4480, d.0.803. Purified by passage through alumina [Traylor and Miksztal *JACS* 109 2770 *1987*].

Hexamethyldisilane *[1450-14-2]* M 146.4, b 111-113°, n 1.422, d 0.729. Grossly impure sample (25% impurities) was purified by repeated spinning band distn. This lowered the impurity level to 500 ppm. The main impurity was identified as 1-hydroxypentamethyldisilane.

Hexamethylenediamine *[124-09-4]* M 116.2, m 42°, b 46-47°/1mm, 84.9°/9mm, 100°/20mm, 204-205°/760mm. Crystd in a stream of nitrogen. Sublimed in a vac.

Hexamethylenediamine dihydrochloride *[6055-52-3]* M 189.2, m 248°. Crystd from water or EtOH.

Hexamethylene glycol *[629-11-8]* M 118.2, m 41.6°. Fractionally crystd from its melt.

Hexamethylenetetramine *[100-97-0]* M 140.2. Crystd from EtOH and stored in a vac.

Hexamethylphosphoric triamide (HMPT) *[680-31-9]* M 179.2, f.p. 7.2°, b 68-70°/1mm, 235°/760mm, n 1.460, d 1.024. The industrial synthesis is usually by treatment of $POCl_3$ with excess of dimethylamine in isopropyl ether. Impurities are water, dimethylamine and its hydrochloride. It is purified by refluxing over BaO or CaO at about 4mm pressure in an atmosphere of nitrogen for several hours, then distd from sodium at the same pressure. The middle fraction (b *ca* 90°) is collected, refluxed over sodium under reduced pressure under nitrogen and distd. It is kept in the dark under nitrogen, and stored in solid CO_2. Can also be stored over 4A molecular sieves.
Alternatively, it is distd under vac from CaH_2 at 60° and crystd twice in a cold room at 0°, seeding the liquid with crystals obtained by cooling in liquid nitrogen. After about two-thirds frozen, the remaining liquid is drained off [Fujinaga, Izutsu and Sakara *PAC* 44 117 *1975*]. For tests of purity see Fujinaga *et al*, in *Purification of Solvents*, ed Coetzee, Pergamon Press, Oxford, *1982*. For efficiency of desiccants

in drying HMPT see Burfield and Smithers [*JOC* **43** 3966 *1978*; Sammes *et al, JCSFT 1* 281 *1986*]. **CARCINOGEN.**

Hexanamide see *n*-caproamide,

n-**Hexane** *[110-54-3]* M 86.2, b 68.7°, n 1.37486, n²⁵ 1.37226, d 0.660. Purification as for *n*-heptane. Modifications include the use of chlorosulphonic acid or 35% fuming H_2SO_4 instead of conc H_2SO_4 in washing the alkane, and final drying and distn from sodium hydride. Unsatd compounds can be removed by shaking the hexane with nitrating acid (58% H_2SO_4, 25% conc HNO_3, 17% water, or 50% HNO_3, 50% H_2SO_4), then washing the hydrocarbon layer with conc H_2SO_4, followed by water, drying, and distg over sodium or *n*-butyl lithium. Also purified by distn under nitrogen from sodium benzophenone ketyl solubilised with tetraglyme. Also purified by chromatography on silica gel followed by distdn [Kajii *et al, JPC* **91** 2791 *1987*].

1,6-Hexanediol *[629-11-8]* M 118.2, m 43-45°. Recrystd from water.

Hexanenitrile see **Capronitrile.**

1-Hexene *[592-41-6]* M 84.2, b 63°, n 1.388, d 0.674,
cis-**2-Hexene** *[7688-21-3]* M 84.2, b 68-70°, n 1,399, d 0.699,
trans-**2-Hexene** *[4050-45-7]* M 84.2, b 65-67°, n 1.390,
trans-**3-Hexene** *[13269-52-8]* M 84.2, b 67-69°, n 1.393, d 0.678. Purified by stirring over Na/K alloy for at least 6hr, then fractionally distd from sodium under nitrogen.

meso-**Hexoestrol** *[84-16-2]* M 270.4, m 185-188°. Crystd from benzene or aq EtOH.

n-**Hexyl alcohol** *[111-27-3]* M 102.2, b 157.5°, n¹⁵ 1.4198, n²⁵ 1.4158, d 0.818. Commercial material usually contains other alcohols which are difficult to remove. A suitable method is to esterify with hydroxybenzoic acid, recrystallise the ester and saponify. [Olivier *Rec Trav chim* **55** 1027 *1936*]. Drying agents include K_2CO_3 and $CaSO_4$, followed by filtration and distn. (Some decomposition to the olefin occurred when Al amalgam was used as drying agent at room temperature, even though the amalgam was removed prior to distn.) If the alcohol is required anhydrous, the redistd material can be refluxed with the appropriate alkyl phthalate or succinate, as described under *Ethanol*.

n-**Hexylamine** *[111-26-2]* M 101.2, b 131°, n 1.419, d 0.765. Dried with, and fractionally distd from, KOH or CaH_2.

n-**Hexyl bromide** *[111-25-1]* M 165.1, b 87-88°/90mm, 155°/743mm, n 1.448, d 1.176. Shaken with H_2SO_4, washed with water, dried with K_2CO_3 and fractionally distd.

n-**Hexyl methacrylate** *[142-09-6]* M 154.2. Purified as for *methyl methacrylate*.

Hexyltrimethylammonium bromide *[2650-53-5]* M 224.3. Recrystd from acetone. Extremely hygroscopic salt.

1-Hexyne *[693-02-7]* M 82.2, b 12.5°/75mm, 71°/760mm, n 1.3989, d 0.7156,
2-Hexyne *[764-35-2]* M 82.2, b 83.8°/760mm, n 1.41382, d 0.73146,
3-Hexyne *[928-49-4]* M 82.1, b 81°/760mm, n 1.4115, 0.7231.
Distd from $NaBH_4$ to remove peroxides. Stood with sodium for 24hr, then fractionally distd under reduced pressure. Also dried by repeated vac transfer into freshly activated 4A molecular sieves, followed by vac transfer into Na/K alloy and stirring for 1hr before fractionally distilling.

Hippuric acid *[495-69-2]* M 178.2, m 187.2°. Crystd from water. Dried over P_2O_5.

Histamine *[51-45-6]* **M 111.2, m 86°(sealed tube), b 167°/0.8mm, 209°/18mm.** Crystd from benzene or chloroform.

Histamine dihydrochloride *[56-92-8]* **M 184.1, m 249-252° (244-245°).** Crystd from aq EtOH.

L-Histidine *[71-00-1]* **M 155.2, m 287°(dec), $[\alpha]_D^{25}$ -39.7° (H_2O), +13.0° (6M HCl).** Likely impurity is arginine. Adsorbed from aq soln on to Dowex 50-H$^+$ ion-exchange resin, washed with 1.5M HCl (to remove other amino acids), then eluted with 4M HCl as the dihydrochloride. Histidine is also purified as the dihydrochloride which is finally dissolved in water, the *p*H adjusted to 7.0, and the free zwitterionic base crystallises out on addition of EtOH.

Histidine dihydrochloride *[1007-42-7]* **M 242.1, m 252°.** Crystd from water or aq EtOH, and washed with acetone, then ethyl ether. Converted to the histidine di-(3,4-dichlorobenzenesulphonate) salt by dissolving 3,4-dichlorobenzenesulphonic acid (1.5g/10ml) in the aqueous histidine soln with warming, and then the soln is cooled in ice. The resulting crystals (m 280° dec) can be recrystd from 5% aq 3,4-dichlorobenzenesulphonic acid, then dried over $CaCl_2$ under vac, and washed with ethyl ether to remove excess reagent. The dihydrochloride can be regenerated by passing the soln through a Dowex-1 (Cl$^-$ form) ion-exchange column. The solid is obtained by evapn of the soln on a steam bath or better in a vac. [Greenstein and Winitz, The Amino Acids **Vol 3**, p 1976 1961.]

L-Histidine monohydrochloride (H_2O) *[7048-02-4]* **M 209.6, $[\alpha]_D^{25}$ +13.0° (6M HCl),**
dl-**Homocysteine** *[6027-13-0]* **M 135.2.** Crystd from aq EtOH.

Homocystine *[626-72-2]* **M 268.4, m 260-265°(dec).** Crystd from water.

Homophthalic acid *[89-51-0]* **M 180.2, m 182-183° (varies with the rate of heating).** Crystd from boiling water (25ml/g). Dried at 100°.

L-Homoserine *[672-15-1]* **M 119.1, m 203°, $[\alpha]_D^{26}$ +18.3°(in 2M HCl).** Likely impurities are *N*-chloroacetyl-L-homoserine, *N*-chloroacetyl-D-homoserine, L-homoserine, homoserine lactone, homoserine anhydride (formed in strong solns of homoserine if slightly acidic). Cyclises to the lactone in strongly acidic soln. Crystd from water by adding 9 volumes of EtOH.

Hordenine *[539-15-1]* **M 165.2, m 117-118°.** Crystd from EtOH or water.

Humulon *[26472-41-3]* **M 362.5, m 65-66.5°.** Crystd from ethyl ether.

Hyamine 1622 *[121-54-0]* **M 448.1** Crystd from boiling acetone after filtering. The ppte was filtered off, washed with ethyl ether and dried for 24hr in a vac desiccator.

Hydantoin *[461-72-3]* **M 100.1.** Crystd from MeOH.

Hydrazine *N,N'*-**dicarboxylic acid diamide** *[110-21-4]* **M 116.1, m 248°,**
4-Hydrazinobenzoic acid *[619-67-0]* **M 152.2, m 217°(dec).** Crystd from water.

1-Hydrazinophthalazine hydrochloride (hydralazine hydrochloride) *[304-20-1]* **M 196.6, m 172-173°.** Crystd from MeOH.

Hydrazobenzene *[122-66-7]* **M 184.2, m 128°.** Crystd from pet ether (b 60-100°) to constant absorption spectrum.

Hydrobenzamide *[92-29-5]* **M 298.4, m 101-102°.** Crystd from abs EtOH or cyclohexane/benzene. Dried under vac over P_2O_5.

dl-**Hydrobenzoin** *[655-48-1]* **M 214.3, m 120°.** Crystd from ethyl ether/pet ether.

meso-**Hydrobenzoin** *[579-43-1]* **214.3, m 139°.** Crystd from EtOH or water.

Hydrocinnamic acid (3-phenylpropionic acid) *[501-52-0]* **M 150.2, m 48-48.5°.** Crystd from benzene, CHCl$_3$ or pet ether (b 40-60°). Dried in a vac.

Hydroquinone *[123-31-9]* **M 110.1, m 175.4, 176.6°.** Crystd from acetone, benzene, EtOH, EtOH/benzene, water or acetonitrile (25g in 30ml), preferably under nitrogen. Dried under vac. [Wolfenden *et al, JACS* **109** 463 *1987*.]

Hydroquinone dimethyl ether see *p*-**dimethoxybenzene.**

Hydroquinone monobenzyl ether see *p*-**benzyloxyphenol.**

Hydroquinone monomethyl ether see *p*-**methoxyphenol.**

Hydroquinone-2-monosulphonate (K salt) *[21799-87-1]* **M 228.3, m 250°(dec).** Recrystd from water.

p-**Hydroxyacetophenone** *[99-93-4]* **M 136.2, m 109°.** Crystd from ethyl ether, aq EtOH or benzene/pet ether.

4'-Hydroxyacetanilide *[103-90-2]* **M 151.2, m 169-170.5°.** Crystd from water.

4-Hydroxyacridine *[18123-20-1]* **M 195.2, m 116.5°.** Crystd from EtOH.

3-Hydroxyanthranilic acid *[548-93-6]* **M 153.1, m >240°(dec),** λ_{max} **298nm, log ε 3000 (0.1M HCl).** Crystd from water. Sublimes below its melting point in a vac.

erythro-**3-Hydroxy-RS-aspartic acid** *[6532-76-9]* **M 149.1.** Likely impurities are 3-chloromalic acid, ammonium chloride, *threo*-3-hydroxyaspartic acid. Crystd from water.

p-**Hydroxyazobenzene** see *p*-**phenylazophenol.**

o-**Hydroxybenzaldehyde** see **salicylaldehyde.**

m-**Hydroxybenzaldehyde** *[100-83-4]* **M 122.1, m 108°.** Crystd from water.

p-**Hydroxybenzaldehyde** *[123-08-0]* **M 122.1, m 115-116°.** Crystd from water (containing some H$_2$SO$_4$). Dried over P$_2$O$_5$ under vac.

m-**Hydroxybenzoic acid** *[99-06-9]* **M 138.1, m 200.8°.** Crystd from abs EtOH.

p-**Hydroxybenzoic acid** *[99-96-7]* **M 138.1, m 213-214°,**
p-**Hydroxybenzonitrile** *[767-00-0]* **M 119.1, m 113-114°.** Crystd from water.

4-Hydroxybenzophenone *[1137-42-4]* **M 198.2, m 135°,**
2-Hydroxybenzothiazole *[934-34-9]* **M 183.1, m 117-118°,**
1-Hydroxybenzotriazole (H$_2$O) *[2592-95-2]* **M 135.1, m 159-160°.** Crystd from aq EtOH or water [Dryland and Sheppard *JCSPT* 125 *1986*].

2-Hydroxybenzyl alcohol *[90-01-7]* **M 124.1, m 87°.** Crystd from water or benzene.

3-Hydroxybenzyl alcohol *[620-24-6]* **M 124.1, m 71°.** Crystd from benzene.

4-Hydroxybenzyl alcohol *[623-05-2]* **M 124.1, m 114-115°.** Crystd from water.

2-Hydroxybiphenyl *[90-43-7]* M 170.2, m 56°, b 145°/14mm, 275°/760mm. Crystd from pet ether.

4-Hydroxybiphenyl *[92-69-3]* M 170.2, m 164-165°, b 305-308°/760mm. Crystd from aq EtOH.

3-Hydroxy-2-butanone *[513-86-0]* M 88.1, b 144-145°, [m 100-105° dimer]. Washed with EtOH until colourless, then with ethyl ether or acetone to remove biacetyl. Air dried by suction and further dried in a vac desiccator.

2-Hydroxycaprylic acid see **2-hydroxyoctadecanoic acid.**

p-**Hydroxycinnamic acid** *[501-98-4]* M 164.2, m 214-215°. Crystd from hot water (charcoal).

4-Hydroxycoumarin *[1076-38-6]* M 162.1, m 206°,
3-(4-Hydroxy-3,5-dimethoxyphenyl)acrylic acid *[2107-59-6]* M 234.1, m 204-205°(dec),
R-2-Hydroxy-3,3-dimethyl-γ-butyrolactone *[79-50-5]* M 130.1, m 89-91°, $[\alpha]_{546}$ -62° (c 3, H_2O). Crystd from water.

4-Hydroxydiphenylamine *[122-37-2]* M 185.2, m 72-73°. Crystd from chlorobenzene/pet ether.

12-Hydroxydodecanoic acid *[505-95-3]* M 216.3, m 86-88°. Crystd from toluene [Sadowik *et al, JACS* 108 7789 *1986*].

2-Hydroxy-4-(*n*-dodecyloxy)benzophenone *[2985-59-3]* M 382.5, m 50-52°. Recryst from *n*-hexane and then 10% (v/v) EtOH in acetonitrile [Valenty *et al, JACS* 106 6155 *1984*].

N-**[2-Hydroxyethyl]ethylenediamine** *[111-41-1]* M 104.1, b 91.2°/5mm, 238-240°/752mm, n 1.485, d 1.030. Distd twice through a Vigreux column. Redistd from solid NaOH, then from CaH_2. Alternatively, converted to the dihydrochloride and recrystd from water. Dried. Mixed with excess of solid NaOH and the free base distd from the mixture. Redistd from CaH_2. [Drinkard, Bauer and Bailar *JACS* 82 2992 *1960*.]

N-**[2-Hydroxyethyl]ethylenediaminetriacetic acid** *[150-39-0]* M 278.3, m 212-214°(dec). Crystd from warm water, after filtering, by addition of 95% EtOH and allowing to cool. The crystals, collected on a sintered-glass funnel, were washed three times with cold abs EtOH, then again crystd from water. After leaching with cold water, the crystals were dried at 100° under vac. [Spedding, Powell and Wheelwright *JACS* 78 34 *1956*.]

N-**Hydroxyethyliminodiacetic acid** *[93-62-9]* M 177.2, m 181°(dec). Crystd from water.

2-Hydroxyethylimino-tris(hydroxymethyl)methane (Mono-Tris) *[7343-51-3]* M 165.2, m 91°. Crystd twice from EtOH. Dried under vac at 25°.

2-Hydroxyethyl methacrylate *[868-77-9]* M 130.1, b 67°/3.5mm, n 1.452, d 1.071. Dissolved in water and extracted with *n*-heptane to remove ethylene glycol dimethacrylate (checked by gas-liquid chromatography) and distd twice under reduced pressure [Strop, Mikes and Kalal *JPC* 80 694 *1976*].

N-**2-Hydroxyethylpiperazine-*N'*-2-ethanesulphonic acid** (HEPES) *[7365-45-9]* M 238.3. Crystd from hot EtOH and water.

3-Hydroxyflavone *[577-85-5]* M 238.2, m 169-170°. Recrystd from MeOH, EtOH or hexane. Also purified by repeated sublimation under high vac, and dried by high vac pumping for at least one hour [Bruker and Kelly *JPC* 91 2856 *1987*].

β-**Hydroxyglutamic acid** *[533-62-0]* M 163.1, m 100°(dec). Crystd from water.

4-Hydroxyindane *[1641-41-1]* **M 134.2, m 49-50°, b 120°/12mm,**
5-Hydroxyindane *[1470-94-6]* **M 134.2, m 55°, b 255°/760mm.** Crystd from pet ether.

2-Hydroxy-5-iodobenzoic acid see **5-iodosalicylic acid.**

α-Hydroxyisobutyric acid see **2-hydroxy-2-methylpropionic acid.**

5-Hydroxy-L-lysine monohydrochloride *[32685-69-1]* **198.7, $[\alpha]_D^{25}$ +17.8° (6M HCl).** Likely impurities are 5-*allo*-hydroxy-(D and L)-lysine, histidine, lysine, ornithine. Crystd from water by adding 2-9 volumes of EtOH stepwise.

4-Hydroxy-3-methoxyacetophenone *[498-02-2]* **M 166.2, m 115°.** Crystd from water, or EtOH/pet ether.

4-Hydroxy-3-methoxycinnamic acid (ferulic acid) *[1135-24-6]* **M 194.2, m 174°.** Crystd from water.

17β-Hydroxy-17α-methyl-3-androsterone *[521-11-9]* **M 304.5°, m 192-193°.** Crystd from ethyl acetate.

3-Hydroxy-4-methylbenzaldehyde *[57295-30-4]* **m 116-117°, b 179°/15mm.** Crystd from water.

dl-**2-Hydroxy-2-methylbutyric acid** *[3739-30-8]* **M 118.1, m 72-73°.** Crystd from benzene, and sublimed at 90°.

dl-**2-Hydroxy-3-methylbutyric acid** *[600-37-3]* **M 118.1, m 86°.** Crystd from ether/pentane.

7-Hydroxy-4-methylcoumarin (4-methylumbelliferone) *[90-33-5]* **M 176.2, m 185-186°.** Crystd from abs EtOH.

5-(Hydroxymethyl)furfural *[67-47-0]* **M 126.1, m 33.5°, b 114-116°/1mm.** Crystd from ethyl ether/pet ether.

3-Hydroxy-3-methylglutaric acid *[503-49-1]* **162.1, m 99-102°.** Recrystd from ethyl ether/hexane and dried under vac at 60° for 1hr.

2-Hydroxymethyl-2-methyl-1,3-propanediol see **1,1,1-tris(hydroxymethyl)ethane.**

dl-**3-Hydroxy-*N*-methylmorphinan** *[297-90-5]* **M 257.4, m 251-253°.** Crystd from anisole + aq EtOH.

5-Hydroxy-2-methyl-1,4-naphthaquinone see **plumbagin.**

6-Hydroxy-2-methyl-1,4-naphthaquinone *[633-71-6]* **M 188.2.** Crystd from aq EtOH. Sublimes on heating.

2-(Hydroxymethyl)-2-nitro-1,3-propanediol *[126-11-4]* **M 151.1, m 174-175°(dec).** Crystd from $CHCl_3$/ethyl acetate or ethyl acetate/benzene.

4-Hydroxy-4-methyl-2-pentanone *[123-42-2]* **M 116.2, b 166°, n 1.4235, n^{25} 1.4213. d 0.932.** Loses water when heated. Can be dried with $CaSO_4$, then fractionally distd under reduced pressure.

17α-Hydroxy-6α-methylprogesterone *[520-85-4]* **M344.5, m 220°, $[\alpha]_D^{25}$ +75°.** Crystd from chloroform.

2-Hydroxy-2-methylpropionic acid *[594-61-6]* **M 104.1, m 79°, b 114°/12mm, 84°/15mm, 212°/760mm.** Distd in steam, crystd from ethyl ether or benzene, sublimed at 50° and dried under vac..

8-Hydroxy-2-methylquinoline *[826-81-3]* **M 159.2, m 74-75°, b 266-267°.** Crystd from EtOH or aq EtOH.

2-Hydroxymyristic acid see **2-hydroxytetradecanoic acid.**

2-Hydroxy-1-naphthaldehyde *[708-06-5]* **M 172.2, m 82°, b 192°/27mm.** Crystd from EtOH (1.5ml/g), ethyl acetate or water.

2-Hydroxy-1-naphthaleneacetic acid *[10441-45-9]* **M 202.2.** Treated with activated charcoal and crystd from EtOH/water (1:9, v/v). Dried under vac, over silica gel, in the dark. Stored in the dark at -20° [Gafni, Modlin and Brand *JPC* **80** 898 *1976*]. Forms a lactone (m 107°) readily.

6-Hydroxy-2-naphthalenepropionic acid *[553-39-9]* **M 216.2, m 180-181°.** Crystd from aq EtOH or aq MeOH.

3-Hydroxy-2-naphthalide *[92-77-3]* **M 263.3, m 248.0-248.5°,**
3-Hydroxy-2-naphtho-4'-chloro-*o*-toluidide *[92-76-2]* **M 311.8, m 243.5-244.5°,**
3-Hydroxy-2-naphthoic-α-naphthalide *[94966-09-2]* **M 314.3, m 217-.5-218.0°,**
3-Hydroxy-2-naphthoic-β-naphthalide *[550-57-2]* **M 305.3, m 243.5-244.5°,** and other naphthol AS derivatives. Crystd from xylene [Schnopper, Broussard and La Forgia *AC* 31 1542 *1959*].

1-Hydroxy-2-naphthoic acid see **1-naphthol-2-carboxylic acid.**

3-Hydroxy-2-naphthoic acid see **3-naphthol-2-carboxylic acid.**

2-Hydroxy-1,4-naphthaquinone *[83-72-7]* **M 174.2, m 192°(dec).** Crystd from benzene.

5-Hydroxy-1,4-naphthaquinone *[481-39-0]* **M 174.2, m 155°.** Crystd from benzene/pet ether.

6-Hydroxynicotinic acid see **2-Hydroxypyridine-5-carboxylic acid.**

2-Hydroxy-5-nitrobenzyl bromide *[772-33-8]* **M 232.0, m 147°.** Crystd from benzene or benzene/ligroin.

4-Hydroxy-2-*n*-nonylquinoline *N*-oxide *[316-66-5]* **M 287.4, m 148-149°.** Crystd from EtOH.

2-Hydroxyoctanoic acid *[617-73-2]* **M 160.2, m 69.5°.** Crystd from EtOH/pet ether or ether/ligroin.

1-Hydroxyphenazine *[528-71-2]* **M 196.2, m 157-158°.** Chromatographed on acidic alumina with benzene/ether. Crystd from benzene/heptane, and sublimed.

2-Hydroxyphenylacetic acid *[614-75-5]* **M 152.2, m 148-149°, b 240-243°/760mm.** Crystd from ether or chloroform.

3-Hydroxyphenylacetic acid *[621-37-4]* **M 152.2, m 137°.** Crystd from benzene/ligroin.

4-Hydroxyphenylacetic acid *[156-38-7]* **M 152.2, m 150-151°.** Crystd from water.

2-(2-Hydroxyphenyl)benzoxazole *[835-64-3]* **M 211.2, m 127°, b 338°/760mm,**
2-(2-Hydroxyphenyl)benzothiazole *[3411-95-8]* **M 227.2, m 132-133°, b 173-179°/3mm.** Recrystd several times from aq EtOH and by sublimation. [Itoh and Fujiwara *JACS* 107 1561 *1985*.]

3-Hydroxy-2-phenylcinchoninic acid *[485-89-2]* **M 265.3, m 206-207°(dec).** Crystd from EtOH.

N-(4-Hydroxyphenyl)-3-phenylsalicylamide *[550-57-2]* M 305.3, m 183-184°. Crystd from aq MeOH.

L-2-Hydroxy-3-phenylpropionic acid (phenyllactic acid) *[20312-36-1]* M 166.2, m 125-126°, $[\alpha]_D^{12}$ - 18.7° (EtOH). Crystd from water, MeOH, EtOH or benzene.

dl-2-Hydroxy-3-phenylpropionic acid *[828-01-3]* M 166.2, m 97-98°, b 148-150°/15mm. Crystd from benzene or chloroform.

3-Hydroxy-2-phenylpropionic acid see **tropic acid**.

3-*p*-Hydroxyphenylpropionic acid *[501-97-3]* M 166.2, m 129-130°. Crystd from ether.

p-Hydroxyphenylpyruvic acid *[156-39-8]* M 180.2, m 220°. Crystd three times from 0.1M HCl/EtOH (4:1, v/v) immediately before use [Rose and Powell *BJ* 87 541 *1963*].

3-β-Hydroxy-5-pregnen-2-one *[145-13-1]* M 316.5, m 189-190°, $[\alpha]_D$ +30° (EtOH), $[\alpha]_{546}$ +34° (c 1, EtOH). Crystd from MeOH.

17α-Hydroxyprogesterone *[604-09-1]* M 330.5, m 222-223°, $[\alpha]_{546}$ +141° (c 2, dioxane), λ_{max} 240nm. Crystd from acetone or EtOH. **Acetate: m 239-240°** and **caproate: m 119-121°** crystallised from CHCl$_3$/MeOH.

21-Hydroxyprogesterone see **11-desoxycorticosterone**.

trans-L-4-Hydroxyproline *[51-35-4]* M 131.1, m 274°, $[\alpha]_D$ -76.0° (c 5, H$_2$O). Crystd from MeOH/EtOH (1:1). Separation from normal *allo*-isomer can be achieved by crystn of the copper salts (see *Biochem Prep* 8 114 *1961*).

4'-Hydroxypropiophenone *[70-70-2]* M 150.2, m 149°. Crystd from water.

2-(α-Hydroxypropyl)piperidine *[24448-89-3]* M 143.2, m 121°, b 226°. Crystd from ether.

7-(2-Hydroxypropyl)theophylline *[603-00-9]* M 238.2, m 135-136°. Crystd from EtOH.

6-Hydroxypurine *[68-94-0]* M 136.1, m 150°(dec). Crystd from hot water. Dried at 105°.

2-Hydroxypyridine *[142-08-5]* M 95.1, m 105-107°, b 181-185°/24mm, ε_{293nm} 5900 (H$_2$O). Distd under vac to remove coloured impurity, then crystd from benzene, CCl$_4$, EtOH or CHCl$_3$/ethyl ether. It can be sublimed under high vacuum. [DePue *et al, JACS* 107 2131 *1985*.]

3-Hydroxypyridine *[109-00-2]* M 95.1, m 129°. Crystd from water or EtOH.

4-Hydroxypyridine *[626-64-2]* M 95.1, m 65°(hydrate), 148.5° (anhydr), b >350°/760mm. Crsytd from water. Loses water of crystn on drying *in vacuo* over H$_2$SO$_4$. Stored over KOH because it is hygroscopic.

2-Hydroxypyridine-5-carboxylic acid *[5006-66-6]* M 139.1, m 304°(dec),
4-Hydroxypyridine-2,6-dicarboxylic acid (chelidamic acid) *[138-60-3]* 183.1, m 254°(dec). Crystd from water.

2-Hydroxypyrimidine *[557-01-7]* M 96.1, m 179-180°. Crystd from EtOH or ethyl acetate.

4-Hydroxypyrimidine *[4562-27-0]* M 96.1, m 164-165°. Crystd from benzene or ethyl acetate.

2-Hydroxypyrimidine hydrochloride *[38353-09-2]* **M 132.5, m 205°(dec)**. Crystd from EtOH.

2-Hydroxyquinoline *[59-31-4]* **M 145.2, m 199-200°**. Crystd from MeOH.

8-Hydroxyquinoline (oxine) *[148-24-3]* **M 145.2, m 75-76°**. Crystd from hot EtOH, acetone, pet ether (b 60-80°) or water. Crude oxine can be purified by pptn of copper oxinate, followed by liberation of free oxine with H_2S or by steam distn after acidification with H_2SO_4. Stored in the dark.

8-Hydroxyquinoline-5-sulphonic acid (H_2O) *[84-88-8]* **M 243.3, m >310°**. Crystd from water or dil HCl (*ca* 2% by weight).

5-Hydroxysalicylic acid *[490-79-9]* **M 154.1, m 204.5-205°**. Crystd from hot water.

trans-**5-Hydroxystilbene** *[6554-98-9]* **M 196.3, m 189°**. Crystd from benzene or acetic acid.

N-**Hydroxysuccinimide** *[6066-32-6]* **M 115.1, m 96-98°**. Recrystd from EtOH/ethyl acetate [Manesis and Goodmen *JOC* **52** 5331 *1987*].

dl-**2-Hydroxytetradecanoic acid** *[2507-55-3]* **M 244.4, m 81-82°**,
R-2-Hydroxytetradecanoic acid *[26632-17-7]* **M 244.4, m 88-2-88.5°**, $[\alpha]_D$ **-31° ($CHCl_3$)**. Crystd from chloroform.

4-Hydroxy-2,2,6,6-tetramethylpiperidine *[2403-88-5]* **M 157.3, m 130-131°**. Crystd from water as hydrate, and crystd from ether as the anhydrous base.

9-Hydroxytriptycene *[73597-16-7]* **M 270.3, m 245-246.5°**. Crystd from benzene/pet ether. Dried at 100° in a vac [Imashiro *et al*, *JACS* **109** 729 *1987*].

5-Hydroxy-L-tryptophan *[4350-09-8]* **M 220.2, m 273°(dec)**, $[\alpha]_D^{22}$ **-32.5°**, $[\alpha]_{546}$ **-73.5° (c 1, H_2O)**. Likely impurities are 5-hydroxy-D-tryptophan and 5-benzyloxytryptophan. Crystd under nitrogen from water by adding EtOH. Stored under nitrogen.

Hydroxyurea *[127-07-1]* **M 76.1, m 70-72° (unstable form), 141°**. Crystd from water by addition of EtOH.

3-Hydroxyxanthone *[3722-51-8]* **M 212.2, m 246°**. Purified by chromatography on SiO_2 gel with pet ether/benzene). Recrystd from benzene or EtOH [Itoh *et al*, *JACS* **107** 4819 *1985*].

α-Hyodeoxycholic acid *[83-49-8]* **M 392.6, m 196-197°**, $[\alpha]_{546}$ **+8° (c 2,EtOH)**. Crystd from ethyl acetate.

Hyoscine (scopolamine, atroscine) *[114-49-8]* **M 321.4, m 59°**, $[\alpha]_D$ **-18° (c 5, EtOH), -28° (c 2, H_2O)**, $[\alpha]_{546}$ **-30° (c 5, $CHCl_3$)**. Crystd from benzene/pet ether. Racemate has m 56-57° (H_2O), 37-38° ($2H_2O$), syrup (anhydr), *l* and *d* isomers can separate as syrups when anhydrous.

Hypericin *[548-04-9]* **M 504.4, m 320°(dec)**. Crystd from pyridine by addition of methanolic HCl.

Hypoxanthine see **6-hydroxypurine**.

Ibogaine *[83-74-9]* M 300.3, m 152-153°. Crystd from aq EtOH.

Imidazole *[288-32-4]* **M 68.1, m 89.5-91°, b 256°.** Crystd from benzene, CCl$_4$, CH$_2$Cl$_2$, EtOH, pet ether, acetone/pet ether and distd deionized water. Dried at 40° under vac over P$_2$O$_5$. Distd at low pressure. Also purified by sublimation or by zone melting. [Caswell and Spiro *JACS* **108** 6470 *1986*.] ^{15}N-imidazole was crystd from benzene [Scholes *et al, JACS* **108** 1660 *1986*].

2-Imidazolidinethione see **ethylene thiourea.**

2-Imidazolidone see **ethylene urea.**

4-(Imidazol-1-yl)acetophenone *[10041-06-2]* **M 186.2, m 104-107°.** Twice recrystd from CH$_2$Cl$_2$/hexane [Collman *et al, JACS* **108** 2588 *1986*].

Iminodiacetic acid *[142-73-4]* **M 133.1, m 225°(dec).** Crystd from water.

2,2'-Iminodiethanol see **diethanolamine.**

1,3-Indandione *[606-23-5]* **M 146.2, m 129-132°.** Recrystd from EtOH [Bernasconi and Paschalis *JACS* **108** 2969 *1986*].

Indane *[496-11-7]* **M 118.1, b 177°, n 1.538, d 0.960.** Shaken with conc H$_2$SO$_4$, then water, dried and fractionally distd.

Indanthrone *[81-77-6]* **M 442.4, m 470-500°.** Crystd repeatedly from 1,2,4-trichlorobenzene.

Indazole *[271-44-3]* **M 118.1, m 147°.** Crystd from water, sublimation under a vac, then pet ether (b 60-80°).

Indene *[95-13-6]* M 116.2, f.p. -1.5°, b 114.5°/100mm, n 1.5763, d 0.994. Shaken with 6M HCl for 24hr (to remove basic nitrogenous material), then refluxed with 40% NaOH for 2hr (to remove benzonitrile). Fractionally distd, then fractionally crystd by partial freezing. The higher-melting portion was converted to its sodium salt by adding a quarter of its weight of sodamide under nitrogen and stirring for 3hr at 120°. Unreacted organic material was distd off at 120°/1mm. The sodium salts were hydrolysed with water, and the organic fraction was separated by steam distn, followed by fractional distn. Before use, the distillate was passed, under nitrogen, through a column of activated silica gel. [Russell *JACS* **78** 1041 *1956*.]

Indigo *[482-89-3]* **M 262.3, and halogen-substituted indigo dyes.** Reduced in alkaline soln with sodium hydrosulphite, and filtered. The filtrate was then oxidised by air, and the resulting ppte was filtered off, dried at 65-70°, ground to a fine powder, and extracted with CHCl$_3$ in a Soxhlet extractor. Evapn of the CHCl$_3$ gave the purified dye. [Brode, Pearson and Wyman *JACS* **76** 1034 *1954*; spectral characteristics are listed.]

Indole *[120-72-9]* **M 117.2, m 52°, 124°/5mm, b 253-254°/760mm.** Crystd from benzene, hexane, water or EtOH/water (1:10). Further purified by sublimation in a vac or zone melting.

Indole-3-acetic acid *[87-51-4]* **M 175.2, m 167-169°,**
Indole-3-butanoic acid *[133-32-4]* **M 203.2, m 124-125°,**
Indole-3-propionic acid *[830-96-6]* **M 189.2, m 134-135°.** Recrystd from EtOH/water [James and Ware *JPC* **89** 5450 *1985*].

(-)-Inosine *[58-63-9]* M 268.2, m 215°, [α]$_{546}$ -76° (c 1, 0.1M NaOH). Crystd from aq 80% EtOH.

i-Inositol *[87-88-8]* M 180.2, m 228°. Crystd from water or aq 50% EtOH. Dried under vac.

meso-Inositol *[87-89-8]* M 180.2, m 223-225°. Crystd from aq EtOH.

Inositol monophosphate *[15421-51-9]* M 260.1, m 195-197°(dec). Crystd from water and EtOH.

Inulin *[9005-80-5]* M (162.14)$_n$. Crystd from water.

Iodoacetamide *[144-48-9]* M 185.0, m *ca* 143°(dec). Crystd from water or CCl_4.

Iodoacetic acid *[64-69-7]* M 160.6, m 78°. Crystd from pet ether (b 60-80°) or $CHCl_3/CCl_4$.

2-Iodoaniline *[615-43-0]* M 219.0, m 60-61°. Distd with steam and crystd from benzene/pet ether.

4-Iodoaniline *[540-37-4]* M 219.0, m 62-63°. Crystd from pet ether (b 60-80°) by refluxing, then cooling in an ice-salt bath freezing mixture. Dried in air. Also crystd from EtOH and dried in a vac for 6hr at 40° [Edidin *et al*, *JACS* **109** 3945 *1987*].

4-Iodoanisole *[696-62-8]* M 234.0, m 51-52°, b 139°/35mm, 237°/726mm. Crystd from aq EtOH.

Iodobenzene *[591-50-4]* M 204.0, b 188°, n^{25} 1.6169, d 1.829. Washed with dil aq $Na_2S_2O_3$, then water. Dried with $CaCl_2$ or $CaSO_4$. Decolorised with charcoal. Distd under reduced pressure and stored with mercury or silver powder.

o-**Iodobenzoic acid** *[88-67-5]* M 248.4, m 162°,
m-**Iodobenzoic acid** *[618-51-9]* M 248.4, m 186.6-186.8°,
p-**Iodobenzoic acid** *[619-58-9]* M 248.4, m 271-272°. Crystd repeatedly from water and EtOH. Sublimed under vac at 100°.

4-Iodobiphenyl *[1591-31-7]* M 280.1,m 113.7-114.3°. Crystd from EtOH/benzene and dried under vac over P_2O_5.

1-Iodobutane see *n*-butyl iodide.

2-Iodobutane *[513-48-4]* M 184.0, b 120.0, n^{25} 1.4973, d 1.50. Purified by shaking with conc H_2SO_4, then washing with water, aq Na_2SO_3 and again with water. Dried with $MgSO_4$ and distd. Alternatively, passed through a column of activated alumina before distn, or treated with elemental bromine, followed by extraction of the free halogen with aq $Na_2S_2O_3$, thorough washing with water, drying and distilling. It is stored over silver powder and distd before use.

1-Iodo-2,4-dinitrobenzene *[709-49-9]* M 294.0, m 88°. Crystd from ethyl acetate.

Iodoethane see ethyl iodide.

Iodoform *[75-47-8]* M 393.7, m 119°. Crystd from MeOH, EtOH or EtOH/ethyl acetate.

1-Iodo-2-methylpropane see isobutyl iodide.

1-Iodo-4-nitrobenzene *[636-98-6]* M 249.0, m 171-172°. Pptd from acetone by addition of water, then recrystd from EtOH.

o-**Iodophenol** *[533-58-4]* M 280.1, m 42°. Crystd from $CHCl_3$ or ethyl ether.

p-**Iodophenol** *[540-38-5]* M 280.1, m 94°. Crystd from pet ether (b 80-100°).

1-Iodopropane see *n*-propyl iodide.

2-Iodopropane see **isopropyl iodide**

3-Iodopropene see **allyl iodide.**

5-Iodosalicylic acid *[119-30-2]* **M 264.0, m 197°.** Crystd from water.

o-**Iodosobenzoic acid** *[304-91-6]* **M 264.0, m >200°.** Crystd from EtOH.

N-**Iodosuccinimide** *[512-12-1]* **M 225.0, m 200-201°.** Crystd from dioxane/CCl$_4$.

α-Iodotoluene see **benzyl iodide.**

p-**Iodotoluene** *[624-31-7]* **M 218.0, m 35°, b 211-212°.** Crystd from EtOH.

3-Iodo-L-tyrosine *[70-78-0]* **M 307.1, m 205-208°(dec), [α]$_D^{25}$ -4.4° (1M HCl).** Likely impurities are tyrosine, diiodotyrosine and iodide. Crystd by soln in dil ammonia, at room temperature, followed by addition of dilute acetic acid to *p*H 6. Stored in the cold.

α-Ionone *[127-41-3]* **M 192.3, b 131°/13mm, n 1.520, d 0.931, [α]$_D^{23}$ +347° (neat).** Purified on a spinning band fractionating column.

β-Ionone *[79-77-6]* **M 192.3, b 150-151°/24mm, n 1.5211, d 0.945, ε$_{296nm}$ 10,700.** Converted to the semicarbazone (**m 149°**) by adding 50g of semicarbazide hydrochloride and 44g of potassium acetate in 150ml of water to a soln of 85g of β-ionone in EtOH. (More EtOH was added to redissolve any β-ionone that pptd.) The semicarbazone crystd on cooling in an ice-bath and was recrystd from EtOH or 75% MeOH to constant m.p. (148-149°). The semicarbazone (5g) was shaken at room temperature for several days with 20ml of pet ether and 48ml of M H$_2$SO$_4$, then the ether layer was washed with water and dil aq NaHCO$_3$, dried and the solvent was evapd. The β-ionone was distd under vac. (The customary steam distn of β-ionone semicarbazone did not increase the purity.) [Young *et al*, *JACS* **66** 855 *1944*.]

Iproniazid phosphate *[305-33-9]* **M 277.2, m 178-179°.** Crystd from water and acetone.

Isatin *[91-56-5]* **M 147.1, m 200°.** Crystd from amyl alcohol.

Isoamyl acetate *[123-92-2]* **M 130.2, b 142.0°, n 1.40535, d 0.871.** Dried with finely divided K$_2$CO$_3$ and fractionally distd.

Isoamyl alcohol *[123-51-3]* **M 88.2, b 132°/760mm, n 1.408, d 0.809.** Dried with K$_2$CO$_3$ or CaSO$_4$, then fractionally distd. If more nearly anhydrous alcohol is required, the distillate can be refluxed with the appropriate alkyl phthalate or succinate as described for *ethanol.*

Isoamyl bromide *[107-82-4]* **M 151.1, f.p. -112°, b 119.2°/737mm, n 1.444, d 1.208.** Shaken with conc H$_2$SO$_4$, washed with water, dried with K$_2$CO$_3$ and fractionally distd.

Isoamyl chloride *[513-36-0]* **M 106.6, b 99°/734mm, n 1.4084, d 0.8704.** Shaken vigorously with 95% H$_2$SO$_4$ until the acid layer no longer became coloured during 12hr, then washed with water, saturated aq Na$_2$CO$_3$, and more water. Dried with MgSO$_4$, filtered and fractionally distd. Alternatively, a stream of oxygen containing 5% of ozone was passed through the chloride for a time, three times longer than was necessary to cause the first coloration of starch iodide paper by the exit gas. Subsequent washing of the liquid with aq NaHCO$_3$ hydrolysed the ozonides and removed organic acids. After drying and filtering, the isoamyl chloride was distd. [Chien and Willard *JACS* **75** 6160 *1953*.]

Isoamyl ether *[544-01-4]* **M 158.3, b 173.4°, n 1.40850, d 0.778.** This is a mixture of 2- and 3-methylbutyl ether. It is purified by refluxing with sodium for 5hr, then distilled under reduced pressure, to remove alcohols. Isoamyl ether can also be dried with $CaCl_2$ and fractionally distd from P_2O_5.

Isoascorbic acid *[89-65-6]* **M 176.1, m 174°(dec), $[\alpha]_D^{25}$ -16.8° (c 2, H_2O).** Crystd from water or dioxane.

dl-**Isoborneol** *[124-76-5]* **M 154.3, m 212° (sealed tube).** Crystd from EtOH or pet ether (b 60-80°). Sublimes in a vac.

Isobutane *[75-28-5]* **M 58.1, b -10.2°, d 0.557.** Olefines and moisture can be removed by passage at 65° through a bed of silica-alumina catalyst which has previously been evacuated at about 400°. Alternatively, water and CO_2 can be taken out by passage through P_2O_5 then asbestos impregnated with NaOH. Treatment with anhydrous $AlBr_3$ at 0° then removes traces of olefines. Inert gases can be separated by freezing the isobutane at -195° and pumping out the system.

Isobutene *[115-11-7]* **M 56.1, b -6.6°/760mm.** Dried by passage through anhydrous $CaSO_4$ at 0°. Purified by freeze-pump-thaw cycles and trap-to-trap distn.

Isobutyl alcohol *[78-83-1]* **M 74.1, b 108°/760mm, n 1.396, d 0.801.** Dried with K_2CO_3, $CaSO_4$ or $CaCl_2$, filtered and fractionally distd. For further drying, the redistd alcohol can be refluxed with the appropriate alkyl phthalate or succinate as described under *ethanol*.

Isobutyl bromide *[78-77-3]* **M 137.0, b 91.2°, n 1.437, d 1.260.** Partially hydrolysed to remove any tertiary alkyl halide, then fractionally distd, washed with conc H_2SO_4, water and aq K_2CO_3, then redistd from dry K_2CO_3. [Dunbar and Hammett *JACS* **72** 109 *1950.*]

Isobutyl chloride *[513-36-0]* **M 92.3, b 68.8°/760mm, n 1.398, d 0.877.** Same methods as described under *isoamyl chloride*.

Isobutylene see **Isobutene.**

Isobutyl formate *[543-27-1]* **M 102.1, b 98.4°, n 1.38546, d 0.885.** Washed with satd aq $NaHCO_3$ in the presence of satd NaCl, until no further reaction occurred, then with satd aq NaCl, dried ($MgSO_4$) and fractionally distd.

Isobutyl iodide *[513-38-2]* **M 184.0, b 83°/250mm, 120°/760mm, n 1.495, d 1.60.** Shaken with conc H_2SO_4, and washed with water, aq Na_2SO_3, and water, dried with $MgSO_4$ and distd. Alternatively, passed through a column of activated alumina before distn. Stored under nitrogen with mercury in a brown bottle or in the dark.

Isobutyl mercaptan see **2-methylpropane-1-thiol.**

Isobutyl vinyl ether *[109-53-5]* **M 100.2, b 108-110°, n 1.398, d 0.768.** Washed three times with equal volumes of aq 1% NaOH, dried with CaH_2, refluxed with sodium for several hours, then fractionally distd from sodium.

Isobutyraldehyde *[78-84-2]* **M 72.1, b 62.0°, n 1.377, d 0.789.** Dried with $CaSO_4$ and used immediately after distn under nitrogen because of the great difficulty in preventing oxidation. Can be purified through its acid bisulphite derivative.

Isobutyramide *[563-83-7]* **M 87.1, m 128-129°, b 217-221°.** Crystd from acetone, benzene, $CHCl_3$ or water, then dried under vac over P_2O_5 or 99% H_2SO_4. Sublimed under vac.

Isobutyric acid *[79-31-2]* **M 88.1. b 154-154.5°, n 1.393, d 0.949.** Distd from $KMnO_4$, then redistd from P_2O_5.

Isobutyronitrile *[78-82-0]* **M 69.1, b 103.6°, n 1.378, d^{25} 0.7650.** Shaken with conc HCl (to remove isonitriles), then with water and aq $NaHCO_3$. After a preliminary drying with silica gel or Linde type 4A molecular sieves, it is shaken or stirred with CaH_2 until hydrogen evolution ceases, then decanted and distd from P_2O_5 (not more than 5g/L, to minimize gel formation). Finally it is refluxed with, and slowly distd from CaH_2 (5g/L), taking precautions to exclude moisture.

Isodurene see **1,2,3,5-tetramethylbenzene.**

L-Isoleucine *[73-32-5]* **M 131.2, m 285-286°(dec), $[\alpha]_D$ +40.6° (6M HCl).** Crystd from water by addition of 4 volumes of EtOH.

Isolysergic acid *[478-95-5]* **M 268.3, m 218°(dec), $[\alpha]_D$ +281° (c 1, pyridine).** Crystd from water.

Isonicotinamide *[1453-82-3]* **M 122.1, m 155.5-156°.** Recrystd from hot water.

Isonicotinic acid *[55-22-1]* **M 123.1, m 320°.** Crystd repeatedly from water. Dried under vac at 110°.

Isonicotinic acid hydrazide *[54-87-3]* **M 137.1, m 172°.** Crystd from 95% EtOH.

1-Isonicotinic acid 2-isopropylhydrazide *[54-92-2]* **M 179.2, m 112.5-113.5°.** Crystd from benzene/pet ether.

1-Isonicotinyl-2-salicylidenehydrazide *[495-84-1]* **M 241.2, m 232-233°.** Crystd from EtOH.

Isonitrosoacetone *[31915-82-9]* **M 87.1, m 69°.** Crystd from ether/pet ether or CCl_4.

Isonitrosoacetophenone *[532-54-7]* **M 149.2, m 126-128°.** Crystd from water.

5-Isonitrosobarbituric acid (violuric acid) *[87-39-8]* **M 175.1, m 245-250°.** Crystd from water or EtOH.

Isononane *[34464-40-9]* **M 128.3, b 142°/760mm.** Passed through columns of activated silica gel and basic alumina (activity 1). Distd under high vac from Na/K alloy.

Isooctane see **2,2,4-trimethylpentane.**

Isopentane see **2-methylbutane.**

Isopentyl- see **isoamyl-.**

Isopentenyl pyrophosphate *[358-71-4]* **M 366.2.** Purified by chromatography on Whatman No 1 paper using *tert*-butyl alcohol/formic acid/water (20:5:8, R_F 0.60) or 1-propanol/ammonia/water (6:3:1. R_F 0.48). Also purified by chromatography on a DEAE-cellulose column or a Dowex-1 (formate form) ion-exchanger using formic acid and ammonium formate as eluents. A further purification step is to convert it to the monocyclohexylammonium salt by passage through a column of Dowex-50 (cyclohexylammonium form) ion-exchange resin. Can also be converted into its lithium salt.

Isophorone *[78-59-1]* **M 138.2, b 94°/16mm, n^{18} 1.4778, d 0.921.** Washed with aq 5% Na_2CO_3 and then distd under reduced pressure, immediately before use. Alternatively, can be purified *via* the semicarbazone. [Erskine and Waight *JCS 3425 1960*.]

Isophthalic acid *[121-91-5]* **M 166.1, m 345-348°.** Crystd from aq EtOH.

Isoprene *[78-79-5]* **M 68.1, b 34.5-35°/762mm, n^{25} 1.4225, d 0.681.** Refluxed with sodium. Distd from sodium or NaBH$_4$ under nitrogen, then passed through a column containing KOH, CaSO$_4$ and silica gel. *tert*-Butylcatechol (0.02% w/w) was added, and the isoprene was stored in this way until redistd before use. The inhibitor (*tert*-butylcatechol) in isoprene can be removed by several washings with dil NaOH and water. The isoprene is then dried over CaH$_2$, distd under nitrogen at atmospheric pressure, and the fraction distilling at 32° is collected. Stored under nitrogen at -15°.

Isopropanol *[67-63-0]* **M 60.1, b 82.5°, n$^{25.8}$ 1.3739, d 0.783.** Isopropyl alcohol is prepared commercially by dissolution of propene in H$_2$SO$_4$, followed by hydrolysis of the sulphate ester. Major impurities are water, lower alcohols and oxidation products such as aldehydes and ketones. Purification of isopropanol follows substantially the same procedure as for *n*-propyl alcohol.
Isopropanol forms a constant-boiling mixture, **b 80.3°**, with water. Most of the water can be removed from this 91% isopropanol by refluxing with CaO (200g/L) for several hours, then distilling. The distillate can be dried further with CaH$_2$, magnesium ribbon, BaO, CaSO$_4$, calcium, anhydrous CuSO$_4$ or Linde type 5A molecular sieves. Distn from sulphanilic acid removes ammonia and other basic impurities. Peroxides [indicated by liberation of iodine from weakly acid (HCl) solns of 2% KI] can be removed by refluxing with solid stannous chloride or with NaBH$_4$ then fractionally distilling. To obtain isopropanol containing only 0.002M of water, sodium (8g/L) has been dissolved in material dried by distn from CaSO$_4$, 35ml of isopropyl benzoate has been added and, after refluxing for 3hr, the alcohol has been distd through a 50-cm Vigreux column. [Hine and Tanabe *JACS* **80** 3002 *1958*.] Other purification steps for isopropanol include refluxing with solid aluminium isopropoxide, refluxing with NaBH$_4$ for 24hr, and the removal of acetone by treatment with, and distn from 2,4-dinitrophenylhydrazine. Peroxides re-form in isopropanol if it is stood for several days.

Isopropenylcyclobutane *[3019-22-5]* **M 98.1.** Purified by preparative chromatography (silicon oil column). Dried with molecular sieves.

Isopropyl acetate *[108-22-5]* **M 102.1, b 88.4°, n 1.3773, d 0.873.** Washed with 50% aq K$_2$CO$_3$ (to remove acid), then with satd aq CaCl$_2$ (to remove any alcohol). Dried with CaCl$_2$ and fractionally distd.

Isopropyl alcohol see **Isopropanol.**

Isopropyl benzene see **cumene.**

Isopropyl bromide *[75-26-3]* **M 123.0, b 0°/69.2mm, 59.4°/760mm, n^{15} 1.42847, n 1.4251, d 1.31.** Washed with 95% H$_2$SO$_4$ (conc acid partially oxidised it) until a fresh portion of acid did not become coloured after several hours, then with water, aq NaHSO$_3$, aq 10% Na$_2$CO$_3$ and again with water. (The H$_2$SO$_4$ can be replaced by conc HCl.) Prior to this treatment, isopropyl bromide has been purified by bubbling a stream of oxygen containing 5% ozone through it for 1hr, followed by shaking with 3% hydrogen peroxide soln, neutralising with aq Na$_2$CO$_3$, washing with distilled water and drying: Alternatively, it has been treated with elemental bromine and stored for 4 weeks, then extracted with aq NaHSO$_3$ and dried with MgSO$_4$. After the acid treatment, isopropyl bromide can be dried with Na$_2$SO$_4$, MgSO$_4$ or CaH$_2$, and fractionally distd.

N-**Isopropylcarbazole** *[1484-09-9]* **M 209.3.** Crystd from isopropanol. Sublimed under vac. Zone refined.

Isopropyl chloride *[75-29-6]* **M 78.5, b 34.8°, n 1.3779, n^{25} 1.3754, d 0.864.** Purified with 95% H$_2$SO$_4$ as described for *isopropyl bromide*, then dried with MgSO$_4$, P$_2$O$_5$ or CaH$_2$, and fractionally distd from Na$_2$CO$_3$ or CaH$_2$. Alternatively, a stream of oxygen containing *ca* 5% ozone has been passed through the chloride for about three times as long as was necessary to obtain the first coloration of starch iodide paper by the exit gas, and the liquid was then washed with NaHCO$_3$ soln to hydrolyse ozonides and remove organic acids before drying and distilling.

Isopropyl ether *[108-20-3]* **M 102.2, b 68.3°, n 1.3688, n^{25} 1.36618, d 0.719.** Common impurities are water and peroxides [detected by the liberation of iodine from weakly acid (HCl) solns of 2% KI]. Peroxides can be removed by shaking with aq Na_2SO_3 or with acidified ferrous sulphate (0.6g $FeSO_4$ and 6ml conc H_2SO_4 in 110ml of water, using 5-10g of soln per L of ether), or aq $NaBH_4$ soln. The ether is then washed with water, dried with $CaCl_2$ and distd. Alternatively, refluxing with $LiAlH_4$ or CaH_2, or drying with $CaSO_4$, then passage through an activated alumina column, can be used to remove water and peroxides. Other dehydrating agents used with isopropyl ether include P_2O_5, sodium amalgam and sodium wire. (The ether is often stored in brown bottles, or in the dark, with sodium wire.) Bonner and Goishi (*JACS* **83** 85 *1961*) treated isopropyl ether with dil sodium dichromate/sulphuric acid soln, followed by repeated shaking with a 1:1 mixture of 6M NaOH and satd $KMnO_4$. The ether was washed several times with water, dil aq HCl, and water, with a final washing with, and storage over, ferrous ammonium sulphate acidified with H_2SO_4. Blaustein and Gryder (*JACS* **79** 540 *1957*), after washing with alkaline $KMnO_4$, then water; treated the ether with ceric nitrate in nitric acid, and again washed with water. Hydroquinone was added before drying with $CaCl_2$ and $MgSO_4$, and refluxing with sodium amalgam (108g Hg/100g Na) for 2hr under nitrogen. The distillate (nitrogen atmosphere) was made 2 x 10^{-5}M in hydroquinone to inhibit peroxide formation (which was negligible if the ether was stored in the dark). Pyrocatechol and resorcinol are alternative inhibitors.

4,4'-Isopropylidenediphenol *[80-05-7]* **M 228.3, m 158°.** Crystd from acetic acid/water (1:1).

Isopropyl iodide *[75-30-9]* **M 170.0, b 88.9°, n 1.4987, d 1.70.** Treated with elemental bromine, followed by extraction of free halogen with aq $Na_2S_2O_3$ or $NaHSO_3$, washing with water, drying ($MgSO_4$ or $CaCl_2$) and distn. (The treatment with bromine is optional.) Other purification methods include passage through activated alumina, or shaking with copper powder or mercury to remove iodine, drying with P_2O_5 and distn. Washing with conc H_2SO_4, water, aq Na_2SO_3, water and aq Na_2CO_3 has also been used. Treatment with silica gel causes some liberation of iodine. Distillations should be carried out at slightly reduced pressure. Purified isopropyl iodide is stored in the dark in the presence of a little mercury.

Isopropyl mercaptan see **propane-2-thiol.**

Isopropyl methyl ether *[598-53-8]* **M 74.1, b 32.5°/777mm, n 1.3576, d^{15} 0.724.** Purified by drying with $CaSO_4$, passage through a column of alumina (to remove peroxides) and fractional distn.

Isopropyl *p*-nitrobenzoate *[13756-40-6]* **M 209.2, m 105-106°.** Dissolved in ethyl ether, washed with aq alkali, then water and dried. Evapn of the ether and recrystn from EtOH gave pure material.

Isopropyl peroxydicarbonate. Crystd from toluene.

***p*-Isopropyl toluene** *[99-87-6]* **M 134.2, b 176.9°/744mm, n 1.4902, d 0.856.** Dried with CaH_2 and fractionally distd. Stored with CaH_2.

Isoquinoline *[119-65-3]* **M 129.2, m 24°, b 120°/18mm, n 1.6148, d 1.0986.** Dried with Linde type 5A molecular sieves or Na_2SO_4 and fractionally distd at reduced pressure. Alternatively, it was refluxed with, and distd from, BaO. Also purified by fractional crystn from the melt and distd from zinc dust. Converted to its phosphate (**m 135°**) or picrate (**m 223°**), which were purified by crystn and the free base recovered and distd. [Packer, Vaughn and Wong *JACS* **80** 905 *1958*.]
The procedure for purifying *via* the picrate comprises the addition of quinoline to picric acid dissolved in the minimum volume of 95% EtOH to yield yellow crystals which are washed with EtOH and air dried before recrystn from acetonitrile. The crystals are dissolved in dimethyl sulphoxide (previously dried over 4A molecular sieves) and passed through a basic alumina column, on which picric acid is adsorbed. The free base in the effluent is extracted with *n*-pentane and distd under vac. Traces of solvent are removed by vapour phase chromatography. [Mooman and Anton *JPC* **80** 2243 *1976*.]

Isovaleric acid *[502-74-2]* **M 102.1, b 176.5°/762mm, n^{15} 1.4064, n 1.40331, d 0.927.** Dried with Na$_2$SO$_4$, then fractionally distd.

L-Isovaline *[595-40-4]* **M 117.2, m *ca* 300° (sublimes in vac), [α]$_D^{18}$ +9° (2M HCl).** Crystd from aq acetone.

Isovanillin see **3-hydroxy-4-methoxybenzaldehyde.**

Isoviolanthrone *[128-64-3]* **M 456.5, m 510-511°(uncorrected).** Dissolved in 98% H$_2$SO$_4$ and pptd by adding water to reduce the acid concentration to about 90%. Sublimes *in vacuo*. [Parkyns and Ubblehode *JCS* 4188 *1960*.]

Itaconic acid *[97-65-4]* **M 130.1, m 165-166°.** Crystd from EtOH, EtOH/water or EtOH/benzene.

Itaconic anhydride *[2170-03-8]* **M 112.1, m 66-68°, b 139-140°/301mm.** Crystd from CHCl$_3$/pet ether.

Janus Red B *[2636-31-9]* M 460.0. Crystd from EtOH/water (1:1 v/v).

Jervine *[469-59-0]* M 425.6, m 241-243°, [α]$_D$ -147° (in EtOH). Crystd from MeOH by addition of water.

Juglone see **2-hydroxy-1,4-naphthaquinone.**

Kerosene *[8008-20-6]* (mixture of hydrocarbons) n 1.443, d 0.75-0.82. Stirred with conc H$_2$SO$_4$ until a fresh portion of acid remains colourless, then washed with water, dried with solid KOH and distd in a Claisen flask. For more complete drying, the kerosene can be refluxed with, and distilled from, sodium.

Ketene *[463-51-4]*, M 42.0, dimer *[674-82-8]*, M 84.1, b 127-130°, n 1.441, d 1.093. Prepared by pyrolysis of acetic anhydride. Purified by passage through a trap at -75° and collected in a liquid-nitrogen-cooled trap. Ethylene was removed by pumping the ethylene in an isopentane-liquid-nitrogen slush pack at -160°. Stored at room temperature in a blackened bulb.

α-Ketoglutaric acid *[328-50-7]* M 146.1, m 111-113°. Crystd from acetone/benzene.

2-Keto-L-gulonic acid *[526-98-7]* M 194.1, m 171°. Crystd from water and washed with acetone.

Khellin *[82-02-0]* M 260.3, m 154-155°, b 180-200°/0.05mm. Crystd from MeOH or ethyl ether.

Kojic acid *[501-30-4]* M 142.1, m 154-155°. Crystd from MeOH (charcoal) by adding ethyl ether. Sublimed at 0.1mm pressure.

Kynurenic acid *[492-27-3]* M 189.1, m 282-283°. Crystd from abs EtOH.

L-Kynurenine *[2922-83-0]* M 208.2, m 190°(dec). Crystd from water.

L-Kynurenine sulphate *[16055-80-4]* M 306.3, m 194°, monohydrate m 178°, [α]$_D^{25}$ +9.6° (H$_2$O). Crystd from water by addition of EtOH.

L-Lactic acid *[79-33-4]* M 90.1, m 52.8°, b 105°/0.1mm, [α]$_D^{15}$ +3.82° (H$_2$O). Purified by fractional distn at 0.1mm pressure, followed by fractional crystn from ethyl ether/isopropyl ether (1:1, dried with sodium). [Borsook, Huffman and Liu *JBC* **102** 449 *1933*.] The solvent mixture, benzene/ethyl ether (1:1) containing 5% pet ether (b 60-80°) has also been used.

Lactobionic acid *[96-82-2]* M 358.3, m 128-130°, [α]$_{546}$ +28° (c 3, after 24hr in H$_2$O). Crystd from water by addition of EtOH.

α-Lactose (H$_2$O) *[16984-38-6]* M 360.3, m 220°(dec), [α]$_D$ +52.3° (c 4.2, H$_2$O). Crystd from water below 93.5°.

Lactulose *[4618-18-2]* M 342.2, m 167-169°(dec), [α]$_{546}$ -57° (c 1, H$_2$O),
Lanatoside A *[17575-20-1]* M 969.1, m 245-248°, [α]$_D$ +32° (EtOH),
Lanatoside B *[17575-21-2]* M 985.1, m 233°(dec), [α]$_D$ +35° (MeOH),
Lanatoside C *[17575-22-3]* M 297.1, m 246-248°, [α]$_D$ +34° (EtOH). Crystd from MeOH.

Lanosterol *[79-63-0]* M 426.7, m 138-140°, [α]$_D$ +62.0° (c 1, CHCl$_3$). Recrystd from anhydrous MeOH. Dried *in vacuo* over P$_2$O$_5$ for 3hr at 90°. Purity checked by proton magnetic resonance.

Lapachol *[84-79-7]* M 226.3, m 140°. Crystd from EtOH or ethyl ether.

dl-Laudanosine *[1699-51-0]* M 357.4, m 114-115°. Crystd from EtOH.

Lauraldehyde (1-dodecanal) *[112-54-9]* M 184.3, b 99.5-100°/3.5mm, n$^{24.7}$ 1.4328. Converted to the addition compound by shaking with satd aq NaHSO$_3$ for 1hr. The ppte was filtered off, washed with ice cold water, EtOH and ether, then decomposed with aq Na$_2$CO$_3$. The aldehyde was extracted into ethyl ether which, after drying and evaporation, gave an oil which was fractionally distd under vac.

Lauric acid (1-dodecanoic acid) *[143-07-7]* M 200.3, m 44.1°, b 141-142°/0.6-0.7mm, 225°/100mm. Vac distd. Crystd from abs EtOH, or from acetone at -25°. Alternatively, purified *via* its methyl ester (b 140.0°/15mm), as described for *capric* acid. Also purified by zone melting.

Lauryl peroxide (dodecyl peroxide) *[105-74-8]* M 398.6, m 53-54°. Crystd from benzene and stored below 0°. Can be **EXPLOSIVE.**

L-Leucine *[61-90-5]* M 131.2, m 293-295°(dec), [α]$_D^{25}$ +15.6° (5M HCl). Likely impurities are isoleucine, valine, and methionine. Crystd from water by adding 4 volumes of EtOH.

Leucomalachite Green *[129-73-7]* M 330.5, m 92-93°. Crystd from 95% EtOH (10ml/g), then from benzene/EtOH, and finally from pet ether.

Lissamine Green B *[3087-16-9]* M 576.6. Crystd from EtOH/water (1:1, v/v).

Lithocholic acid *[434-13-9]* M 376.6, m 184-186°, [α]$_D$ +33.8° (c 1.5, EtOH). Crystd from EtOH or acetic acid.

Lumazine see 2,4(1*H*,3*H*)-pteridinedione.

Lumichrome *[1086-80-2]* M 242.2, m >290°. Recrystd twice from glacial acetic acid and dried at 100° in a vac.

Luminol *[521-31-3]* M 177.2, m 329-332°. Dissolved in KOH soln, treated with Norit (charcoal), filtered and pptd with conc HCl. [Hardy, Sietz and Hercules *Talanta* 24 297 1977.] Stored in the dark in an inert atmosphere, because its structure changes during its luminescence. It has been recrystd from 0.1M KOH [Merenyi *et al, JACS* 108 77716 1986].

dl-Lupinane *[10248-30-3]* M 169.3, m 98-99°. Crystd from acetone.

Lupulon *[468-28-0]* M 414.6, m 92-94°. Crystd from 90% MeOH.

Lutein *[127-40-2]* M 568.9, m 196°, $\varepsilon_{1cm}^{1\%}$ 1750 (423nm), 2560 (446nm), 2340 (477.5nm) in EtOH; λ$_{max}$ in CS$_2$ 446, 479 and 511nm. Crystd from MeOH (copper-coloured prisms) or from ethyl ether by adding MeOH. Also purified by chromatography on columns of magnesia or calcium hydroxide, and crystd from CS_2/EtOH. May be purified *via* the dipalmitate ester. Stored in the dark, in an inert atmosphere.

Lutidine (mixture). For the preparation of pure 2,3-, 2,4- and 2,5-lutidine from commercial "2,4- and 2,5-lutidine" see Coulson *et al, JCS* 1934 *1959*, and Kyte, Jeffery and Vogel *JCS* 4454 *1960*.

2,3-Lutidine *[583-61-9]* M 107.2, f.p. -14.8°, b 160.6°, n 1.50857, d 0.9464. Steam distd from a soln containing about 1.2 equivalents of 20% H_2SO_4, until *ca* 10% of the base has been carried over with the non-basic impurities. The acid soln was then made alkaline, and the base was separated, dried over NaOH or BaO, and fractionally distd. The distd lutidine was converted to its urea complex by stirring 100g with 40g of urea in 75ml of water, cooling to 5°, filtering at the pump, and washing with 75ml of water. The complex dissolved in 300ml of water was steam distd. until the distillate gave no turbidity with a little solid NaOH. The distillate was then treated with excess solid NaOH, and the upper layer was removed: the aqueous layer was then extracted with ethyl ether. The upper layer and the ether extract were combined, dried (K_2CO_3), and distd through a short column. Final purification was by fractional crystn using partial freezing. [Kyte, Jeffery and Vogel *JCS* 4454 *1960*.]

2,4-Lutidine *[108-47-4]* M 107.2, b 157.8°, n 1.50087, n^{25} 1.4985, d 0.9305. Dried with Linde type 5A molecular sieves, BaO or sodium, and fractionally distd. The distillate (200g) was heated with benzene (500ml) and conc HCl (150ml) in a Dean and Stark apparatus on a water bath until water no longer separated, and the temperature just below the liquid reached 80°. When cold, the supernatant benzene was decanted and the 2,4-lutidine hydrochloride, after washing with a little benzene, was dissolved in water (350ml). After removing any benzene by steam distn, an aq soln of NaOH (80g) was added, and the free lutidine was steam distd. It was isolated by saturating the distillate with solid NaOH, and distd through a short column. The pptn cycle was repeated, then the final distillate was partly frozen in an apparatus at -67.8-68.5° (cooled by acetone/CO_2). The crystals were then melted and distd. [Kyte, Jeffery and Vogel *JCS* 4454 *1960*.] Alternative purifications are *via* the picrate [Clarke and Rothwell *JCS* 1885 *1960*], or the hydrobromide [Warnhoff *JOC* 27 4587 *1962*]. The latter is pptd from a soln of lutidine in benzene by passing dry HBr gas: the salt is recrystd from $CHCl_3$/methyl ethyl ketone, then decomposed with NaOH, and the free base is extracted into ethyl ether, dried, evaporated and the residue distd.

2,5-Lutidine *[589-93-5]* M 107.2, m -15.3°, b 156.7°/759mm, n^{25} 1.4982, d 0.927. Steam distd from a soln containing 1-2 equivalents of 20% H_2SO_4 until about 10% of the base had been carried over with the non-basic impurities, then the acid soln was made alkaline, and the base was separated, dried with NaOH and fractionally distd twice. Dried with sodium and fractionally distd through a Todd column packed with glass helices.

2,6-Lutidine *[108-48-5]* M 107.2, m -59°, b 144.0°, n 1.49779, d 0.92257. Likely contaminants include 3- and 4-picoline (similar boiling points). However, they are removed by using BF_3, with which they react preferentially, by adding 4ml of BF_3 to 100ml of dry fractionally distd 2,6-lutidine and redistilling. Distn of commercial material from $AlCl_3$ (14g per 100ml) can also be used to remove picolines (and water). Alternatively, lutidine (100ml) can be refluxed with ethyl benzenesulphonate (20g) or ethyl *p*-toluenesulphonate (20g) for 1hr, then the upper layer is cooled, separated and distd. The distillate is refluxed with BaO or CaH_2, then fractionally distd, through a glass helices-packed column.
2,6-Lutidine can be dried with KOH or sodium, or by refluxing with (and distilling from) BaO, prior to distn. For purification via its picrate, 2,6-lutidine, dissolved in abs EtOH, is treated with an excess of warm ethanolic picric acid. The ppte is filtered off, recrystd from acetone (to give m 163-164.5°), and partitioned between ammonia and $CHCl_3$/ethyl ether. The organic soln, after washing with dil aq KOH, is dried with Na_2SO_4 and fractionally distd. [Warnhoff *JOC* 27 4587 *1962*.] Alternatively, 2,6-lutidine can be purified *via* its urea complex, as described under 2,3-lutidine. Other purification procedures include azeotropic distn with phenol [Coulson *et al, J Appl Chem (London)* 2 71 *1952*], fractional crystn by partial freezing, and vapour-phase chromatography using a 180-cm column of

polyethylene glycol-400 (Shell, 5%) on Embacel (May and Baker) at 100°, with argon as carrier gas [Bamford and Block *JCS* 4989 *1961*].

3,5-Lutidine *[591-22-0]* M 107.2, f.p. -6.3°, b 172.0°/767mm, n 1.50613, n^{25} 1.5035, d 0.9419. Dried with sodium and fractionally distd through a Todd column packed with glass helices. Dissolved (100ml) in dil HCl (1:4) and steam distd until 1L of distillate was collected. Excess conc NaOH was added to the residue which was again steam distd. The base was extracted from the distillate, using ethyl ether. The extract was dried with K_2CO_3, and distd. It was then fractionally crystd by partial freezing.

Lycopene *[502-65-8]* M 536.9, m 172-173°, $\varepsilon_{1cm}^{1\%}$ 2250 (446nm), 3450 (472nm), 3150 (505nm) in pet ether. Crystd from CS_2/MeOH, ethyl ether/pet ether, or acetone/pet ether, and purified by column chromatography on deactivated alumina, $CaCO_3$, calcium hydroxide or magnesia. Stored in the dark, in an inert atmosphere.

Lycorine *[476-28-8]* M 552.9, m 275-280°(dec). Crystd from EtOH.

Lycoxanthin *[19891-74-8]* M 268.3, m 173-174°, $\varepsilon_{1cm}^{1\%}$ 3360 (472.5nm), also λ_{max} 444 and 503nm in pet ether. Crystd from ethyl ether/light petroleum, benzene/pet ether or CS_2. Purified by chromatography on columns of $CaCO_3$, $Ca(OH)_2$ or deactivated alumina, washing with benzene and eluting with 3:1 benzene/MeOH. Stored in the dark, in an inert atmosphere, at -20°.

Lysergic acid *[82-58-6]* M 268.3, m 240°(dec), $[\alpha]_D$ +40° (pyridine). Crystd from water.

L-Lysine *[56-87-1]* M 146.2, m >210°(dec). Crystd from aq EtOH.

L-Lysine dihydrochloride *[657-26-1]* M 219.1, m 193°, $[\alpha]_D^{25}$ +25.9° (5M HCl). Crystd from MeOH, in the presence of excess HCl, by adding ethyl ether.

L-Lysine monohydrochloride *[657-27-2]* M 182.7, $[\alpha]$ as above. Likely impurities are arginine, D-lysine, 2,6-diaminoheptanedioic acid and glutamic acid. Crystd from water at *p*H 4-6 by adding 4 volumes of EtOH. Above 60% relative humidity it forms a dihydrate.

β-D-Lyxose *[1114-34-7]* M 150.1, m 118-119°, $[\alpha]_D$ -14° (c 4, H_2O). Crystd from EtOH or aq 80% EtOH. Dried under vac at 60°, and stored in a vac desiccator over P_2O_5 or $CaSO_4$.

Malachite Green (carbinol) *[510-13-4]* **M 346.4, m 112-114°**. The oxalate was recrystd from hot water and dried in air. The carbinol was pptd from the oxalate (1g) in distd water (100ml) by adding M NaOH (10ml). The ppte was filtered off, recrystd from 95% EtOH containing a little dissolved KOH, then washed with ether, and crystd from pet ether. Dried in vac at 40°. An acid soln (2 x 10^{-5}M in 6 x 10^{-5}M H_2SO_4) rapidly reverted to the dye. [Swain and Hedberg *JACS* **72** 3373 *1950*.]

Z-Maleamic acid *[557-24-4]* **M 115.1, m 172-173°(dec)**. Crystd from EtOH.

Maleic acid *[110-16-7]* **M 116.1, m 143.5°**. Crystd from acetone/pet ether (b 60-80°) or hot water. Dried at 100°.

Maleic anhydride *[108-31-6]* **M 98.1, m 54°, b 94-96°/20mm, 199°/760mm**. Crystd from benzene, $CHCl_3$, CH_2Cl_2 or CCl_4. Sublimed under reduced pressure. [Skell *et al, JACS* **108** 6300 *1986*.]

Maleic hydrazide *[123-3-1]* **M 112.1, m 144°(dec)**. Crystd from water.

Maleuric acid *[105-61-3]* **M 158.1, m 167-168°(dec)**. Crystd from hot water.

dl-**Malic acid** *[6915-15-7]* **M 134.1, m 128-129°**. Crystd from acetone, then from acetone/CCl_4, or from ethyl acetate by adding pet ether (b 60-70°). Dried at 35° under 1mm pressure to avoid formation of the anhydride.

L-Malic acid *[617-48-1]* **M 134.1, m 104.5-106°, $[\alpha]_D$ -2.3°** (c 8.5, H_2O). Crystd (charcoal) from ethyl acetate/pet ether (b 55-56°), keeping the temperature below 65°. Or, dissolved by refluxing in fifteen parts of anhydrous ethyl ether, decanted, concentrated to one-third volume and crystd at 0°, repeatedly to constant melting point.

Malonamide *[108-13-4]* **M 102.1, m 170°**. Crystd from water.

Malonic acid *[141-82-2]* **M 104.1, m 136°**. Crystd from benzene/ethyl ether (1:1) containing 5% of pet ether (b 60-80°), washed with ethyl ether, then recrystd from water or acetone. Dried under vac over conc H_2SO_4.

Malononitrile *[109-77-3]* **M 66.1, m 32-34°, b 220°/760mm**. Crystd from water, EtOH, benzene or chloroform. Distd from, and stored over, P_2O_5. [Bernasconi *et al, JACS* **107** 7692 *1985*; Gratenhuis *JACS* **109** 8044 *1987*.]

Maltol *[118-71-8]* **M 126.1, m 161-162°**. Crystd from $CHCl_3$ or aq 50% EtOH. Volatile in steam. It can be readily sublimed in a vac.

Maltose (H_2O) *[63-63-53-7]* **M 360.3, m 118°**. Purified by chromatography from aq soln on to a charcoal/Celite (1:1) column, washed with water to remove glucose and other monosaccharides, then eluted with aq 75% EtOH. Crystd from water, aq EtOH or EtOH containing 1% nitric acid. Dried as the monohydrate at room temperature under vac over H_2SO_4 or P_2O_5.

S-Mandelic acid *[17199-29-0]* **M 152.2, m 133°, $[\alpha]_{546}$ +188°** (c 5, H_2O),
dl-**Mandelic acid** *[61-72-3]* **M 152.2, m 118°**. Purified by Soxhlet extraction with benzene (about 6ml/g), allowing the extract to crystallise. Dried at room temperature under vac.

D-Mannitol *[69-65-8]* **M 182.2, m 166.1°, $[\alpha]_{546}$ + 29°** (c 10, after 1hr in 8% borax soln). Crystd from EtOH or distd water and dried at 100°.

Mannitol hexanitrate *[130-39-2]* **M 452.2, m 112-113°.** Crystd from EtOH. **EXPLOSIVE on concussion.**

α-D-Mannose *[3458-58-4]* **M 180.2, m 132°, $[\alpha]_D$ +14.1° (c 4, H_2O).** Crystd repeatedly from EtOH or aq 80% EtOH, then dried under vac over P_2O_5 at 60°.

Margaric acid see **heptadecanoic acid.**

Meconic acid (3-hydroxy-γ-pyrone-2,6-dicarboxylic acid) *[497-59-6]* **M 200.1, m 100° (-H_2O).** Crystd from water and dried at 100° for 20min.

Melamine *[108-78-1]* **M 126.1, m 353°.** Crystd from water or dil aq NaOH.

D(+)-Melezitose (H_2O) *[597-12-6]* **M 540.5, m 153-154°(dec), 2H_2O m 160°(dec), $[\alpha]_D$ +88° (c 4, H_2O).** Crystallises from water as the dihydrate, then dried at 110° (anhydrous).

D(+)-Melibiose (H_2O) *[585-99-9]* **M 360.3, m 84-85°, $[\alpha]_D$ +135° (c 5, after 10hr H_2O).** Crystallises as a hydrate from water or aq EtOH.

p-**Menta-1,5-diene** see **α-phellandrene.**

(-)-Menthol *[2216-51-5]* **M 156.3, m 44-46.5°, $[\alpha]_D$ 50° (c 10, EtOH).** Crystd from $CHCl_3$, pet ether or EtOH/water.

Meprobamate *[57-53-4]* **246.3, m 104-106°.** Crystd from hot water.

2-Mercaptobenzimidazole *[583-39-1]* **M 150.2, m 302-304°.** Crystd from aq EtOH or aq ammonia.

2-Mercaptobenzothiazole *[149-30-4]* **M 167.2, m 182°.** Crystd repeatedly from 95% EtOH, or purified by incomplete pptn by dil H_2SO_4 from a basic soln, followed by several crystns from acetone/water or benzene.

2-Mercaptoethylamine (cysteamine) *[60-23-1]* **M 77.2, m 97-98.5°.** Sublimed under vac, and stored under nitrogen.

2-Mercaptoimidazole *[872-35-5]* **M 100.1, m 221-222°.** Crystd from water.

2-Mercapto-1-methylimidazole *[60-56-0]* **M 114.2, m 145-147°.** Crystd from EtOH.

2-Mercaptopurine (H_2O) *[6112-76-1]* **M 170.2, m >315°(dec).** Crystd from pyridine (30ml/g), washed with pyridine, then triturated with water (25ml/g), adjusting to *p*H 5 by adding M HCl. Recrystd by heating, then cooling, the soln. Filtered, washed with water and dried at 110°. Has also been crystd from water (charcoal).

8-Mercaptoquinoline (2H_2O, thioxine) *[491-33-8]* **M 197.3, m 58-59°.** Easily oxidised in air to give diquinolyl-8,8'-disulphide (which is stable). It is more convenient to make 8-mercaptoquinoline by reduction of the material. [Nakamura and Sekido *Talanta* **17** 515 *1970*.]

Mercaptosuccinic acid see **thiomalic acid.**

Mesaconic acid *[498-24-8]* **M 130.1, m 204-205°.** Crystd from water or EtOH [Katakis *et al*, *JCSDT* 1491 *1986*].

Mescaline sulphate [2-(3,4,5-trimethoxyphenyl)ethylamine sulphate] *[5967-42-0]* **M 309.3, m 183-184°.** Crystd from water.

Mesitoic acid see **2,4,6-trimethylbenzoic acid.**

Mesitylene (1,3,5-trimethylbenzene) *[108-67-8]* **M 120.2, m -44.7°, b 99.0-99.8°/100mm, 166.5-167°/760mm, m 1.4962, n^{25} 1.4967, d 0.865.** Dried with $CaCl_2$ and distd from sodium in a glass helices packed column. Treated with silica gel and redistd. Alternative purifications include vapour-phase chromatography, or fractional distn followed by azeotropic distn with 2-methoxyethanol (which is subsequently washed out with water), drying and fractional distn. More exhaustive purification uses sulphonation by dissolving in two volumes of conc H_2SO_4, precipitating with four volumes of conc HCl at 0°, washing with conc HCl and recrystallising from $CHCl_3$. The mesitylene sulphonic acid is hydrolysed with boiling 20% HCl and steam distd. The separated mesitylene is dried ($MgSO_4$ or $CaSO_4$) and distd. It can also be fractionally crystd.

Mesityl oxide *[141-79-7]* **M 98.2, b 112°/760mm, n^{24} 1.4412, d 0.854.** Purified *via* the semicarbazone (m 165°). [Erskine and Waight *JCS* 3425 *1960.*]

Metalphthalein (H_2O) *[2411-89-4]* **M 636.6, m 186°.** Dissolved in sodium acetate and fractionally pptd with HCl. This removed unsubstituted and monosubstituted cresol phthaleins (which separated at lower acidities). Washed with cold water, dried to mononhydrate at 30° *in vacuo.*

Metanilic acid *[121-47-1]* **M 173.2, decomposes on heating.** Crystd from water (as the hydrate), under CO_2 in a semi-darkened room. (The soln is photosensitive.) Dried over 90% H_2SO_4 in a vac desiccator.

α-Methacraldehyde *[78-85-3]* **M 68.1, b 68.4°.** Fractionally distd under nitrogen through a short Vigreux column. Stored in sealed ampoules. (Slight polymerisation may occur.)

Methacrylamide *[79-39-0]* **M 85.1, m 111-112°.** Crystd from benzene or ethyl acetate and dried under vac at room temperature.

Methacrylic acid *[79-41-4]* **M 86.1, b 72°/14mm, 160°/760mm, n 1.431, d 1.015.** Aq methacrylic acid (90%) was satd with NaCl (to remove the bulk of the water), then the organic phase was dried with $CaCl_2$ and distd under vac. Polymerisation inhibitors include 0.25% *p*-methoxyphenol, 0.1% hydroquinone, or 0.05% *N,N'*-diphenyl-*p*-phenylenediamine.

Methacrylic anhydride *[760-93-0]* **M 154.2, b 65°/2mm, m 1.454, d 1.040.** Distd at 2mm pressure, immediately before use, in the presence of hydroquinone.

Methacrylonitrile *[126-98-7]* **M 67.1, b 90.3°, m 1.4007. n^{30} 1.3954, d 0.800.** Washed (to remove inhibitors such as *p-tert*-butylcatechol) with satd aq $NaHSO_3$, 1% NaOH in satd NaCl and then with satd NaCl. Dried with $CaCl_2$ and fractionally distd under nitrogen to separate from impurities such as methacrolein and acetone.

Methadone hydrochloride *[1095-90-5]* **M 345.9, m 241-242°.** Crystd from EtOH.

Methane *[74-82-8]* **M 16.0, m -184°, b -164°/760mm, -130°/6.7atm, d^{-164} 0.466 (air 1).** Dried by passage over $CaCl_2$ and P_2O_5, then passed through a Dry-ice trap and fractionally distd from a liquid-nitrogen trap. Oxygen can be removed by prior passage in a stream of hydrogen over reduced copper oxide at 500°, and higher hydrocarbons can be removed by prechlorinating about 10% of the sample: the hydrocarbons, chlorides and HCl are readily separated from the methane by condensing the sample in the liquid-nitrogen trap and fractionally distilling it. Methane has also been washed with conc H_2SO_4, then solid NaOH and then 30% NaOH soln. It was dried with $CaCl_2$, then P_2O_5, and condensed in a trap at liquid air temperature then transferred to another trap cooled in liquid nitrogen. CO_2, O_2, N_2 and higher hydrocarbons can be removed from methane by adsorption on charcoal. [Eiseman and Potter *J Res Nat Bur Stand* **58** 213 *1957.*]

Methanesulphonic acid *[76-75-2]* M 96.1, m 20°, b 134.5-135°/3mm, n 1.432, d 1.483. Dried, either by azeotropic removal of water with benzene or toluene, or by stirring 20g of P_2O_5 with 500ml of the acid at 100° for 0.5hr/ Then distd under vac and fractionally crystd by partial freezing. Sulphuric acid, if present, can be removed by prior addition of $Ba(OH)_2$ to a dilute soln, filtering off the $BaSO_4$ and concentrating under reduced pressure, and is sufficiently pure for most applications.

Methanesulphonyl chloride *[124-63-0]* M 114.5. b 55°/11mmm, n 1.452, d 1.474. Distd from P_2O_5 under vac.

Methanol *[67-56-1]* M 32.0, b 64.5°, n^{15} 1.33057, n^{25} 1.32663, d^{15} 0.79609, d^{25} 1.32663. Almost all methanol is now obtained synthetically. Likely impurities are water, acetone, formaldehyde, ethanol, methyl formate and traces of dimethyl ether, methylal, methyl acetate, acetaldehyde, carbon dioxide and ammonia. Most of the water (down to about 0.01%) can be removed by fractional distn. Drying with CaO is unnecessary and wasteful. Anhydrous methanol can be obtained from "absolute" material by passage through Linde type 4A molecular sieves, or by drying with CaH_2, $CaSO_4$, or with just a little more sodium than required to react with the water present; in all cases the methanol is then distd. Two treatments with sodium reduces the water content to about 5 X 10^{-5}%. {Friedman, Gill and Doty *JACS* **83** 4050 *1961*.} Lund and Bjerrum [*Ber* **64** 210 *1931*] warmed clean dry magnesium turnings (5g) and iodine (0.5g) with 50-75ml of "absolute" methanol in a flask until the iodine disappeared and all the magnesium was converted to methoxide. Up to 1L of methanol was added and, after refluxing for 2-3hr, it was distd off, excluding moisture from the system. Redistn from tribromobenzoic acid removes basic impurities and traces of magnesium oxides, and leaves conductivity-quality material. The method of Hartley and Raikes [*JCS* **127** 524 *1925*] gives a slightly better product. This consists of an initial fractional distn, followed by distn from aluminium methoxide, and then ammonia and other volatile impurities are removed by refluxing for 6hr with freshly dehydrated $CuSO_4$ (2g/L) while dry air is passed through: the methanol is finally distd. (The aluminium methoxide is prepared by warming with aluminium amalgam (3g/L) until all the aluminium has reacted. The amalgam is obtained by warming pieces of sheet aluminium with a soln of $HgCl_2$ in dry methanol.) This treatment also removes aldehydes.
If acetone is present in the methanol, it is usually removed prior to drying. Bates, Mullaly and Hartley [*JCS* 401 *1923*] dissolved 25g of iodine in 1L of methanol and then poured the soln, with constant stirring, into 500ml of M NaOH. Additn of 150ml of water pptd iodoform. The soln was stood overnight, filtered, then boiled under reflux until the odour of iodoform disappeared, and fractionally distd. (This treatment also removes formaldehyde.) Morton and Mark [*IECAE* **6** 151 *1934*] refluxed methanol (1L) with furfural (50ml) and 10% NaOH soln (120ml) for 6-12hr, the refluxing resin carrying down with it the acetone and other carbonyl-containing impurities. The alcohol was then fractionally distd. Evers and Knox [*JACS* **73** 1739 *1951*], after refluxing 4,5L of methanol for 24hr with 50g of magnesium, distd off 4L of it, which they then refluxed with $AgNO_3$ for 24hr in the absence of moisture or CO_2. The methanol was again distd, shaken for 24hr with activated alumina before being filtered through a glass sinter and distd under nitrogen in an all-glass still. Material suitable for conductivity work was obtained.
Variations of the above methods have also been used. For example, a sodium hydroxide soln containing iodine has been added to methanol and, after standing for 1day, the soln has been poured slowly into about a quarter of its volume of 10% $AgNO_3$, shaken for several hours, then distd. Sulphanilic acid has been used instead of tribromobenzoic acid in Lund and Bjerrum's method. A soln of 15g of magnesium in 500ml of methanol has been heated under reflux, under nitrogen, with hydroquinone (30g), before degassing and distilling the methanol, which was subsequently stored with magnesium (2g) and hydroquinone (4g per 100ml). Refluxing for about 12hr removes the bulk of the formaldehyde from methanol: further purification has been obtained by subsequent distn, refluxing for 12hr with dinitrophenylhydrazine (5g) and H_2SO_4 (2g/L), and again fractionally distilling.
A simple purification procedure consists of adding 2g of $NaBH_4$ to 1.5L methanol, gently bubbling with argon and refluxing for a day at 30°, then adding 2g of freshly cut sodium (washed with methanol) and refluxing for 1day before distilling. The middle fraction is taken. [Jou and Freeman *JPC* **81** 909 *1977*.]

β-Methazone M 392.5, m 231-136°(dec), λ_{max} 238nm (log ε 4.18). Crystd from ethyl acetate.

dl-Methionine *[59-51-8]* M 149.2, m 281°(dec). Crystd from hot water.

L-Methionine *[63-68-3]* M 149.2, m 283°(dec), [α]$_D^{25}$ +21.2° (0.2M HCl). Crystd from aq EtOH.

dl-Methionine sulphoxide *[454-41-1]* M 165.2, m >240°(dec). Likely impurities are *dl*-methionine sulphone and *dl*-methionine. Crystd from water by adding EtOH in excess.

Methoxyacetic acid *[625-45-6]* M 90.1, b 97°/13-14mm, n 1.417, d 1.175. Fractionally crystd by repeated partial freezing, then fractionally distd under vac through a vacuum-jacketed Vigreux column 20cm long.

p-Methoxyacetophenone *[100-06-1]* M 150.2, m 39°, b 139°/15mm, 264°/736mm. Crystd from ethyl ether/pet ether.

Methoxyamine hydrochloride *[593-56-6]* M 83.5, m 151-152°. Crystd from abs EtOH or EtOH by addition of ethyl ether. [Kovach *et al*, *JACS* 107 7360 1985.]

3-Methoxybenzanthrone *[3688-79-7]* M 274.3. Crystd from benzene.

p-Methoxybenzene *[2396-60-3]* M 212.3, m 54-56°. Crystd from EtOH.

m-Methoxybenzoic acid *[586-38-9]* M 152.2, m 110°. Crystd from EtOH/water.

p-Methoxybenzoic acid *[100-09-4]* M 152.2, m 184.0-184.5°. Crystd from EtOH, water, EtOH/water or toluene.

"Methoxychlor", 1,1-Bis(*p*-methoxyphenyl)-2,2,2-trichloroethane (dimorphic) *[72-43-5]* M 345.7, m 78-78.2°, or 86-88°. Freed from 1,1-bis(*p*-chlorophenyl)-2,2,2-trichloroethane by crystn from EtOH.

trans-p-Methoxycinnamic acid *[830-09-1]* M 178.2, m 173.4-174.8°. Crystd from MeOH to constant melting point and UV spectrum.

2-Methoxyethanol *[109-86-4]* M 76.1, b 124.4°, n 1.4017, d 0.964. Peroxides can be removed by refluxing with stannous chloride or by filtration under slight pressure through a column of activated alumina. 2-Methoxyethanol can be dried with K_2CO_3, $CaSO_4$, $MgSO_4$ or silica gel, with a final distn from sodium. Aliphatic ketones (and water) can be removed by making the solvent 0.1% in 2,4-dinitrophenylhydrazine and allowing to stand overnight with silica gel before fractionally distilling.

β-Methoxyethylamine *[109-85-3]* M 75.1, b 94°, n 1.407, d 0.874. An aq 70% soln was dehydrated by azeotropic distn with benzene and the amine was distd twice.

6-Methoxy-1-indanone *[13623-25-1]* M 162.2, m 151-153°. Crystd from MeOH, then sublimed.

5-Methoxyindole *[1006-94-6]* M 147.2, m 55°, b 176-178°/17mm. Crystd from cyclohexane or pet ether.

1-Methoxynaphthalene *[2216-69-5]* M 158.2, b 268.4-268.5°, n 1.621, d 1.094. Fractionally distd from CaH_2.

2-Methoxynaphthalene *[93-04-9]* M 158.2, m 73.0-73.6°, b 273°/760mm. Fractionally distd under vac. Crystd from abs EtOH, aq EtOH, benzene or *n*-heptane, and dried under vac in an Abderhalden pistol or distd *in vacuo*. [Kikuchi *et al*, *JPC* 91 574 1987.]

4-Methoxynitrobenzene see **nitroanisole**.

1-Methoxy-4-nitronaphthalene *[4900-63-4]* **M 203.2, m 85°.** Purified by chromatography on silica gel and recrystd from MeOH [Bunce *et al, JOC* **52** 4214 *1987*].

p-**Methoxyphenol** *[150-76-5]* **M 124.1, m 54-55°, b 243°.** Crystd from benzene, pet ether or water, and dried under vac over P_2O_5 at room temperature. Sublimes *in vacuo*. [Wolfenden *et al, JACS* **109** 463 *1987*.]

(*m*-**Methoxyphenyl)acetic acid** *[1798-09-0]* **M 166.2, m 71.0-71.2°.** Crystd from water (charcoal), or aq EtOH.

(*p*-**Methoxyphenyl)acetic acid** *[104-01-8]* **M 166.2, m 85-87°.** Crystd from EtOH/water.

5-(*p*-Methoxyphenyl)-1,2-dithiole-3-thione *[42766-10-9]* **M 240.2, m 111°.** Crystd from butyl acetate.

N-(*p*-**Methoxyphenyl)-*p*-phenylenediamine** *[101-64-4]* **M 214.3, m 102°, b 238°/12mm.** Crystd from ligroin.

8-Methoxypsoralen *[298-81-7]* **M 216.2, m 148°.** Crystd from EtOH/ether or benzene/pet ether.

4-Methoxystyrene *[637-69-4]* **M 134.2, b 41-42°/0.5mm, n 1.5622. d 1.009.** Distd from CaH_2 and stored under argon at -10° [Hall *et al, JOC* **52** 5528 *1987*].

N-**Methylacetamide** *[79-16-3]* **M 73.1, m 30°, b 70-71°/2.5-3mm.** Fractionally distd under vac, then fractionally crystd twice from its melt. Impurities include acetic acid, methyl amine and water. For detailed purification procedure, see Knecht and Kolthoff, *Inorg Chem* **1** 195 *1962*.
Although *N*-methylacetamide is commercially available it is often extensively contaminated with acetic acid, methylamine, water and an unidentified impurity. The recommended procedure is to synthesise it in the laboratory by direct reaction. The gaseous amine is passed into hot glacial acetic acid, to give a partially aqueous soln of methylammonium acetate which is heated to *ca* 130° to expel water. Chemical methods of purification such as extraction by pet ether, treatment with H_2SO_4, K_2CO_3 or CaO can be used but are more laborious.
Tests for purity include the Karl Fischer titration for water; this can be applied directly. Acetic acid and methylamine can be detected polarographically.
In addition to the above, purification of *N*-methylacetamide can be achieved by fractional freezing, including zone melting, repeated many times, or by chemical treatment with vacuum distn under reduced pressures. For details of zone melting techniques, see Knecht in **Recommended Methods for Purification of Solvents and Tests for Impurities**, Coetzee ed., Pergamon Press *1982*.

N-**Methylacetanilide** *[579-10-2]* **M 149.2, m 102-104°.** Crystd from water, ether or light petroleum (b 80-100°).

Methyl acetate *[79-20-9]* **M 74.1, b 56.7-57.2°. n 1.36193, n^{25} 1.3538, d 0.934.** Methanol in methyl acetate can be detected by measuring solubility in water. At 20°, the solubility of methyl acetate in water is *ca* 35g per 100ml, but 1% MeOH confers miscibility. Methanol can be removed by conversion to methyl acetate, using refluxing for 6hr with acetic anhydride (85ml/L), followed by fractional distn. Acidic impurities can be removed by shaking with anhydrous K_2CO_3 and distilling. An alternative treatment is with acetyl chloride, followed by washing with conc NaCl and drying with CaO or $MgSO_4$. (Solid $CaCl_2$ cannot be used because it forms a crystalline addition compound.) Distn from copper stearate destroys peroxides. Free alcohol or acid can be eliminated from methyl acetate by shaking with strong aq Na_2CO_3 or K_2CO_3 (three times), then with aq 50% $CaCl_2$ (three times), satd aq NaCl (twice), drying with K_2CO_3 and distn from P_2O_5.

p-Methylacetophenone *[122-00-9]* M 134.2, m 22-24°, b 93.5°/7mm, 110°/14mm, n 1.5335, d 1.000. Impurities, including the *o*- and *m*-isomers, were removed by forming the semicarbazone which, after repeated crystn, was hydrolysed to the ketone. [Brown and Marino *JACS* **84** 1236 *1962*.] Also purified by distn under reduced pressure, followed by low temperature crystn from isopentane.

Methyl acrylate *[96-33-3]* M 86.1, b 80°, n 1.4040, d 0.9535. Washed repeatedly with aq NaOH until free from inhibitors (such as hydroquinone), then washed with distd water, dried (CaCl$_2$) and fractionally distd under reduced pressure in an all-glass apparatus. Sealed under nitrogen and stored at 0° in the dark. [Bamford and Han *JCSFT 1* **78** 855 *1982*.]

1-Methyladamantane *[768-91-2]* M 150.2,
2-Methyladamantane *[700-56-1]* M 150.2. Purified by zone melting.

Methylal see dimethoxymethane.

2-Methylalanine see α-aminoisobutyric acid.

Methylamine (gas) *[74-89-5]* M 31.1, b -7.55°/719mm. Dried with sodium or BaO.

Methylamine hydrochloride *[593-51-1]* M 67.5, m 231.8-233.4°, b 225-230°/15mm. Crystd from *n*-butanol, abs EtOH or MeOH/CHCl$_3$. Washed with CHCl$_3$ to remove traces of dimethylamine hydrochloride. Dried under vac first with H$_2$SO$_4$ then P$_2$O$_5$. Deliquescent, stored in a desiccator over P$_2$O$_5$.

1-Methylaminoanthraquinone *[82-38-2]* M 237.3, m 166.5°. Crystd to constant melting point from butan-1-ol, then crystd from EtOH. It can be sublimed under vac.

N-Methyl-*o*-aminobenzoic acid (*N*-methylanthranilic acid) *[119-68-6]* M 151.2, m 178.5°. Crystd from water or EtOH.

p-Methylaminophenol sulphate *[55-55-0]* M 344.4, m 260°(dec). Crystd from MeOH.

N-Methylaniline *[100-61-8]* M 107.2, b 57°/4mm, 81-82°/14mm, n 1.570, d 0.985. Dried with KOH pellets and fractionally distd under vac. Acetylated and the acetyl derivative was recrystd to constant melting point (m 101-102°), then hydrolysed with aq HCl and distd from zinc dust under reduced pressure. [Hammond and Parks *JACS* **77** 340 *1955*.]

o-, *m*- and *p*-Methylaniline see *o*-, *m*- and *p*-toluidine.

N-Methylaniline hydrochloride *[2739-12-0]* M 143.7, m 123.0-123.1°. Crystd from dry benzene/CHCl$_3$ and dried under vac.

Methyl *p*-anisate *[121-98-2]* M 166.2, m 48°. Crystd from EtOH.

4-Methyl anisole *[104-93-8]* M 122.2, b 175-176°, n 1.512, d$_{15}$^{15} 0.9757. Dissolved in ethyl ether, washed with M NaOH, water, dried (Na$_2$CO$_3$), evapd and the residue distd under vac.

2-Methylanthracene *[613-12-7]* M 192.3, m 204-206°,
4-Methylanthracene *[779-02-2]* M 192.3, m 77-79°, b 196-197°/12mm, d 1.066. Chromatographed on silica gel with cyclohexane as eluent and recrystd from EtOH [Werst *JACS* **109** 32 *1987*].

N-Methylanthranilic acid see *N*-methyl-*o*-aminobenzoic acid.

2-Methylanthraquinone *[84-54-8]* M 222.3, m 176°. Crystd from EtOH, then sublimed.

Methylarenes (see also pentamethyl- and hexamethyl- benzenes). Recrystd from EtOH and sublimed in vac [Schlesener *et al, JACS* **106** 7472 *1984*].

Methyl benzoate *[93-58-3]* M 136.2, b 104-105°/39mm, 199.5°/760mm, n^{15} 1.52049, n 1.51701, d 1.087. Washed with dil aq NaHCO$_3$, then water, dried with Na$_2$SO$_4$ and fractionally distd under reduced pressure.

p-**Methylbenzophenone** *[134-84-9]* M 196.3, m 57°. Crystd from MeOH and pet ether.

Methyl-1,4-benzoquinone *[553-79-9]* M 122.1, m 68-69°. Crystd from heptane or EtOH, dried rapidly (vac over P$_2$O$_5$) and stored under vac.

Methyl benzoylformate *[15206-55-0]* M 164.2, m 246-248°. Purified by radial chromatography (ethyl ether/hexane, 1:1), and dried at 110-112° at 6mm pressure. [Meyers and Oppenlaender *JACS* **108** 1989 *1986*.]

2-Methyl-3,4-benzphenanthrene *[652-04-0]* M 242.3, m 70°. Crystd from EtOH.

dl-α-**Methylbenzyl alcohol** *[13323-81-4]* M 122.2, b 60.5-61.0°/3mm. Dried with MgSO$_4$ and distd under vac.

p-**Methylbenzyl alcohol** see *p*-**tolyl carbinol.**

p-**Methylbenzyl bromide** *[104-81-4]* M 185.1, m 35°, b 218-220°/760mm. Crystd from pentane.

p-**Methylbenzyl chloride** *[104-82-5]* M 140.6, b 80°/2mm, n 1.543, d 1.085. Dried with CaSO$_4$ and fractionally distd under vac.

Methyl benzylpenicillinate *[653-89-4]* M 348.3, m 97°, [α]$_D$ +328° (c 1, MeOH). Crystd from CCl$_4$.

Methylbixin *[26585-94-4]* M 408.5, m 163°. Crystd from EtOH/CHCl$_3$.

Methyl bromide *[74-83-9]* M 94.9, b 3.6°. Purified by bubbling through conc H$_2$SO$_4$, followed by passage through a tube containing glass beads coated with P$_2$O$_5$. Also purified by distn from AlBr$_3$ at -80°, by passage through a tower of KOH pellets and by partial condensation.

Methyl *o*-bromobenzoate *[610-94-6]* M 215.1, b 122°/17mm, 234-244°/760mm. Soln in ether is washed with 10% aq Na$_2$CO$_3$, water, then dried and distd.

Methyl *p*-bromobenzoate *[619-42-1]* M 215.1, m 79.5-80.5°. Crystd from MeOH.

2-Methyl-1,3-butadiene see isoprene.

2-Methylbutane *[78-78-4]* M 72.2, b 27.9°, n 1.35373, n^{25} 1.35088, d 0.621. Stirred for several hours in the cold with conc H$_2$SO$_4$ (to remove olefinic impurities), then washed with water, aq Na$_2$CO$_3$ and water again. Dried with MgSO$_4$ and fractionally distd using a Todd column packed with glass helices. Material transparent down to 180nm was obtained by distilling from sodium wire, and passing through a column of silica gel which had previously been dried in place at 350° for 12hr before use. [Potts *JPC* **20** 809 *1952*.]

2-Methyl-1-butanol *[dl 137-32-6]* *[l 1565-80-6]* M 88.2, b 128.6°, n^{25} 1.4082, d 0.809. Refluxed with CaO, distd, refluxed with magnesium and again fractionally distd. A small sample of highly purified material was obtained by fractional crystn after conversion into a suitable ester such as the trinitrophthalate or the 3-nitrophthalate. The latter was converted to the cinchonine salt in acetone and recrystd from $CHCl_3$ by adding pentane. The salt was saponified, extracted with ether, and fractionally distd. [Terry *et al, J Chem Eng Data* **5** 403 *1960.*]

2-Methyl-2-butanol see *tert*-**amyl alcohol.**

3-Methyl-1-butanol *[123-51-3]* M 88.2, b 128°/750mm, 132°/760mm, n^{15} 1.4085, n 1.4075, d^{15} 0.8129. Dried by heating with CaO and fractionally distilling, then heating with BaO and redistilling. Alternatively, boiled with conc KOH, washed with dil H_3PO_4, and dried with K_2CO_3, then anhydrous $CuSO_4$, before fractionally distilling. It is separated from 2-methyl-1-butanol by fractional distn, fractional crystn and preparative gas chromatography.

3-Methyl-2-butanol *[598-75-4]* M 88.2, b 111.5°, n 1.4095, n^{25} 1.4076, d 0.807. Refluxed with magnesium, then fractionally distd.

3-Methyl-2-butanone *[563-80-4]* M 86.1, b 93-94°/752mm, n 1.410, d 0.818. Refluxed with a little $KMnO_4$. Fractionated on a spinning-band column. Dried with $CaSO_4$ and distd.

2-Methyl-2-butene *[513-35-9]* M 70.1, f.p. -133.8°, b 38.4°/760mm, n^{15} 1.3908, d^{15} 0.66708, d 0.6783, d^{25} 0.65694. Distd from sodium.

1-Methylbutyl- see *sec*-**amyl-.**

Methyl *n*-butyrate *[623-42-7]* M 102.1, b 102.3°/760mm, n 1.389, d 0.898. Treated with anhydrous $CuSO_4$, then distd under dry nitrogen.

Methyl carbamate *[598-55-0]* M 75.1, m 54.4-54.8°. Crystd from benzene.

9-Methylcarbazole *1484-12-4]* M 181.2, m 89°. Purified by zone melting.

Methyl carbitol see **diethylene glycol monomethyl ether.**

4-Methylcatechol *[452-86-8]* M 124.1, m 68°, b 112°/3mm, 241°/760mm. Crystd from high-boiling pet ether and distd in a vac.

Methylcellosolve see **2-methoxyethanol.**

Methyl chloride *[74-87-3]* M 50.5, b -24.1°. Bubbled through a sintered-glass disc dipping into conc H_2SO_4, then washed with water, condensed at low temperature and fractionally distd. Has been distd from $AlCl_3$ at -80°. Alternatively, passed through towers containing $AlCl_3$, soda-lime and P_2O_5, then condensed and fractionally distd. Stored as a gas.

Methyl chloroacetate *[96-34-4]* M 108.5, b 129-130°, n 1.423, d 1.230. Shaken with satd aq Na_2CO_3 (three times), aq 50% $CaCl_2$ (three times), satd aq NaCl (twice), dried (Na_2SO_4) and fractionally distd.

3-Methylcholanthrene *[56-49-5]* M 268.4, m 179-180°. Crystd from benzene and ethyl ether.

Methyl cyanide see **acetonitrile.**

Methyl cyanoacetate *[105-34-0]* M 99.1, f.p. -13°, b 205°, n 1.420, d 1.128. Purified by shaking with 10% Na_2CO_3 soln, washing well with water, drying with anhydrous Na_2SO_4, and distilling.

Methylcyclohexane *[108-87-2]* M 98.2, b 100.9°. n 1.4231, n^{25} 1.42058, d^{25} 0.7650. Passage through a column of activated silica gel gives material transparent down to 220nm. Can also be purified by passage through a column of activated basic alumina, or by azeotropic distn with MeOH, followed by washing out the MeOH with water, drying and distilling. Methylcyclohexane can be dried with $CaSO_4$,

CaH$_2$ or sodium. Has also been purified by shaking with a mixture of conc H$_2$SO$_4$ and HNO$_3$ in the cold, washing with water, drying with CaSO$_4$ and fractionally distilling from potassium. Percolation through a Celite column impregnated with 2,4-dinitrophenylhydrazine, phosphoric acid and water (prepared by grinding 0.5g DNPH with 6ml 85% H$_3$PO$_4$, then mixing with 4ml of distd water and 10g of Celite) removes carbonyl-containing impurities.

2-Methylcyclohexanol *[583-59-5]* M 114.2, b 65°/20mm, 167.6°/760mm, n 1.46085, d 0.922, *cis*- and *trans*-3-Methylcyclohexanol *[591-23-1]* M 114.2, b 69°/16mm, 172°/760mm, n 1.45757, n$^{25.5}$ 1.45444, d 0.930. Dried with Na$_2$SO$_4$ and distd under vac.

4-Methylcyclohexanone *[589-92-4]* M 112.2, b 165.5°/743mm, n 1.44506, d 0.914. Dried with CaSO$_4$, then fractionally distd.

1-Methylcyclohexene *[591-49-1]* M 107.4-108°/760mm, n 1.451, d 0.813. Freed from hydroperoxides by passing through a column containing basic alumina or refluxing with cupric stearate, filtered and fractionally distd from sodium.

Methylcyclopentene *[96-37-7]* M 84.2, b 71.8°, n 1.40970, n^{25} 1.40700, d 0.749. Purification procedures include passage through columns of silica gel (prepared by heating in nitrogen to 350° prior to use) and activated basic alumina, distn from sodium-potassium alloy, and azeotropic distn with MeOH, followed by washing out the methanol with water, drying and distilling. It can be stored with CaH$_2$ or sodium.

3'-Methyl-1,2-cyclopentenophenanthrene *[549-38-2]* M 232.3, m 126-127°. Crystd from acetic acid.

S-Methyl-L-cysteine *[1187-84-4]* M 135.2, m 207-211°, [α]^{26} -32.0° (H$_2$O). Likely impurities are cysteine and S-methyl-*dl*-cysteine. Crystd from water by adding 4 volumes of EtOH.

5-Methylcytosine *[554-01-8]* M 125.1, m 270°(dec). Crystd from water.

Methyl decanoate *[110-42-9]* M 186.3, b 114°/15mm, 224°/760mm, n 1.426, d 0.874. Passed through alumina before use.

Methyl 2,4-dichlorophenoxyacetate *[1928-38-7]* M 235.1, m 43°, b 119°/11mm. Crystd from MeOH.

***m*-Methyl-*N,N*-dimethylaniline** *[121-72-2]* M 135.2, b 72-74°/5mm, 215°/760mm, *p*-Methyl-*N,N*-dimethylaniline *[99-97-8]* M 135.2, b 76.5, 77.5°/4mm, 211°/760mm. Refluxed for 3hr with 2gram-equivalents of acetic anhydride, then fractionally distd under reduced pressure.

Methyl dodecanoate *[111-82-0]* M 214.4, m 5°, b 141°/15mm, n^{50} 1.4199, d 0.870. Passed through alumina before use.

***N*-Methyleneaminoacetonitrile** *[109-82-0]* M 68.1, m 129°. Crystd from EtOH or acetone.

Methylene-bis-acrylamide see **bis-acrylamide**.

***p,p'*-Methylene-bis-(*N,N*-dimethylaniline)** *[101-61-1]* M 254.4, m 89.5°. Crystd from 95% EtOH (charcoal) (*ca* 12ml/g).

Methylene Blue *[61-73-4]* M 319.9, ε$_{654}$ 94,000 (EtOH), ε$_{664}$ 81,000 (H$_2$O). Crystd from 0.1M HCl (16ml/g), the crystals were separated by centrifugation, washed with chilled EtOH and ethyl ether and dried under vac. Crystd from 50% aq EtOH, washed with abs EtOH, and dried at 50-55° for 24hr. Also crystd from benzene-MeOH (3:1). Salted out with NaCl from a commercial conc aq soln, then crystd from water, dried at 100° in an oven for 8-10hr.

Methylene chloride see **dichloromethane**.

4,4'-Methylenedianiline see *p,p'*-**diaminodiphenylmethane.**

3,4-Methylenedioxyaniline *[14268-66-7]* **M 137.1, m 45-46°, b 144°/14mm.** Crystd from pet ether.

3,4-Methylenedioxycinnamic acid *[2373-80-0]* **M 192.2, m 243-244°(dec).** Crystd from glacial acetic acid.

5,5'-Methylenedisalicylic acid *[122-25-8]* **M 372.3, m 238°(dec).** Crystd from acetone and benzene.

Methylene Green *[2679-01-8]* **M 364.9.** Crystd three times from water (18ml/g).

Methylene iodide see **diiodomethane.**

Methyl ether *[115-10-6]* **M 46.1, b -63.5°/96.5mm.** Dried by passing over alumina and then BaO, or over CaH_2, followed by fractional distn at low temperatures.

N-**Methyl ethylamine hydrochloride** *[624-60-2]* **M 95.6, m 126-130°.** Crystd from abs EtOH or ethyl ether.

Methyl ethyl ketone see **2-butanone.**

N-**Methyl formamide** *[123-39-7]* **M 59.1, m -3.5°, b 100.5°/25mm, n^{25} 1.4306, d 1.005.** Dried with molecular sieves for 2days, then distd under reduced pressure through a column packed with glass helices. Fractionally crystd by partial freezine and the solid portion was vac distd.

Methyl formate *[107-31-3]* **M 60.1, b 31.5°, n^{15} 1.34648, n 1.34332, d 0.971.** Washed with strong aq Na_2CO_3, dried with solid Na_2CO_3 and distd from P_2O_5. (Procedure removes free alcohol or acid.)

2-Methylfuran *[534-22-5]* **M 82.1, b 62.7-62.8°/731mm, n 1.436, d 0.917.** Washed with acidified satd ferrous sulphate soln (to remove peroxides), separated, dried with $CaSO_4$ or $CaCl_2$, and fractionally distd from KOH immediately before use. To reduce the possibility of spontaneous polymerisation, addition of about one-third of its volume of heavy mineral oil to 2-methylfuran prior to distn has been recommended.

Methyl gallate *[99-24-1]* **M 184.2, m 202°,**
N-**Methylglucamine** *[6284-40-8]* **M 195.2, m 128-129°, $[\alpha]_{546}$ -19.5°(c 2, H_2O),**
Methyl α-D-glucosamine *[97-30-3]* **M 194.2, m 165°, $[\alpha]_D^{25}$ +157.8° (c 3.0, H_2O).** Crystd from MeOH.

α-**Methylglutaric acid** *[617-62-8]* **M 146.1, m 79°,**
β-**Methylglutaric acid** *[626-51-7]* **M 146.1, m 87°.** Crystd from distd water, then dried under vac over conc H_2SO_4.

Methylglyoxal *[78-98-8]* **M 72.1, b ca 72°/760mm.** Commercial 30% (w/v) aq soln was diluted to about 10% and distd twice, taking the fraction boiling below 50°/20mm Hg. (This treatment does not remove lactic acid).

Methyl Green *[82-94-0]* **M 458.5.** Crystd from hot water.

1-Methylguanine *[938-85-2]* **M 165.2, m >300°(dec).** Crystd from 50% aq acetic acid.

7-Methylguanine *[578-76-7]* **M 165.2.** Crystd from water.

2-Methylhexane *[591-76-4]* **M 100.2, b 90.1°, n 1.38485, n^{25} 1.38227, d 0.678,**

3-Methylhexane *[589-34-4]* M 100.2, b 91.9°, n 1.38864, n^{25} 1.38609, d 0.687. Purified by azeotropic distn with MeOH, then washed with water (to remove the MeOH), dried over type 4A molecular sieves and distd.

Methyl hexanoate *[106-70-7]* M 130.2, b 52°/15mm, 150°/760mm, n 1.410, d 0.885. Passed through alumina before use.

Methylhydrazine *[60-34-4]* M 46.1, b 87°/745mm, n 1.436, d 0.876. Dried with BaO, then vac distd. Stored under nitrogen.

Methyl hydrazinocarboxylate *[6294-89-9]* M 90.1, m 70-73°. To remove impurities the material was melted and pumped under vac until the vapours were spectroscopically pure {Caminati *et al, JACS* **108** 4364 *1986*].

2-Methyl-4-hydroxyazobenzene *[1435-88-7]* M 212.2, m 100-101°,
3-Methyl-4-hydroxyazobenzene *[62-48-1]* M 212.2, m 125-126°. Crystd from hexane.

Methyl 4-hydroxybenzoate *[99-76-3]* M 152.2, m 127.5°. Fractionally crystd from its melt, recrystd from benzene, then from benzene/MeOH and dried over $CaCl_2$ in a vac desiccator.

Methyl 3-hydroxy-2-naphthoate *[883-99-8]* M 202.2, m 73-74°. Crystd from MeOH (charcoal) containing a little water.

N-**Methylimidazole** *[616-47-7]* M 82.1, b 81-84°/27mm, 197-198°/760mm, n 1.496, d 1.032. Dried with sodium metal and then distd. Stored at 0° under dry argon.

2-Methylimidazole *[693-98-1]* M 82.1, m 140-141°, b 267°/760mm,
4-Methylimidazole *[822-36-6]* M 82.1, m 47-48°, b 263°/760mm. Recrystd from benzene or pet ether.

2-Methylindole *[95-20-5]* M 131.2, m 61°,
3-Methylindole *[83-34-1]* M 131.2, m 95°. Crystd from benzene. Purified by zone melting.

Methyl iodide *[74-88-4]* M 141.9, b 42.8°, n 1.5315, d 2.281. Deteriorates rapidly with liberation of iodine if exposed to light. Usually purified by shaking with dil aq $Na_2S_2O_3$ or $NaHSO_3$ until colourless, then washed with water, dil aq Na_2CO_3, and more water, dried with $CaCl_2$ and distd. It is stored in a brown bottle away from sunlight in contact with a small amount of mercury, powdered silver or copper. (Prolonged exposure of mercury to methyl iodide forms methylmercuric iodide.) Methyl iodide can be dried further using $CaSO_4$ or P_2O_5. An alternative purification is by percolation through a column of silica gel or activated alumina, then distn. The soln can be degassed by using a repeated freeze-pump-thaw cycle.

Methyl isobutyl ketone see **4-methyl-2-pentanone.**

Methyl isopropyl ketone see **3-methyl-2-butanone.**

N-**Methyl maleimide** *[930-88-1]* M 111.1, m 94-96°. Crystd three times from ethyl ether.

Methylmevalonic acid *[516-05-2]* M 118.1, m 135°(dec). Crystallises as the hydrate from water.

Methyl methacrylate *[80-62-6]* M 100.1, f.p. -50°, b 46°/100mm, n 1.4144, d 0.937. Washed twice with aq 5% NaOH (to remove inhibitors such as hydroquinone) and twice with water. Dried with $CaCl_2$, Na_2CO_3, Na_2SO_4 or $MgSO_4$, then with CaH_2 under nitrogen at reduced pressure. The distillate is stored at low temperatures and redistd before use. Prior to distn, inhibitors such as β-naphthylamine (0.2%) or di-β-naphthol are sometimes added. Also purified by boiling for 12-14hr with dry powdered KOH, then filtering and distilling under reduced pressure in oxygen-free nitrogen to remove carbonyl compounds. Alternatively it was washed for several hours with 2M NaOH to remove quinol stabiliser. The organic layer was removed and the methacrylate layer was washed several times with distd water and dried over anhydrous Na_2SO_4 [Shizuka *et al, JPC* **89** 320 *1985*] or, washed twice with 5% aq NaOH soln, then with 5% aq H_3PO_4 soln and finally with satd NaCl soln. It was dried for 24hr over anhydrous $CaSO_4$, distd at 0.1mm Hg at room temperature and stored at -30° (Albeck et al, *JCSFT1* 1 1488 *1978*].

α-Methylmethionine *[562-48-1]* **M 163.0, m 283-284°.** Crystd from aq EtOH.

S-Methyl-L-methionine chloride *[1115-84-0]* **M 199.5, $[\alpha]_D^{23}$ +33° (0.2M HCl).** Likely impurities are methionine, methionine sulphoxide and methionine sulphone. Crystd from water by adding a large excess of EtOH. Stored in a cool, dry place, protected from light.

***N*-Methylmorpholine** *[109-02-4]* **M 101.2, b 116-117°/764mm, n 1.436, d 0.919.** Dried by refluxing with BaO or sodium, then fractionally distd through a helices-packed column.

1-Methylnaphthalene *[90-12-0]* **M 142.2, f.p. -30°, b 244.6°, n 1.6108, d 1/021.** Dried for several days with $CaCl_2$ or by prolonged refluxing with BaO. Fractionally distd through a glass helices-packed column from sodium. Purified further by soln in MeOH and pptn of its picrate complex by adding to a satd soln of picric acid in MeOH. The picrate, after crystn to constant melting point (**m** 140-141°) from MeOH., was dissolved in benzene and extracted with aq 10% LiOH until the extract was colourless. Evaporation of the benzene under vac gave 1-methylnaphthalene [Kloetzel and Herzog *JACS* **72** 1991 *1950*]. However, neither the picrate nor the styphnate complexes satisfactorily separates 1- and 2-methylnaphthalenes. To achieve this, 2-methylnaphthalene (10.7g) in 95% EtOH (50ml) has been pptd with 1,3,5-trinitrobenzene (7.8g) and the complex has been crystd from MeOH to **m** 153-153.5° (**m** of the 2-methyl isomer is 124°). [Alternatively, 2,4,7-trinitrofluorenone in hot glacial acetic acid could be used, and the derivative (**m** 163-164°) recrystd from glacial acetic acid.] The 1-methylnaphthalene was regenerated by passing a soln of the complex in dry benzene through a 15-in column of activated alumina and washing with benzene/pet ether (b 35-60°) until the coloured band of the nitro compound had moved down near the end of the column. The complex can also be decomposed using tin and acetic-hydrochloric acids, followed by extraction with ethyl ether and benzene; the extracts were washed successively with dil HCl, strongly alkaline sodium hypophosphite, water, dil HCl and water. [Soffer and Stewart *JACS* **74** 567 *1952*.] It can be purified from anthracene by zone melting.

2-Methylnaphthalene *[91-57-6]* **M 142.2, m 34.7-34.9°, b 129-130°/25mm.** Fractionally crystd repeatedly from its melt, then fractionally distd under reduced pressure. Crystd from benzene and dried under vac in an Abderhalden pistol. Purified *via* its picrate (**m** 114-115°) as described for 1-methylnaphthalene.

6-Methyl-2-naphthol *[17579-79-2]* **M 158.2, m 128-129°,**
7-Methyl-2-naphthol *[26593-50-0]* **M 158.2, m 118°.** Crystd from EtOH or ligroin and sublimed *in vacuo*.

2-Methyl-1,4-naphthoquinone (vitamin K_3) *[58-27-5]* **M 172.2, m 106°.** Crystd from EtOH or pet ether (b 60-80°).

Methyl 1-naphthyl ether *[2216-69-5]* **M 158.2, b 90-91°/2mm, n^{26} 1.6210, d 1.095.** Steam distd from alkali. The distillate was extracted with ethyl ether. After drying the extract and evaporating the ethyl ether, the methyl naphthyl ether was distd under reduced pressure.

Methyl nitrate *[598-58-3]* **M 77.0, b 65°/760mm, d^5 1.2322, d^{15} 1.2167, d^{25} 1.2032.** Distd at -80°. The middle fraction was subjected to several freeze-pump-thaw cycles. **VAPOUR EXPLODES ON HEATING.**

Methyl nitrite *[624-91-9]* M 61.0, b -12°, d^{15} (liq) 0.991. Condensed in a liquid nitrogen trap. Distd under vac, first trap containing dry Na_2CO_3 to free it from acid impurities then into further Na_2CO_3 traps before collection.

N-**Methyl-4-nitroaniline** *[100-15-2]* M 152.2, m 152.2°. Crystd from aq EtOH.

2-Methyl-5-nitroaniline *[99-55-8]* M 152.2, m 109°. Acetylated, and the acetyl derivative crystd to constant melting point, then hydrolysed with 70% H_2SO_4 and the free base regenerated by treatment with ammonia [Bevan, Fayiga and Hirst *JCS* 4284 *1956*].

4-Methyl-3-nitroaniline *[119-32-4]* M 152.2, m 81.5°. Crystd from hot water (charcoal), then ethanol and dried in a vac desiccator.

Methyl 3-nitrobenzoate *[618-95-1]* M 181.2, m 78°. Crystd from MeOH (1g/ml).

Methyl 4-nitrobenzoate *[619-50-1]* M 181.2, m 95-95.5°. Dissolved in ethyl ether, then washed with aq alkali, the ether was evaporated and the ester was recrystd from EtOH.

2-Methyl-2-nitro-1,3-propanediol *[77-49-6]* M 135.1, m 145°. Crystd from *n*-butanol.

2-Methyl-2-nitro-1-propanol *[76-39-1]* M 119.1, m 87-88°. Crystd from pet ether.

N-**Methyl-4-nitrosoaniline** *[10595-51-4]* M 136.2, m 118°. Crystd from benzene.

N-**Methyl-*N*-nitroso-*p*-toluenesulphonamide** (diazald) *[80-11-5]* M 214.2, m 62°. Crystd from benzene by addition of pet ether.

3-Methyloctane *[2216-33-3]* M 128.3, b 142-144°/760mm, n 1.407, d 0.719. Passed through a column of silica gel [Klassen and Ross *JPC* 91 3668 *1987*].

Methyl octanoate (methyl caprylate) *[111-11-5]* M 158.2, b 83°/15mm, 193-194°/760mm, n 1.419, d 0.877. Passed through alumina before use.

Methyl oleate *[112-62-9]* M 296.5, f.p. -19.9°, b 217°/16mm, n 1.4522, d 0.874. Purified by fractional distn under reduced pressure, and by low temperature crystn from acetone.

Methyl Orange *[547-58-0]* M 327.3. Crystd twice from hot water, then washed with a little EtOH followed by ethyl ether.

3-Methyl-2-oxazolidone *[19836-78-3]* M 101.1, m 15°, b 88-91°/1mm, n 1.455, d 1.172. Purified by successive fractional freezing, then dried in a dry-box over Linde type 4A molecular sieves for 2days.

Methylpentane (mixture of isomers). Passage through a long column of activated silica gel (or alumina) gave material transparent down to 200nm.

2-Methylpentane *[107-83-5]* M 86.2, b 60.3°, n 1.37145, n^{25} 1.36873, d 0.655. Purified by azeotropic distn with MeOH, followed by washing out the MeOH with water, drying ($CaCl_2$, then sodium), and distn. [Forziati *et al*, *J Res Nat Bur Stand* 36 129 *1946*.]

3-Methylpentane *[96-14-0]* M 86.2, b 63.3°, n 1.37652, n^{25} 1.37384, d 0.664. Purified by azeotropic distn with MeOH, as for 2-methylpentane. Purified for ultraviolet spectroscopy by passage through columns of silica gel or alumina activated by heating for 8hr at 210° under a stream of nitrogen. Has also been treated with conc (or fuming) H_2SO_4, then washed with water, aq 5% NaOH, water again, then dried ($CaCl_2$, then sodium), and distd through a long, glass helices-packed column.

2-Methyl-2,4-pentanediol *[107-41-5]* M 118.2, b 107.5-108.5°/25mm, n^{25} 1.4265, d 0.922. Dried with Na$_2$SO$_4$, then CaH$_2$ and fractionally distd under reduced pressure through a packed column, taking precautions to avoid absorption of water.

2-Methyl-1-pentanol *[105-30-6]* M 102.2, b 65-66°/60mm, 146-147°/760mm, n 1.420, d 0.827. Dried with Na$_2$SO$_4$ and distd.

4-Methyl-2-pentanol *[108-11-2]* M 102.2, b 131-132°, n 1.413, d 0.810. Washed with aq NaHCO$_3$, dried and distd. Further purified by conversion to the phthalate ester by adding 120ml of dry pyridine and 67g of phthalic anhydride per mole of alcohol, purifying the ester and steam distilling it in the presence of NaOH. The distillate was extracted with ether, and the extract was dried and fractionally distd. [Levine and Walti *JBC* **94** 367 *1931*.]

3-Methyl-3-pentanol carbamate (Emylcamate) *[78-28-4]* M 145.2, m 56-58.5°. Crystd from 30% EtOH.

4-Methyl-2-pentanone *[108-10-1]* M 100.2, b 115.7°, n 1.3958, n^{25} 1.3938, d 0.801. Refluxed with a little KMnO$_4$, washed with aq NaHCO$_3$, dried with CaSO$_4$ and distd. Acidic impurities were removed by passage through a small column of activated alumina.

2-Methyl-1-pentene *[763-29-1]* M 84.2, b 61.5-62°, n 1.395, d 0.680. Water was removed, and peroxide formation prevented, by several vac distns from sodium, followed by storage with sodium-potassium alloy.

cis-**4-Methyl-2-pentene** *[691-38-3]* M 84.2, m -134.4°, b 57.7-58.5°, n 1.388, d 0.672,
trans-**4-Methyl-2-pentene** *[674-76-0]* M 84.2, m -140.8°, b 58.5°, n 1.389, d 0.669. Dried with CaH$_2$, and distd.

3-Methyl-1-propyn-3-ol carbamate *[302-66-9]* M 141.2, m 55.8-57°. Crystd from ether/pet ether or cyclohexane.

5-Methyl-1,10-phenanthroline *[3002-78-6]* M 194.2, m 113°(anhydr). Crystd from benzene/pet ether.

N-**Methylphenazonium methosulphate** *[299-11-6]* M 306.3, m 155-157°. Crystd from EtOH.

N-**Methylphenothiazine** *[1207-72-3]* M 213.2, α-form m 99.3° and b 360-365°, β-form m 78-79°. Recrystn (three times) from EtOH gave α-form (prisms). Recrystn from EtOH/benzene gave the β-form (needles). Also purified by vac sublimation and carefully dried in a vacuum line. Also crystd from toluene and stored in the dark [Guarr *et al, JACS* **107** 5104 *1985;* Olmsted *et al, JACS* **109** 3297 *1987*].

4-Methylphenylacetic acid *[622-47-9]* M 150.2, m 94°. Crystd from heptane.

1-Methyl-1-phenylhydrazine sulphate *[33008-18-3]* M 218.2. Crystd from hot water by addition of hot EtOH.

3-Methyl-1-phenyl-5-pyrazolone *[89-25-8]* M 174.2, m 127°. Crystd from hot water, or EtOH/water (1:1).

N-**Methyl-2-phenylsuccinimide** see **phensuximide.**

N-**Methylphthalimide** *[550-44-7]* M 161.1, m 133.8°. Recrystd from abs EtOH.

2-Methylpiperazine *[109-07-9]* M 100.2, m 66°. Purified by zone melting.

3-Methylpiperidine *[626-56-2]* M 99.2, b 125°/763mm, n^{25} 1.4448, d 0.846. Purified *via* the hydrochloride (m 172°). [Chapman, Isaacs and Parker *JCS* 1925 *1959*.]

4-Methylpiperidine *[626-58-4]* M 99.2, b 124.4°/755mm, n^{25} 1.4430, d 0.839. Purified *via* the hydrochloride (m 189°). Freed from 3-methylpyridine by zone melting.

2-Methylpropane-1,2-diamine *[811-93-8]* M 88.2, b 47-48°/17mm. Dried with sodium for 2days, then distd under reduced pressure from sodium.

2-Methylpropane-1-thiol *[513-44-0]* M 90.2, b 41.2°/142mm, n^{25} 1.43582. Dissolved in EtOH, and added to 0.25M Pb(OAc)$_2$ in 50% aq EtOH. The pptd lead mercaptide was filtered off, washed with a little EtOH, and impurities were removed from the molten salt by steam distn. After cooling, dil HCl was added dropwise to the residue, and the mercaptan was distd directly from the flask. Water was separated from the distillate, and the mercaptan was dried (Na$_2$CO$_3$) and distd under nitrogen. [Mathias *JACS* 72 1897 *1950*.]

2-Methylpropane-2-thiol *[75-66-1]* M 90.2, b 61.6°/701mm, n^{25} 1.41984, d^{25} 0.79426. Dried for several days with CaO, then distd from CaO. Purified as for *2-methylpropane-1-thiol*.

2-Methyl-1-propanol *[78-83-1]* M 74.1, b 107.9°, n^{15} 1.39768, n^{25} 1.3939, d 0.804. Dried by refluxing with CaO and BaO for several hours, followed by treatment with calcium or aluminium amalgam, then fractional distn from sulphanilic or tartaric acids. More exhaustive purifications involve formation of phthalate or borate esters. Heating with phthalic anhydride gives the acid phthalate which, after crystn to constant melting point (m 65°) from pet ether, is hydrolysed with aq 15% KOH. The alcohol is distd as the water azeotrope and dried with K$_2$CO$_3$, then anhydrous CuSO$_4$, and finally magnesium turnings, followed by fractional distn. [Hückel and Ackermann *J prakt Chem* 136 15 *1933*.] The borate ester is formed by heating the dried alcohol for 6hr in an autoclave at 160-175° with a quarter of its weight of boric acid. After several fractional distns under vac the ester is hydrolysed by heating for a short time with aq alkali and the alcohol is dried with CaO and distd. [Michael, Scharf and Voigt *JACS* 38 653 *1916*.]

2-Methyl-2-propanol see *tert*-butyl alcohol.

N-**Methylpropionamide** *[1187-58-2]* M 87.1, f.p. -30.9°, b 103°/12-13mm, n^{25} 1.4356, d 0.934. A colourless, odourless, neutral liquid at room temperature with a high dielectric constant. The amount of water present can be determined directly by Karl Fischer titration; g.l.c. and nmr have been used to detect unreacted propionic acid. Commercial material of high quality is available, probably from the condensation of anhydrous methylamine with 50% excess of propionic acid. Rapid heating to 120-140° with stirring favours the reaction by removing water either directly or as the ternary xylene azeotrope. The quality of the distillate improves during the distn.
The propionamide can be dried over CaO. Water and unreacted propionic acid were removed as their xylene azeotropes. It was vac dried. Material used as an electrolyte solvent (specific conductance less than 10^{-6} ohm^{-1}cm^{-1}) was obtained by fractional distn under reduced pressure, and was stored over BaO or molecular sieves because it readily absorbs moisture from the atmosphere on prolonged storage. [Hoover *PAC* 37 581 *1974*; **Recommended Methods for Purification of Solvents and Tests for Impurities**, Coetzee. ed, Pergamon Press, *1982.*]

Methyl propionate *[554-12-1]* M 88.1, b 79.7°. Washed with satd aq NaCl, then dried with Na$_2$CO$_3$ and distd from P$_2$O$_5$. (This removes any free acid and alcohol.) It has also been dried with anhydrous CuSO$_4$.

Methyl *n*-propyl ether *[557-17-5]* M 74.1, b 39°, n^{14} 1.3602, d 0.736. Dried with CaSO$_4$, then passed through a column of alumina (to remove peroxides) and fractionally distd.

Methyl *n*-propyl ketone *[107-87-9]* **M 86.1, b 102.4°, n 1.3903, d 0.807.** Refluxed with a little KMnO₄, dried with CaSO₄ and distd. It was converted to its bisulphite addition compound by shaking with excess satd aq NaHSO₃ at room temperature, cooling to 0°, filtering, washing with ethyl ether and drying. Steam distillation gave a distillate from which the ketone was recovered, washed with aq NaHCO₃ and distd water, dried (K₂CO₃) and fractionally distd. [Waring and Garik *JACS* **78** 5198 *1956*.]

2-Methylpyrazine *[109-08-0]* **M 94.1, b 136-137°, n 1.505, d 1.025.** Purified *via* the picrate. [Wiggins and Wise *JCS* 4780 *1956*.]

2-Methylpyridine (2-picoline) *[109-06-8]* **M 93.1, b 129.4°, n 1.50102, d 0.9444.** Biddiscombe and Handley [*JCS* 1957 *1954*] steam distd a boiling soln of the base in 1,2 equivalents of 20% H₂SO₄ until about 10% of the base had been carried over, along with non-basic impurities. Excess aq NaOH was then added to the residue, the free base was separated, dried with solid NaOH and fractionally distd. 2-Methylpyridine can also be dried with BaO, CaO, CaH₂, LiAlH₄, sodium or Linde type 5A molecular sieves. An alternative purification is *via* the ZnCl₂ adduct, which is formed by adding 2-methylpyridine (90ml) to a soln of anhydrous ZnCl₂ (168g) and 42ml conc HCl in abs EtOH (200ml). Crystals of the complex are filtered off, recrystd twice from abs EtOH (to give **m** 118.5-119.5°), and the free base is liberated by addition of excess aq NaOH. It is steam distd, and solid NaOH added to the distillate to form two layers, the upper one of which is then dried with KOH pellets, stored for several days with BaO and fractionally distd. Instead of ZnCl₂, HgCl₂ (430g in 2.4L of hot water) can be used. The complex, which separates on cooling, can be dried at 110° and recrystd from 1% HCl (to **m** 156-157°).

3-Methylpyridine (3-picoline) *[108-99-6]* **M 93.1, m -18.5°, b 144°/767mm, n 1.5069, d 0.957.** In general, the same methods of purification that are described for *2-methylpyridine* can be used. However, 3-methylpyridine often contains 4-methylpyridine and 2,6-lutidine, neither of which can be removed satisfactorily by drying and fractionation, or by using the ZnCl₂ complex. Biddiscombe and Handley [*JCS* 1957 *1954*], after steam distn as for *2-methylpyridine*, treated the residue with urea to remove 2,6-lutidine, then azeotropically distd with acetic acid (the azeotrope had **b** 114.5°/712mm), and recovered the base by adding excess of aq 30% NaOH, drying with solid NaOH and carefully fractionally distilling. The distillate was then fractionally crystd by slow partial freezing. An alternative treatment [Reithof *et al*, *IECAE* **18** 458 *1946*] is to reflux the crude base (500ml) for 20-24hr with a mixture of acetic anhydride (125g) and phthalic anhydride (125g) followed by distn until phthalic anhydride begins to pass over. The distillate was treated with NaOH (250g in 1.5L of water) and then steam distd. Addition of solid NaOH (250g) to this distillate (*ca* 2L) led to the separation of 3-methylpyridine which was removed, dried (K₂CO₃, then BaO) and fractionally distd. (Subsequent fractional freezing would probably be advantageous.)

4-Methylpyridine (4-picoline) *[108-89-4]* **M 93.1, m 4.25°, b 145.0°/765mm, n 1.5058, d 0.955.** Can be purified as for *2-methylpyridine*. Biddescombe and Handley's method for 3-methylpyridine is also applicable. Lidstone [*JCS* 242 *1940*] purified *via* the oxalate (**m** 137-138°) by heating 100ml of 4-methylpyridine to 80° and adding slowly 110g of anhydrous oxalic acid, followed by 150ml of boiling EtOH. After cooling and filtering, the ppte was washed with a little EtOH, then recrystd from EtOH, dissolved in the minimum quantity of water and distd with excess 50% KOH. The distillate was dried with solid KOH and again distd. Hydrocarbons can be removed from 4-methylpyridine by converting the latter to its hydrochloride, crystallising from EtOH/ethyl ether, regenerating the free base by adding alkali and distilling. As a final purification step, 4-methylpyridine can be fractionally crystd by partial freezing to effect a separation from 3-methylpyridine. Contamination by 2,6-lutidine is detected by its strong absorption at 270nm.

4-Methylpyridine 1-oxide *[1003-67-4]* **M 109.1, m 184°.** Crystd from acetone/ether.

***N*-Methylpyrrole** *[96-54-8]* **M 81.1, b 115-116°/756mm, n 1.487, d 0.908.** Dried with CaSO₄, then fractionally distd from KOH immediately before use.

1-Methyl-2-pyrrolidinone *[872-50-4]* **M 99.1, f.p. -24.4, b 202°/760mm, n 1.471,d 1.033.** Dried by removing water as benzene azeotrope. Fractionally distd at 10 torr through a 100-cm column packed with glass helices.

2-Methylquinoline (quinaldine) *[91-63-4]* **M 143.2, b 86-87°/1mm, 155°/14mm, 246-247°/760mm, n 1.6126, d 1.058.** Dried with Na_2SO_4 or by refluxing with BaO, then fractionally distd under reduced pressure. Redistd from zinc dust. Purified by conversion to its phosphate (m 220°) or picrate (m 192°) from which after recrystn, the free base was regenerated. [Packer, Vaughan and Wong *JACS* **80** 905 *1958*.] Its $ZnCl_2$ complex can be used for the same purpose.

4-Methylquinoline (lepidine) *[491-35-0]* **M 143.2, b 265.5°, n 1.61995, d 1.084.** Refluxed with BaO, then fractionally distd. Purified *via* its recrystd dichromate salt (m 138°). [Cumper, Redford and Vogel *JCS* 1176 *1962*.]

6-Methylquinoline *[91-62-3]* **M 143.2, b 258.6°, n 1.61606, d 1.067.** Refluxed with BaO, then fractionally distd. Purified *via* its recrytd $ZnCl_2$ complex (m 190°). [Cumper, Redford and Vogel *JCS* 1176 *1962*.]

7-Methylquinoline *[612-60-2]* **M 143.2, M 143.2, m 38°, b 255-260°, n 1.61481, d 1.052.** Purified *via* its dichromate complex (m 149°, after five recrystns from water). [Cumper, Redford and Vogel *JCS* 1176 *1962*.]

8-Methylquinoline *[611-321-5]* **M 143.2, b 122.5°/16mm, 247.8°/760mm, n 1.61631, d 1.703.** Purified as for 2-methylquinoline. The phosphate and picrate have m 158° and m 201° respectively.

2-Methyl-8-quinolinol see **8-hydroxy-2-methylquinoline.**

Methyl Red *[493-52-7]* **M 269.3, m 181-182°.** Extracted with boiling toluene in a Soxhlet flask. The crystals which separated on slow cooling to room temperature are filtered off, washed with a little toluene and recrystd from glacial acetic acid, benzene or toluene followed by pyridine/water. Alternatively, dissolved in aq 5% $NaHCO_3$ soln, and pptd from hot soln by dropwise addition of aq HCl. Repeated until the extinction coefficients did not increase.

Methyl salicylsalicylate *[580-02-9]* **M 194.2, m 51-52°.** Crystd from pet ether.

α-Methylstyrene (monomer) *[98-83-9]* **M 118.2, b 57°/15mm, n 1.5368, d 0.910.** Washed three times with aq 10% NaOH (to remove inhibitors such as quinol), then six times with distd water, dried with $CaCl_2$ and distd under vac. The distillate is kept under nitrogen, in the cold, and redistd if kept for more than 48hr before use. It can also be dried with CaH_2.

trans-**β-Methylstyrene** *[873-66-5]* **M 118.2, b 176°/760mm, n 1.5496, d 0.910.** Distd under nitrogen from powdered NaOH through a Vigreux column, and passed through activated neutral alumina before use [Wong *et al, JACS* **109** 3428 *1987*].

Methylsuccinic acid *[498-21-5]* **M 132.1, m 115.0°.** Crystd from water.

Methyl tetradecanoate (methyl myristate) *[134-10-7]* **M 382.7, m 18.5°, b 155-157°/7mm.** Passed through alumina before use.

2-Methyltetrahydrofuran *[96-47-9]* **M 86.1, b 80.0°, n 1.405, d 0.853.** Likely impurities are 2-methylfuran, methyldihydrofurans and hydroquinone (stabiliser, which is removed by distn under reduced pressures). It was washed with 10% aq NaOH, dried, vac distd from CaH_2, passed through freshly activated alumina under nitrogen, and refluxed over sodium metal under vac. Stored over sodium. [Ling and Kevan *JPC* **80** 592 *1976*.] Vac distd from sodium, and stored with sodium-potassium alloy. (Treatment removes water and prevents the formation of peroxides.) Alternatively, it can be freed from peroxides by treatment with ferrous sulphate and sodium bisulphate, then solid KOH, followed by drying with, and distilling from, sodium, or type 4A molecular sieves under argon. It may be difficult to remove benzene if it is present as an impurity (can be readily detected by its ultraviolet absorption in the 249-268nm region). [Ichikawa and Yoshida *JPC* **88** 3199 *1984*.]

N-Methylthioacetamide *[5310-10-1]* M 89.1, m 59°. Recrystd from benzene.

3-(Methylthio)aniline see S-methyl-L-cysteine.

3-Methylthiophen *[616-44-4]* M 98.2, b 111-113°. n 1.531,d 1.024. Dried with Na_2SO_4, then distd from sodium.

6-Methyl-2-thiouracil *[56-04-2]* M 142.2, m 330°(dec). Crystd from a large volume of water.

17α-Methyltostesterone *[58-18-4]* M 302.5, m 164-165°, $[\alpha]_{546}$ +87° (c 1, dioxane). Crystd from hexane/benzene.

Methyltricaprylylammonium chloride see Aliquat 336.

2-Methyltricycloquinazoline *[2642-52-6]* M 334.4. Purified by vac sublimation. CARCINOGEN.

N-Methyltryptophan (L-abrine) *[526-31-8]* M 218.3, m 295°(dec), $[\alpha]_D^{21}$ +44.4° (c 2.8, 0.5M HCl). Crystd from water.

dl-5-Methyltryptophan *[951-55-3]* M 218.3, m 275°(dec). Crystd from aq EtOH.

6-Methyluracil *[626-48-2]* M 126.1, m 270-280°(dec), λ_{max} 260$_{nm}$ logε 3.97. Crystd from EtOH or acetic acid.

3-Methyluric acid *[39717-48-1]* M 182.1, m >350°,
7-Methyluric acid *[30409-21-3]* M 182.1, m >380°,
9-Methyluric acid *[30345-24-5]* M 182.1, m >400°. Crystd from water.

Methyl vinyl ketone *[78-94-4]* M 70.1, b 62-68°/400mm, 79-80°/760mm, n 1.413, d 0.845. Forms an 85% azeotrope with water. After drying with K_2CO_3 and $CaCl_2$ (with cooling), the ketone is distd at low pressures.

Methyl vinyl sulphone *[3680-02-2]* M 106.1, b 116-118°/20mm, n 1.461, d 1.215. Passed through a column of alumina, then degassed and distd on a vac line and stored at -190° until required.

Methyl Violet *[8004-87-3]* M 394.0. Crystd from abs EtOH by pptn with ethyl ether during cooling in an ice-bath. Filtered off and dried at 105°.

Methyl viologen dichloride see *N,N'*-dimethyl-4,4'-dipyridyl chloride (paraquat dichloride).

1-Methylxanthine *[6136-37-4]* M 166.1, m >360°,
3-Methylxanthine *[1076-22-5]* M 166.1, m >360°,
7-Methylxanthine *[552-62-5]* M 166.1, m >380°(dec),
8-Methylxanthine *[17338-96-4]* M 166.1, m 292-293°(dec),
9-Methylxanthine *[1198-33-0]* M 166.1, m 384°(dec). Crystd from water.

Metrazole *[56-95-5]* M 138.2, m 61°. Crystd from ethyl ether. Dried under vac over P_2O_5.

Mevalonic acid lactone *[674-26-0]* M 130.2, m 28°, b 145-150°/5mm. Purified *via* the dibenzyl-ethylenediammonium salt (m 124-125°) [Hofmann *et al, JACS* **79** 2316 *1957*], or by chromatography on paper or on Dowex-1 (formate) column. [Bloch *et al*, *JBC* **234** 2595 *1959*]. Stored as DBED salt, or as the lactone in a sealed container at 0°.

Mevalonic acid 5-phosphate *[1189-94-2]* **M 228.1.** Purified by conversion to the tricyclohexylammonium salt (**m** 154-156°) by treatment with cyclohexylamine. Crystd from water/acetone at -15°. Alternatively, the phosphate was chromatographed by ion-exchange or paper (Whatman No 1) in a system isobutyric acid/ammonia/water (66:3:30; R_F 0.42). Stored as the cyclohexylammonium salt.

Mevalonic acid 5-pyrophosphate *[1492-08-6]* **M 258.1.** Purified by ion-exchange chromatography on Dowex-1 formate {Bloch *et al*, *JBC* **234** 2595 *1959*], DEAE-cellulose [Skilletar and Kekwick, *Anal Biochem* **20** 171 *1967*], on by paper chromatography [Rogers *et al*, *BJ* **99** 381 *1966*]. Likely impurities are ATP and mevalonic acid phosphate. Stored as a dry powder or as a slightly alkaline (*p*H 7-9) soln at -20°.

Michler's ketone *[90-94-8]* **M 268.4, m 179°.** Dissolved in dil HCl, filtered and pptd by adding ammonia (to remove water-insoluble impurities such as benzophenone). Then crystd from EtOH or pet ether. [Suppan *JCSFT1* **71** 539 *1975*.] It was also purified by dissolving in benzene, then washed with water until the aq phase was colourless. The benzene was evaporated off and the residue recrystd three times from benzene and EtOH [Hoshino and Kogure *JPC* **72** 417 *1988*].

Milling Red SWB *[6459-94-5]* **M 832.8,**
Milling Yellow G *[51569-18-7]*, Salted out three times with sodium acetate, then repeatedly extracted with EtOH. [McGrew and Schneider *JACS* **72** 2547 *1950*.] See entry under Chlorazol Sky Blue FF.

Monensin *[17090-79-8]* **M 670.9.** Purified by chromatography.

Monobutyl urea *[592-31-4]* **M 116.2, m 96-98°,**
Monoethyl urea *[625-52-5]* **M 88.1, m 92-95°,**
Monomethyl urea *[598-50-5]* **M 74.1, m 93-95°.** Crystd from EtOH/water, then dried under vac at room temperature.

Monopropyl urea *[627-06-5]* **M 102.1, m 110°.** Crystd from EtOH.

Morin (hydrate) *[480-16-0]* **M 302.2, m 289-292°.** Stirred at room temperature with ten times its weight of abs EtOH, then left overnight to settle. Filtered, and evapd under a heat lamp to one-tenth its volume. An equal volume of water was added, and the pptd morin was filtered off, dissolved in the minimum amount of EtOH and again pptd with an equal volume of water. The ppte was filtered, washed with water and dried at 110° for 1hr. (Yield *ca* 2.5%.) [Perkins and Kalkwarf *AC* **28** 1989 *1956*.]

Morphine (H$_2$O) *[57-27-2]* **M 302.2, m 230°(dec),** $[\alpha]_D^{23}$ **-130.9° (MeOH).** Crystd from MeOH.

Morpholine *[110-91-8]* **M 87.1, f.p. -4.9°, b 128.9°, n 1.4540, n^{25} 1.4533, d 1.0007.** Dried with KOH, fractionally dist, then refluxed with sodium, and again fractionally distd. Dermer and Dermer [*JACS* **59** 1148 *1937*] pptd as the oxalate by adding slowly to slightly more than 1 molar equivalent of oxalic acid in EtOH. The ppte was filtered and recrystd twice from 60% EtOH. Addition of the oxalate to conc aq NaOH regenerated the base, which was separated and dried with solid KOH, then sodium, before being fractionally distd.

2-(*N*-Morpholino)ethanesulphonic acid (MES) *[4432-31-9]* **M 213.3, m >300°(dec), pKa 6.95.** Crystd from hot EtOH containing a little water.

Mucic acid see **galactaric acid.**

Mucochloric acid *[87-56-9]* **M 169.0, m 124-126°.** Crystd twice from water (charcoal).

trans,trans-**Muconic acid** *[3588-17-8]* **M 142.1, m 300°,**
Muramic acid (H$_2$O) *[1114-41-6]* **M 251.2, m 152-154°(dec), [α]$_D$ +155° to +110° (in H$_2$O).** Crystd from water.

Murexide *[3051-09-0]* **M 284.2, m >300°, λ$_{max}$ 520nm (ε 12,000).** The sample may be grossly contaminated with uramil, alloxanthine, etc. Difficult to purify. It is better to synthesise it from pure alloxanthine [Davidson *JACS* **58** 1821 *1936*]. Crystd from water.

Myristic acid *[544-63-8]* **M 228.4, m 58°.** Purified *via* the methyl ester (**b** 153-154°/10mm, **n^{25}** 1.4350), as for capric acid. [Trachtman and Miller *JACS* **84** 4828 *1962*.] Also purified by zone melting. Crystd from pet ether.

Naphthacene (2,3-benzanthracene) [92-24-0] M 228.3, m 340-341°. Crystd from EtOH or benzene. Dissolved in sodium-dried benzene and passed through a column of alumina. The benzene was evaporated under vac, and the chromatography was repeated using fresh benzene. Finally, the naphthacene was sublimed under vac. [Martin and Ubblehode *JCS* 4948 *1961*.]

2-Naphthaldehyde [66-99-9] M 156.2, m 59°, b 260°/19mm. Distilled with steam and crystd from water or EtOH.

Naphthalene [91-20-3] M 128.2, m 80.3°, b 87.5°/10mm, 218.0°, n^{85} 155898, d 1.0253, d^{100} 0.9625. Crystd one or more times from the following solvents: EtOH, MeOH, CCl_4, benzene, glacial acetic acid, acetone or ethyl ether, followed by drying at 60° in an Abderhalden drying apparatus. Also purified by vac sublimation and by fractional crystn from its melt. Other purification procedures include refluxing in EtOH over Raney Ni, and chromatography of a CCl_4 soln on alumina with benzene as eluting solvent. Baly and Tuck [*JCS* 1902 *1908*] purified naphthalene for spectroscopy by heating with conc H_2SO_4 and MnO_2, followed by steam distn (repeating the process), and formation of the picrate which, after recrystn, was decomposed and the naphthalene was steam distd. It was then crystd from dil EtOH. It can be dried over P_2O_5 under vac. Also purified by sublimation and subsequent crystn from cyclohexane. Alternatively, it has been washed at 85° with 10% NaOH to remove phenols, with 50% NaOH to remove nitriles, with 10% H_2SO_4 to remove organic bases, and with 0.8g $AlCl_3$ to remove thianaphthalenes and various alkyl derivatives. Then it was treated with 20% H_2SO_4, 15% Na_2CO_3 and finally distd. [Gorman *et al*, *JACS* 107 4404 *1985*.]
Zone refining purified naphthalene from anthracene, 2,4-dinitrophenylhydrazine, methyl violet, benzoic acid, methyl red, chrysene, pentacene and indoline.

Naphthalenediol see **dihydroxynaphthalene**.

Naphthalene-2,5-disulphonic acid [92-41-1] M 288.2. Crystd from conc HCl.

Naphthalene Scarlet Red 4R [2611-82-7] M 623.5. Dissolved in the minimum quantity of boiling water, filtered and enough EtOH was added to ppte *ca* 80% of the dye. This process was repeated until a soln of the dye in aq 20% pyridine had a constant extinction coefficient.

Naphthalene-1-sulphonic acid [85-47-2] M 208.2, m ($2H_2O$) 90°, (anhydrous) 139-140°. Crystd from conc HCl and twice from water.

Naphthalene-2-sulphonic acid [120-18-3] M 208.2, m 91°. Crystd from conc HCl.

Naphthalene-2-sulphonyl chloride [93-11-8] M 226.7, m 79°. Crystd (twice) from benzene/pet ether (1:1 v/v).

Naphthalene-2-thiol [91-60-1] M 160.2, m 81-82°. Crystd from EtOH.

1,8-Naphthalic anhydride [81-84-5] M 198.2, m 274°. Extracted with cold aq Na_2CO_3 to remove free acid, then crystd from acetic anhydride.

Naphthamide [2243-82-5] M 171.2, m 195°. Crystd from EtOH.

Naphthazarin (5,8-dihydroxy-1,4-naphthoquinone) [475-38-7] M 190.2, m 225-230°. Recrystd from hexane and purified by vac sublimation. [Huppert *et al*, *JPC* 89 5811 *1985*.]

Naphthionic acid see **4-Aminonaphthalene-1-sulphonic acid**.

α-Naphthoic acid *[86-55-5]* M 172.2, m 162.5-163.0°. Crystd from toluene (3ml/g) (charcoal), pet ether (b 80-100°), or aq 50% EtOH.

β-Naphthoic acid *[93-09-4]* M 172.2, m 184-185°. Crystd from EtOH (4ml/g), or aq 50% EtOH. Dried at 100°.

α-Naphthol *[90-15-3]* M 144.2, m 95.5-96°. Sublimed, then crystd from aq MeOH (charcoal), aq 25% or 50% EtOH, benzene, cyclohexane, heptane, CCl_4 or boiling water. Dried over P_2O_5 under vac. [Shizuka *et al, JACS* **107** 7816 *1985*.]

β-Naphthol *[135-19-3]* M 144.2, m 122.5-123.5°. Crystd from aq 25% EtOH (charcoal), water, benzene, toluene or CCl_4, e.g. by repeated extraction with small amounts of EtOH, followed by dissolution in a minimum amount of EtOH and pptn with distd water, then drying over P_2O_5 under vac. Has also been dissolved in aq NaOH, and pptd by adding acid (repeated several times), then pptd from benzene by addition of heptane. Final purification can be by zone melting or sublimation *in vacuo*. [Bardez *et al, JPC* **89** 5031 *1985*; Kikuchi *et al, JPC* **91** 574 *1987*.]

α-Naphtholbenzein *[6948-88-5]* M 392.5, m 122-125°. Crystd from EtOH, aq EtOH or glacial acetic acid.

1-Naphthol-2-carboxylic acid *[86-48-6]* M 188.2, m 203-204°. Successively crystd from EtOH/water, ethyl ether and acetonitrile, with filtration through a column of charcoal and Celite. [Tong and Glesmann *JACS* **79** 583 *1957*]

2-Naphthol-3-carboxylic acid *[92-70-6]* M 188.2, m 222-223°. Crystd from water or acetic acid.

1,2-Naphthaquinone *[524-42-5]* M 158.2, m 140-142°(dec). Crystd from ether (red needles) or benzene (orange leaflets).

1,4-Naphthaquinone *[130-15-4]* M 158.2, m 125-125.5°. Crystd from ethyl ether (charcoal). Steam distd. Crystd from benzene or aq EtOH. Sublimed in a vac.

β-Naphthoxyacetic acid *[120-23-0]* M 202.2, m 156°. Crystd from hot water or benzene.

β-Naphthoyltrifluoroacetone *[893-33-4]* M 254.1. Crystd from EtOH.

Naphthvalene *[34305-47-0]* M 104.1. Purified by chromatography on alumina and eluting with pentane [Abelt *et al, JACS* **107** 4148 *1985*].

1-Naphthyl acetate *[830-81-9]* M 186.2, m 45-46°. Chromatographed on silica gel.

2-Naphthyl acetate *[1523-11-1]* M 186.2, m 71°. Crystd from pet ether (b 60-80°) or dil aq EtOH.

1-Naphthylacetic acid *[86-87-3]* M 186.2, m 132°. Crystd from EtOH or water.

2-Naphthylacetic acid *[581-96-4]* M 186.2, m 143.1-143.4°. Crystd from water or benzene.

α-Naphthylamine *[134-32-7]* M 143.2, m 50.8-51.2°, b 160°. Sublimed at 120° in a stream of nitrogen, then crystd from pet ether (b 60-80°), or abs EtOH then ethyl ether. Dried under vac in an Abderhalden pistol. Has also been purified by crystn of its hydrochloride from water, followed by liberation of the free base and distn; finally purified by zone melting. **CARCINOGEN.**

β-Naphthylamine *[91-59-8]* M 143.2, m 113°. Sublimed at 180° in a stream of nitrogen. Crystd from hot water or benzene. Dried under vac in an Abderhalden pistol. **CARCINOGEN.**

α-Naphthylamine hydrochloride *[552-46-5]* M 179.7. Crystd from water (charcoal).

1-Naphthylamine-4-sulphonic acid *[84-86-6]* M 223.3, m >300°(dec),
1-Naphthylamine-5-sulphonic acid *[84-89-9]* M 223.3,
2-Naphthylamine-1-sulphonic acid *[81-16-3]* M 223.3. Crystd under nitrogen from boiling water and dried in a steam oven.

2-Naphthylamine-6-sulphonic acid *[93-00-5]* M 223.3. Crystd from a large volume of hot water.

2-Naphthylethylene *[827-54-3]* M 154.2, m 66°, b 95-96°/2.1mm, 135-137°/18mm. Crystd from aq EtOH.

***N*-(α-Naphthyl)ethylenediamine dihydrochloride** *[1465-25-4]* M 291.2, m 188-190°. Crystd from water.

β-Naphthyl lactate *[93-43-6]* M 216.2. Crystd from EtOH.

Naphthyl methyl ether see **methoxynaphthalene.**

2-(β-Naphthyloxy)ethanol *[93-20-9]* M 188.2, m 76.7°. Crystd from benzene/pet ether.

***N*-1-Naphthylphthalamic acid** *[132-66-1]* M 291.3, m 203°. Crystd from EtOH.

β-Naphthyl salicyclate *[613-78-5]* M 264.3, m 95°,
α-Naphthyl thiourea *[86-88-4]* M 202.2. Crystd from EtOH.

1-Naphthyl urea *[6950-84-1]* M 186.2, m 215-220°,
2-Naphthyl urea *[13114-62-0]* M 186.2, m 219-220°. Crystd from EtOH.

1,5-Naphthyridine *[254-79-5]* M 130.1, m 75°, b 112°/15mm. Purified by repeated sublimation.

Narcein *[131-28-2]* M 445.4, m 176-177° (145° anhydrous). Crystd from water (as trihydrate).

Narigenine *[480-41-1]* M 272.3, m 251°. Crystd from aq EtOH.

Naringin *[10236047-2]* M 580.5, m 171° (2H$_2$O), [α]$_D^{19}$ -90° (c 1, EtOH), [α]$_{546}$ -107° (c 1, EtOH). Crystd from water. Dried at 110°(to give the dihydrate).

Neopentane (2,2-dimethylpropane) *[463-82-1]* M 72.2, b 79.3°, n 1.38273, d 0.6737. Purified from isobutene by passage over conc H$_2$SO$_4$ or P$_2$O$_5$, and through silica gel.

Neopentyl glycol see **2,2-dimethyl-1,3-propanol.**

D(+)-Neopterin *[2009-64-5]* M 253.2, m >300°(dec), [α]$_{546}$ +64.5° (c 0.14, in 0.1M HCl). Purified as biopterin.

Neostigmine bromide *[114-80-7]* M 303.2, m 176°(dec). Crystd from EtOH/ethyl ether. (*Highly Toxic.*)

Neostigmine methyl sulphate *[51-60-5]* M 334.4, m 142-145°. Crystd from EtOH. (*Highly Toxic.*).

Nerolidol *[142-50-7]* M 222.4, m of semicarbazide 134-135°. Purified by thin layer chromatography on plates of kieselguhr G [McSweeney *J Chromatog* 17 183 *1965*] or silica gel plates impregnated with AgNO$_3$, using 1,2-dichloromethane/CHCl$_3$/ethyl acetate/propanol (10:10:1:1) as solvent system. Also by gas/liquid chromatography on butanediol succinate (20%) on Chromosorb W. Stored in a cool place, in an inert atmosphere, in the dark.

Neutral Red (Basic Red 5, CI 50040) *[553-24-2]* **M 288.8, m 290°(dec).** Crystd from benzene/MeOH (1:1).

New Methylene Blue N (CI 927) *[6586-05-6]* **M 416.1.** Crystd from benzene/MeOH (3:1).

Nicotinaldehyde thiosemicarbazone *[3608-75-1]* **M 180.2, m 222-223°.** Crystd from water.

Nicotinamide *[98-92-0]* **M 122.1, m 128-131°.** Crystd from benzene.

Nicotinic acid (niacin) *[59-67-6]* **M 123.1, m 232-234°.** Crystd from benzene.

Nicotinic acid hydrazide *[553-53-7]* **M 137.1, m 158-159°.** Crystd from aq EtOH or benzene.

Nile Blue A *[3625-57-8]* **M 415.5, m 138°(dec).** Crystd from pet ether.

Ninhydrin *[485-47-2]* **M 178.1, m 241-243°(dec).** Crystd from hot water (charcoal). Dried under vac and stored in a sealed brown container.

Nioxime see **cyclohexanedione dioxime.**

Nisin *[1414-45-5]* **M 3354.0.** Crystd from EtOH.

Nitrioltriacetatic acid *[139-13-9]* **191.1, m 247°(dec).** Crystd from water. Dried at 110°.

2,2',2"-Nitrilotriethanol hydrochloride see **triethanolamine hydrochloride.**

Nitrin *[63363-93-9]* **M 227-229°.** Crystd from acetone.

2-Nitroacetanilide *[552-32-9]* **M 180.2, m 93-94°.** Crystd from water.

4-Niroacetanilide *[104-04-1]* **M 180.2, m 217°.** Pptd from 80% H_2SO_4 by adding ice, then washed with water, and crystd from EtOH. Dried in air.

3-Nitroacetophenone *[121-89-1]* **M 165.2, m 81°, b 167°/18mm, 202°/760mm.** Distilled in steam and crystd from EtOH.

4-Nitroacetophenone *[100-19-6]* **M 165.2, m 80-81°, b 145-152°/760mm.** Crystd from EtOH or aq EtOH.

3-Nitroalizarin *[568-93-4]* **M 285.2, m 244°(dec).** Crystd from acetic acid.

o-**Nitroaniline** *[88-74-4]* **M 138.1, m 72.5-73.0°.** Crystd from hot water (charcoal), then crystd from water, aq 50% EtOH, or EtOH, and dried in a vac desiccator. Has also been chromatographed on alumina, then recrystd from benzene.

m-**Nitroaniline** *[99-09-2]* **M 138.1, m 114°.** Purified as for *o*-nitroaniline. **Warning: it is absorbed through the skin.**

p-**Nitroaniline** *[100-01-6]* **M 138.1, m 148-148.5°.** Purified as for *o*-nitroaniline. Also crystd from acetone. Freed from *o*- and *m*-isomers by zone melting and sublimation.

o-**Nitroanisole** *[91-23-6]* **M 153.1, f.p. 9.4°, b 265°/737mm, n 1.563, d 1.251.** Purified by repeated vac distn in the absence of oxygen.

p-**Nitroanisole** *[100-17-4]* **M 153.1, m 54°.** Crystd from pet ether or hexane and dried *in vacuo.*

9-Nitroanthracene *[602-60-8]* **M 223.2, m 142-143°.** Purified by recrystn from EtOH or MeOH. Further purified by sublimation or tlc.

5-Nitrobarbituric acid *[480-68-2]* **M 173.1, m 176°.** Crystd from water.

o-**Nitrobenzaldehyde** *[552-89-6]* **M 151.1, m 44-45°, b 120-144°/3-6mm.** Crystd from toluene (2-2.5ml/g) by addition of pet ether (b 40-60°)(7ml/ml of soln). Can also be distd at reduced pressures.

m-**Nitrobenzaldehyde** *[99-61-6]* **M 151.1, m 58°,**
p-**Nitrobenzaldehyde** *[555-16-8]* **M 151.1, m 106°.** Crystd from water or EtOH/water, then sublimed twice at 2mm pressure at a temperature slightly above its melting point.

Nitrobenzene *[98-95-3]* **M 123.1, f.p. 5.8°, b 84-86.5°/6.5-8mm, 210.8°/760mm, n^{15} 1.55457, n 1.55257, d 1.206.** Common impurities include nitrotoluene, dinitrothiophen, dinitrobenzene and aniline. Most impurities can be removed by steam distn in the presence of dil H_2SO_4, followed by drying with $CaCl_2$, and shaking with, then distilling at low pressure from BaO, P_2O_5, $AlCl_3$ or activated alumina. It can also be purified by fractional crystn from abs EtOH (by refrigeration). Another purification process includes extraction with aq 2M NaOH, then water, dil HCl, and water, followed by drying ($CaCl_2$, $MgSO_4$ or $CaSO_4$) and fractional distn under reduced pressure. The pure material is stored in a brown bottle, in contact with silica gel or CaH_2. It is very hygroscopic.

4-Nitrobenzene-azo-resorcinol (magneson II) *[74-39-5]* **M 259.2, m 199-200°.** Crystd from EtOH.

4-Nitrobenzhydrazide *[606-26-8]* **M 181.1, m 213-214°.** Crystd from EtOH.

2-Nitrobenzoic acid *[552-16-9]* **M 167.1, m 146-148°.** Crystd from benzene (twice), *n*-butyl ether (twice), then water (twice). Dried and stored in a vac desiccator. [Le Noble and Wheland *JACS* **80** 5397 *1958.*] Has also been crystd from EtOH/water.

3-Nitrobenzoic acid *[121-92-6]* **M 167.1, m 143-143.5°,**
4-Nitrobenzoic acid *[62-23-7]* **M 167.1, m 241-242°.** Crystd from benzene, water, EtOH (charcoal), glacial acetic acid or MeOH/water. Dried and stored in a vac desiccator.

4-Nitrobenzoyl chloride *[122-04-3]* **M 185.6, m 75°. b 155°/20mm.** Crystd from dry pet ether (b 60-80°) or CCl_4. Distilled under vac.

4-Nitrobenzyl alcohol *[619-73-8]* **M 153.1, m 93°.** Crystd from EtOH.

4-Nitrobenzyl bromide *[100-11-8]* **M 216.0, m 98.5-99.0°.** Recrystd four times from abs EtOH, then twice from cyclohexane/hexane/benzene (1:1:1), followed by vac sublimation at 0.1mm and a final recrystn from the same solvent mixture. [Lichtin and Rao *JACS* **83** 2417 *1961.*] Has also been crystd from pet ether (b 80-100°, 10ml/g, charcoal). It slowly decomposes even when stored in a desiccator in the dark.

m-**Nitrobenzyl chloride** *[619-23-8]* **M 171.6, m 45°.** Crystd from pet ether (b 90-120°).

p-**Nitrobenzyl chloride** *[100-14-1]* **M 171.6, m 72.5-73°.** Crystd from CCl_4, dry ethyl ether, 95% EtOH or *n*-heptane, and dried under vac.

p-**Nitrobenzyl cyanide** *[555-21-5]* **M 162.2, m 117°.** Crystd from EtOH.

4-(4-Nitrobenzyl)pyridine *[1083-48-3]* **M 214.2, m 70-71°.** Crystd from cyclohexane.

2-Nitrobiphenyl *[86-00-0]* **M 199.2, m 36.7°.** Crystd from EtOH (seeding required). Sublimed under vac.

3-Nitrocinnamic acid *[555-68-0]* **M 193.2, m 200-201°.** Crystd from benzene or EtOH.

4-Nitrocinnamic acid *[619-89-6]* **M 193.2, m 143°** (*cis*), **286°**(*trans*). Crystd from water.

N-**Nitrosodiethanolamine** *[1116-54-7]* **M 134.4.** Purified by dissolving the amine (0.5g) in 1-propanol (10ml) and 5g of anhydrous Na_2SO_4 added with stirring. After standing for 1-2hr, it was filtered and passed through a chromatographic column packed with AG 50W x 8 (H^+ form, a strongly acidic cation exchanger). The eluent and washings were combined and evapd to dryness at 35°. [Fukuda *et al, AC* **53** 2000 *1981*.] **Possible carcinogen.**

4-Nitrodiphenylamine *[836-30-6]* **M 214.2, m 133-134°.** Crystd from EtOH.

Nitrodurene *[38899-21-7]* **M 179.2, m 53-55°, b 143-144°/10mm.** Crystd from EtOH, MeOH, acetic acid, pet ether or chloroform.

Nitroethane *[79-24-3]* **M 75.1, b 115°, n 1.3920, n^{25} 1.39015, d 1.049.** Purified as described for *nitromethane.* A spectroscopic impurity has been removed by shaking with activated alumina, decanting and rapidly distilling.

2-Nitrofluorene *[607-57-8]* **M 211.2, m 156°.** Crystd from aq acetic acid.

Nitroguanine *[556-88-7]* **M 104.1, m 232°**(dec). Crystd from water (20ml/g).

5-Nitroindole *[6146-52-7]* **M 162.1, m 141-142°.** Decolorised (charcoal) and recrystd twice from aq EtOH.

Nitromesitylene *[603-71-4]* **M 165.2, m 44°, b 255°.** Crystd from EtOH.

Nitromethane *[75-52-5]* **M 61.0, f.p. -28.5°, b 101.3°, n 1.3819, n^{30} 1.37730, d 1.13749, d^{30} 1.12398.** Nitromethane is generally manufactured by gas-phase nitration of methane. The usual impurities include aldehydes, nitroethane, water and small amounts of alcohols. Most of these can be removed by drying with $CaCl_2$ or by distn to remove the water/nitromethane azeotrope, followed by drying with $CaSO_4$. Phosphorus pentoxide is not suitable as a drying agent. [Wright *et al, JCS* 199 *1936*.] The purified material should be stored by dark bottles, away from strong light, in a cool place. Purifications using extraction are commonly used. For example, Van Looy and Hammett [*JACS* **81** 3872 *1959*] mixed about 150ml of conc H_2SO_4 with 1L of nitromethane and allowed it to stand for 1 or 2days. The solvent was washed with water, aq Na_2CO_3, and again with water, then dried for several days with $MgSO_4$, filtered again with $CaSO_4$. It was fractionally distd before use. Smith, Fainberg and Winstein [*JACS* **83** 618 *1961*] washed successively with aq $NaHCO_3$, aq $NaHSO_3$, water, 5% H_2SO_4, water and dil $NaHCO_3$. The solvent was dried with $CaSO_4$, then percolated through a column of Linde type 4A molecular sieves, followed by distn from some of this material (in powdered form). Buffagni and Dunn [*JCS* 5105 *1961*] refluxed for 24hr with activated charcoal while bubbling a stream of nitrogen through the liquid. The suspension was filtered, dried (Na_2SO_4) and distd, then passed through an alumina column and redistd. It has also been refluxed over CaH_2, distd and kept under argon over 4A molecular sieves.
Can be purified by zone melting or by distn under vac at 0°, subjecting the middle fraction to several freeze-pump-thaw cycles. An impure sample containing higher nitroalkanes and traces of cyanoalkanes was purified (on the basis of its NMR spectrum) by crystn from ethyl ether at -60° (cooling in Dry-ice)[Parrett and Sun *J Chem Educ* **54** 448 *1977*].
Fractional crystn was more effective than fractional distn from Drierite in purifying nitromethane for conductivity measurements. [Coetzee and Cunningham *JACS* **87** 2529 *1965*.] Specific conductivities around 5 x 10^{-9} $ohm^{-1}cm^{-1}$ were obtained.

Nitron *[487-88-7]* M 312.4, m 189°(dec). Crystd from EtOH or chloroform.

1-Nitronaphthalene *[86-57-7]* M 173.2, m 57.3-58.3°, b 30-40°/0.01mm. Fractionally distd under reduced pressure, then crystd from EtOH, aq EtOH or heptane. Chromatographed on alumina from benzene/pet ether. Sublimes *in vacuo*.

2-Nitronaphthalene *[581-89-5]* M 173.2, m 79°, b 165°/15mm. Crystd from aq EtOH and sublimed in a vac.

1-Nitro-2-naphthol *[550-60-7]* M 189.2, m 103°. Crystd (repeatedly) from benzene/pet ether (b 60-80°)(1:1).

2-Nitro-1-naphthol *[607-24-9]* M 189.2, m 127-128°. Crystd (repeatedly) from EtOH.

5-Nitro-1,10-phenanthroline *[4199-88-6]* M 225.2, m 197-198°. Crystd from benzene/pet ether, until anhydrous.

2-Nitrophenol *[88-75-5]* M 139.1, m 44.5-45.5°. Crystd from EtOH/water, water, EtOH, benzene or MeOH/pet ether (b 70-90°). Can be steam distd. Petrucci and Weygandt [*AC* 33 275 *1961*] crystd from hot water (twice), then EtOH (twice), followed by fractional crystn from the melt (twice), drying over $CaCl_2$ in a vac desiccator and then in an Abderhalden drying pistol.

3-Nitrophenol *[554-84-7]* M 139.1, m 96°, b 160-165°/12mm. Crystd from water, $CHCl_3$, CS_2, EtOH or pet ether (b 80-100°), and dried under vac over P_2O_5 at room temperature. Can also be distd at low pressure.

4-Nitrophenol *[100-02-7]* M 139.1, m 113-114°. Crystd from water (which may be acidified, e.g. *N* H_2SO_4 or 0.5*N* HCl), EtOH, aq MeOH, $CHCl_3$, benzene or pet ether, then dried under vac over P_2O_5 at room temperature. Can be sublimed at 60°/10^{-4}mm.

2-Nitrophenoxyacetic acid *[1878-87-1]* M 197.2, m 150-159°. Crystd from water.

p-**Nitrophenyl acetate** *[830-03-5]* M 181.2. Recrystd from abs EtOH [Moss *et al, JACS* 108 5520 *1986*].

3-Nitrophenylacetic acid *[3740-52-1]* M 181.2, m 120°. Crystd from EtOH/water.

4-Nitrophenylacetic acid *[104-03-0]* M 181.2, m 80.5°. Crystd from EtOH/water (1:1), then from sodium-dried ethyl ether and dried over P_2O_5 under vac.

4-Nitrophenylacetonitrile *[555-21-5]* M 162.2, m 116-117°. Crystd from EtOH.

4-(4-Nitrophenylazo)resorcinol see **4-nitrobenzene-azo-resorcinol**.

4-Nitro-1,2-phenylenediamine *[99-56-9]* M 153.1, m 201°. Crystd from water.

4-Nitrophenylhydrazine *[100-16-3]* M 153.1, m 158°(dec). Crystd from EtOH.

3-Nitrophenyl isocyanate *[3320-87-4]* M 164.1, m 52-54°,
4-Nitrophenyl isocyanate *[100-28-7]* M 164.1, m 53°. Crystd from pet ether (b 28-38°).

4-Nitrophenyl trifluoroacetate *[4195-17-9]* M 223.2, m 93-95°. Recrystd from $CHCl_3$/hexane [Margolis *et al, JBC* 253 7891 *1078*].

2-Nitrophenylpropiolic acid *[16619-65-1]* M 191.1, m 157°(dec). Crystd from water.

4-Nitrophenyl urea *[556-10-5]* M 181.2, m 238°. Crystd from EtOH and hot water.

3-Nitrophthalic acid *[603-11-2]* M 211.1, m 216-218°. Crystd from hot water (1.5ml/g). Air dried.

4-Nitrophthalic acid *[610-27-5]* M 211.1, m 165°. Crystd from ether or ethyl acetate.

3-Nitrophthalic anhydride *[641-70-3]* M 193.1, m 164°. Crystd from benzene, benzene/pet ether, acetic actic or acetone. Dried at 100°.

1-Nitropropane *[108-03-2]* M 89.1, b 131.4°, n 1.40161, n^{25} 1.39936, d 1.004,
2-Nitropropane *[79-46-9]* M 89.1, b 120.3°, n 1.3949, n^{25} 1.39206, d 0.989. Purified as *nitromethane*.

5-Nitro-2-*n*-propoxyaniline *[553-79-7]* 196.2, m 47.5-48.5°. Crystd from *n*-propyl alcohol/pet ether.

5-Nitroquinoline *[607-34-1]* M 174.2, m 70°. Crystd from pentane, then from benzene.

8-Nitroquinoline *[706-35-2]* M 174.2, m 88-89°. Crystd from hot water, MeOH, EtOH or EtOH/ethyl ether (3:1).

4-Nitroquinoline 1-oxide *[56-57-5]* M 190.2, m 157°. Recrystd from aq acetone [Seki *et al*, *JPC* **91** 126 *1987*].

2-Nitroresorcinol *[601-89-8]* M 155.1, m 81-81°. Crystd from aq EtOH.

4-Nitrosalicylic acid *[619-19-1]* M 183.1, m 277-288°. Crystd from water.

5-Nitrosalicylic acid *[96-97-9]* M 183.1, m 233°. Crystd from acetone (charcoal), then twice more from acetone alone.

N-**Nitrosodiphenylamine** *[156-10-5]* M 198.2, m 144-145°(dec). Crystd from benzene.

1-Nitroso-2-naphthol *[131-91-9]* M 173.2, m 110.4-110.8°. Crystd from pet ether (b 60-80°, 7.5ml/g).

2-Nitroso-1-naphthol *[132-53-6]* M 173.2, m 158°(dec). Purified by recrystn from pet ether (b 60-80°) or by dissolving in hot EtOH, followed by successive addition of small volumes of water.

4-Nitroso-1-naphthol *[605-60-7]* 173.2, m 198°. Crystd from benzene.

2-Nitroso-1-naphthol-4-sulphonic acid (3H$_2$O) *[3682-32-4]* M 316.3, m 142-146°(dec). Crystd from dil HCl soln. Crystals were dried over CaCl$_2$ in a vac desiccator. Also purified by dissolution in aq alkali and pptn by addition of water.

4-Nitrosophenol *[104-91-6]* M 123.1, m >124°(dec). Crystd from xylene.

N-**Nitroso-*N*-phenylbenzylamine** *[612-98-6]* M 212.2, m 58°. Crystd from abs EtOH and dried in air.

β-**Nitrostyrene** *[102-96-5]* M 149.2, m 60°. Crystd from abs EtOH, or three times from benzene/pet ether (b 60-80°) (1:1).

4-Nitrostyrene *[100-13-0]* M 149.2, m 20.5-21°. Crystd from CHCl$_3$/hexane. Purified by addition of MeOH to ppte the polymer, then crystd at -40° from MeOH. Also crystd from EtOH [Bernasconi *et al*, *JACS* **108** 4541 *1986*].

2-Nitro-4-sulphobenzoic acid *[552-23-8]* M 247.1, m 111°. Crystd from dil HCl.

2-Nitrotoluene *[88-72-2]* M 137.1, m -9.55° (α-form), -3.85° (β-form), b 118°/16mm, 222.3°/760mm, n 1.545, d 1.163. Crystd (repeatedly) from abs EtOH by cooling in a Dry-ice/alcohol mixture, Further purified by passage of an alcoholic soln through a column of alumina.

3-Nitrotoluene *[99-08-1]* M 137.1, m 16°, b 113-114°/15mm, 232.6°, n 1.544, d 1.156. Dried with P_2O_5 for 24hr, then fractionally distd under reduced pressure. [Org. Synth **Vol I** 416 *1948*.]

4-Nitrotoluene *[99-89-0]* M 137.1, m 52°. Crystd from EtOH, MeOH/water, EtOH/water (1:1) or MeOH. Air dried, then dried in a vac desiccator over H_2SO_4. [Wright and Grilliom *JACS* **108** 2340 *1986*.]

Nitrourea *[556-89-8]* M 105.1, m 158.4-158.8°(dec). Crystd from EtOH/pet ether.

Nonactin *[6833-84-7]* M 737.0, m 147-148°, [α] 0 ±2° (c 1.2, $CHCl_3$). Crystd from MeOH.

n-**Nonane** *[111-84-2]* M 126.3, b 150.8°, n 1.40542, n^{25} 1.40311, d 0.719. Fractionally distd, then stirred with successive volumes of conc H_2SO_4 for 12hr each until no further colouration was observed in the acid layer. Then washed with water, dried with $MgSO_4$ and fractionally distd. Alternatively, it was purified by azeotropic distn with 2-ethoxyethanol, followed by washing out the alcohol with water, drying and distilling. [Forziati *et al, J Res Nat Bur Stand* **36** 129 *1946*.]

Nonanoic acid see **pelargonic acid.**

2,5-Norbornadiene *[121-46-0]* M 92.1, b 89°, n 1.4707- d 0.854. Purified by distn from activated alumina [Landis and Halpern *JACS* **109** 1746 *1987*].

Norbornylene *[498-66-8]* M 94.2, m 44-46°, b 96°. Refluxed over Na, and distd [Gilliom and Grubbs *JACS* **108** 733 *1986*]. Also purified by sublimation *in vacuo* onto an ice-cold finger [Woon *et al, JACS* **108** 7990 *1986*].

Norcamphor (bicyclo[2.2.1]heptan-2-one) *[497-38-1]* M 110.2, m 94-95°. Crystd from water.

Norcholanic acid *[511-18-2]* M 346.5, m 177°. Crystd from acetic acid.

Norcodeine *[467-15-2]* M 285.3, m 185°. Crystd from acetone or ethyl acetate.

Nordihydroguaiaretic acid *[500-38-9]* M 302.4, m 184-185°. Crystd from dil acetic acid.

Norleucine *[R: 327-56-0] [S: 327-57-1]* M 117.2, m 301° [α]$_{546}$ ±28° (c 5, 5M HCl); *[RS: 616-06-8]* m 297-300°. Crystd from water.

Norvaline *[R: 2031-12-9] [S: 6600-40-4]* M 117.2, m 305°(dec), [α]$_{546}$ ±25° (c 10, 5M HCl). Crystd from aq EtOH or water.

Novobiocin *[303-81-1]* M 612.6, two forms m 152-156° and m 174-178°, $\lambda_{max.}$330nm (acid EtOH), 305nm (alk EtOH). Crystd from EtOH and stored in the dark. The sodium salt can be crystd from MeOH, then dried at 60°/0.5mm. [Sensi, Gallo and Chiesa, *AC* **29** 1611 *1957*.]

Nuclear Fast Red *[6409-77-4]* M 357.3, m >290°(dec). λ_{max} 518nm. A soln of 5g of the dye in 250ml of warm 50% EtOH was cooled to 15° for 36hr, then filtered on a Büchner funnel, washed with EtOH until the washings were colourless, then with 100ml of ethyl ether and dried over P_2O_5. [Kingsley and Robnett *AC* **33** 552 *1961*.]

Nylon powder. Pellets were dissolved in ethylene glycol under reflux. Then pptd as a white powder on addition of EtOH at room temperature. This was washed with EtOH and dried at 100° under vac.

n-**Octacosane** *[630-02-4]* **M 394.8, m 62.5°.** Purified by forming its adduct with urea, washing and crystallising from acetone/water. [McCubbin *TFS* **58** 2307 *1962*.] Crystd from hot, filtered isopropyl ether soln (10ml/g).

n-**Octadecane** *[593-45-3]* **M 254.5, m 28.1°, b 173.5°/10mm, 316.1°/760mm, n 1.4390. d 0.7768.** Crystd from acetone and distd under reduced pressure from sodium.

n-**Octadecanoic acid** *[124-07-2]* **M 144.2, m 16.7-17°, b 144-145°/27mm, n 1.428, d 0.911.** Fractionally crystd by partial freezing. Dried with Linde type 4A molecular sieves and fractionally distd under reduced pressure.

1-Octadecanol see *n*-octadecyl alcohol.

Octadecyl acetate *[822-23-1]* **M 312.5, m 32.6°.** Distd under vac, then crystd from ethyl ether/MeOH.

n-**Octadecyl alcohol** *[112-92-5]* **M 270.5, m 61°, b 153-154°/0.3mm.** Crystd from MeOH, or dry ethyl ether and benzene, then fractionally distd under reduced pressure. Purified by column chromatography. Freed from cetyl alcohol by zone melting.

Octadecyl ether *[6297-03-6]* **M 523.0, m 59.4°.** Vac distd, then crystd from MeOH/benzene.

Octadecyltrimethylammonium bromide *[1120-02-1]* **M 392.5.** Recrystd from EtOH. Dried in a vac desiccator.

2,3,7,8,12,13,17,18-Octaethylporphin *[2683-82-1]* **M 534.8.** Chromatographed on SiO_2 using $CHCl_3$ as eluent.

Octafluoropropane (profluorane) *[76-19-7]* **M 188.0, b -38°.** Purified for pyrolysis studies by passage through a copper vessel containing CoF_3 at about 270°, then fractionally distd. [Steunenberg and Cady *JACS* **74** 4165 *1952*.]

1,2,3,4,6,7,8,9-Octahydroanthracene *[1079-71-6]* **M 186.3, m 78°.** Crystd from EtOH, then zone refined.

Octamethylcyclotetrasiloxane *[556-67-2]* **M 296.6, m 17.3°, b 175-176°, n 1.396, d 0.957.** Purified by zone melting.

Octan-1,8-diol *[629-41-4]* **M 146.2, m 59-61°, b 172°/0.2mm.** Recrystd from EtOH.

Octan-4,5-diol see **dipropylene glycol.**

n-**Octane** *[111-65-9]* **M 114.2, b 126.5°, n 1.39743, n^{25} 1.39505, d 0.704.** Extracted repeatedly with conc H_2SO_4 or chlorosulphonic acid, then washed with water, dried and distd. Also purified by azeotropic distn with EtOH, followed by washing with water to remove the EtOH, drying and distilling. For further details, see n-*heptane*. Also purified by zone melting.

1-Octanethiol *[111-88-6]* **M 146.3, b 86°/15mm, 197-200°/760mm. n 1.4540, d 0.8433.** Passed through a column of alumina [Battacharyya *et al, JCSFT1* **82** 135 *1986*].

Octaphenylcyclotetrasiloxane *[546-56-5]* **M 793.2, m 201-202°, b 330-34°/760mm.** Crystd from benzene/EtOH or glacial acetic acid.

1-Octene *[111-66-0]* **M 112.2, b 121°/742mm, n 1.4087, d 0.716,**

(*trans*)-2-Octene *[13389-42-9]* **M 112.2, b 124-124.5°/760mm, n 1.4132, d 0.722.** Distd under nitrogen from sodium. [Removes water and peroxides.] Peroxides can also be removed by percolation through dried, acid washed alumina. Stored under nitrogen in the dark. [Strukul and Michelin *JACS* **107** 7563 *1985.*]

n-Octyl alcohol *[111-87-5]* **M 130.2, b 98°/19mm, 195.3°/760mm, n 1.43018, d 0.828.** Fractionally distd under reduced pressure. Dried with sodium and again fractionally distd. or refluxed with boric anhydride and distd (**b** 195-205°/5mm), the distillate being neutralised with NaOH and again fractionally distd. Also purified by distn from Raney nickel and by preparative glc.

n-Octylammonium 9-anthanilate *[88020-99-9]* **M 351.5, m 134-135°.** Recrystd several times from ethyl acetate.

n-Octylammonium hexadecanoate *[88020-97-7]* **M 385.7, m 52-53°.** Purified by several recrystns from *n*-hexane or ethyl acetate. The solid was then washed with cold anhydrous ethyl ether, and dried *in vacuo* over P_2O_5.

n-Octylammonium octadecanoate *[32580-92-0]* **M 413.7, m 56-57°.** Purified as for the *hexanoate* above.

n-Octylammonium tetradecanoate *[544-61-8]* **M 358.6, m 46-48°.** Purified as for the *hexanoate* above.

4-Octylbenzoic acid *[3575-31-3]* **M 234.3, m 99-100°.** Crystd from EtOH has **m** 139°; crystd from aq EtOH has **m** 99-100°.

n-Octyl bromide *[111-83-1]* **M 193.1, b 201.5°, n^{25} 1.4503, d 1.118.** Shaken with H_2SO_4, washed with water, dried with K_2CO_3 and fractionally distd.

4-(*tert*-Octyl)phenol *[140-66-9]* **M 206.3, m 85-86°.** Crystd from *n*-hexane.

1-Octyne *[629-05-0]* **M 110.2, b 126.2°/760mm, n^{25} 1.1.4159, d 0.717.** Distd from $NaBH_4$ to remove peroxides.

β-Oestradiol *[50-28-2]* **M 272.4, m 178°, $[\alpha]^{18}$ +78° (EtOH).** Crystd from 80% aq EtOH.

β-Oestradiol benzoate *[50-50-0]* **M 376.5, m 194-195°, $[\alpha]_{546}$ +70° (c 2, dioxane),**
Oestrone *[53-16-7]* **M 270.4, m 256°.** Crystd from EtOH.

Oleic acid *[112-80-1]* **M 282.5, m 16°, b 360°(dec), n^{30} 1.4571, d 0.891.** Purified by fractional crystn from its melt, followed by molecular distn at 10^{-3}mm, or by conversion to its methyl ester, the free acid can be crystd from acetone at -40° to -45° (12ml/g). For purification by the use of lead and lithium salts, see Keffler and McLean [*JCS Ind (London)* **54** 176T *1935*].
Purification based on direct crystn from acetone is described by Brown and Shinowara [*JACS* **59** 6 *1937*].

Oleyl alcohol *[142-28-2]* **M 268.5, b 182-184°/1.5mm, $n^{27.5}$ 1.4582, d 0.847.** Purified by fractional crystn at -40° from acetone, then distd under vac.

Opianic acid (2-formyl-4,5-dimethylbenzoic acid) *[519-05-1]* **M 210.2, m 150°.** Crystd from water.

Orcinol *[504-15-4]* **M 124.2, m 107.5°, (H_2O) m 59-61°.** Crystd from $CHCl_3$/benzene (2:3).

L-Ornithine *[70-26-8]* **M 132.2, m 140°, $[\alpha]_D^{25}$ +16° (c 0.5, H_2O).** Crystd from water containing 1mM EDTA (to remove metal ions).

L-Ornithine monohydrochloride *[3184-13-2]* **M 168.6, $[\alpha]_D^{25}$ +28.3° (5M HCl).** Likely impurities are citrulline, arginine and D-ornithine. Crystd from water by adding 4 volumes of EtOH.

Orotic acid (H_2O) *[50887-69-9]* **M 174.1, m 235-346°(dec).** Crystd from water.

Orthanilic acid (2-aminobenzenesulphonic acid) *[88-21-1]* **M 173.2, m >300°(dec).** Crystd from aq soln, containing 20ml of conc HCl per L, then crystd from distd water.

Ouabain *[630-60-4]* **M 728.8, m 180°(dec), $[\alpha]_{546}$ -30° (c 1, H_2O).** Crystd from water. Dried at 130°. Stored in the dark.

Oxalic acid ($2H_2O$) *[630-60-4]* **M 90.0, m 101.5° ; *[anhydrous 144-62 -7]* m 189.5°.** Crystd from dist water. Dried in vac over H_2SO_4. The anhydrous acid can be obtained by drying at 100° overnight.

Oxaloacetic acid *[328-42-7]* **M 132.1, m 160°(decarboxylates).** Crystd from boiling ethyl acetate, or from hot acetone by addition of hot benzene.

Oxamide *[471-46-5]* **M 88.1, m >320°(dec).** Crystd from water. Ground. Dried in an oven at 150°.

Oxamycin see **R-4-amino-3-isoxazolidone.**

2-Oxazolidone *[497-25-6]* **M 89-90°.** Crystd from benzene.

2-Oxaloglutaric acid *[328-50-7]* **M 146.1, m 114°.** Crystd repeatedly from acetone/benzene.

2-Oxohexamethyleneimine see **ε-caprolactam.**

Oxalylindigo *[2533-00-8]* **M 316.3.** Recrystd twice from nitrobenzene and dried by heating *in vacuo* for several hours. [Sehanze *et al, JACS* **108** 2646 *1986*].

Oxetane (1.3-trimethylene oxide) *[503-30-0]* **M 58.1, b 48°/760mm, n 1.395, d 0.892.** Distd from sodium metal. Also purified by preparative gas chromatography using a 2m silica gel column.

Oxine Blue *[3733-85-5]* **M 369.4, m 134-135°.** Recrystd from EtOH. Dried over H_2SO_4.

2,2'-Oxydiethanol see **diethylene glycol.**

Palmitic acid *[57-10-3]* **M 256.4, m 63-64°.** Crystd from EtOH. Purified *via* the methyl ester (**b 193-194°/12mm, n^{35} 1.4359**) as for *capric acid*, or by zone melting.

R-Pantothenic acid *[867-81-2]* **M 241.2, m 122-124°, [α]$_D^{25}$ +27° (c 5, H$_2$O).** Crystd from EtOH.

Papain see Chapter 6.

Paraffin (oil) *[8012-95-1]* **n 1.482, d 0.880.** Treated with fuming H$_2$SO$_4$, then washed with water and dil aq NaOH, then percolated through activated silica gel.

Paraffin Wax. Melted in the presence of NaOH, washed with water until all of the base had heen removed. The paraffin was allowed to solidify after each wash. Finally, 5g of paraffin was melted by heating on a water-bath, then shaken for 20-30min with 100ml of boiling water and fractionally crystd.

Paraldehyde *[30525-89-4]* **M 132.2, m 12.5°, 124°, n 1.407, d 0.995.** Washed with water and fractionally distd.

Patulin *[149-29-1]* **M 154.1, m 110°.** Crystd from ethyl ether or chloroform. (*Highly Toxic*).

Pavatrine hydrochloride *[548-65-2]* **M 333.7, m 143-144°.** Recrystd from isopropanol, and dried over P$_2$O$_5$ under vac.

Pectic acid see Chapter 6.

Pectin see Chapter 6.

Pelargonic acid (nonanoic acid) *[112-05-0]* **M 158, m 15°, b 98.9°/1mm, 225°/760mm.** Esterified with ethylene glycol and distd. (This removes dibasic acids as undistillable residues.) The acid was regenerated by hydrolysing the ester.

Penicillic acid *[90-65-3]* **M 158.2, m 58-64° (H$_2$O), 83-84° (anhydrous).** Crystd from water as the monohydrate, or from pet ether.

Pentaacetyl-α-D-glucopyranose *[604-68-2]* **M 390.4, m 112°, [α]$_{546}$ +119° (c 5, CHCl$_3$),**
Pentaacetyl-β-D-glucopyranose *[604-69-3]* **M 390.4, m 131°, [α]$_{546}$ +5°(c 5, CHCl$_3$).** Crystd from EtOH.

Pentabromoacetone *[79-49-2]* **M 452.6, m 76°.** Crystd from ethyl ether or EtOH.

Pentabromophenol *[608-71-9]* **M 488.7, m 229°.** Purified by crystn (charcoal) from toluene then from CCl$_4$. Dried for 2 weeks at *ca* 75°.

1-Pentacene *[13360-61-7]* **M 278.4, m 300°.** Crystd from benzene.

Pentachloroethane (pentalin) *[76-01-7]* **M 202.3, b 69°/37mm, 152.2°/64mm, 162.0°, n^{15} 1.50542, d 1.678.** Usual impurities include trichloroethylene. Partially decomposes if distd at atmospheric pressure. Drying with CaO, KOH or sodium is unsatisfactory because HCl is split off. It can be purified by steam distn, or by washing with conc H$_2$SO$_4$, water, and then aq K$_2$CO$_3$, drying with solid K$_2$CO$_3$ or CaSO$_4$, and fractionally distd under reduced pressure.

Pentachloronitrobenzene *[82-68-8]* **M 295.3, m 146°.** Crystd from EtOH.

Pentachlorophenol *[87-86-5]* **M 266.3, m 190-191°.** Twice crystd from toluene/EtOH. Sublimed *in vacuo*.

Pentachlorothiophenol *[133-49-3]* **M 282.4.** Crystd from benzene.

Pentadecanoic acid *[1002-84-2]* **242.4, m 51-53°, b 158°/1mm, 257°/760mm, d^{80} 0.8424.** Purification as for hexadecanoic acid.

Penta-1,3-diene *[cis: 1574-41-0], [trans: 2004-70-8]* **M 68.1, b 42°, n 1.4316, d 0.680,**
Penta-1,4-diene *[591-93-5]* **M 68.1, b 25.8-26.2°/756mm, n 1.3890, d 0.645.** Distd from $NaBH_4$. Purified by preparative gas chromatography. [Reimann *et al*, *JACS* **108** 5527 *1986*.]

Penta-2,4-dione see **acetylacetone**.

Pentaerythritol *[115-77-5]* **M 136.2, m 260.5°.** Refluxed with an equal vol of MeOH, then cooled and the ppte dried at 90°. Crystd from dil aq HCl. Sublimed under vac at 200°.

Pentaerythritol tetraacetate *[597-71-7]* **M 304.3, m 78-79°.** Crystd from hot water, then leached with cold water until the odour of acetic acid was no longer detectable.

Pentaerythrityl laurate *[13057-50-6]* **M 864.6, m 50°.** Crystd from pet ether.

Pentaerythrityl tetranitrate. *[78-11-5]* **M 316.2, m 140.1°.** Crystd from acetone or acetone/EtOH. **EXPLOSIVE.**

Pentaethylenehexamine *[4067-16-7]* **M 232.4.** Fractionally distd twice at 10-20mm, the fraction boiling at 220-250° being collected. Its soln in MeOH (40ml in 250ml) was cooled in an ice-bath and conc HCl was added dropwise with stirring. About 50ml was added, and the pptd hydrochloride was filtered off, washed with acetone and ethyl ether, then dried in a vac desiccator. [Jonassen *et al*, *JACS* **79** 4279 *1957*.]

2,2,3,3,3-Pentafluoropropan-1-ol *[422-05-9]* **M 150.1, b 80°, n 1.288, d 1.507.** Shaken with alumina for 24hr, dried with anhydrous K_2CO_3, and distd, collecting the middle fraction (b 80-81°) and redistilling.

2',3,4',5,7-Pentahydroxyflavone see **Morin**.

Pentamethylbenzene *[700-12-9]* **M 148.3, m 53.5-55.1°.** Successively crystd from abs EtOH, toluene and MeOH, and dried under vac. [Rader and Smith *JACS* **84** 1443 *1962*.] It has also been crystd from benzene or aq EtOH, and sublimed.

1,5-Pentamethylenetetrazole *[54-95-5]* **M 138.2, m 60-61°, b 194°/12mm.** Crystd from ethyl ether.

n-**Pentane** *[109-66-0]* **m 72.2, b 36.1°, n 1.35748, n^{25} 1.35472, d 0.626.** Stirred with successive portions of conc H_2SO_4 until there was no further coloration during 12hr, then with 0.5*N* $KMnO_4$ in 3M H_2SO_4 for 12hr, washed with water and aq $NaHCO_3$. Dried with $MgSO_4$ or Na_2SO_4, then P_2O_5 and fractionally distd through a column packed with glass helices. It was also purified by passage through a column of silica gel, followed by distn and storage with sodium hydride. An alternative purification is by azeotropic distn with MeOH, which is subsequently washed out from the distillate (using water), followed by drying and distn. For removal of carbonyl-containing impurities, see n-*heptane*.
Also purified by fractional freezing (*ca* 40%) on a copper coil through which cold air was passed, then washed with conc H_2SO_4 and fractionally distd.

2,4-Pentanedione see **acetylacetone**.

Pentane-1-thiol *[110-66-7]* **M 104.2, b 122.9°/697.5mm, d^{25} 0.8375.** Dissolved in aq 20% NaOH, then extracted with a small amount of ethyl ether. The soln was acidified slightly with 15% H_2SO_4, and the thiol was distd out, dried with $CaSO_4$ or $CaCl_2$, and fractionally distd under nitrogen. [Ellis and Reid *JACS* **54** 1674 *1932*.]

Pentan-1-ol see *n*-amyl alcohol.

Pentan-2-ol *[6032-29-7]* M 88.2, b 119.9°, n 1.41787, n^{25} 1.4052, d 0.810.
Pentan-3-ol *[684-02-1]* M 88.2, b 116.2°, n^{25} 1.4072, d 0.819. Refluxed with CaO, distd, refluxed with magnesium and again fractionally distd.

Pentan-3-one *[96-22-0]* M 86.1, b 101.7°, n 1.39240, n^{25} 1.39003, d 0.813. Refluxed with $CaCl_2$ for 2hr, then left overnight with fresh $CaCl_2$, filtered and distd.

Pentaquine monophosphate *[5428-64-8]* M 395.6, m 189-190°. Crystd from 95% EtOH.

Pent-2-ene (mixed isomers) *[109-68-2]* M 70.1, b 36.4°, n 1.38003, n^{25} 1.3839, d 0.650. Refluxed with sodium wire, then fractionally distd twice through a Fenske column.

cis-**Pent-2-ene** *[627-20-3]* M 70.1, b 37.1°, n 1.3830, n^{25} 1.3798, d 0.657. Dried with sodium wire and fractionally distd, or purified by azeotropic distn with MeOH, followed by washing out the MeOH with water, drying and distilling. Also purified by chromatography through silica gel and alumina [Klassen and Ross *JPC* **91** 3668 *1987*].

trans-**Pent-2-ene** *[646-04-8]* M 70.1, b 36.5°, n 1.3793, d 0.6482. It was treated as above and washed with water, dried over anhydrous Na_2CO_3, and fractionally distd. The middle cut was purified by two passes of fractional melting.

Pentobarbital *[76-74-4]* M 226.4. Soln of the sodium salt in 10% HCl was prepared and the acid was extracted by addition of ether. Then purified by repeated crystn from $CHCl_3$. [Bucket and Sandorfy *JPC* **88** 3274 *1984*.]

Pentyl- see *tert*-butyl.

tert-**Pentyl** *see tert*-amyl.

neo-**Pentyl alcohol** see 2,2-dimethyl-1-propanol.

Pent-2-yne *[627-21-4]* M 68.1, b 26°/2.4mm, n^{25} 1.4005, d 0.710. Stood with, then distd at low pressure from, sodium or $NaBH_4$.

Pepsin see Chapter 6.

Perbenzoic acid *[93-59-4]* M 138.1, m 41-43°. Crystd from benzene. Readily sublimed.

Perchlorobutadiene *[87-68-3]* M 260.8, b 144.1°/100mm, 210-212°/760mm, n 1.5556, d 1.683. Washed with four or five 1/10th volumes of MeOH (or until the yellow colour has been extracted), then stirred for 2hr with H_2SO_4, washed with distd water until neutral and filtered through a column of P_2O_5. Distd under reduced pressure through a packed column. [Rytner and Bauer *JACS* **82** 298 *1960*.]

Perfluorobutyric acid *[375-22-4]* M 214.0, m -17.5°, b 120°/735mm, n^{16} 1.295, d 1.651. Fractionally distd twice in an Oldershaw column with an automatic vapour-dividing head, the first distn in the presence of conc H_2SO_4 as a drying agent.

Perfluorocyclobutane *[115-25-3]* M 200.0, m -40°, b -5°, d^{-20} 1.654, d^0 1.72. Purified by trap-to-trap distn, retaining the middle portion.

Perfluorocyclohexane *[355-68-0]* **M 300.1, m 51° (sublimes), b 52°.** Extracted repeatedly with MeOH, then passed through a column of silica gel (previously activated by heating at 250°).

Perfluoro-1,3-dimethylcyclohexane *[335-27-3]* **M 400.1, b 101°, n 1.300, d 1.829.** Fractionally distd, then 35ml was sealed with about 7g KOH pellets in a borosilicate glass ampoule and heated at 135° for 48hr. The ampoule was cooled and opened, and the liquid was resealed with fresh KOH in another ampoule and heated as before. This process was continued until no further decomposition was observed. The substance was then washed with distd water, dried (CaSO$_4$) and distd. [Grafstein *AC* 26 523 *1954*.]

Perfluoroheptane *[335-57-9]* **M 388.1, b 99-101°, d^{25} 1.72006.** Purified as for *perflurodimethylhexane*. Other procedures include shaking with H$_2$SO$_4$, washing with water, drying with disodium for 48hr and fractionally distilling. Alternatively, it has been refluxed for 24hr with satd acid KMnO$_4$ (to oxidise and remove hydrocarbons), then neutralised, steam distd, dried with P$_2$O$_5$, and passed slowly through a column of dry silica gel. It has been purified by fractional crystn, using partial freezing.

Perfluoro-*n*-hexane *[355-42-0]* **M 338.1, m -4°, b 58-60°, d 1.684.** Purified by fractional freezing. The methods described for *perfluoroheptane* should be applicable here.

Perfluoro(methylcyclohexane) *[355-02-2]* **M 350.1, b 76.3°, d^{25} 1.78777.** Refluxed for 24hr with satd acid KMnO$_4$ (to oxidise and remove hydrocarbons), then neutralised, steam distd, dried with P$_2$O$_5$ and passed slowly through a column of dry silica gel. [Glew and Reeves *JPC* 60 615 *1956*.] Also purified by percolation through a 1m neutral activated alumina column, and ^{1}H-impurities checked by nmr.

Perfluorononane *[375-96-2]* **M 488.1.** Purified as for *perfluorodimethylcyclohexane*.

Perfluoropropane *[76-19-7]* **M 188.0.** Purified by several trap-to-trap distns.

Perfluorotripropylamine *[338-83-0]* **M 521.1.** Purified as for *perfluorodimethylcyclopropane*.

Pericyazine *[2622-26-6]* **M 365.4.** Recrystd from a satd soln in cyclohexane.

Perylene *[198-55-0]* **M 252.3, m 273-274°.** Purified by silica-gel chromatography of its recrystd picrate. [Ware *JACS* 83 4374 *1961*.] Crystd from benzene, toluene or EtOH and sublimed in a flow of oxygen-free nitrogen. [Gorman *et al, JACS* 107 4404 *1985*; Johansson *et al, JACS* 109 7374 *1987*.]

Petroleum ether *[8032-32-4]* **b 35-60°, n 1.363, d 0.640.** Shaken several times with conc H$_2$SO$_4$, then 10% H$_2$SO$_4$ and conc KMnO$_4$ (to remove unsatd, including aromatic, hydrocarbons) until the permanganate colour persists. Washed with water, aq Na$_2$CO$_3$ and again with water. Dried with CaCl$_2$ or Na$_2$SO$_4$, and distd. It can be dried further using CaH$_2$ or sodium wire. Passage through a column of activated alumina, or treatment with CaH$_2$ or sodium, removes peroxides. For the elimination of carbonyl-containing impurities without using permanganate, see *n-heptane*. These procedures could be used for all fractions of pet ethers.

R(-)-α-Phellandrene *[4221-98-1]* **M 136.2, b 61°/11mm, 175-176°/760mm, n 1.471, d 0.838.** Purified by gas chromatography on an Apiezon column.

Phenacetin see *p*-acetophenetidine.

Phenacylamine hydrochloride *[5468-37-1]* **M 171.6, m 194°(dec).** Recrystd from 2-propanol [Castro *JACS* 108 4179 *1986*].

Phenacyl bromide see **2-bromoacetophenone**.

Phenanthrene *[85-01-8]* **M 178.2, m 98°.** Likely contaminants include, anthracene, carbazole, fluorene and other polycyclic hydrocarbons. Purified by distn from sodium, boiling with maleic anhydride in xylene, crystn from acetic acid, sublimation and zone melting. Has also been recrystd repeatedly from EtOH benzene or pet ether (b 60-70°), with subsequent drying under vac over P_2O_5 in an Abderhalden pistol. Feldman, Pantages and Orchin [*JACS* 73 4341 *1951*] separated from most of the anthracene impurity by refluxing phenanthrene (671g) with maleic anhydride (194g) in xylene (1.25L) under nitrogen for 22hr, then filtered. The filtrate was extracted with aq 10% NaOH, the organic phase was separated, and the solvent was evaporated. The residue, after stirring for 2hr with 7g of sodium, was vac distd, then recrystd twice from 30% benzene in EtOH, then dissolved in hot glacial acetic acid (2,2ml/g), slowly adding an aq soln of CrO_3 (60g in 72ml H_2O added to 2.2L of acetic acid), followed by slow addition of conc H_2SO_4 (30ml). The mixture was refluxed for 15min, diluted with an equal volume of water and cooled. The ppte was filtered off, washed with water, dried and distd., then recrystd twice from EtOH. Further purification is possible by chromatography from $CHCl_3$ soln on activated alumina, with benzene as eluent, and by zone refining.

Phenanthrene-9-aldehyde *[4707-71-5]* **M 206.3, m 102.2-103°/12mm.** Crystd from EtOH and sublimed at 95-98°/0.07mm.

9,10-Phenanthrenequinone *[84-11-7]* **M 208.2, m 208°.** Crystd from dioxane or 95% EtOH and dried under vac.

Phenanthridine *[229-87-8]* **M 179.2, m 106.5°.** Purified *via* the $HgCl_2$ addition compound formed when phenanthridine (20g) in 1:1 HCl (100ml) was added to aq $HgCl_2$ (60g in 3L), and the mixture was heated to boiling. Conc HCl was then added until all of the solid had dissolved. The compound separated on cooling, and was decomposed with strong aq NaOH (*ca* 5M). Phenanthridine was extracted with ethyl ether and crystd from pet ether (b 80-100°) or ethyl acetate. [Cumper, Ginman and Vogel *JCS* 45218 *1962*.] Also purified by zone melting.

1,10-Phenanthroline *[66-71-7]* **M 198.2, m 108-110° (H_2O), 118° (anhydrous).** Crystd as its picrate (m 191°) from EtOH, then the free base was liberated, dried at 78°/8mm over P_2O_5 and crystd from pet ether (b 80-100°). [Cumper, Ginman and Vogel *JCS* 1188 *1962*.] It can be purified by zone melting. Also crystd from hexane, benzene/pet ether (b 40-60°) or sodium-dried benzene, dried and stored over H_2SO_4. The monohydrate is obtained by crystn from aq EtOH or ethyl acetate.

4,7-Phenanthroline-5,6-dione *[84-12-8]* **M 210.2, m 295°(dec).** Crystd from MeOH.

Phenazine *[92-82-0]* **M 180.2, m 171°.** Crystd from EtOH, $CHCl_3$ or ethyl acetate, after pretreatment with activated charcoal. It can be sublimed *in vacuo*, and zone refined.

Phenazine monosulphate *[299-11-6]* **M 306.3, m 155-157° (or 198° dec on rapid heating).** Crystd from EtOH (charcoal).

Phenethylamine *[64-04-0]* **M 121.2, b 87°/13mm, n 1.535, d 0.962.** Distd from CaH_2, under reduced pressure, just before use.

Phenethyl bromide *[103-63-9]* **M 185.1, b 92°/11mm, n 1.557, d 1.368.** Washed with conc H_2SO_4, water aq 10% Na_2CO_3 and water again, then dried with $CaCl_2$ and fractionally distd just before use.

Phenethyl urea *[2158-04-5]* **M 164.2, m 173-174°.** Crystd from water.

Phenetole *[103-73-1]* **M 122.2, b 60°/9mm, 77.5°/31mm, 170.0°/760mm, n 1.50735, n^{25} 1.50485, d 0.967.** Small quantities of phenol can be removed by shaking with NaOH, but this is not a very likely contaminant of commercial material. Fractional distn from sodium, at low pressures, probably gives adequate purification. It can be dissolved in ethyl ether and washed with 10% NaOH (to remove phenols), then water. The ethereal soln was evapd and the phenetole fractionally distd under vac.

Phenocoll hydrochloride (*p*-phenetidine HCl) *[536-10-6]* **M 230.7, m 234°.** Crystd from water. Sublimes *in vacuo*.

Phenol *[108-95-2]* **M 94.1, m 40.9°, b 85.5-86.0°/20mm, 180.8°/760mm, n^{41} 1.54178, n^{46} 1.53957, d 1.06.** Steam was passed through a boiling soln containing 1mole of phenol and 1.5-2.0moles of NaOH in 5L of water until all non-acidic naterial had distd. The residue was cooled, acidified with 20% (v/v) H$_2$SO$_4$, and the phenol was separated, dried with CaSO$_4$ and fractionally distd under reduced pressure. It was then fractionally crystd several times from its melt, [Andon *et al, JCS* 5246 *1960*.] Purification *via* the benzoate has been used by Berliner, Berliner and Nelidow [*JACS* 76 507 *1954*.] The benzoate was crystd from 95% EtOH, then hydrolysed to the free phenol by refluxing with two equivalents of KOH in aq EtOH until the soln became homogeneous. It was acidified with HCl and extracted with ethyl ether. The ether layer was freed from benzoic acid by thorough extraction with aq NaHCO$_3$, and, after drying and removing the ether, the phenol was distd.

Phenol has also been crystd from a 75% w/w soln in water by cooling to 11° and seeding with a crystal of the hydrate. The crystals were centrifuged off, rinsed with cold water (0-2°) satd with phenol, and dried. It can be crystd from pet ether [Berasconi and Paschalis *JACS* 108 2969 *1986*].

Draper and Pollard [*Science* 109 448 *1949*] added 12% water, 0.1% aluminium (can also use zinc), and 0.05% NaHCO$_3$ to phenol, and distd at atmospheric pressure until the azeotrope was removed, The phenol was then distd at 25mm. Phenol has also been dried by distn from the benzene soln to remove the water-benzene azeotrope and the excess benzene, followed by distn of the phenol at reduced pressure under nitrogen. Processes such as this are probably adequate for analytical grade phenol which has as its main impurity water. Phenol has also been crystd from pet ether/benzene or pet ether (b 40-60°). Purified material is stored in a vac desiccator over P$_2$O$_5$ or CaSO$_4$.

Phenol-2,4-disulphonic acid *[96-77-5]* **M 254.2.** Crystd from EtOH/ethyl ether.

Phenolphthalein *[787-09-8]* **M 319.2, m 263°.** Dissolved in EtOH (7ml/g), then diluted with eight volumes of cold water. Filtered. Heated on a water-bath to remove most of the alcohol and the pptd phenolphthalein was filtered off and dried under vac.

Phenolphthalol *[81-92-5]* **M 306.3, m 201-202°.** Crystd from aq EtOH.

Phenosafranine *[81-93-6]* **M 322.8, λ$_{max}$ 530nm (H$_2$O).** Crystd from dil HCl.

Phenothiazine *[92-84-2]* **M 199.3, m 184-185°.** Crystd from benzene or toluene (charcoal) after boiling for 10min under reflux. Filtered on a suction filter. Dried in an oven at 100°, then in a vac desiccator over paraffin chips. Also twice recrystd from water and dried in an oven at 100° for 8-10hr.

Phenoxazine *[135-67-1]* **M 199.2, m 156°.** Crystd from EtOH and sublimed *in vacuo*.

Phenoxyacetic acid *[122-59-8]* **M 152.2, m 98-99°.** Crystd from water or aq EtOH.

4-Phenoxyaniline *[139-59-3]* **M 185.2, m 95°.** Crystd from water.

Phenoxybenzamine *[59-96-1]* **M 303.5, hydrochloride** *[63-92-3]* **M 340.0, m 137.5-140°.** Crystd from EtOH/ethyl ether.

2-Phenoxybenzoic acid *[2243-42-7]* **M 214.2, m 113°, b 355°/760mm,**
3-Phenoxybenzoic acid *[3739-38-6]* **M 214.2, m 145°.** Crystd from aq EtOH.

2-Phenoxypropionic acid *[940-31-8]* **M 166.2, m 115-116°, b 105-106°/5mm, 265-266°/758mm.** Crystd from water.

Phensuximide *[86-34-0]* **M 189.2, m 71-73°.** Crystd from hot 95% EtOH.

Phenylacetamide *[103-81-1]* **M 135.2, m 158.5°.** Crystd repeatedly from abs EtOH. Dried under vac over P_2O_5.

Phenyl acetate *[122-79-2]* **M 136.2, b 78°/10mm, n^{22} 1.5039, d 1.079.** Freed from phenol and acetic acid by washing (either directly or as a soln in pentane) with aq 5% Na_2CO_3, then with satd aq $CaCl_2$, drying with $CaSO_4$ or Na_2SO_4, and fractional distn at reduced pressure.

Phenylacetic acid *[103-82-2]* **M 136.2, m 76-77°, b 140-150°/20mm.** Crystd from pet ether (b 40-60°), isopropyl alcohol, aq 50% EtOH or hot water. Dried under vac. It can be distd under reduced pressure

Phenylacetone *[103-79-9]* **M 134.2, b 69-71°/3mm, n 1.516, d 1.00.** Converted to the semicarbazone and crystd three times from EtOH (to m 186-187°). The semicarbazone was hydrolysed with 10% phosphoric acid and the ketone was distd. [Kumler, Strait and Alpen *JACS* **72** 1463 *1950*.]

Phenylacetonitrile see **benzyl cyanide.**

4'-Phenylacetophenone *[92-91-1]* **M 196.3, m 120.3-121.2°, b 196-210°/18mm.** Crystd from EtOH. Can also be distd under reduced pressure.

Phenylacetylene *[536-74-3]* **M 102.1, b 75°/80mm, n^{25} 1.5463, d 0.930.** Distd through a spinning band column. Should be filtered through a short column of alumina before use [Collman *et al*, *JACS* **108** 2988 *1986*].

***dl*-Phenylalanine** *[150-30-1]* **M 165.2, m 162°.** Crystd from water and dried under vac over P_2O_5.

L-Phenylalanine *[63-91-2]* **M 165.2, m 280°(dec), $[\alpha]_D^{25}$ -34.0° (c 2, H_2O).** Likely impurities are leucine, valine, methionine and tyrosine. Crystd from water by adding 4 volumes of EtOH. Dried under vac over P_2O_5. Also crystd from satd refluxing aq solns at neutral *p*H, or 1:1 (v/v) EtOH/water soln, or conc HCl.

3-Phenylallyl chloride (cinnamyl chloride) *[E: 18685-01-3][Z: 18684-06-1]* **M 152.6, b 92-93°/3mm.** Distd under vac three times from K_2CO_3.

Phenyl 4-aminosalicylate *[133-11-9]* **M 229.2, m 153°.** Crystd from isopropanol.

4-Phenylanisole *[361-37-6]* **M 184.2, m 89.9-90.1°.** Crystd from benzene/pet ether. Dried under vac in an Abderhalden pistol.

9-Phenylanthracene *[602-55-1]* **M 254.3, 153-154°.** Chromatographed on alumina in benzene and crystd from acetic acid.

***N*-Phenylanthranilic acid** *[91-40-7]* **M 213.2, 182-183°.** Crystd from EtOH (5ml/g) or acetic acid (2ml/g) by adding hot water (1ml/g).

2-Phenyl-1-azaindolizine *[56983-95-0]* **M 194.2, m 140°.** Crystd from EtOH or benzene/pet ether.

***p*-Phenylazoaniline** see *p*-aminoazobenzene,

***p*-Phenylazobenzoyl chloride** *[104-24-5]* **M 244.7, m 93°.** Crystd from pet ether (b 60-80°).

4-Phenylazodiphenylamine see **benzeneazodiphenylamine.**

1-Phenylazo-2-naphthol (Sudan I) *[842-07-9]* **M 248.3, m 131°.** Crystd from EtOH.

4-Phenylazo-α-naphthylamine *[131-22-6]* **M 247.3.** Crystd from cyclohexane.

1-Phenylazo-2-napthylamine *[85-84-7]* **M 247.3, m 99-100°.** Crystd from abs EtOH or glacial acetic acid.

4-Phenylazophenacyl bromide **M 317.3, m 103-104°.** Purified on a column of silica gel, using pet ether/ethyl ether (9:1 v/v) as solvent.

4-Phenylazophenol *[1689-82-3]* **M 198.2, m 155°.** Crystd from benzene or 95% EtOH.

Phenyl benzoate *[93-99-2]* **M 198.2, m 69.5°, b 198-199°.** Crystd from EtOH using *ca* twice the volume needed for complete soln at 69°.

Phenyl-1,4-benzoquinone *[363-03-1]* **M 184.2, m 114-115°.** Crystd from heptane or pet ether (b 60-70°) and sublimed *in vacuo*. [Carlson and Miller *JACS* **107** 479 *1985*.]

N-**Phenylbenzylamine** see **benzylaniline.**

1-Phenylbiguanide *[102-02-3]* **M 177.2, m 144-146°.** Crystd from water or toluene.

1-Phenyl-1,3-butanedione see **benzoylacetone.**

Phenylbutazone *[50-33-9]* **M 308.4, m 105°.** Crystd from EtOH.

trans-**4-Phenyl-3-buten-2-one** see **benzalacetone.**

2-Phenylbutyramide *[90-26-6]* **M 163.2, m 86°.** Crystd from water.

4-Phenylbutyric acid *[1821-12-1]* **M 164.2, m 50°.** Crystd from pet ether (b 40-60°).

o-**(Phenylcarbamoyl)-1-scopolamine methobromide** **M 518.4, m 200.5-201.5°(dec).** Crystd from 95% EtOH.

9-Phenylcarbazole *[1150-62-5]* **M 243.3, m 94-95°.** Crystd from EtOH or isopropanol and sublimed *in vacuo*.

Phenyl cinnamate *[2757-04-2]* **M 224.3, m 75-76°, b 205-207°/15mm.** Crystd from EtOH (2ml/g). It can also be distd under reduced pressure.

α-Phenylcinnamic acid *[91-48-5]* **M 224.3, m 174°(*cis*), m 138-139°(*trans*).** Crystd from ether/pet ether.

α-Phenyl-*p*-cresol see *p*-**benzylphenol.**

o-**Phenylenediamine** *[95-54-5]* **M 108.1, m 100-101°.** Crystd from aq 1% sodium hydrosulphite (charcoal), washed with ice-water and dried in a vac desiccator, or sublimed *in vacuo*. It has been purified by recrystn from toluene and zone refined [Anson *et al*, *JACS* **108** 6593 *1986*]. Purification by refluxing a CH$_2$Cl$_2$ solution containing charcoal was also carried out followed by evapn and recrystn [Koola and Kochi *JOC* **52** 4545 *1987*.]

p-**Phenylenediamine** *[106-50-3]* **M 108.1, m 140°.** Crystd from EtOH or benzene, and sublimed *in vacuo*.

o-**Phenylenediamine dihydrochloride** *[615-28-1]* **M 181.1, m 180°.** Crystd from dil HCl (60ml conc HCl, 40ml water, with 2g stannous chloride), after treatment of the hot soln with charcoal by adding an equal vol of conc HCl and cooling in an ice-salt mixture. The crystals were washed with a small amount of conc HCl and dried in a vac desiccator over NaOH.

2-Phenyl-1,3-diaza-azulene *[2161-31-1]* **M 187.5.** Recrystd three times from deaerated cyclohexane in the dark.

Phenyl disulphide see **diphenyl disulphide.**

dl-**1-Phenylethanol** *[13323-81-4]* **M 122, b 106-107°/22-23mm, n^{25} 1.5254, d 1.01.** Purified *via* its hydrogen phthalate. [See Houssa and Kenyon *JCS* 2260 *1930*.] Shaken with a soln of ferrous sulphate, and the alcohol layer was washed with distd water and fractionally distd.

2-Phenylethanol *[60-12-8]* **M 122.1, b 215-217°, d 1.020.** Purified by shaking with a soln of ferrous sulphate, and the alcohol layer was washed with distd water and fractionally distd.

Phenyl ether *[101-84-8]* **M 170.2, m 27.0°, n$^{30.7}$ 1.57596, d 1.074.** Crystd from 90% EtOH. Melted, washed with 3M NaOH and water, dried with CaCl$_2$ and fractionally distd under reduced pressure. Fractionally crystd from its melt and stored over P$_2$O$_5$.

p-α-**Phenylethylphenol** *[1988-89-2]* **M 198.3, m 56.0-56.3°.** Crystd from pet ether.

5-(α-Phenylethyl)semioxamazide *[93-95-8]* **M 207.1, m 167-168° (*l*-), 157° (*dl*-).** Crystd from EtOH.

9-Phenyl-3-fluorone *[975-17-7]* **M 320.3, m >300°(dec), λ$_{max}$ 462nm (ε 4.06 x 10^4, in 1M HCl aq EtOH).** Recrystd from warm, acidified EtOH by addition of ammonia. The crude material (1g) can be extracted with EtOH (50ml) in a Soxhlet apparatus for 10hr to remove impurities. Impurities can be detected by paper electrophoresis. [Petrova *et al, Anal Lett* **5** 695 *1972*.]

L-α-Phenylglycine *[2935-35-5]* **M 151.2, m 305-310°, [α]$_{546}$ +185° (c 1, M HCl).** Crystd from EtOH.

Phenylglycine-*o*-carboxylic acid *[612-42-0]* **M 195.2, m 208°.** Crystd from hot water (charcoal).

Phenylglyoxaldoxime see **isonitrosoacetophenone.**

Phenylhydrazine *[100-63-0]* **M 108.1, m 23°, b 137-138°/18mm, 241-242°/760mm, n 1.607, d 1.10.** Purified by chromatography, then crystd from pet ether (b 60-80°)/benzene. [Shaw and Stratton *JCS* 5004 *1962*.]

Phenylhydrazine hydrochloride *[59-88-1]* **M 144.5, m 244°.** One litre of boiling EtOH was added to 100g of phenylhydrazine hydrochloride dissolved during 1-3hr (without heating) in 200ml of warm water (60-70°). The soln was filtered off, while still hot, through Whatman No 2 filter paper and cooled in a refrigerator. The ppte was collected on a medium sintered-glass filter and recrystd twice this way, then washed with cold EtOH, dried thoroughly and stored in a stoppered brown bottle. [Peterson, Karrer and Guerra *AC* **29** 144 *1957*.] Hough, Powell and Woods [*JCS* 4799 *1956*] boiled the hydrochloride with three times its weight of water, filtered hot (charcoal), added one-third volume of conc HCl and cooled to 0°. The crystals were washed with acetone, and dried over P$_2$O$_5$ under vac. The salt has also been crystd from 95% EtOH.

Phenylhydroxylamine *[100-65-2]* **M 109.1, m 82°.** Crystd from water.

2-Phenyl-1,3-indandione *[83-12-5]* **M 222.2, m 149-151°,**
2-Phenylindolizine *[25379-20-8]* **M 193.2, m 214°(dec).** Crystd from EtOH.

Phenylisocyanate *[103-71-9]* M 119.1, b 45-47°/10mm, n 1.536, d 1.093. Distd under reduced pressure from P_2O_5.

3-Phenyllactic acid see **2-hydroxy-3-phenylpropionic acid.**

Phenyl methanesulphonate *[16156-59-5]* M 172.1, m 61-62°. Crystd from MeOH.

2-Phenylnaphthalene *[612-94-2]* M 204.3, m 103-104°. Chromatographed on alumina in benzene and crystd from aq EtOH.

N-Phenyl-1-naphthylamine *[90-30-2]* M 219.3, m 63.7-64.0°. Crystd from EtOH, pet ether or benzene/EtOH. Dried under vac in an Abderhalden pistol.

N-Phenyl-2-naphthylamine *{135-88-6]* M 219.3, m 107.5-108.5°. Crystd from EtOH, MeOH, glacial acetic acid or benzene/hexane.

4-Phenylphenacyl bromide *[135-73-9]* M 275.2, m 126°. Crystd (charcoal) from EtOH (15ml/g), or ethyl acetate/pet ether (b 90-100°.

4-Phenylphenol (4-hydroxybiphenyl) *[92-69-3]* M 170.2, m 166-167°. Crystd from benzene, EtOH or EtOH/water, and vac dried in a desiccator over $CaCl_2$. [Buchanan *et al, JACS* **108** 7703 *1986.*]

Phenylphosphonous acid *[121-70-0]* M 141.1, m 71°. Crystd from hot water.

2-Phenylpropanal *[93-53-8]* M 134.2, b 206°/760mm, n 1.5183, d 1.001. May contain up to 15% of acetophenone. Purified *via* the bisulphite addition compound [Lodge and Heathcock *JACS* **109** 3353 *1987*].

Phenyl-2-propanone see **phenylacetone.**

Phenylpropiolic acid *[637-44-5]* M 146.2, m 137.8-138.4°. Crystd from benzene, CCl_4 or aq EtOH.

α-Phenylpropionic acid *[492-37-5]* M 150.2, m 49°. Crystd from pet ether (b 40-60°).

3-Phenylpropyl bromide *[637-59-2]* M 199.1, b 110°/12mm, 128-129°/29mm, d 1.31. Washed successively with conc H_2SO_4, water, 10% aq Na_2CO_3 and again with water, then dried with $CaCl_2$ and fractionally distd just before use.

Phenyl 2-pyridyl ketoxime *[1826-28-4]* M 198.2, m 151-152°,
6-Phenylquinoline *[162-95-3]* M 205.3, m 110.5-111.5°. Crystd from EtOH (charcoal).

2-Phenylquinoline-4-carboxylic acid *[132-60-5]* M 249.3, m 215°. Crystd from EtOH (*ca* 20ml/g).

Phenyl salicylate *[118-55-8]* M 214.2, m 41.8-42.6°. Fractionally crystd from its melt, then crystd from benzene.

2-Phenylsalicylic acid *[304-06-3]* M 214.3, m 186-187.5°. Dissolved in *ca* 1 equivalent of satd aq Na_2CO_3, filtered and pptd by adding 0.8 equivalents of M HCl. Crystd from ethylene dichloride (charcoal), and sublimed at 0.1mm. [Brooks, Eglington and Norman *JCS* 661 *1961.*]

1-Phenylsemicarbazide *[103-03-7]* M 151.2, m 172°,
4-Phenylsemicarbazide *[537-47-3]* M 151.2, m 122°,
Phenylsuccinic acid *[635-51-8]* M 194.2, m 168°,
1-Phenyl-2-thiourea *[103-85-5]* M 152.1, m 154°. Crystd from water.

1-Phenyl-5-sulphanilamidopyrazole *[526-08-9]* M 314.3, m 179-183°,
1-Phenylthiosemicarbazide *[645-48-7]* M 167.2, m 200-201°(dec),
4-Phenylthiosemicarbazide *[5351-69-9]* M 167.2, m 140°. Crystd from EtOH.

Phenyltoloxamine hydrochloride *[6152-43-8]* M 291.8, m 119-120°. Crystd from isobutyl methyl ketone.

Phenyl 4-toluenesulphonate *[640-60-8]* M 248.2, m 94.5-95.5°. Crystd from MeOH or glacial acetic acid.

Phenyl 4-tolylcarbonate *[13183-20-5]* M 228.2, m 67°. Purified by preparative glc with 20% Apiezon on Embacel, and sublimed *in vacuo*.

4-Phenyl-1,2,4-triazole-3,5-diol *[15988-11-1]* M 175.2, m 207-209°. Crystd from water.

4-Phenylurazole see **4-phemyl-1,2,4-triazole-3,5-diol.**

Phenylurea *[64-10-8]* M 136.2, m 148°. Crystd from boiling water (10ml/g). Dried in a steam oven at 100°.

Phloretic acid see **3-*p*-hydroxyphenylpropionic acid.**

Phloretin *[60-82-2]* M 274.3, m 264-271°(dec). Crystd from aq EtOH.

Phlorizin (2H$_2$O) *[60-81-1]* M 472.5, m 110°, [α]$_{546}$ -62° (c 3.2, EtOH). Crystd as dihydrate from water.

Phloroacetophenone (2H$_2$O) *[480-66-0]* M 186.2, m 218-219°. Crystd from hot water (35ml/g).

Phloroglucinol (2H$_2$O) *[6099-90-7]* M 126.1, m 217-219°, 117° (anhydrous). Crystd from water, and stored in the dark under nitrogen.

Phorone *[504-20-1]* M 138.2, m 28°, b 197°/743mm. Crystd repeatedly from EtOH.

"Phosphine" (a dye, CI 793). Crystd from benzene/EtOH.

Phthalazine *[253-52-1]* M 130.2, m 90-91°. Crystd from ethyl ether or benzene, and sublimed under vac.

Phthalazine-1,4-dione *[1445-69-8]* M 162.2. Twice recrystd from 0.1M KOH [Merenyi *et al, JACS* **108** 7716 *1986*].

Phthalazone *[119-39-1]* M 146.2, m 183-184°, b 337°/760mm. Crystd from water and sublimed *in vacuo*.

Phthalein complexon *[2411-89-4]* M 654.6. *o*-Cresolphthalein is a contaminant and is one of the starting materials. It can be removed by dissolving the reagent in water and adding a 3-fold excess of sodium acetate and fractionally precipitating it by dropwise addition of HCl to the clear filtrate. The pure material gives a single spot on paper chromatography (eluting solvent EtOH/water/phenol, 6:3:1; and developing with NaOH). [Anderre *et al, Helv Chim Acta* **37** 113 *1954*.]

Phthalhydrazide see **phthalazine-1,4-dione.**

o-**Phthalic acid** *[88-99-3]* M 166.1, m 211-211.5°. Crystd from water.

Phthalic anhydride *[85-44-9]* **M 148.1, m 132°, b 295°.** Distd under reduced pressure. Purified from the acid by extraction with hot $CHCl_3$, filtered and evaporated. The residue was crystd from $CHCl_3$, CCl_4 or benzene. Fractionally crystd from its melt. Dried under vac at 100°. [Saltiel *JACS* **108** 2674 *1986*.]

Phthalide *[87-41-2]* **M 134.1, m 72-73°.** Crystd from water (75ml/g) and dried in air on filter paper.

Phthalimide *[85-41-6]* **M 147.1, m 235°.** Crystd from EtOH (20ml/g) (charcoal), or by sublimation.

Phthalimidoglycine *[4702-13-0]* **M 205.2, m 192-193°.** Crystd from water or EtOH.

Phthalonitrile *[91-15-6]* **M 128.1, m 141°.** Crystd from EtOH or benzene. Can also be distd under high vac.

Phthalylsulphacetamide *[131-69-1]* **M 362.3, m 196°.** Crystd from water.

Phthiocol *[483-55-6]* **M 188.1, m 173-174°.** Crystd from ethyl ether/pet ether.

Physalien *[144-67-2]* **M 1044, m 98.5-99.5°, $\varepsilon_{1cm}^{1\%}$ 1410 (449nm), 1255 (478nm) in hexane.** Purified by chromatography on water-deactivated alumina, using hexane/ethyl ether (19:1) to develop the column. Crystd from benzene/EtOH. Stored in the dark, in inert atmosphere, at 0°.

Physostigmine (Eserine) *[57-47-6]* **M 275.4, m 105-106o, $[\alpha]^{546}$ -91° (c 1.7, $CHCl_3$).** Crystd from ethyl ether or benzene.

Phytoene *[540-04-5]* **M 544.9, $\varepsilon_{1cm}^{1\%}$ 850 (287nm) in hexane, λ_{max} 275, 287 and 297nm nm.** Purified by chromatography on columns of magnesia-Supercel or alumina [Rabourn *et al*, *Arch Biochem Biophys* **48** 267 *1954*]. Stored as a soln in pet ether under nitrogen at -20°.

Phytofluene *[27664-65-9]* **M 542.9, $\varepsilon_{1cm}^{1\%}$ 1350 (348nm) in pet ether, λ_{max} 331, 348, 267.** Purified by chromatography on partially deactivated alumina [Kushwaha *et al*, *JBC* **245** 4708 *1970*]. Stored as a soln in pet ether under nitrogen at -20°.

Picein *[530-14-3]* **M 298.3, m 195-196°.** Crystd from MeOH or (as monohydrate) from water.

Picene *[213-14-3]* **M 278.3, m 364°.** Crystd from isopropylbenzene/xylene. Can also be sublimed.

Picoline see **methylpyridine.**

Picolinic acid *[98-98-6]* **M 123.1, m 138°.** Crystd from water or benzene.

α-Picolinium chloride *[14401-91-3]* **M 129.6, m 200°.** 1:1 Mixture of α-picoline and HCl, distd at 275°. Then vac sublimed at 91-91.5°.

N-**Picolinoylbenzimidazole M 173.3, m 105-107°.** Recrystd three times from hexane [Fife and Przystas *JACS* **108** 4631 *1986*].

Picric acid *[88-89-1]* **M 229.1, m 122-123°.** Crystd first from acetic acid then acetone, toluene, $CHCl_3$, aq 30% EtOH, 95% EtOH, MeOH or water. Dried in a vac oven at 80° for 2hr. Alternatively, dried over $Mg(ClO_4)_2$ or fused and allowed to freeze under vac three times. Because it is **EXPLOSIVE**, picric acid should be stored moistened with water, and only small portions should be dried at any one time. The dried acid should **NOT** be heated.

Picrolonic acid *[550-74-3]* **M 264.2, m 120°(dec).** Crystd from water or EtOH.

Picrotoxin *[124-87-8]* M 602.6, m 203°, $[\alpha]_{546}$ -40° (c 1, EtOH). Crystd from water.

Picryl chloride *[88-880]* M 226.3, m 83°. Crystd from $CHCl_3$ or EtOH.

Picryl iodide *[4436-27-5]* M 340.0, m 164-165°. Crystd from benzene.

Pimelic acid *[111-16-0]* M 160.2, m 105-106°. Crystd from water or from benzene containing 5% ethyl ether.

Pinacol (hexahydrate) *[6091-58-3]* M 194.3, m 46.5°, b 59°/4mm. Distd then crystd repeatedly from water.

Pinacol (anhydrous) *[76-09-5]* M 118.1, m 41.1°, b 172°. The hydrate is rendered anhydrous by azeotropic distn of water with benzene. Recrystd from benzene or toluene/pet ether, abs EtOH or dry ethyl ether. Recrystn from water gives the hexahydrate.

Pinacolone oxime *[2475-93-6]* M 115.2, m 78°. Crystd from aq EtOH.

Pinacyanol chloride *[2768-90-3]* M 388.9, m 270°(dec). Crystd from EtOH/ethyl ether.

R-α-Pinene *[7785-70-8]* M 136.2, b 61°/30mm, 156.2°/760mm, n^{15} 1.4634, n 1.4658, d 0.858, $[\alpha]_D^{25}$ +47.3°,
S-α-Pinene *[7785-26-4]* M 136.2, b 155-156°/760mm, n 1.4634, d 0.858, $[\alpha]_D$ -47.2°. Isomerised by heat, acids and certain solvents. Should be distd under reduced pressure under nitrogen and stored in the dark. Purified *via* the nitrosochloride [Waterman *et al, Rec Trav Chim Pays-Bas* **48** 1191 *1929*]. For purification of optically active forms see Lynn [*JACS* **91** 361 *1919*].
Small quantities (0.5ml) have been purified by glc using helium as carrier gas and a column at 90° packed with 20 wt% of polypropylene sebacate on a Chromosorb support. Larger quantities were fractionally distd under reduced pressure in a column packed with stainless steel gauze spirals. Material could be dried with CaH_2 or sodium, and stored in a refrigerator: $CaSO_4$ and silica gel were not satisfactory because they induced spontaneous isomerisation. [Bates, Best and Williams *JCS* 1521 *1962*.]

dl-**Pipecolinic acid** *[4043-87-2]* M 129.1, m 264°. Crystd from water.

Piperazine *[110-85-0]* M 86.1, m 110-112°, 44° (hexahydrate *142-63-2*) b 125-130°/760mm. Crystd from EtOH or anhydrous benzene, and dried at 0.01mm. It can be sublimed under vac and purified by zone melting.

Piperazine-N,N'-bis(2-ethanesulphonic acid) (PIPES) *[5625-37-6]* M 302.4. Crystd from boiling water (maximum solubility is about 1g/L) or as described for ADA, pK_a^{20} 7.85.

Piperazine dihydrochloride (H_2O) *[6094-40-2]* M 177.1, m 82.5-83.5°. Crystd from aq EtOH. Dried at 110°.

Piperazine-2,5-dione *[106-57-0]* M 114.1, m 309-310°. Crystd from water.

Piperazine phosphate (H_2O) *[18534-18-4]* M 197.6. Crystd twice from water, air-dried and stored for several days over Drierite. The salt dehydrates slowly if heated at 70°.

Piperic acid *[136-72-1]* M 218.2, m 217°. Crystd from EtOH. Protect from light.

Piperidine *[110-89-4]* **M 85.2, f.p. -9°, b 35.4°/40mm, 106°/760mm, n 1.4535, n²⁵ 1.4500, d 0.862.** Dried with BaO, KOH, CaH₂, or sodium, and fractionally distd (optionally from sodium, CaH₂, or P₂O₅). Purified from pyridine by zone melting.

dl-**Piperidine-2-carboxylic acid** see **pipecolinic acid.**

Piperidinium hydrochloride *[6091-44-7]* **M 121.6, m 244-245°.** Crystd from EtOH/ethyl ether in the presence of a small amount of HCl.

Piperidinium nitrate *[6091-45-8]* **M 145.2, m 110°.** Crystd from acetone/ethyl acetate.

Piperine *[94-62-2]* **M 285.4, m 129-129.5°.** Crystd from EtOH or benzene/ligroin.

Piperonal *[120-57-0]* **M 150.1, m 37°, b 140°/15mm, 263°/760mm.** Crystd from aq 70% EtOH or EtOH/water.

Piperonylic acid *[94-53-1]* **M 166.1, m 229°.** Crystd from EtOH or water.

Pivalic acid *[75-98-9]* **M 102.1, m 35.4°, b 71-73°/0.1mm.** Fractionally distd under reduced pressure, then fractionally crystd from its melt. Recrystd from benzene.

Plumbagin *[481-42-5]* **M 188.1, m 78-79°.** Crystd from aq EtOH.

Polyacrylonitrile *[25014-41-9].* Pptd from dimethylformamide by addition of MeOH.

Polybrene see **1,5-Dimethyl-1,5-diazaundecamethylene polymethobromide.**

Poly(diallyldimethylammonium) chloride. Pptd from water in acetone, and dried in vac for 24hr. [Hardy and Shriner *JACS* **107** 3822 *1985*.]

Polyethylene *[9002-88-4].* Crystd from thiophen-free benzene and dried over P₂O₅ under vac.

Polygalacturonic acid see **pectic acid.**

Polymethyl acrylate *[9002-21-8].* Pptd from a 2% soln in acetone by addition of water.

Polystyrene *[9003-53-6].* Pptd repeatedly from CHCl₃ or toluene soln by addition of MeOH. Dried *in vacuo* [Miyasaka *et al, JPC* **92** 249 *1988*].

Polystyrenesulphonic acid (sodium salt) *[25704-18-1].* Purified by repeated pptn of the sodium salt from aq soln by MeOH, with subsequent conversion to the free acid by passage through an Amberlite IR-120 ion-exchange resin. [Kotin and Nagasawa *JACS* **83** 1026 *1961*.]
Also purified by passage through cation and anion exchange resins in series (Rexyn 101 cation exchange resin and Rexyn 203 anion exchange resin), then titrated with NaOH to *p*H 7. The sodium form of polystyrenesulphonic acid pptd by addition of 2-propanol. Dried in a vac oven at 80° for 24hr, finally increasing to 120° prior to use. [Kowblansky and Ander *JPC* **80** 287 *1976*.]

Polyvinyl acetate *[9003-20-7].* Pptd from acetone by addition of *n*-hexane.

Poly(*N*-vinylcarbazole) *[25067-59-8].* Pptd seven times from tetrahydrofuran with MeOH, with a final freeze-drying from benzene. Dried under vac.

Polyvinyl chloride *[9002-81-2].* Pptd from cyclohexanone by addition of MeOH.

Poly(4-vinylpyridine) *[25232-41-1]* M (105.1)$_n$. Purified by repeated pptn from solns in EtOH and dioxane, and then EtOH and Ethyl acetate. Finally, freeze-dried from *tert*-butanol.

Poly(N-vinylpyrrolidone) *[9003-39-8]* M (111.1)$_n$, crosslinked *[25249-54-1]* m >300°. Purified by dialysis, and freeze-dried. Also by pptn from CHCl$_3$ soln by pouring into ether. Dried in a vac over P$_2$O$_5$. For the crosslinked polymer purification is by boiling for 10min in 10% HCl and then washing with glass-distd water until free from Cl ions. Final Cl ions were removed more readily by neutralising with KOH and continued washing.

Pentachrome Azure Blue B. Crystd from MeOH.

Pontacyl Carmine 2G *[3734-67-6]* M 510.4,
Pontacyl Light Yellow GX *[6359-98-4]* M 552.3. Salted out three times with sodium acetate, then repeatedly extracted with EtOH. See *Chlorazol Sky Blue FF*. [McGrew and Schneider *JACS* 72 2547 *1950*.]

Prednisone *[53-03-2]* M 358.5, m 238°(dec), [α]$_D$ +168° (c 1, dioxane), λ$_{max}$ 238nm (log ε 4.18) in MeOH. Crystd from acetone/hexane.

Pregnane *[24909-91-9]* M 300.5, m 83.5°, [α]$_D$ +21° (CHCl$_3$). Crystd from MeOH.

5β-Pregnane-3α,20α-diol *[80-92-2]* M 320.5, m 243-244°, [α]$_{564}$ +31° (c 1, EtOH). Crystd from acetone.

5β-Pregnane-3α,20β-diol *[80-91-1]* M 320.5, m 244-246°, [α]$_{564}$ +22° (c 1, EtOH). Crystd from EtOH.

Pregnenolone see **3β-hydroxy-5-pregnen-20-one.**

Prehnitine see **1,2,3,4-trimethylbenzene.**

Procaine *[59-46-1]* M 236.3, m 51° (dihydrate), 61° (anhydrous). Crystd as the dihydrate from aq EtOH and as anhydrous material from pet ether or ethyl ether. The latter is hygroscopic.

Proclavine (3,6-diaminoacridine) *[92-62-6]* M 209.2, m 284-286°. Crystd from aq MeOH.

Proflavine see **3,6-diaminoacridine hydrochloride.**

Progesterone *[57-83-0]* M 314.5, m 128.5°, [α]$_{546}$ +220° (c 2, dioxane). Crystd from EtOH. When crystd from pet ether m is 121°, λ$_{max}$ 240nm, log ε 4.25 (EtOH).

L-Proline *[147-85-3]* M 115.1, m 215-220°(dec)(D-isomer), 220-222°(dec) (L-form), 205°(dec)(DL-isomer), [α]$_D^{25}$ (H$_2$O, L-isomer). Likely impurity are hydroxyproline. Purified *via* its picrate which was crystd twice from water, then decomposed with 40% H$_2$SO$_4$. The picric acid was extracted with ethyl ether, the H$_2$SO$_4$ was pptd with Ba(OH)$_2$, and the filtrate evapd. The residue was crystd from hot abs EtOH [Mellan and Hoover *JACS* 73 3879 *1951*] or EtOH/ether. Hygroscopic. Stored in a desiccator.

Prolycopene *[2361-24-2]* M 536.5, m 111°, λ$_{max}$ 443.5, 470nm in pet ether. Purified by chromatography on deactivated alumina [Kushwaha *et al*, *JBC* 245 4708 *1970*]. Crystd from pet ether. Stored in the dark, in an inert atmosphere at -20°.

L-Prolylglycine *[2578-57-6]* M 172.2. Crystd from water at 50-60° by addition of EtOH.

Proneurosporene *[10467-46-6]* M 538.9, λ$_{max}$ 408, 432, 461 nm, ε$_{1cm}^{1\%}$ 2040 (432nm) in hexane. Purified by chromatography on deactivated alumina [Kushwaha *et al*, *JBC* 245 4708 *1970*]. Stored in the dark, in an inert atmosphere at 0°.

Prodiene see **allene**

Propane *[74-98-6]* **M 44.1, m -189.7, b -42.1°/760mm, n 1.2898, d 0.5005.** Purified by bromination of the olefinic contaminants. Propane was treated with bromine for 30min at 0°. Unreacted bromine was quenched, and the propane was distd through two -78° traps and collected at -196° [Skell *et al, JACS* **108** 6300 *1986*].

Propane-1,2-diamine *[78-90-0]* **M 74.1, b 120.5°, n 1.446, d 0.868.** Purified by azeotropic distn with toluene. [Horton, Thomason and Kelly *AC* **27** 269 *1955*.]

Propane-1,2-diol *[57-55-6]* **M 76.1, b 104°/32mm, n 1.433, d 1.040.** Dried with Na_2SO_4, decanted and distd under reduced pressure.

Propane-1,3-diol *[504-63-2]* **M 76.1, b 110-122°/12mm, $n^{18.5}$ 1.4398, d 1.053.** Dried with K_2CO_3 and distd under reduced pressure. More extensive purification involved conversion with benzaldehyde to 2-phenyl-1,3-dioxane (m 47-48°) which was subsequently decomposed by shaking with 0.5M HCl (3ml/g) for 15min and standing overnight at room temperature. After neutralisation with K_2CO_3, the benzaldehyde was removed by distn and the diol was recovered from the remaining aq soln by continuous extraction with $CHCl_3$ for 1day. The extract was dried with K_2CO_3, the $CHCl_3$ was evapd and the diol was distd. [Foster, Haines and Stacey *Tetrahedron* **16** 177 *1961*.]

Propane-1-thiol *[1120-71-4]* **M 76.1, b 65.3°/702mm, n^{25} 1.43511, d^{25} 0.83598,**
Propane-2-thiol *[75-33-2]* **M 76.1, b 49.8°/696mm, n^{25} 1.42154, d^{25} 0.80895.** Purified by soln in aq 20% NaOH, extraction with a small amount of benzene and steam distn until clear. After cooling, the soln was acidified slightly with 15% H_2SO_4, and the thiol was distd out, dried with anhydrous $CaSO_4$ or $CaCl_2$, and fractionally distd under nitrogen. [Mathias and Filho *JPC* **62** 1427 *1958*.] Also purified by liberation of the mercaptan by adding dil HCl to the residue remaining after steam distn. After direct distn from the flask, and separation of the water, the mercaptan was dried (Na_2SO_4) and distd under nitrogen.

1,2,3-Propanetricarboxylic acid see **tricarballic acid.**

Propan-1-ol see *n*-**propyl alcohol.**

Propan-2-ol see **isopropyl alcohol.**

Propargyl alcohol *[107-19-7]* **M 56.1, b 54°/57mm, 113.6°/760mm, n 1.432, d 0.947.** Commercial material contains a stabiliser. An aq soln of propargyl alcohol can be concentrated by azeotropic distn with butanol or butyl acetate. Dried with K_2CO_3 and distd under reduced pressure, in the presence of about 1% succinic acid, through a glass helices-packed column.

Propene *[115-07-1]* **M 42.1, m -185.2°, b -47.8°/750mm, n^{-71} 1.357, d 0.519.** Purified by freeze-pump-thaw cycles and trap-to-trap distn.

p-**(1-Propenyl)phenol** *[539-12-8]* **M 134.2, m 93-94°.** Crystd from water.

β-**Propiolactone** *[57-57-8]* **M 72.1, b 83°/45mm, n^{25} 1.4117, d 1.150.** Fractionally distd under reduced pressure, from sodium. **CARCINOGEN.**

Propionaldehyde *[123-38-6]* **M 58.1, b 48.5-48.7°, n 1.3733, n^{25} 1.37115, d 0.804.** Dried with $CaSO_4$ or $CaCl_2$, and fractionally distd under nitrogen or in the presence of a trace of hydroquinone (to retard oxidation). Blacet and Pitts [*JACS* **74** 3382 *1952*] repeatedly vac distd the middle fraction until no longer gave a solid polymer when cooled to -80°. It was stored with $CaSO_4$.

Propionamide *[79-05-0]* **M 73.1, m 79.8-80.8°.** Crystd from acetone, benzene, $CHCl_3$, water or acetone/water, then dried in a vac desiccator over P_2O_5 or conc H_2SO_4.

Propionic acid *[79-09-4]* **M 74.1, b 141°, n 1.3865, n^{25} 1.3843, d 0.992.** Dried with Na_2SO_4 or by fractional distn, then redistd after refluxing with a few crystals of $KMnO_4$. An alternative purification uses the conversion to the ethyl ester, fractional distn and hydrolysis. [Bradbury *JACS* **74** 2709 *1952*.] Propionic acid can also be heated for 0.5hr with an amount of benzoic anhydride equivalent to the amount of water present (in the presence of CrO_3 as catalyst), followed by fractional distn. [Cham and Israel *JCS* 196 *1960*.]

Propionic anhydride *[123-62-6]* **M 130.2, b 67°/18mm, 168°/780mm, n 1.012, d 1.407.** Shaken with P_2O_5 for several nimutes, then distd.

Propionitrile *[107-12-0]* **M 55.1, b 97.2°, n^{15} 1.36812, n^{30} 1.36132, d 1.407.** Shaken with 1:5 dil HCl, or with conc HCl until the odour of isonitrile has gone, then washed with water, and aq K_2CO_3. After a preliminary drying with silica gel or Linde type 4A molecular sieves, it is stirred with CaH_2 until hydrogen evolution ceases, then decanted and distd from P_2O_5 (not more than 5g/L, to minimise gel formation). Finally, it is refluxed with, and slowly distd from CaH_2 (5g/L), taking precautions to exclude moisture.

n-**Propyl acetate** *[109-60-4]* **M 102.1, b 101.5°, n 1.38442, d 0.887.** Washed with satd aq $NaHCO_3$ until neutral, then with satd aq NaCl. Dried with $MgSO_4$ and fractionally distd.

n-**Propyl alcohol** *[71-23-8]* **M 60.1, b 97.2°, m 1.385, d^{25} 0.79995.** The main impurities in *n*-propyl alcohol are usually water and 2-propen-1-ol, reflecting the commercial production by hydration of propene. Water can be removed by azeotropic distn either directly (azeotrope contains 28% water) or by using a ternary system, e.g. by adding benzene. Alternatively, for gross amounts of water, refluxing over CaO for several hours is suitable, followed by distn and a further drying. To obtain more nearly anhydrous alcohol, suitable drying agents are firstly NaOH, $CaSO_4$ or K_2CO_3, then CaH_2, aluminium amalgam, magnesium activated with iodine, or a small amount of sodium. Alternatively, the alcohol can be refluxed with *n*-propylsuccinate or phthalate in a method similar to the one described under EtOH. Allyl alcohol is removed by adding bromine (15ml/L) and then fractionally distilling from a small amount of K_2CO_3. Propionaldehyde, also formed in the bromination, is removed as the 2,4-dinitrophenylhydrazone. *n*-Propyl alcohol can be dried down to 20ppm of water by passage through a column of pre-dried molecular sieves (type A, K^+ form, heated for 3hr at 300°) in a current of nitrogen. Distn from sulphanilic or tartaric acids removes impurities.
Albrecht [*JACS* **82** 3813 *1960*] obtained spectroscopically pure material by heating with charcoal to 50-60°, filtering and adding 2,4-dinitrophenylhydrazine and a few drops of conc H_2SO_4. After standing for several hours, the mixture was cooled to 0°, filtered and vac distd. Gold and Satchell [*JCS* 1938 *1963*] heated *n*-propyl alcohol with 3-nitrophthalic anhydride at 76-110° for 15hr, then recrystd the resulting ester from water, benzene/pet ether (b 100-120°)(3:1), and benzene. The ester was hydrolysed under reflux with aq 7.5M NaOH for 45min under nitrogen, followed by distn (also under nitrogen). The fraction (b 87-92°) was dried with K_2CO_3 and stirred under reduced pressure in the dark over 2,4-dinitrophenylhydrazine, then freshly distilled. Also purified by adding 2g $NaBH_4$ to 1.5L alcohol, gently bubbling with argon and refluxing for 1day at 50°. Then added 2g of freshly cut sodium (washed with propanol) and refluxed for one day. Distd, taking the middle fraction [Jou and Freeman *JPC* **81** 909 *1977*].

n-**Propylamine** *[107-10-8]* **M 59.1, b 48.5°, n 1.38815, d 0.716.** Distd from zinc dust, at reduced pressure, in an atmosphere of nitrogen.

n-**Propyl bromide.** *[106-94-5]* **M 123.0, b 71.0°, n^{15} 1.43695, n^{25} 1.43123, d 1.354.** Likely contaminants include *n*-propyl alcohol and isopropyl bromide. The simplest purification procedure uses drying with $MgSO_4$ or $CaCl_2$ (with or without a preliminary washed of the bromide with aq $NaHCO_3$, then water), followed by fractional distn away from bright light. Chien and Willard [*JACS* **79** 4872 *1957*] bubbled a stream of oxygen containing 5% ozone through *n*-propyl bromide for 1hr, then shook with 3% hydrogen peroxide soln, neutralised with aq Na_2CO_3, washed with distd water and dried. Then followed vigorous stirring with 95% H_2SO_4 until fresh acid did not discolour within 12hr. The propyl bromide was separated, neutralised, washed dried with $MgSO_4$ and fractionally distd. The centre cut was stored in the dark. Instead of ozone, Schuler and McCauley [*JACS* **79** 821 *1957*] added bromine and stored for 4 weeks, the bromine then being extracted with aq $NaHSO_3$ before the sulphuric acid treatment was applied. Distd. Further purified by preparative gas chromatography on a column packed with 30% SE-30 (General Electric ethylsilicone rubber) on 42/60 Chromosorb P at 150° and 40psi, using helium. [Chu *JPC* **41** 226 *1964*.]

n-**Propyl chloride** *[540-54-5]* M 78.5, b 46.6°, n 1.3880, d 0.890. Dried with $MgSO_4$ and fractionally distd. More extensively purified using extraction with H_2SO_4 as for *n*-propyl bromide. Alternatively, Chien and Willard [*JACS* **75** 6160 *1953*] passed a stream of oxygen containing about 5% ozone through the *n*-propyl chloride for three times as long as was needed to cause the first coloration of starch iodide paper by the exit gas. After washing with aq $NaHCO_3$ to hydrolyse ozonides and remove organic acids, the chloride was dried with $MgSO_4$ and fractionally distd.

1-Propyl-3-(*p*-chlorobenzenesulphonyl) urea *[94-20-2]* M 260.7, m 127-129°. Crystd from aq EtOH.

Propylene carbonate *[108-32-7]* M 102.1, b 110°/0.5-1mm, 238-239°/760mm, n 1.423, d 1.204. Manufactured by reaction of 1,2-propylene oxide with CO_2 in the presence of a catalyst (quaternary ammonium halide). Contaminants include propylene oxide, carbon dioxide, 1,2- and 1,3-propanediols, allyl alcohol and ethylene carbonate. It can be purified by percolation through molecular sieves (Linde 5A, dried at 350° for 14hr under a stream of argon), followed by distn under vac. [Jasinski and Kirkland *AC* **39** 163 *1967*.] It can be stored over molecular sieves under an inert gas atmosphere. When purified in this way it contains less than 2ppm water. Activated alumina and dried CaO have been also used as drying agents prior to fractional distn under reduced pressure. It has been dried with 3A molecular sieves and distd under nitrogen in the presence of *p*-toluenesulphonic acid. Then redistilled and the middle fraction collected.

Propylenediamine see **propane-1,2-diamine.**

Propyleneglycol see **propane-1,2-diol.**

dl-**Propylene oxide** *[75-56-9]* M 58.1, b 34.5°, n 1.3664, d 0.829. Dried with Na_2SO_4 or CaH_2, and fractionally distd through a column packed with glass helices after refluxing with sodium, CaH_2, or KOH pellets.

n-**Propyl ether** *[111-43-3]* M 102.2, b 90.1°, n^{15} 1.38296, n 1.3803, d 0.740. Purified by drying with $CaSO_4$, by passage through an alumina column (to remove peroxides), and by fractional distn.

Propyl formate *[110-74-7]* M 88.1, b 81.3°, n 1.3779, d 0.9058. Distd, then washed with satd aq NaCl, and with satd aq $NaHCO_3$ in the presence of solid NaCl, dried with $MgSO_4$ and fractionally distd.

n-**Propyl gallate** *[121-79-9]* M 212.2, m 150°. Crystd from aq EtOH.

n-**Propyl iodide** *[107-08-4]* M 170.0, b 102.5°, n 1.5041, d 1.745. Should be distd at reduced pressure to avoid decomposition. Dried with $MgSO_4$ or silica gel and fractionally distd. Stored under nitrogen with mercury in a brown bottle. Prior to distn, free iodine can be removed by shaking with copper powder or by washing with aq $Na_2S_2O_3$ and drying. Alternatively, the *n*-propyl iodide can be treated with bromine, then washed with aq $Na_2S_2O_3$ and dried. See also *n-butyl iodide.*

n-**Propyl propionate** *[106-36-5]* M 120.2, b 122°, n 1.393, d 0.881. Treated with anhydrous $CuSO_4$, then distd under nitrogen.

Propyne *[74-99-7]* M 40.1, m -101.5°, b -23.2°/760mm, n^{-40} 1.3863, d^{-50} 0.7062. Purified by preparative gas chromatography.

2-Propyn-1-ol see **propargyl alcohol.**

Protocatechualdehyde *[139-85-5]* M 138.1, m 153°. Crystd from water or toluene.

Protopine *[130-86-9]* M 353.4, m 208°. Crystd from EtOH/chloroform.

Protoporphyrin *[553-12-8]* M 562.7, m >300°. Crystd from ethyl ether.

Pseudocumene see **1,2,4-trimethylbenzene.**

S,S-Pseudoephedrine *[90-82-4]* M 165.2, m 118-119°, $[\alpha]_D$ +53.0° (EtOH), +40.0° (H$_2$O). Crystd from dry ethyl ether, or from water and dried in a vac desiccator.

Pseudoephedrine hydrochloride *[347-78-8]* M 210.7, m 181-182°,
Pseudoisocyanine iodide, Crystd from EtOH.

Pteridine *[91-18-9]* M 132.2, m 139.5-140°. Crystd from EtOH, benzene, *n*-hexane, *n*-heptane or pet ether. It sublimes at 120-130°/20mm. Stored at 0°, in the dark; turns green in the presence of light.

2,4-(1H,3H)-Pteridinedione (H$_2$O) *[487-21-8]* M 182.1, m >350°. Crystd from water.

Pterin (2-aminopteridin-4(3H)-one) *[2236-60-4]* M 163.1, m >300°. It was dissolved in hot 1% aq ammonia, filtered, and an equal volume of hot 1M aq formic acid was added. The soln was allowed to cool at 0-2° overnight. The solid was collected and washed with distd water several times by centrifugation and dried *in vacuo* over P$_2$O$_5$ overnight, and then at 100° overnight.

Pterocarpin *[524-97-0]* M 298.3, m 165°, $[\alpha]_D$ -220°. Crystd from EtOH.

Pteroic acid *[119-24-4]* M 312.3, m >300°(dec). Crystd from dil HCl.

R(+)-Pulegone *[89-82-7]* M 152.2, b 69.5°/5mm, n 1.4849, d 0.935, $[\alpha]_{546}$ +23.5°(neat). Purified *via* the semicarbazone. [Erskine and Waight *JCS 3425 1960.*]

Purine *[120-73-0]* M 120.1, m 216-217°. Crystd from toluene or EtOH.

Purpurin *[81-54-9]* M 256.2, m 253-256°. Crystd from aq EtOH. Dried at 100°.

Purpurogallin *[569-77-7]* M 220.2, m 274° (rapid heating). Crystd from acetic acid.

Pyocyanine *[573-77-3]* M 210.2, m 133°. Crystd from water.

Pyrazine *[290-37-9]* M 80.1, m 47°, b 115.5-115.8°. Distd in steam and crystd from water. Purified by zone melting.

Pyrazinecarboxamide *[98-96-4]* M 123.1, m 189-191° (sublimes slowly at 159°). Crystd from water or EtOH.

Pyrazinecarboxylic acid *[98-97-5]* M 124.1, m 225-229°(dec). Crystd from water.

Pyrazine-2,3-dicarboxylic acid *[89-01-0]* M 168.1, m 183-185°(dec). Crystd from water. Dried at 100°.

Pyrazole *[288-13-1]* **M 68.1, m 70°.** Crystd from pet ether, cyclohexane, or water. [Barszcz *et al*, *JCSDT* 2025 *1986*.]

Pyrazole-3,5-dicarboxylic acid *[3112-31-0]* **M 174.1, m 287-289°**(dec). Crystd from water or EtOH.

Pyrene *[129-00-0]* **M 202.3, m 149-150°.** Crystd from EtOH, glacial acetic acid, benzene or toluene. Purified by chromatography of CCl_4 solns on alumina, with benzene or *n*-hexane as eluent. [Backer and Whitten *JPC* **91** 865 *1987*.] Also zone refined, and purified by sublimation. Marvel and Anderson [*JACS* **76** 5434 *1954*] refluxed pyrene (35g) in toluene (400ml) with maleic anhydride (5g) for 4days, then added 150ml of aq 5% KOH and refluxed for 5hr with occasional; shaking. The toluene layer was separated, washed thoroughly with water, concentrated to about 100ml and allowed to cool. Crystalline pyrene was filtered off and recrystd three times from EtOH or acetonitrile. [Chu and Thomas *JACS* **108** 6270 *1986*; Russell *et al*, *AC* **50** 2961 *1986*.] The material was free from anthracene derivatives. Another purification step involved passage of pyrene in cyclohexane through a column of silica gel. It can be sublimed in a vac and zone refined. [Kano *et al*, *JPC* **89** 3748 *1985*.]

Pyrene-1-aldehyde *[3029-19-4]* **M 230.3, m 125-126°.** Recrystd three times from aq EtOH.

1-Pyrenebutyric acid *[3443-45-6]* **M 288.4, m 184-186°.** Crystd from benzene, EtOH or EtOH/water (7:3 v/v). Dried over P_2O_5. [Chu and Thomas *JACS* **108** 6270 *1986*.]

1-Pyrenecarboxylic acid *[3029-19-4]* **M 230.3, m 126-127°.** Crystd from benzene or 95% EtOH.

1-Pyrenesulphonic acid *[26651-23-0]* **M 202.2.** Crystd from EtOH/water.

1,3,6,8-Pyrenetetrasulphonic acid *[6528-53-6]* **M 522.2.** Crystd from water.

Pyridine *[110-86-1]* **M 79.1, f.p. -41.^{8}, b 115.6°, n 1.51021, d 0.9831.** Likely impurities are water and amines such as the picolines and lutidines. Pyridine is hygroscopic and is miscible with water and organic solvents. It can be dried with solid KOH, NaOH, CaO, BaO or sodium, followed by fractional distn. Other methods of drying include standing with Linde type 4A molecular sieves, CaH_2 or $LiAlH_4$, azeotropic distn of the water with toluene or benzene, or treated with phenylmagnesium bromide in ether, followed by evaporation of the ether and distn of the pyridine. A recommended [Lindauer and Mukherjee *PAC* **27** 267 *1971*] method dries pyridine over solid KOH (20g/Kg) for 2weeks, and fractionally distils the supernatant over Linde type 5A molecular sieves and solid KOH. The product is stored under CO_2-free nitrogen. Pyridine can be stored in contact with BaO, CaH_2 or molecular sieves. Non-basic materials can be removed by steam distilling a soln containing 1.2 equivalents of 20% H_2SO_4 or 17% HCl until about 10% of the base has been carried over along with the non-basic impurities. The residue is then made alkaline, and the base is separated, dried with NaOH and fractionally distd.

Alternatively, pyridine can be treated with oxidising agents. Thus pyridine (800ml) has been stirred for 24hr with a mixture of ceric sulphate (20g) and anhydrous K_2CO_3 (15g), then filtered and fractionally distd. Hurd and Simon [*JACS* **84** 4519 *1962*] stirred pyridine (135ml), water (2.5L) and $KMnO_4$ (90g) for 2hr at 100°, then stood for 15hr before filtering off the pptd manganese oxides. Addition of solid KOH (*ca* 500g) caused pyridine to separate. It was decanted, refluxed with CaO for 3hr and distd.

Separation of pyridine from some of its homologues can be achieved by crystn of the oxalates. Pyridine is pptd as its oxalate by adding it to the stirred soln of oxalic acid in acetone. The ppte is filtered, washed with cold acetone, and pyridine is regenerated and isolated. Other methods are based on complex formation with $ZnCl_2$ or $HgCl_2$. Heap, Jones and Speakman [*JACS* **43** 1936 *1921*] added crude pyridine (1L) to a soln of $ZnCl_2$ (848g) in 730ml of water, 346ml of conc HCl and 690ml of 95% EtOH. The crystalline ppte of $ZnCl_2.2$(pyridine) was filtered off, recrystd twice from abs EtOH, then treated with a conc NaOH soln, using 26.7g of solid NaOH to 100g of the complex. The ppte was filtered off, and the pyridine was dried with NaOH pellets and distd. Similarly, Kyte, Jeffery and Vogel [*JCS* 4454 *1960*] added pyridine (60ml) in 300ml of 10% (v/v) HCl to a soln of $HgCl_2$ (405g) in hot water (2.3L). On cooling, crystals of pyridine-$HgCl_2$ (1:1) complex separated and were filtered off, crystd from 1% HCl (to **m** 178.5-179°), washed with a little EtOH and dried at 110°. The free base was liberated by addition of excess aq NaOH and separated by steam distn. The distillate was satd with solid KOH, and the upper layer was removed, dried further with KOH, then BaO and distd. Another possible purification step is fractional crystn by partial freezing.

Small amounts of pyridine have been purified by vapour-phase chromatography, using a 180-cm column of polyethyleneglycol-400 (Shell 5%) on Embacel (May and Baker) at 100°, with argon as carrier gas. The Karl Fischer titration can be used for determining water content. A colour test for pyrrole as a contaminant is described by Biddiscombe *et al* [*JCS* 1957 *1954*].

Pyridine-2-aldehyde *[1121-60-4]* **M 107.1, b 81.5°/25mm, n 1.535, d 1.121,**
Pyridine-3-aldehyde *[500-22-1]* **M 107.1, b 89.5°/14mm, n 1.549, d 1.141,**
Pyridine-4-aldehyde *[872-85-5]* **M 107.1, b 79.5°/12mm, n 1.544, d 1.137.** Sulphur dioxide was bubbled into a soln of 50g in 250ml of boiled out water, under nitrogen, at 0°, until pptn was complete. The addition compound was filtered off rapidly and, after washing with a little water, it was refluxed in 17% HCl (200ml) under nitrogen until a clear soln was obtained. Neutralisation with $NaHCO_3$ and extraction with ether separated the aldehyde which was recovered by drying the extract, then distilling twice, under nitrogen. [Kyte, Jeffery and Vogel *JCS* 4454 *1960*.]

Pyridine-2-aldoxime *[873-69-8]* **M 122.1, m 113°,**
Pyridine-3-aldoxime *[1193-92-6]* **M 122.1, m 150°,**
Pyridine-4-aldoxime *[696-54-8]* **M 122.1, m 129°.** Crystd from water.

Pyridine-2-carboxylic acid see **picolinic acid.**

Pyridine-3-carboxylic acid see **nicotinic acid.**

Pyridine-4-carboxylic acid see **isonicotinic acid.**

2,6-Pyridinedialdoxime *[2851-68-5]* **M 165.1, m 212°.** Crystd from water.

Pyridine-2,5-dicarboxylic acid *[100-26-5]* **M 167.1, m 254°.** Crystd from dil HCl.

Pyridine-2,6-dicarboxylic acid see **dipicolinic acid.**

Pyridine-3,4-dicarboxylic acid *[490-11-9]* **M 167.1, m256°.** Crystd from dil aq HCl.

Pyridine hydrochloride *[628-13-7]* **M 115.6, m 144°, b 218°.** Crystd from $CHCl_3$/ethyl acetate and washed with ethyl ether.

Pyridine *N*-oxide *[694-59-7]* **M 95.1, m 67°.** Purified by vac sublimation.

Pyridinium bromide perbromide *[39416-48-3]* **M 319.8, m 130°.** Crystd from acetic acid.

Pyridoxal hydrochloride *[65-12-5]* **M 203.6, m 176-180°(dec).** Dissolved in water, the *p*H was adjusted to 6 with NaOH and set aside overnight to crystallise. The crystals were washed with cold water, dried in a vac desiccator over P_2O_5 and stored in a brown bottle at room temperature. [Fleck and Alberty *JPC* **66** 1678 *1962*.]

Pyridoxamine hydrochloride *[5103-96-8]* **M 241.2, m 226-227°(dec).** Crystd from hot MeOH.

Pyridoxine hydrochloride (vitamin B$_6$)   *[58-56-0]* **M 205.7, m 209-210°(dec).** Crystd from EtOH/acetone.

1-(2-Pyridylazo)-2-naphthol (PAN) *[85-85-8]* **M 249.3, m 140-142°.** Purified by repeated crystn from MeOH. It can also be purified by sublimation under vac. Purity can be checked by tlc using a mixed solvent (pet ether, ethyl ether, EtOH; 10:10:1) on a silica gel plate.

4-(2-Pyridylazo)resorcinol (PAR) *[1141-59-9]* **M 215.2, m >195o(dec), kmax 415nm, ε 2,59 x 104 (pH 6-12).** Purified as the sodium salt by recrystn from 1:1 EtOH/water. Purity can be checked by tlc using a silica gel plate and a mixed solvent (*n*-BuOH:EtOH:2M NH$_3$; 6:2:2).

Pyridyldiphenyltriazine *[1046-56-6]* **M 310.4, m 191-192°.** Purified by repeated recrystn from EtOH/dimethylformamide.

Pyrocatechol see **catechol.**

Pyrogallol *[87-66-1]* **M 126.1, m 136°.** Crystd from EtOH/benzene.

L-Pyroglutamic acid *[98-66-1]* **M 129.1, m 162-164°, [α]^{546} -10° (c 5, H$_2$O).** Crystd from EtOH by addition of pet ether.

Pyromellitic acid *[89-05-4]* **M 254.2, m 276°.** Dissolved in 5.7 parts of hot dimethylformamide, decolorised and filtered. The ppte obtained on cooling was separated and air dried, the solvent being removed by heating in an oven at 150-170° for several hours. Crystd from water.

Pyromellitic dianhydride *[89-32-7]* **M 218.1, m 286°.** Crystd from ethyl methyl ketone or dioxane. Dried, and sublimed *in vacuo*.

Pyronin B *[2150-48-3]* **M 358.9.** Crystd from EtOH.

Pyronin Y, CI 739 *[92-32-0]* **M 302.8.** Commercial material contained a large quantity of zinc. Purified by dissolving 1g in 50ml of hot water containing 5g NaEDTA. Cooled to 0°, filtered, evapd to dryness and the residue was extracted with EtOH. The soln was evaporated to 5-10ml, filtered, and the dye pptd by addition of excess of dry ethyl ether. It was centrifuged and the crystals were washed with dry ether. The procedure was repeated, then the product was dissolved in CHCl$_3$, filtered and evapd. The dye was stored in a vacuum.

Pyrrole *[109-97-7]* **M 67.1, b 129-130°, n 1.5097, d 0.966.** Dried with NaOH, CaH$_2$ or CaSO$_4$. Fractionally distd under reduced pressure from sodium or CaH$_2$. Stored under nitrogen. Redistd immediately before use.

Pyrrolidine *[123-75-1]* **M 71.1, b 87.5-88.5°, n 1.443, d 0.860.** Dried with BaO or sodium, then fractionally distd, under nitrogen, through a Todd column packed with glass helices.

2-Pyrrolidone-5-carboxylic acid (D-pyroglutamic acid) *[4042-36-8]* **M 129.1, m 182-183°, [α]$_D$ +10.7° (H$_2$O).** Crystd from EtOH/pet ether.

Pyruvic acid *[127-17-3]* **M 88.1, m 13°, b 65°/10mm.** Distd twice, then fractionally crystd by partial freezing.

p-**Quaterphenyl** *[135-70-6]* **M 306.4, m 312-314°.** Recrystd from dimethyl sulphoxide at *ca* 50°.

Quercetin (2H$_2$O) *[6151-25-3]* **M 338.3, m *ca* 315°(dec).** Crystd from aq EtOH and dried at 100°.

Quercitrin *[117-39-3]* **M 302.2, m 168°.** Crystd from aq EtOH and dried at 135°.

Quinaldic acid *[93-10-7]* **M 173.2, m 156-157°.** Crystd from benzene.

Quinaldine see **2-methylquinoline.**

Quinalizarin *[81-61-8]* **M 272.2.** Crystd from acetic acid or nitrobenzene. It can be sublimed under vac.

Quinazoline *[253-82-7]* **M 130.2, m 48.0-48.5°, b 120-121°/17-18mm.** Purified by passage through an activated alumina column in benzene or pet ether (b 40-60°). Distd under reduced pressure, sublimed under vac and crystd from pet ether. [Armarego *J Appl Chem* **11** 70 *1961*.]

Quinhydrone *[106-34-3]* **M 218.2, m 168°.** Crystd from water heated to 65°, then dried in a vac desiccator.

1R,3R,4R,5R-Quinic acid *[77-95-2]* **M 192.3, m 172°(dec), [α]$_{546}$ -51° (c 20, H$_2$O).** Crystd from water.

Quinidine *[56-54-2]* **M 324.4, m 171°, [α]$_{546}$ +301.1° (CHCl$_3$ contg 2.5% (v/v) EtOH).** Crystd from benzene or dry CHCl$_3$/pet ether (b 40-60°), discarding the initial, oily crop of crystals. Dried under vac at 100° over P$_2$O$_5$.

Quinine *[130-95-0]* **M 324.4, m 177°(dec), [α]$_{546}$ -160° (c 1, CHCl$_3$).** Crystd from abs EtOH.

Quinine bisulphate *[804-63-7]* **M 422.4, m 160° (anhydrous).** Crystd from 0.1M H$_2$SO$_4$, forms heptahydrate when crystd from water

Quinine sulphate (2H$_2$O) *[6591-63-5]* **M 783.0, m 205°.** Crystd from water. dried at 110°.

Quinizarin *[81-64-1]* **M 240.2, m 200-202°.** Crystd from glacial acetic acid.

Quinol see **hydroquinone.**

Quinoline *[91-22-5]* **M 129.2, m -16°, b 111.5°, 236°/758mm, n 1.625, d 1.0937.** Dried with Na$_2$SO$_4$ and vac distd from zinc dust. Also dried by boiling with acetic anhydride, then fractionally distilling. Calvin and Wilmarth [*JACS* **78** 1301 *1956*] cooled redistd quinoline in ice and added enough HCl to form its hydrochloride. Diazotization removed aniline, the diazo compound being broken down by warming the soln to 60°. Non-basic impurities were removed by ether extraction. Quinoline was liberated by neutralising the hydrochloride with NaOH, then dried with KOH and fractionally distd at low pressure. Addition of cuprous acetate (7g/L of quinoline) and shaking under hydrogen for 12hr at 100° removed impurities due to the nitrous acid treatment. Finally the hydrogen was pumped off and the quinoline was distd. Other purification procedures depend on conversion to the phosphate (**m 159°**, pptd from MeOH soln, filtered, washed with MeOH, then dried at 55°) or the picrate (**m 201°**) which, after crystn were reconverted to the amine.
The method using the picrate [Packer, Vaughan and Wong *JACS* **80** 905 *1958*] is as follows: quinoline is added to picric acid dissolved in the minimum volume of 95% EtOH, giving yellow crystals which were washed with EtOH, air-dried and crystd from acetonitrile. These were dissolved in dimethyl sulphoxide (previously dried over 4A molecular sieves) and passed through basic alumina, on which the

picric acid is adsorbed. The free base in the effluent is extracted with *n*-pentane and distd under vac. Traces of solvent can be removed by vapour-phase chromatography. [Moonaw and Anton *JPC* **80** 2243 *1976*]. The ZnCl$_2$ and dichromate complexes have also been used. [Cumper, Redford and Vogel *JCS* **1176** *1962*.]

2-Quinolinealdehyde *[5470-96-2]* **M 157.2, m 71°**. Steam distd. Crystd from water. Protected from light.

8-Quinolinecarboxylic acid *[86-59-9]* **M 173.2, m 186-187.5°**. Crystd from water.

Quinoline ethiodide *[634-35-5]* **M 285.1, m 158-159°**. Crystd from aq EtOH.

Quinolinol see **hydroxyquinoline.**

Quinoxaline *[91-19-0]* **M 130.2, m 28° (anhydr), 37°(H$_2$O), b 108-110°/0.1mm, 140°/40mm**. Crystd from pet ether. Crystallises as the monohydrate on addition of water to a pet ether soln.

Quinoxaline-2,3-dithiol *[1199-03-7]* **M 194.1, m 345°(dec)**. Purified by repeated dissolution in alkali and re-pptn by acetic acid.

p-**Quinquephenyl** *[61537-20-0]* **M 382.5, m 388.5°**. Recrystd from boiling dimethyl sulphoxide (b 189°, lowered to 110°). The solid obtained on cooling was filtered off and washed repeatedly with toluene, then with conc HCl. The final material was washed repeatedly with hot EtOH. It was also recrystd from pyridine, then sublimed *in vacuo*.

Quinuclidine *[100-76-5]* **M 111.2, m 158°(sublimes)**. Crystd from ethyl ether.

D-Raffinose (5H$_2$O) *[512-69-6]* M 594.5, m 80°, [α]$_{546}$ +124° (c 10, H$_2$O). Crystd from aq EtOH.

Rauwolscine hydrochloride *[6211-32-1]* **M 390.0, m 278-280°**. Crystd from water.

RDX see **cyclotrimethylenetrinitramine.**

Reductic acid *[80-72-8]* **M 114.1, m 213°**. Crystd from ethyl acetate.

Rescinnamine *[24815-24-5]* **M 634.7, m 238-239°(vac), [α]$_D$ -87-97° (c 1, CHCl$_3$)**. Crystd from benzene.

Reserpic acid *[83-60-3]* **M 400.5, m 241-243°**. Crystd from MeOH.

Reserpine *[50-55-5]* **M 608.7, m 262-263°, [α]$_{546}$ -148° (c 1, CHCl$_3$)**. Crystd from aq acetone.

Resorcinol *[108-46-3]* M 110.1, m 111.2-111.6°. Crystd from benzene, toluene or benzene/ethyl ether.

Resorufin *[635-78-9]* M 213.2. Washed with water and recrystd from EtOH several times.

Retene *[483-65-8]* M 234.3, m 99°. Crystd from EtOH.

Retinal *[116-31-4]* M 284.5, $\varepsilon_{1cm}^{1\%}$ (*all-trans*) 1530 (381 nm), (*13-cis*) 1250 (375 nm) in EtOH. Separated from retinol by column chromatography on water-deactivated alumina. Eluted with 1-2% acetone in hexane, or on tlc plates of silica gel G development with 1:1 ether/hexane. Crystd from pet ether or *n*-hexane, or as the semicarbazone from EtOH. Six isomers are reported. It is stored in the dark, in an inert atmosphere, at 0°.

Retinoic acid *[302-79-4]* M 300.4, $\varepsilon_{1cm}^{1\%}$ (*all-trans*) 1500 (350nm), (*13-cis*) 1320 (354 nm), (*9-cis*) 1230 (345 nm), (*9,13-di-cis*) 1150 (346 nm) in EtOH. Crystd from MeOH, EtOH, isopropyl alcohol, or as methyl ester from MeOH. Purified by column chromatography on silicic acid columns, eluting with a small amount of EtOH in hexane. Stored in the dark, in an inert atmosphere, at 0°.

Retinol *[68-26-8]* M 286.5, $\varepsilon_{1cm}^{1\%}$ (*all-trans*) 1832 (325 nm),(*13-cis*) 1686 (328nm), (*11-cis*) 1230 (319 nm), (*9-cis*) 1480 (323 nm), (*9,13-di-cis*) 1379 (324 nm), (*11-13-di-cis*) 908 (311 nm) in EtOH. Purified by chromatography on columns of water-deactivated alumina eluting with 3-5% acetone in hexane. Separation of isomers is by tlc plates on silica gel G, developed with pet ether (low boiling)/methyl heptanone (11:2). Stored in the dark, under nitrogen, at 0°, as in ethyl ether, acetone or ethyl acetate. [See Gunghaly *et al*, *Arch Biochem Biophys* **38** 75 *1952*.]

Retinyl acetate *[127-49-9]* M 328.5, m 57°. Separated from retinol by column chromatography, then crystd from MeOH. See Kofler and Rubin [*Vitamin Hormones* **18** 315 *1960*] for review of purification methods. Stored in the dark, under inert atmosphere, at 0°.

Retinyl palmitate *[79-81-2]* M 524.9, $\varepsilon_{1cm}^{1\%}$ (*all-trans*) 1000 (325 nm) in EtOH. Separated from retinol by column chromatography on water-deactivated alumina with hexane containing a very small percentage of acetone. Also chromatographed on tlc silica gel G, using pet ether/isopropyl ether/acetic acid/water (180:20:2:5) or pet ether/acetonitrile/acetic acid/water (190:10:1:15) to develop the chromatogram. Then recrystd from propylene.

Rhamnetin *[90-19-7]* M 316.3, m >300°(dec). Crystd from EtOH.

L-α-Rhamnose (H_2O) *[3615-41-6]* M 182.2, m 105°, $[\alpha]_D^{15}$ +9.1° (c 5, H_2O). Crystd from water or EtOH.

Rhodamine B, CI 749 *[81-88-9]* M 442.5. Major impurities are partially dealkylated compounds not removed by crystn. Purified by chromatography, using ethyl acetate/isopropanol/ammonia (0.888 sg)(9:7:4, R_F 0.75 on Kieselgel G). Crystd from conc soln in MeOH by slow addition of dry ethyl ether. Stored in the dark.

Rhodamine B chloride *[81-88-9]* M 479.0. Crystd from EtOH containing a drop of conc HCl by slow addition of ten volumes of dry ethyl ether. The solid was washed with ether and air dried. The dried material has also been extracted with benzene to remove oil-soluble material prior to recrystn.

Rhodamine 6G *[989-38-8]* M 479.3. Crystd from MeOH or EtOH, and dried in a vac oven.

Rhodanine *[141-84-4]* M 133.2, m 168.5° (capillary). Crystd from glacial acetic acid or water.

Riboflavin *[83-88-5]* M 376.4, m 295-300°(dec), $[\alpha]_D$ -9.8° (H_2O), -125° (c 5, 0.05N NaOH). Crystd from 2M acetic acid, then extracted with $CHCl_3$ to remove lumichrome impurity. [Smith and Metzler *JACS* **85** 3285 *1963*.] Has also been crystd from water.

Riboflavin-5'-phosphate (Na salt, $2H_2O$) *[130-40-5]* **M 514.4.** Crystd from acidic aq soln.

Ribonucleic acid see Chapter 6.

α-D-Ribose *[50-69-1]* **M 150.1, m 90°, $[\alpha]_{546}$ -24° (after 24hr, c 10, H_2O).** Crystd from aq 80% EtOH, dried under vac at 60° over P_2O_5 and stored in a vac desiccator.

Ricinoleic acid *[141-22-0]* **M 298.5, m 7-8° (α-form), 5.0° (γ-form), n 1.4717.** Purified as methyl acetylricinoleate [Rider *JACS* **53** 4130 *1931*], fractionally distilling at 180-185°/0.3mm, then 87g of this ester was refluxed with KOH (56g), water (25ml), and MeOH (250ml) for 10min. The free acid was separated, crystd from acetone at -50°, and distd in small batches, b 180°/0.005mm. [Bailey *et al, JCS* 3027 *1957*.]

Rosaniline (Fuchsin) *[632-99-5]* **M 323.8, λ_{max} 544 nm.** Crystd from water. Dried *in vacuo* at 40°.

Rose Bengal *[11121-48-5]* **M 1017.7.** Purified chromatographically on silica tlc ring using a 35:65 mix of EtOH/acetone as eluent.

Rubijervine *[79-58-3]* **413.6.** Crystd from EtOH. It has solvent of crystn.

Rubrene *[517-51-1]* **M 532.7, m >320°.** Recrystd from benzene (under red light).

(+)-Rutin *[153-18-4]* **M 610.5, m 188-189, $[\alpha]_{546}$ +13° (c 5, EtOH).** Crystd from MeOH or water/EtOH, air dried, then dried for several hours at 110°.

Saccharic acid *[81-07-2]* M 183.2, m 125-126°. Crystd from 95% EtOH.

Saccharin *[87-73-0]* M 210.4, m 229°. Crystd from water.

Safranine O *[477-73-6]* M 350.9, λ_{max} 530nm. Crystd from benzene/MeOH (1:1) or water. Dried under vac over H_2SO_4.

D(-)-Salicin *[138-52-3]* M 286.3, m 204-208°, $[\alpha]_D^{25}$ -63.5° (c *ca* 3, H_2O). Crystd from EtOH.

Salicylaldehyde *[90-02-8]* M 122.1, b 93°/25mm, 195-197°/760mm, n 1.574, d 1.167. Pptd as the bisulphite addition compound by pouring the aldehyde slowly and with stirring into a 25% soln of $NaHSO_3$ in 30% EtOH, then standing for 30min. The ppte, after filtering at the pump, and washing with EtOH, was decomposed with aq 10% $NaHCO_3$, and the aldehyde was extracted into ethyl ether, dried with Na_2SO_4 or $MgSO_4$, and distd, under reduced pressure. Alternatively, salicylaldehyde can be pptd as its copper complex by adding it to warm, satd soln of copper acetate, shaking and then standing in ice. The ppte was filtered off, washed thoroughly with EtOH, then with ethyl ether, and decomposed with 10% H_2SO_4, the aldehyde was extracted into ethyl ether, dried and distd. It has also been purified by repeated vac distn, and by dry column chromatography on Kiesel gel G [Nishiya *et al, JACS* **108** 3880 *1986*].

Salicylaldoxime *[94-67-7]* M 137.1, m 57°. Crystd from $CHCl_3$/pet ether (b 40-60°).

Salicylamide *[65-45-2]* M 137.1, m 142-144°. Crystd from water or repeatedly from chloroform [Nishiya *et al, JACS* **108** 3880 *1986*].

Salicylanilide *[97-17-2]* M 213.2, m 135°. Crystd from water.

Salicylhydroxamic acid *[89-73-6]* M 153.1, m 179-180°(dec). Crystd from acetic acid.

Salicylic acid *[69-72-7]* M 138.1, m 159-160°. Crystd from hot water, abs MeOH, or cyclohexane. Air dried. It can also be sublimed in a vac.

Sarcosine *[107-97-1]* M 89.1, m 212-213°(dec). Crystd from abs EtOH.

Sarcosine anhydride *[5076-82-4]* M 142.2, m 146-147°. Crystd from water, EtOH or ethyl acetate. Dried in vac at room temperature.

Scopolamine see **hyoscine**.

Scopoletin *[92-61-5]* M 192.2, m 206°. Crystd from water or acetic acid.

Secobarbital *[76-73-3]* M 238.4. A soln of the salt in 10% HCl was pptd and the acid form was extracted by the addition of ether. Then purified by repeated crystn from $CHCl_3$. [Buchet and Sandorfy *JPC* **88** 3274 *1984*.]

Sebacic acid *[111-20-6]* M 202.3, m 134.5°. Purified *via* the disodium salt which, after crystn from boiling water (charcoal), was again converted to the free acid. The free acid was crystd repeatedly from hot distd water and dried under vac.

Selenopyronine *[85051-91-8]* M 365.8, λ_{max} 571 (ϵ 81,000). Purified as the hydrochloride from hydrochloric acid [Fanghanel *et al, JPC* **91** 3700 *1987*].

Selenourea *[630-10-4]* M 123.0, m 214-215°(dec). Crystd from water under nitrogen.

Semicarbazide hydrochloride *[563-41-7]* **M 111.5, m 175°.** Crystd from aq 75% EtOH and dried under vac over CaSO$_4$. Also crystd from a mixture of 3.6 mole % MeOH and 6.4 mole % of water. [Kovach *et al, JACS* **107** 7360 *1985.*]

Sennoside A *[81-27-6]* **M 862.7,**
Sennoside B *[128-57-4]* **M 962.7.** Crystd from aq acetone.

L-Serine *[56-45-1]* **M 105.1, m 228°(dec), [α]$_D^{25}$ +14.5° (1M HCl), [α]$_{546}$ +16° (c 5, 5M HCl).** Likely impurity is glycine. Crystd from water by adding 4 volumes of EtOH. Dried. Stored in a desiccator.

Serotonin creatinine sulphate (H$_2$O) *[61-47-1]* **M 405.4, m 220°(dec).** Crystd (as monohydrate) from water.

Shikimic acid *[138-59-0]* **M 174.2, m 190°, [α]$_{546}$ -210° (c 2, H$_2$O),**
Sinomenine hydrochloride *[6080-33-7]* **M 365.9, m 231°.** Crystd from water.

Sitosterols *[12002-39-0],* **M 414.7.** Crystd from EtOH.

β-Sitosterol *[83-46-5]* **M 414.7, m 136-137°, [α]$_{546}$ -42° (c 2, CHCl$_3$).** Crystd from MeOH. Also purified by zone melting.

Skatole see **3-methylindole.**

Sinapinic acid see **3-(4-hydroxy-3,5-dimethoxyphenyl)acrylic acid.**

Skellysolve A is essentially *n*-pentane, b 28-30°,
Skellysolve A is essentially *n*-hexane, b 60-68°,
Skellysolve C is essentially *n*-heptane, b 90-100°,
Skellysolve D is mixed heptanes, b 75-115°,
Skellysolve E is mixed octanes, b 100-140°,
Skellysolve F is pet ether, b 30-60°,
Skellysolve G is pet ether, b 40-75°,
Skellysolve H is hexanes and heptanes, b 69-96°,
Skellysolve L is essentially octanes, b 95-127°. For methods of purification, see **petroleum ether.**

Smilagenin *[126-18-1]* **416.6, m 185°.** Crystd from acetone.

Solanidine *[80-78-4]* **M 397.6, m 218-219°.** Crystd from CHCl$_3$/MeOH.

Solanine-S *[51938-42-2]* **M 884.1, m 284°(dec).** Crystd from aq 85% EtOH.

Solasodine *[126-17-0]* **M 413.6, m 202°.** Crystd (as monohydrate) from aq 80% EtOH.

Solasonine *[19121-58-5]* **M 884.0, m 279°.** Crystd from aq 80% dioxane.

Solochrome Violet R *[2092-55-9]* **M 367.3.** Converted to the monosodium salt by pptn with sodium acetate/acetic acid buffer of *p*H 4, then purified as described for *Chlorazole Sky Blue FF*. Dried at 110°. It is hygroscopic. [Coates and Rigg *TFS* **57** 1088 *1961.*]

Sorbic acid *[110-44-1]* **M 112.1, m 134°.** Crystd from boiling water.

Sorbitol *[50-70-4]* **M 182.2, m 89-93° (hemihydrate), 110-111° (anhydrous), [α]$_{546}$ -1.8° (c 10, H$_2$O).** Crystd (as hemihydrate) several times from EtOH/water (1:1), then dried by fusing and storing over MgSO$_4$.

Spirilloxanthin *[34255-08-8]* **M 596.9, m 216-218°,** λ_{max} **463, 493, 528 nm,** $\varepsilon_{1cm}^{1\%}$ **2680 (493 nm), in pet ether (b 40-70°).** Crystd from $CHCl_3$/pet ether, acetone/pet ether, benzene/pet ether or benzene. Purified by chromatography on a column of $CaCO_3$/$Ca(OH)_2$ mixture or deactivated alumina. [Polgar *et al, Arch Biochem Biophys* **5** 243 *1944*.] Stored in the dark in an inert atmosphere, at -20°.

Squalene *[111-01-3]* **M 422.8, f.p. -5.4°, b 213°/1mm, n 1.4905, d^{25} 0.8670.** Crystd repeatedly from acetone (1.4ml of acetone per ml) by cooling in a Dry-ice bath, washing the crystals with cold acetone, then freezing the squalene from the solvent under vac. The squalene was further purified by passage through a column of silica gel. It has also been chromatographed on activated alumina, using pet ether as eluent. Dauben *et al* [*JACS* **74** 4321 *1952*] purified squalene *via* its hexachloride. See also Capstack *et al* [*JBC* **240** 3258 *1965*] and Krishna *et al* [*Arch Biochem Biophys* **114** 200 *1966*].

Starch *[9005-84-9]* **M (162.1)n.** Defatted by Soxhlet extraction with ethyl ether or 95% EtOH. For fractionation of starch into "amylose" and "amylopectin" fractions, see Lansky, Kooi and Schoch [*JACS* **71** 4066 *1949*].

Stearic acid *[57-11-4]* **M 284.5, m 71.4°.** Crystd from acetone, acetonitrile, EtOH (5 times), aq MeOH, ethyl methyl ketone or pet ether (b 60-90°), or by fractional pptn by dissolving in hot 95% EtOH and pouring into distd water, with stirring. The ppte, after washing with distd water, was dried under vac over P_2O_5. It has also been purified by zone melting. [Tamai *et al, JPC* **91** 541 *1987*].

Stearyl alcohol see *n*-octadecyl alcohol.

Stigmasterol *[83-48-7]* **M 412.7, m 170°, $[\alpha]_D^{22}$ -51° ($CHCl_3$), $[\alpha]_{546}$ -59° (c 2, $CHCl_3$).** Crystd from hot EtOH. Dried in vac over P_2O_5 for 3hr at 90°. Purity was checked by NMR.

cis-**Stilbene** *[645-49-8]* **M 180.3, b 145°/12mm.** Purified by chromatography on alumina using hexane and distd under vac. (The final product contains *ca* 0.1% of the *trans*-isomer). [Lewis *et al, JACS* **107** 203 *1985*; Saltiel *JPC* **91** 2755 *1987*.]

trans-**Stilbene** *[103.30-0]* **M 180.3, m 125.9°, b 305-307°/744mm, d 0.970.** Purified by vac distn. (The final product contains about 1% of the *cis* isomer). Crystd from EtOH. Purified by zone melting. [Lewis *et al, JACS* **107** 203 *1985*; Bollucci *et al, JACS* **109** 515 *1987*; Saltiel *JPC* **91** 2755 *1987*.]

(-)-Strychnine *[57-24-9]* **M 334.4, m 268°, $[\alpha]_{546}$ -139° (c 1, $CHCl_3$).** Crystd as the hydrochloride from water, then neutralised with ammonia.

Styphnic acid *[82-71-3]* **M 245.1, m 179-180°.** Crystd from ethyl acetate. **[Explodes violently on rapid heating.]**

Styrene *[100-42-5]* **M 104.2, b 41-42°/18mm, 145.2°/760mm, n 1.5469, n^{25} 1.5441, d 0.907.** Styrene is difficult to purify and keep pure. Usually contains added inhibitors (such as a trace of hydroquinone). Washed with aq NaOH to remove inhibitors (e.g. *tert*-butanol), then with water, dried for several hours with $MgSO_4$ and distd at 25° under reduced pressure in the presence of an inhibitor (such as 0.005% *p*-*tert*-butylcatechol). It can be stored at -78°. It can also be stored and kept anhydrous with Linde thype 5A molecular sieves, CaH_2, $CaSO_4$, BaO or sodium, being fractionally distd, and distd in a vacuum line just before use. Alternatively styrene (and its deuterated derivative) were passed through a neutral alumina column before use [Woon *et al, JACS* **108** 7990 *1986*; Collman *JACS* **108** 2588 *1986*].

(±)-Styrene glycol (±-1-phenyl-1,2-ethanediol) *[93-56-1]* **M 138.2, m 67-68°.** Crystd from pet ether.

Styrene oxide *[96-09-3]* **M 120.2, b 84-86°/16.5mm, n 1.535, d 1.053.** Fractional distn at reduced pressure does not remove phenylacetaldehyde. If this material is present, the styrene oxide is treated with hydrogen under 3 atmospheres pressure in the presence of platinum oxide. The aldehyde, but not the oxide, is reduced to β-phenylethanol) and separation is now readily achieved by fractional distn. [Schenck and Kaizermen *JACS* **75** 1636 *1953*.]

Suberic acid *[505-48-6]* **M 174.2, m 141-142°.** Crystd from acetone.

Succinamide *[110-14-5]* **M 116.1, m 262-265°(dec).** Crystd from hot water.

Succinic acid *[110-15-6]* **M 118.1, m 185-185.5°.** Washed with ethyl ether. Crystd from acetone, distd water, or *tert*-butanol. Dried under vac over P_2O_5 or conc H_2SO_4. Also purified by conversion to the disodium salt which, after crystn from boiling water (charcoal), is treated with mineral acid to regenerate the succinic acid. The acid is then recrystd and vac dried.

Succinic anhydride *[108-30-5]* **M 100.1, m 119-120°.** Crystd from redistd acetic anhydride or $CHCl_3$, then filtered, washed with ethyl ether and dried under vac.

Succinimide *[123-56-8]* **M 99.1, m 124-125°.** Crystd from EtOH (1ml/g) or water.

Succinonitrile *[110-61-2]* **M 80.1, m 57.9°, b 108°/1mm, 267°/760mm.** Purified by vac sublimation, also crystd from acetone.

D(+)-Sucrose *[57-50-1]* **M 342.3, m 186-188°, $[\alpha]_{546}$ +78° (c 10, H_2O).** Crystd from water.

Sucrose diacetate hexaisobutyrate. Melted and, while molten, treated with $NaHCO_3$ and charcoal, then filtered.

D-Sucrose octaacetate *[126-14-7]* **M 678.6, m 83-85°, $[\alpha]_{546}$ +70° (c 1, $CHCl_3$).** Crystd from EtOH.

Sudan III, CI 248 *[85-86-9]* **M 352.4, m 199°(dec), λ_{max} 354, 508 nm.** Crystd from EtOH, EtOH/water or benzene/abs EtOH (1:1).

Sudan IV *[85-83-6]* **M 380.5, m 184°.** Crystd from EtOH/water or acetone/water.

Sudan Yellow *[824-07-9]* **M 248.3, m 135°.** Crystd from EtOH.

Sulphaguanidine *[57-67-0]* **M 214.2, m 189-190°.** Crystd from hot water (7ml/g).

Sulphamethazine *[57-68-1]* **M 278.3, m 198-200°.** Crystd from dioxane.

Sulphanilamide *[63-74-1]* **M 172.2, m 166°.** Crystd from water or EtOH.

Sulphanilic acid *[121-57-3]* **M 173.2.** Crystd (as dihydrate) from boiling water. Dried at 105° for 2-3hr, then over 90% H_2SO_4 in a vac desiccator.

Sulphapyridine *[144-83-2]* **M 349.2, m 193°.** Crystd from 90% acetone and dried at 90°.

o-**Sulphobenzoic acid** (H_2O) *[632-25-7]* **M 202.2, m 68-69°.** Crystd from water.

o-**Sulphobenzoic acid** (monoammonium salt) *[6939-89-5]* **M 219.5.** Crystd from water.

o-**Sulphobenzoic anhydride** *[81-08-3]* **M 184.2, m 128°, b 184-186°/18mm.** Crystd from dry benzene. It can be distd under vac.

Sulpholane *[126-33-0]* **M 120.2, m 28.5°, b 153-154°/18mm, 285°/760mm, n^{30} 1.4820, d 1.263.** Prepared commercially by Diels-Alder reaction of 1,3-butadiene and sulphur dioxide, followed by Raney nickel hydrogenation. The principle impurities are water, 3-sulpholene, 2-sulpholene and 2-

isopropyl sulpholanyl ether. It is dried by passage through a column of molecular sieves. Distd under reduced pressure through a column packed with stainless steel helices. Again dried with molecular sieves and distd. [Cram *et al, JACS* **83** 3678 *1961*; Coetzee *PAC* **49** 211 *1977*.]

Also, it was stirred at 50° and small portions of solid $KMnO_4$ were added until the colour persisted during 1hr. Dropwise addition of MeOH then destroyed the excess $KMnO_4$, the soln was filtered, freed from potassium ions by passage through an ion-exchange column and dried under vac. It has also been vac distd from KOH pellets. It is *hygroscopic*. [See Sacco *et al, JPC* **80** 749 *1976; JCSFT 1* **73** 1936 *1977*; **74** 2070 *1978*; *TFS* **62** 2738 *1966*.] Coetzee has reviewed the methods of purification of sulpholane, and also the removal of impurities. [Coetzee in *Recommended Methods of Purification of Solvents and Tests for Impurities*, Coetzee ed. Pergamon Press, 1982.]

5-Sulphosalicylic acid *[5965-83-3]* **M 254.2, m 108-110°.** Crystd from water. Alternatively, it was converted to the monosodium salt which was crystd from water and washed with a little water, EtOH and then ethyl ether. The free acid was recovered by acidification.

Syringaldehyde *[134-96-3]* **M 182.2, m 113°.** Crystd from pet ether.

D(-)-Tagatose *[87-81-0]* M 180.2, m 134-135°, $[\alpha]^{546}$ -6.5° (c 1, H_2O). Crystd from aq EtOH.

d- Tartaric acid *[147-71-7]* M 150.1, m 169.5-170° (2S,3S-form, natural) $[\alpha]_{546}^{o}$ -15° (c 10, H_2O); m 208° (2RS,3RS-form). Crystd from distd water or benzene/ethyl ether containing 5% of pet ether (b 60-80°) (1:1). Soxhlet extraction with ethyl ether has been used to remove an impurity absorbing at 265nm. It has also been crystd from abs EtOH/hexane, and dried in a vac for 18hr [Kornblum and Wade *JOC* **52** 5301 *1987*].

*meso-*Tartaric acid *[147-73-9]* M 150.1, m 139-141°. Crystd from water, washed with cold MeOH and dried at 60° under vac.

Taurocholic acid *[81-24-3]* M 515.6, m 125°(dec). Crystd from EtOH/ethyl ether.

Terephalaldehyde *[623-27-8]* M 134.1, m 116°, b 245-248°/771mm. Crystd from water.

Terephthalic acid *[100-21-0]* M 166.1, sublimes >300° without melting. Purified *via* the sodium salt which, after crystn from water, was reconverted to the acid by acidification with mineral acid.

Terephthaloyl chloride *[100-20-9]* M 203.0, m 80-82°. Crystd from dry hexane.

*o-*Terphenyl *[84-15-1]* M 230.3, m 57-58°,
*m-*Terphenyl *[92-06-8]* M 230.3, m 88-89°. Crystd from EtOH. Purified by chromatography of CCl_4 solns on alumina, with pet ether as eluent, followed by crystn from pet ether (b 40-60°) or pet ether/benzene. They can also be distd under vac.

*p-*Terphenyl *[92-94-4]* M 230.3, m 212.7°. Crystd from nitrobenzene or trichlorobenzene. It was purified by chromatography on alumina in a darkened room, using pet ether, and then crystallizing from pet ether (b 40-60°) or pet ether/benzene.

Terpin hydrate *[2451-01-6]* M 190.3, m 105.5° (*cis*), 156-158° (*trans*). Crystd from water or EtOH.

2,2':6',2"-Terpyridyl *[1148-79-4]* M 233.3, m 91-92°. Crystd from ethyl ether, toluene or from pet ether, then aq MeOH, followed by vac sublimation at 90°.

Terramycin *[79-57-2]* M 248.4, sinters at 182°, melts at 184-185°(dec), $[\alpha]_D$ -196.6° (equilibrium in 0.1M HCl), -2.1° (equilibrium in 0.1M NaOH). Crystd (as dihydrate) from water or aq EtOH.

Terric acid *[121-40-4]* M 154.1, m 127-127.5°. Crystd from benzene or hexane. Sublimed *in vacuo*.

Testosterone *[58-22-0]* M 288.4, m 155°, $[\alpha]_{546}$ +130° (c 1, dioxane). Crystd from aq acetone.

Testosterone propionate *[57-85-2]* M 344.5, m 118-122°, $[\alpha]_{546}$ +100° (c 1, dioxane). Crystd from aq EtOH.

2,3,4,6-Tetraacetyl-α-D-glucopyranosyl bromide see **acetobromoglucose.**

Tetra-*n*-amylammonium bromide *[866-97-7]* M 278.6, m 100-101°. Purified by crystn from acetone/ether mixtures, and dried *in vacuo* at 60° for 2 days.

Tetra-*n*-amylammonium iodide *[2498-20-6]* M 425.5, m 135-137°. Crystd from EtOH and dried at 35° under vac. Also purified by dissolving in acetone and pptd by adding ethyl ether; and dried at 50° for 2 days.

Tetrabenazine *[58-46-8]* M 317.4, m 127-128°. Crystd from MeOH.

3',3'',5'.5''-Tetrabromo-*m*-cresolsulphophthalein see **bromocresol green.**

1,1,2,2,-Tetrabromoethane *[79-27-6]* M 345.7, f.p. 0.0°, b 243.5°, n 1.63533, d 2.965. Washed successively with conc H_2SO_4 (three times) and water (three times), dried with K_2CO_3 and $CaSO_4$ and distd.

Tetrabromophenolphthalein ethyl ester *[1176-74-5]* M 662.0. Crystd from benzene, dried at 120° and kept under vac.

3',3'',5',5''-Tetrabromophenolsulphonephthalein see **bromophenol blue.**

Tetra-*n*-butylammonium bromide *[1643-19-2]* M 322.4, m 119.6°. Crystd from benzene (5ml/g) at 80° by adding hot *n*-hexane (three volumes) and allowing to cool. Dried over P_2O_5 or $Mg(ClO_4)_2$, under vac. The salt is *very hygroscopic*. It can also be crystd from ethyl acetate or dry acetone by adding ethyl ether and dried *in vacuo* at 60° for 2 days.. It has been crystd from acetone by addition of ethyl ether. So hygroscopic that all manipulations should be carried out in a dry-box. Purified by pptn of a satd soln in dry CCl_4 by addition of cyclohexane or by recrystn from ethyl acetate, then heating in vac to 75° in the presence of P_2O_5. [Symons *et al, JCSFT 1* **76** 2251 *1908*]. Also crystd from CH_2Cl_2/ethyl ether [Blau and Espenson *JACS* **108** 1962 *1986*].

Tetra-*n*-butylammonium chloride *[1112-67-0]* M 295.9. Crystd from acetone by addition of ethyl ether. *Very hygroscopic.*

Tetra-*n*-butylammonium fluoroborate *[429-42-5]* M 329.3, m 161-163°. Recrystd from ethyl acetate/pentane in dry acetonitrile. [Hartley and Faulkner *JACS* **107** 3436 *1985*.]

Tetra-*n*-butylammonium hexafluorophosphate *[3109-27-8]* M 387.5, m 239-241°. Recrystd from satd EtOH/water and dried for 10hr in vac at 70°. It was also recrystd three times from abs EtOH and dried for 2 days in a drying pistol under vac at boiling toluene temperature [Bedard and Dahl *JACS* **108** 5933 *1986*].

Tetra-*n*-butylammonium hydrogen sulphate *[32503-27-8]* M 339.5, m 171-172°. Crystd from acetone.

Tetra-*n*-butylammonium iodide *[311-28-4]* M 369.4, m 146°. Crystd from toluene/pet ether (see entry for the corresponding bromide), acetone, ethyl acetate, EtOH/ethyl ether, nitromethane, aq EtOH or water. Dried at room temperature under vac. It has also been dissolved in MeOH/acetone (1:3, 10ml/g), filtered and allowed to stand at room temperature to evaporate to *ca* half its original volume. Distd water (1ml/g) was then added, and the ppte was filtered off and dried. It was also dissolved in acetone, pptd by adding ether and dried in vac at 90° for 2 days. Crystd from CH_2Cl_2/pet ether or hexane, or anhydrous MeOH and stored over P_2O_5. [Chau and Espenson *JACS* **108** 1962 *1986*.]

Tetrabutylammonium nitrate *[1941-27-1]* M 304.5, m 119°. Crystd from benzene (7ml/g) or EtOH., dried in a vac over P_2O_5 at 60° for 2 days.

Tetra-*n*-butylammonium perchlorate *[1923-70-2]* M 341.9°, m 210°(dec). Crystd from EtOH, ethyl acetate, from *n*-hexane or ethyl ether/acetone mixture, ethyl acetate or hot CH_2Cl_2. Dried in vac at room temperature over P_2O_5 for 24hr. [Anson *et al, JACS* **106** 4460 *1984*; Ohst and Kochi *JACS* **108** 2877 *1986*; Collman *et al, JACS* **108** 2916 *1986*; Blau and Espenson *JACS* **108** 1962 *1986*; Gustowski *et al,JACS* **108** *1986*; Ikezawa and Kutal *JOC* **52** 3299 *1987*.]

Tetra-*n*-butylammonium picrate *[914-45-4]* M 490.6, m 89°. Crystd from EtOH and dried under vac.

Tetrabutylammonium tetrabutylborate (Bu$_4$N$^+$ Bu$_4$B$^-$) *[23231-91-6]* M 481.7, m 109.5°. Dissolved in MeOH or acetone, and crystd by adding distd water. Dried in vac at 70°. It has also been successively recrystd from isopropyl ether, isopropyl ether/acetone (50:1) and isopropyl ether/EtOH (50:1) for 10hr, then isopropyl ether/acetone for 1hr, and dried at 65° under reduced pressure for 1 week. [Kondo *et al, JCSFT 1* 76 812 *1980*.]

Tetrabutylammonium tetrafluoroborate *[429-42-5]* M 329.3, m 160-162°. Recrystd from ethyl acetate, and dried at 80° under vac [Detty and Jones *JACS* 109 5666 *1987*].

1,2,4,5-Tetrachloroaniline *[634-83-3]* M 230.9, m 119-120°,
2,3,5,6-Tetrachloroaniline *[3481-20-7]* M 230.9, m 107-108°. Crystd from EtOH.

1,2,3,4-Tetrachlorobenzene *[634-66-2]* M 215.9, m 45-46°, b 254°/760mm,
1,2,3,5-Tetrachlorobenzene *[634-90-2]* M 215.9, m 51°, b 246°/760mm. Crystd from EtOH.

1,2,4,5-Tetrachlorobenzene *[95-94-3]* M 215.9, m 139.5-140.5°, b 240°/760mm. Crystd from EtOH, ether, benzene, benzene/EtOH or carbon disulphide.

Tetrachloro-*o*-benzoquinone *[2435-53-2]* M 245.9, m 130°. Crystd from glacial acetic acid.

1,1,2,2-Tetrachloro-1,2-difluoroethane *[72-12-0]* M 203.8, f.p. 26.0°, b 92.8°/760mm. Purified as for trichlorotrifluoroethane.

sym-**Tetrachloroethane** *[79-34-5]* M 167.9, b 146.2°, n^{15} 1.49678, d 1.588. Stirred, on a steam-bath, with conc H$_2$SO$_4$ until a fresh portion of acid remained colourless. The organic phase was then separated, distd in steam, dried (CaCl$_2$ or K$_2$CO$_3$), and fractionally distd.

Tetrachloroethylene *[127-18-4]* M 165.8, b 121.2°, n^{15} 1.50759, n 1.50566, d^{15} 1.63109, d 1.623. It decomposes under similar conditions to CHCl$_3$, to give phosgene and trichloroacetic acid. Inhibitors of this reaction include EtOH, ethyl ether and thymol (effective at 2-5ppm). Tetrachloroethylene should be distd under a vac (to avoid phosgene formation), and stored in the dark out of contact with air. It can be purified by washing with 2M HCl until the aq phase no longer becomes coloured, then with water, drying with Na$_2$CO$_3$, Na$_2$SO$_4$, CaCl$_2$ or P$_2$O$_5$, and fractionally distilling just before use. 1,1,2-Trichloroethane and 1,1,1,2-tetrachloroethane can be removed by counter-current extraction with EtOH/water.

Tetrachloro-*N*-methylphthalimide *[14737-80-5]* M 298.9, m 209.7°. Crystd from abs EtOH.

2,3,4,6-Tetrachloronitrobenzene *[879-39-0]* M 260.9, m 42°,
2,3,5,6-Tetrachloronitrobenzene *[117-18-0]* M 260.9, m 99-100°. Crystd from aq EtOH.

2,3,4,5-Tetrachlorophenol *[4901-51-3]* M 231,9, m 116-117°,
2,3,4,6-Tetrachlorophenol *[58-90-2]* M 231.9, m 70°, b 150°/15mm,
2,3,5,6-Tetrachlorophenol *[935-95-5]* M 231.9, m 115°. Crystd from ligroin.

Tetrachlorophthalic anhydride *[117-08-8]* M 285.9, m 255-257°. Crystd from chloroform or benzene, then sublimed.

2,3,4,6-Tetrachloropyridine *[14121-36-9]* M 216.9, m 74-75°, b 130-135°/16-20mm. Crystd from 50% EtOH.

Tetracosane *[646-31-1]* M 338.7, m 54°, b 243-244°/15mm. Crystd from ether.

Tetracosanoic acid *[557-59-5]* M 368.7, m 87.5-88°. Crystd from acetic acid.

1,2,4,5-Tetracyanobenzene *[712-74-3]*, **M 178.1, m 270-272° (280°).** Crystd from EtOH and sublimed *in vacuo*. [Lawton and McRitchie *JOC* **24** 26 *1959*; Bailey *et al*, *Tetrahedron* **19** 161 *1963*.]

Tetracyanoethylene *[670-54-2]* **M 128.1, m 199-200° (sealed tube).** Crystd from chlorobenzene, dichloroethane, or methylene dichloride [Hall *et al*, *JOC* **52** 5528 *1987*]. Stored at 0° in a desiccator over NaOH pellets. (It slowly evolves HCN on exposure to moist air.) It can also be sublimed at 120° under vac. Also purified by repeated sublimation at 120-130°/0.5mm. [Frey *et al*, *JACS* **107** 748 *1985*; Traylor and Miksztal *JACS* **109** 2778 *1987*].

7,7,8,8-Tetracyanoquinodimethane *[1518-16-7]* **M 204.2, m 287-290°(dec).** Recrystd from redistd dried acetonitrile.

Tetracycline *[60-54-8]* **M 444.4, m 172-174°(dec), $[\alpha]_{546}$ +270° (c 1, MeOH).** Crystd from toluene.

Tetradecane *[629-59-4]* **M 198.4, m 6°, b 122°/10mm, 252-254°, n 1.429, d 0.763.** Washed successively with 4M H_2SO_4 and water. Dried over $MgSO_4$ and distd several times under reduced pressure [Poë *et al*, *JACS* **108** 5459 *1986*].

Tetradecanoic acid see **myristic acid.**

1-Tetradecanol *[112-72-1]* **M 214.4, m 39-39.5°, b 160°/10mm, 170-173°/20mm.** Crystd from aq EtOH. Purified by zone melting.

Tetradecyl ether *[5412-98-6]* **M 410.7.** Distd under vac and then crystd repeatedly from MeOH/benzene.

Tetradecyltrimethylammonium bromide *[1119-97-7]* **M 336.4, m 244-249°.** Crystd from acetone or a mixture of acetone and >5% MeOH. Washed with ethyl ether and dried in a vac oven at 60°. [Dearden and Wooley *JPC* **91** 2404 *1987*.]

Tetraethoxymethane *[62695-86-7]* **M 192.2, b 159°.** Dried with Na_2SO_4 and distd.

Tetraethylammonium bromide *[71-91-0]* **M 210.2, m 284°(dec).** Recrystd from EtOH , $CHCl_3$ or ethyl ether, or, recrystd from acetonitrile, and dried over P_2O_5 under reduced pressure for several days. Also recrystd from EtOH/ethyl ether (1:2), ethyl acetate, water or boiling MeOH/acetone (1:3) or by adding equal volume of acetone and allowing to cool. Dried at 100° *in vacuo* for 12 days, and stored over P_2O_5.

Tetraethylammonium chloride *[56-34-8]* **M 165.7.** Crystd from EtOH by adding ethyl ether, from warm water by adding EtOH and ethyl ether, from dimethylacetamide or from CH_2Cl_2 by addition of ethyl ether. Dried over P_2O_5 in vac for several days. Also crystd from acetone/CH_2Cl_2/hexane (2:2:1) [Blau and Espenson *JACS* **108** 1962 *1986*; White and Murray *JACS* **109** 2576 *1987*].

Tetraethylammonium iodide *[68-05-3]* **M 257.2, m >300°(dec).** Crystd from acetone/MeOH, EtOH/water, dimethylacetamide or ethyl acetate/EtOH (19:1). Dried under vac at 50° and stored over P_2O_5.

Tetraethylammonium perchlorate *[2567-83-1]* **M 229.7.** Crystd repeatedly from water, aq MeOH, acetonitrile or acetone, and dried at 70° under vac for 24hr. [Cox *et al*, *JACS* **106** 5965 *1984*; Liu *et al*, *JACS* **108** 1740 *1986*; White and Murray *JACS* **109** 2576 *1987*.] Also twice crystd from ethyl acetate/95% EtOH (2:1) [Lexa *et al*, *JACS* **109** 6464 *1987*].

Tetraethylammonium picrate *[741-03-7]* **M 342.1.** Purified by successive crystns from water or 95% EtOH followed by drying in vac at 70°.

Tetraethylammonium tetrafluoroborate *[429-06-1]* **M 217.1.** Recrystd three times from a mixture of ethyl acetate/hexane (5:1) or MeOH/pet ether, then stored at 95° for 48hr under vac [Henry and Faulkner *JACS* **107** 3436 *1985*; Huang *et al, AC* **58** 2889 *1986*].

Tetraethylammonium tetraphenylborate *[12099-10-4]* **M 449.4.** Recrystd from aq acetone. Dried in a vac oven at 60° for several days. *Similarly for the propyl and butyl homologues.*

Tetraethyl 1,1,2,2-ethanetetracarboxylate *[632-56-4]* **M 318.3, m 73-74°.** Twice recrystd from EtOH by cooling to 0°.

Tetraethylene glycol dimethyl ether *[143-24-8]* **M 222.3, b 105°/1mm, n 1.435, d 1.010.** Stood with CaH_2, $LiAlH_4$ or sodium, and distd when required.

Tetraethylenepentamine *[112-57-2]* **M 189.3, b 169-171°/0.05mm, n 1.506, d 0.999.** Distd under vac. Purified *via* its pentachloride, nitrate or sulphate. Jonassen, Frey and Schaafsma [*JPC* **61** 504 *1957*] cooled a soln of 150g of the base in 300ml of 95% EtOH, and added dropwise 180ml of conc HCl, keeping the temperature below 20°. The white ppte was filtered, crystd three times from EtOH/water, then washed with ethyl ether and dried by suction. Reilley and Holloway [*JACS* **80** 2917 *1958*], starting with a similar soln cooled to 0°, added slowly (keeping the temperature below 10°) a soln of 4.5g-moles of HNO_3 in 600ml of aq 50% EtOH (also cooled to 0°). The ppte was filtered by suction, recrystd five times from aq 5% HNO_3, then washed with acetone and abs EtOH and dried at 50°. [For purification *via* the sulphate see Reilley and Vavoulis (*AC* **31** 243 *1959*), and for an additional purification step using the Schiff base with benzaldehyde see Jonassen *et al, JACS* **79** 4279 *1957*.]

1,1,2,2-Tetrafluorocyclobutane *[374-12-9]* **M 128.1.** Purified by preparative gas chromatography using a 2m x 6mm(i.d.) column packed with β,β'-oxydipropionitrile on Chromosorb P at 33°. [Conlin and Fey *JCSFT 1* **76** 322 *1980*.]

Tetrafluoro-1,3-dithietane *[1717-50-6]* **M 164.1, m -6°, b 47-48°/760mm, n^{25} 1.3908, d^{25} 1.6036.** Purified by preparative gas chromatography or by distn through an 18in spinning band column. Also purified by shaking vigorously *ca* 40ml with 25ml of 10% NaOH, 5ml of 30% H_2O_2 until the yellow colour disappeared The larger layer was separated, dried over silica gel to give a colourless liquid boiling at 48°. It had a single line at -1.77ppm in the NMR spectrum. [Middleton, Howard and Sharkey, *JOC* **30** 1375 *1965*.]

2,2,3,3-Tetrafluoropropanol *[76-37-9]* **M 132.1, b 106-106.5°.** Tetrafluoropropanol (450ml) was added to a soln of 2.25g of $NaHSO_3$ in 90ml of water, shaken vigorously and stood for 24hr. The fraction distilling at or above 99° was refluxed for 4hr with 5-6g of KOH and rapidly distd, followed by a final fractional distn. [Kosower and Wu *JACS* **83** 3142 *1961*.] Alternatively, shaken with alumina for 24hr, dried overnight with anhydrous K_2CO_3 and distd, taking the middle fraction (b 107-108°).

Tetera-*n*-heptylammonium bromide *[4368-51-8]* **M 490.7, m 89-91°.** Crystd from *n*-hexane, then dried in a vac oven at 70°.

Tetra-*n*-heptylammonium iodide *[3535-83-9]* **M 537.7.** Crystd from EtOH.

Tetra-*n*-hexylammonium bromide *[4328-13-6]* **M 434.6, m 99-100°.** Washed with ether, and dried in a vac at room temperature for 3 days.

Tetra-*n*-hexylammonium chloride *[5922-92-9]* **M 390.1.** Crystd from EtOH.

Tetra-*n*-hexylammonium iodide *[2138-24-1]* **M 481.6, m 99-101°.** Washed with ethyl ether and dried at room temperature *in vacuo* for 3 days.

Tetrahexylammonium perchlorate *[4656-81-9]* **M 454.1, m 104-106°.** Crystd from acetone and dried *in vacuo* at 80° for 24hr.

Tetrahydrofuran *[109-99-9]* **M 72.1, b 65.4°, n^{25} 1.4040, d 0.888.** It is obtained commercially by catalytic hydrogenation of furan from pentosan-containing agricultural residues. It was purified by refluxing with, and distilling from LiAlH$_4$ which removes water, peroxides, inhibitors and other impurities. Peroxides can also be removed by passage through a column of activated alumina, or by treatment with aq ferrous sulphate and sodium bisulphate, followed by solid KOH. In both cases, the solvent is then dried and fractionally distd from sodium. Lithium wire or vigorously stirred molten potassium have also been used for this purpose. CaH$_2$ has also been used as a drying agent.
Several methods are available for obtaining the solvent almost anhydrous. Ware [*JACS* **83** 1296 *1961*] dried vigorously with sodium-potassium alloy until a characteristic blue colour was evident in the solvent at Dry-ice/cellosolve temperatures. The solvent was kept in contact with the alloy until distd for use. Worsfold and Bywater [*JCS* 5234 *1960*], after refluxing and distilling from P$_2$O$_5$ and KOH, in turn, refluxed the solvent with sodium-potassium alloy and fluorenone until the green colour of the disodium salt of fluorenone was well established. [Alternatively, instead of fluorenone, benzophenone, which forms a blue ketyl, can be used.] The tetrahydrofuran was then fractionally distd, degassed and stored above CaH$_2$. *p*-Cresol or hydroquinone inhibit peroxide formation. The method described by Coetzee and Chang [*PAC* **57** 633 *1985*] for 1,4-dioxane also applies here.

1,2,3,4-Tetrahydronaphthalene see **tetralin.**

l-**Tetrahydropalmatine** **M 355.4, m 148-149°, [α]$_D$ 291° (EtOH).** Crystd from MeOH by addition of water [see *JCS,C* 530 *1967*].

Teterahydropyran *[142-68-7]* **M 86.1, b 88.0°, n 1.4202, d 0.885.** Dried with CaH$_2$, then passed through a column of silica gel to remove olefinic impurities and fractionally distd. Freed from peroxides and moisture by refluxing with sodium, then distilling from LiAlH$_4$. Alternatively, peroxides can be removed by treatment with aq ferrous sulphate and sodium bisulphate, followed by solid KOH, and fractional distn from sodium.

Tetrahydrothiophen *[110-01-0]* **M 88.2, m -96°, b 14.5°/1omm, 120.9°/760mm, n 1.5289, d 0.997.** Crude material was purified by crystn of the mercuric chloride complex to a constant melting point. It was then regenerated, washed, dried, and fractionally distd. [Whitehead *et al, JACS* **73** 3632 *1951*.]. It has been dried over Na$_2$SO$_4$ and distd in a vac [Roberts and Friend *JACS* **108** 7204 *1986*].

Tetrahydroxy-p-benzoquinone (2H$_2$O) *[5676-48-2]* **M 208.1.** Crystd from water.

Tetrakis(dimethylamine)ethylene *[996-70-3]* **M 300.2, b 60°/1mm, n 1.4817, d 0.861.** Impurities include tetramethylurea, dimethylamine, tetramethylethanediamine and tetramethyloxamide. It was washed with water while being flushed with nitrogen to remove dimethylamine, dried over molecular sieves, then passed through a silica gel column (previously activated at 400°) under nitrogen. Degassed on a vacuum line by distn from a trap at 50° to one at -70°. Finally, it was stirred over sodium-potassium alloy for several days. [Holroyd *et al, JPC* **89** 4244 *1985*.]

Tetralin *[119-64-2]* **M 132.2, n 65-66°/5mm, 207.6°/760mm, n 1.5413, d 0.968.** It was washed with successive portions of conc H$_2$SO$_4$ until the acid layer no longer became coloured, then washed with aq 10% Na$_2$CO$_3$, and then distd water. Dried (CaSO$_4$ or Na$_2$SO$_4$), filtered, refluxed and fractionally distd at under reduced pressure from sodium or BaO. It can also be purified by repeated fractional freezing. Bass [*JCS* 3498 *1964*] freed tetralin, purified as above, from naphthalene and other impurities by conversion to ammonium tetralin-6-sulphonate. Conc H$_2$SO$_4$ (150ml) was added slowly to stirred tetralin (272ml) which was then heated on a water bath for about 2hr to give complete soln. The warm mixture, when poured into aq NH$_4$Cl soln (120g in 400ml water), gave a white ppte which, after filtering off, was crystd from boiling water, washed with 50% aq EtOH and dried at 100°. Evapn of its boiling aq soln on a steam bath removed traces of naphthalene. The pure salt (229g) was mixed with conc H$_2$SO$_4$ (266ml) and steam distd from an oil bath at 165-170°. An ether extract of the distillate was washed with aq Na$_2$SO$_4$, and the ether was evapd, prior to distilling the tetralin from sodium. Tetralin has also been purified *via* barium tetralin-6-sulphonate, conversion to the sodium salt and decomposition in 60% H$_2$SO$_4$ using superheated steam.

Tetralin hydroperoxide *[771-29-9]* **M 164.2, m 56°.** Crystd from hexane.

Tetramethylammonium bromide *[64-20-0]* **M 154.1, sublimes with dec >230°.** Crystd from EtOH, EtOH/ethyl ether, MeOH/acetone, water or from acetone/MeOH (4:1) by adding an equal volume of acetone. It was dried at 110° under reduced pressure or at 140° for 24hr.

Tetramethylammonium chloride *[57-75-0]* **M 109.6, m >230°(dec).** Crystd from EtOH, EtOH/CHCl$_3$, EtOH/ethyl ether, acetone/EtOH (1:1), isopropanol or water. Traces of the free amine can be removed by washing with CHCl$_3$.

Tetramethylammonium hydroxide (2H$_2$O) *[10424-65-4]* **M 181.2, m 63°(dec).** Freed from chloride ions by passage through an ion-exchange column (Amberlite IRA-400, prepared in its OH⁻ form by passing 2M NaOH until the effluent was free from chloride ions, then washed with distd water until neutral). A modification, to obtain carbonate-free hydroxide, uses the method of Davies and Nancollas [*Nature* 165 237 *1950*].

Tetramethylammonium iodide *[75-58-1]* **M 201.1, m >230°(dec).** Crystd from water or 50% EtOH, EtOH/ethyl ether, ethyl acetate, or from acetone/MeOH (4:1) by adding an equal volume of acetone. Dried in a vac desiccator.

Tetramethylammonium perchlorate *[2537-36-2]* **M 173.6, m >300 °(dec).** Crystd from acetone and dried *in vacuo* at 60° for several days.

Tetramethylammonium tetraphenylborate *[15525-13-0]* **M 393,3.** Recrystd from acetone, acetone/CCl$_4$ and from acetone/1,2-dichloroethane. Dried over P$_2$O$_5$ in vac, or in a vac oven at 60° for several days.

1,2,3,4-Teteramethylbenzene *[488-23-3]* **M 134.2, m -6.3°, b 79.4°/10mm, 204-205°/760mm, n 1.5203, d 0.905.** Dried over sodium and distd under reduced pressure.

1,2,3,5-Tetramethylbenzene (isodurene) *[527-53-7]* **M 134.2, m -23.7°, b 74.4°/10mm, 198°/760mm, n 1.5130, d 0.890.** Refluxed over sodium and distd under reduced pressure.

1,2,3,5-Tetramethylbenzene see **durene.**

N,N,N',N'-**Tetramethylbenzidine** *[366-29-0]* **M 240.4, m 195.4-195.6°.** Crystd from EtOH or pet ether, then from pet ether/benzene, and sublimed in a vac. [Guarr *et al, JACS* 107 5104 *1985*.] Dried under vac in an Abderhalden pistol, or carefully on a vacuum line.

2,2,4,4-Tetramethylcyclobutan-1,3-dione *[933-52-8]* **M 140.2, m 114.5-114.9°.** Crystd from benzene and dried under vac over P$_2$O$_5$ in an Abderhalden pistol.

p,p'-**Tteramethyldiaminophenylmethane** *[101-61-1]* **M 254.4, m 89-90°.** Crystd from EtOH (2ml/g).

3,3,4,4-Tetramethyldiazetidine dioxide **M 144.2.** Purified by recrystn from MeOH.

Tetramethylene glycol see **1,4-butanediol.**

Tetramethylenesulphone see **sulpholane.**

Tetramethylene sulphoxide (tetrahydrothiophen 1-oxide) *[1600-44-8]* **M 104.2, b 235-237°, n 1.525, d 1.175.** Shaken with BaO for 4 days, then distd from CaH$_2$ under reduced pressure.

N,N,N',N'-**Tetramethylethylenediamine** (TEMED) *[110-18-9]* **M 116.2, b 122°, n^{25} 1.4153, d 1.175.** Partially dried with molecular sieves (Linde type 4A), and distd in vac from butyl lithium. This treatment removes all traces of primary and secondary amines and water. [Hay, McCabe and Robb *JCSTF 1* **68** 1 *1972*.] Or, dried with KOH pellets. Refluxed for 2hr with one-sixth its weight of *n*-butyric anhydride (to remove primary and secondary amines) and fractionally distd. Refluxed with fresh KOH, and distd under nitrogen. [Cram and Wilson *JACS* **85** 1245 *1963*.] Also distd from sodium.

Tetramethylethylenediamine dihydrochloride *[7677-21-8]* **M 198.2.** Crystd from 98% EtOH/conc HCl.

1.1.3.3-Tetramethylguanidine *[80-70-6]* **M 115.2, b 159-160°, n 1.470, d 0.917.** Refluxed over granulated BaO, then fractionally distd.

N,N,N',N'-**Tetramethyl-1,4-phenylenediamine** *[100-22-1]* **M 164.3, m 51°, b 260°/760mm.** Crystd from pet ether or water. It can be sublimed or dried carefully in a vac line, and stored in the dark under nitrogen. Also recrystd from its melt.

N,N,N',N'-**Tetramethyl-1,4-phenylenediamine dihydrochloride** *[637-01-4]* **M 237.2, m 222-224°.** Crystd from isopropyl or *n*-butyl alcohols, satd with HCl. Treated with aq NaOH to give the free base which was filtered, dried and sublimed in a vac. [Guarr *et al, JACS* **107** 5104 *1985*.]

2,2,6,6-Tetramethylpiperidinyl-1-oxy (TEMPO) *[2564-83-2]* **M 156.3, m 36-38°.** Purified by sublimation (33°, water aspirator) [Hay and Fincke *JACS* **109** 8012 *1987*].

Tetramethylthiuram disulphide [bis(dimethylthiocarbamyl)-disulphide] *[137-26-8]* **M 240.4, m 146-148°.** Crystd (three times) from boiling CHCl$_3$, then recrystd from boiling CHCl$_3$ by adding EtOH dropwise to initiate pptn, and allowed to cool. Finally it was pptd from cold CHCl$_3$ by adding EtOH (which retained the monosulphide in soln). [Ferington and Tobolsky *JACS* **77** 4510 *1955*.]

1,1,3,3-Tetramethyl urea *[632-22-4]* **M 116.2, f.p. -1.2°, b 175.2°/760mm, n 1.453, d 0.969.** Dried over BaO and distd under nitrogen.

Tetramethyl uric acid *[2309-49-1]* **M 224.2, m 228°.** Crystd from water.

1,3,5,5-Tetranitrohexahydropyrimidine *[81360-42-1]* **M 270.1.** Crystd five times from EtOH.

Tetranitromethane *[509-14-8]* **M 196.0, m 14.2°, b 21-23°/23mm, 126°/760mm, n 1.438, d 1.640.** Shaken with dil NaOH, washed, steam distd, dried with Na$_2$SO$_4$ and fractionally crystd by partial freezing. The melted crystals were dried with MgSO$_4$ and fractionally distd under reduced pressure. Shaken with a large volume of dil NaOH until no absorption attributable to the nitroform anion is observable in the water. Then washed with distd water, and distd at room temperature by passing a stream of air or nitrogen through the liquid and condensing in a trap at -80°. It can be dried with MgSO$_4$ or Na$_2$SO$_4$, fractionally crystd from the melt, and fractionally distd under reduced pressure.

Tetra(*p*-nitrophenyl)ethylene *[47797-98-8]* **M 512.4.** Crystd from dioxane and dried at 150°/0.1mm.

4,7,13,18-Tetraoxa-1,10-diazabicyclo[8.5.5]eicosane (Cryptand 211) *[31250-06-3]* **M 288.1.** Redistd, dried under high vac over 24hr, and stored under nitrogen.

1,7,10,16-Tetraoxa-4,13-diazacyclooctadecane (4,13-diaza-18-crown-6) *[23978-55-4]* **M 262.3, m 118-116°**. Twice recrystd from benzene/n-heptane, and dried for 24hr under high vacuum [D'Aprano and Sesta *JPC* **91** 2415 *1987*].

Tetrapentylammonium bromide *[866-97-7]* **M 378.5, m 100-101°**. Crystd from pet ether, benzene or acetone/ether and dried in vac at 40-50°.

Tetraphenylethylene *[632-51-9]* **M 332.4, m 223-224°, b 415-425°/760mm**. Crystd from dioxane or from EtOH/benzene. Sublimed under high vac.

Tetraphenylhydrazine *[632-52-0]* **M 336.4, m 147°**. Crystd from 1:1 CHCl$_3$/toluene or CHCl$_3$/EtOH. Stored in a refrigerator, in the dark.

trans-**1,1,4,4-Tetraphenyl-2-methylbutadiene** *[20411-57-8]* **M 372.5**. Crystd from EtOH.

Tetraphenylphosphonium chloride *[2001-45-8]* **M 374.9, m 273-275°**. Crystd from acetone. Dried at 70° under vac. Also recrystd from a mixture of 1:1 or 1:2 dichloroethane/pet ether, the solvents having been dried under anhydrous K$_2$CO$_3$. The purified salt was dried at room temperature under vac for 3 days, and at 170° for a further three days. *Extremely hygroscopic.*

5,10,15,20-Tetraphenylporphyrin (TPP) *[917-23-7]* **M 614.7, λ_{max} 482nm**. Purified by chromatography on neutral (Grade I) alumina, and recrystd from CH$_2$Cl$_2$/MeOH [Yamashita *et al, JPC* **91** 3055 *1987*].

Tetra-n-propylammonium bromide *[1941-30-6]* **M 266.3, m >280°(dec)**. Crystd from ethyl acetate/EtOH (9:1), acetone or MeOH. Dried at 110° under reduced pressure.

Tetra-n-propylammonium iodide *[631-40-3]* **M 313.3, m >280°(dec)**. Purified by crystn from EtOH, EtOH/ethyl ether (1:1), EtOH/water or aq acetone. Dried at 50° under vac. Stored over P$_2$O$_5$ in a vac desiccator.

Tetra-n-propylammonium perchlorate *[15780-02-6]* **M 285.8, m 239-241°**. Crystd from acetonitrile/water (1:4.v/v), or conductivity water. Dried in an oven at 60° for several days, or dried under vac over P$_2$O$_5$.

5,10,15,20-Tetra-4'-pyridinylporphyrin *[16834-13-2]* **M 618.7**. Purified by chromatography on alumina (neutral, Grade I), followed by recrystn from CH$_2$Cl$_2$/MeOH [Yamashita *et al, JPC* **91** 3055 *1987*].

Tetrathiafulvalene *[31366-25-3]* **M 204.4, m 122-124°**. Recrystd from cyclohexane/hexane under an argon atmosphere [Kauzlarich *et al, JACS* **109** 4561 *1987*].

1,2,3,4-Tetrazole *[288-94-8]* **M 70.1, m 156°**. Crystd from EtOH, sublimed under high vac at *ca* 120° (*care should be taken due to possible* explosion).

TETREN see **tetraethylenepentamine**.

Thapsic acid see **1,14-hexadecanedioic acid**.

Thebaine *[115-37-7]* **M 311.4, m 193°, $[\alpha]_D^{25}$ -219° (EtOH)**. Sublimed at 170-180°.

2-Thenoyltrifluoroacetone *[326-91-0]* **M 222.2**. Crystd from hexane or benzene. (Aq solns slowly decompose.)

2-Thenylamine *[27757-85-3]* **M 113.1, b 78.5°/15mm**. Distd under reduced pressure (nitrogen), from BaO, through a column packed with glass helices.

Theobromine *[83-67-0]* **M 180.2, m 337°,**
Theophylline *[58-55-9]* **M 180.2, m 270-274°.** Crystd from water.

Thevetin *[11018-93-2]* **M 858.9, softens at 194°, m 210°.** Crystd (as trihydrate) from isopropanol. Dried at 100°/10mm to give the hemihydrate (*very hygroscopic*).

Thianthrene *[92-85-3]* **M 216.3, m 158°.** Crystd from acetone (charcoal), acetic acid or EtOH. Sublimed under vac.

ε-[2-(4-Thiazolidone)]hexanoic acid M 215.3, m 140°. Crystd from water, acetone or MeOH.

4-(2-Thiazolylazo)resorcinol *[2246-46-0]* **M 221.2, m 200-202°(dec), λ_{max} 500 nm.** Dissolved in alkali, extracted with ethyl ether, and re-pptd with dil HCl. The purity was checked by tlc on silica gel using pet ether/ethyl ether/EtOH (10:10:1) as the mobile phase.

Thietane *[287-27-4]* **M 74.1, b 93.8-94.2°/752mm, n^{23} 1.5059, d^{23} 1.0283.** Purified by preparative gas chromatography on a dinonyl phthalate column.

Thioacetamide *[62-55-5]* **M 75.1, m 112-113°.** Crystd from abs ethyl ether or benzene. Dried at 70° in vac and stored over P_2O_5 at 0° under nitrogen. (*Develops an obnoxious odour on keeping*, and absorption at 269mm decreases, hence it should be freshly crystd before use.)

Thioacetanilide *[677-53-6]* **M 151.2, m 75-76°.** Crystd from water and dried in a vac desiccator.

Thiobarbituric acid *[504-17-6]* **M 144.2, m 235°(dec).** Crystd from water.

Thiobenzanilide *[636-04-4]* **M 213.2, m 101.5-102°.** Crystd from MeOH at Dry-ice temperature.

Thiocarbanilide see *sym*-diphenylthiourea.

Thiochrome *[92-35-3]* **7M 262.3, m 227-228°.** Crystd from chloroform.

Thiocresol see **toluenethiol**.

6,8-Thioctic acid *[1077-28-7]* **M 206.3, m 45-47.5° (R-isomer), 60-61° (RS-form).** Crystd from cyclohexane.

Thiodiglycollic acid *[123-93-3]* **M 150.2, m 129°,**
β,β'-Thiodipropionic acid *[111-17-1]* **M 178.2.** Crystd from water.

Thioflavine T *[2390-54-7]* **M 318.9.** Crystd from benzene/EtOH (1:1).

Thioformamide *[115-08-2]* **M 61.0, m 29°.** Crystd from ethyl acetate or ether/pet ether.

Thioglycollic acid *[68-11-1]* **M 92.1, b 95-96°/8mm, n 1.505, d 1.326.** Mixed with an equal volume of benzene, the benzene then being distd to dehydrate the acid. After heating to 100° to remove most of the benzene, the residue was distd under vac and stored in sealed ampoules at 3°. [Eshelman *et al, AC* **22** 844 *1960*.]

Thioguanosine *[85-31-4]* **M 299.3, m 230-231°(dec).** Crystd (as hemihydrate) from water.

Thioindigo *[522-75-8]* **M 296.2, m >280°.** Adsorbed on silica gel from CCl_4/benzene (3:1), eluted with benzene, crystd from $CHCl_3$ and dried at 60-65°. [Wyman and Brode *JACS* **73** 1487 *1951*.] This paper also gives details of purification of other thioindigo dyes.

Thiomalic acid *[70-49-5]* **M 150.2, m 153-154°.** Extracted from aq soln several times with ethyl ether, and the aq soln freeze-dried.

Thio-Michler's Ketone *[1226-46-6]* **M 284.6,** λ_{max} **457 nm (**ε **2.92 x 10^4 in 30% aq *n*-propanol).** Purified by recrystn from hot EtOH or by triturating with a small volume of $CHCl_3$, followed by filtration and washing with hot EtOH [Terbell and Wystrade *JPC* **68** 2110 *1964*].

Thionanthone *[492-22-8]* **M 212.3, m 210-213°.** Recrystd from benzene [Ikezawa *et al, JACS* **108** 1589 *1986*].

2-Thionaphthol *[91-60-1]* **M 160.2, m 81°, b 153.5°/15mm, 286°/760mm.** Crystd from EtOH.

Thionin *[581-64-6]* **M 263.8,** ε_{590} **6.2 x 10^4 M^{-1}cm^{-1}.** The standard biological stain is highly pure. It can be crystd from water or 50% EtOH, then chromatographed on alumina using $CHCl_3$ as eluent [Shepp, Chaberek and McNeil *JPC* **66** 2563 *1962*]. Dried overnight at 100° and stored in vac. The hydrochloride can be crystd from 50% EtOH or dil HCl and aq *n*-butanol. Purified also by column chromatography and washed with $CHCl_3$ and acetone. Dried in vac at room temperature.

Thiophane (tetrahydrothiophen) *[110-01-0]* **M 88.2, b 40.3°/39.7mm, n 1.504, d 1.001.** Distd from sodium.

Thiophen *[110-02-1]* **M 84.1, f.p. -38.5°, b 84.2°, n 1.52890, n^{30} 1.5223, d 1.525.** The simplest purification procedure is to dry with solid KOH, or reflux with sodium, and fractionally distd through a glass-helices packed column. More extensive treatments include an initial wash with aq HCl, then water, drying with $CaSO_4$ or KOH, and passage through columns of activated silica gel or alumina. Fawcett and Rasmussen [*JACS* **67** 1705 *1945*] washed thiophen successively with 7M HCl, 4M NaOH, and distd water, dried with $CaCl_2$ and fractionally distd. Benzene was removed by fractional crystn by partial freezing, and the thiophen was degassed and sealed in Pyrex flasks. [Also a method is described for recovering the thiophen from the benzene-enriched portion.]

2-Thiophenaldehyde *[98-03-3]* **M 112.2, m b 106°/30mm, m 1.222, d 1.593.** Washed with 50% HCl and distd under reduced pressure just before use.

Thiophen-2-acetic acid *[1918-77-0]* **M 142.2, m 76°,**
Thiophen-3-acetic acid *[6964-21-2]* **M 142.2, m 79-80°.** Crystd from ligroin.

Thiophen-2-carboxylic acid *[527-72-0]* **M 128.2, m 129-130°,**
Thiophen-3-carboxylic acid *[88-31-1]* **M 128.1, m 137-138°.** Crystd from water.

Thiophenol see **benzenethiol**

Thiopyronine *[2412-14-8]* **M 318.9,** λ_{max} **564nm (**ε **78,500) H$_2$O.** Purified as the hydrochloride by recrystn from hydrochloric acid. [Fanghanel *et al, JPC* **91** 3700 *1987*.]

Thiosalicylic acid *[147-93-3]* **M 154.2, m 164-165°.** Crystd from hot EtOH (4ml/g), after adding hot distd water (8ml/g) and boiling with charcoal. The hot soln was filtered, cooled, the solid collected and dried in vac over P_2O_5.

Thiosemicarbazide *[79-19-6]* **M 91.1, m 181-183°.** Crystd from water.

1-Thiosorbitol *[24531-57-5]* **M 198.2, m 92-93°.** Crystd from EtOH.

Thiothienoyltrifluoroacetone *[4552-64-1]* **M 228.2, m 61-62°.** Easily oxidised and has to be purified before use. This may be by recrystd from benzene or by dissolution in pet ether, extraction into 1M NaOH soln, acidification of the aqueous phase with 1-6M HCl soln, back extraction into pet ether and final evapn of the solvent. The purity can be checked by tlc. It was stored in ampoules under nitrogen at 0° in the dark. [Muller and Rother *Anal Chim Acta* **66** 49 *1973*.]

Thiouracil *[141-90-2]* M 128.2, m 240°(dec). Crystd from water or EtOH.

Thiourea *[62-56-6]* M 76.1, m 179°. Crystd from abs EtOH, MeOH, acetonitrile or water. Dried under vac over H_2SO_4 at room temperature.

Thioxanthene-9-one *[492-22-8]* M 212.3, m 209°. Crystd from $CHCl_3$ and sublimed *in vacuo*.

Thiram see **bis(dimethylcarbamyl)disulphide**.

L-Threonine *[72-19-5]* M 119.1, m 251-253°, $[\alpha]_D^{26}$ -28.4° (H_2O). Likely impurities are *allo*-threonine and glycine. Crystd from water by adding 4 volumes of EtOH. Dried and stored in a desiccator.

Thymidine *[50-89-5]* M 242.2, m 185°. Crystd from ethyl acetate.

Thymine *[65-71-4]* M 126.1, m 326°. Crystd from ethyl acetate or water. Purified by preparative (2mm thick) tlc plates of silica gel, eluting with ethyl acetate/isopropanol/water (75:16:9, v/v; R_F 0.75). Spot localised by uv lamp, cut from plate, placed in MeOH, shaken and filtered through a millipore filter, then rotary evapd. [Infante *et al, JCSFT 1* **68** 1586 *1973*.]

Thymolphthalein complexone *[1913-93-5]* M 720.8, m 190°(dec). Purification as for phthalein complexone except that it was synthesised from thymolphthalein instead of cresolphthalein.

S-Thyroxine *[57-48-9]* M 776.9, m 235°, $[\alpha]_D^{22}$ +26° (EtOH/1M aq HCl; 1:1). Likely impurities are tyrosine, iodotyrosine, iodothyroxines and iodide. Dissolved in dil ammonia at room temperature, then crystd by adding dil acetic acid to *p*H 6.

Tiglic acid *[80-59-1]* M 100.1, m 63.5-64°, b 198.5°. Crystd from water.

Tinuvin P *[50936-05-5]*. Recrystd from *n*-heptane [Woessner *et al, JPC* **81** 3629 *1985*].

***dl*-α-Tocopherol** (vitamin E) *[59-02-9]* M 430.7, $\varepsilon_{1cm}^{1\%}$ 74.2 at 292 nm in MeOH. Dissolved in anhydrous MeOH (15ml/g) cooled to -6° for 1hr, then chilled in a Dry-ice/acetone bath, crystn being induced by scratching with a glass rod.

Tolan see **diphenylacetylene**.

***o*-Tolidine** *[119-93-7]* M 212.3, m 131-132°. Dissolved in benzene, percolated through a column of activated alumina and crystd from benzene/pet ether.

***p*-Tolualdehyde** *[104-87-0]* M 120.2, b 83-85°/0.1mm, 199-200°/760mm, n 1.548, d 1.018. Steam distd, dried with $CaSO_4$ and fractionally distd.

***o*-Toluamide** *[527-85-5]* M 135.2, m 141°. Crystd from hot water (10ml/g) and dried in air.

Toluene *[108-88-3]* M 92.1, b 110.6°, n 1.49693, n^{25} 1.49413, d^{10} 0.87615, d^{25} 0.86231. Dried with $CaCl_2$, CaH_2 or $CaSO_4$, and dried further by standing with sodium, P_2O_5 or CaH_2. It can be fractionally distd from sodium or P_2O_5. Unless specially purified, toluene is likely to be contaminated with methylthiophens and other sulphur containing impurities. These can be removed by shaking with conc H_2SO_4, but the temperature must be kept below 30° if sulphonation of toluene is to be avoided. A typical procedure consists of shaking toluene twice with cold conc H_2SO_4 (100ml of acid per L), once with water, once with aq 5% $NaHCO_3$ or NaOH, again with water, then drying successively with $CaSO_4$ and P_2O_5, with final distn from P_2O_5 or over $LiAlH_4$ after refluxing for 30min. Alternatively, the treatment with NaHCO$_3$ can be replaced by boiling under reflux with 1% sodium amalgam. Sulphur compounds can also be removed by prolonged shaking of the toluene with mercury, or by two distns from AlCl$_3$, the distillate then being washed with water, dried with K$_2$CO$_3$ and stored with sodium wire. Other purification procedures include refluxing and distn of sodium dried toluene from diphenylpicrylhydrazyl, and from SnCl$_2$ (to ensure freedom from peroxides). It has also been co-distd with 10% by volume of ethyl methyl ketone, and again fractionally distd. [Brown and Pearsall *JACS* **74** 191 *1952*.] For removal of carbonyl impurities see *benzene*. Toluene has been purified by distn under nitrogen in the presence of sodium benzophenone ketyl. Toluene has also been dried with MgSO$_4$, after the sulphur impurities have been removed, and then fractionally distd from P$_2$O$_5$ and stored in the dark [Tabushi *et al, JACS* **107** 4465 *1985*]. Toluene can be purified by passage through a tightly packed column of Fuller's earth.

Toluene-2,4-diamine (*asym*-xylidine) *[95-80-7]* M 122.2, m 99°, b 148-150°/8mm, 292°/760mm. Recrystd from water containing a vary small amount of sodium dithionite (to prevent air oxidation), and dried under vac.

o-**Toluenesulphonamide** *[88-19-7]* M 171.2, m 155.5^o,
p-**Toluenesulphonamide** *[70-55-3]* M 171.2, m 138°. Crystd from hot water, then from EtOH.

p-**Toluenesulphonic acid** *[6192-52-5]* M 190.2, m 38° (anhydrous), m 105-107° (monohydrate). Purified by pptn from a satd soln at 0° by introducing HCl gas. Also crystd from conc HCl, then crystd from dil HCl (charcoal) to remove benzenesulphonic acid. It has been crystd from EtOH/water. Dried in a vac desiccator over solid KOH and CaCl$_2$. *p*-Toluenesulphonic acid can be dehydrated by azeotropic distn with benzene or by heating at 100° for 4hr under water-pump vac. The anhydrous acid can be crystd from benzene, CHCl$_3$, ethyl acetate, anhydrous MeOH, or from acetone by adding a large excess of benzene. It can be dried under vac at 50°.

p-**Toluenesulphonyl chloride** *[98-59-9]* M 190.7, m 69°, b 146°/15mm. Material that has been standing for a long time contains tosic acid and HCl and has m *ca* 65-68°. It is purified by dissolving (10g) in the minimum volume of CHCl$_3$ (*ca* 25ml) filtered, and diluted with five volumes (i.e. 125ml) of pet ether (b 30-60°) to precipitate impurities. The soln is filtered, clarified with charcoal and concentrated to 40ml by evaporation. Further evaporation to a very small volume gave 7g of white crystals which were analytically pure, m 67.5-68.5^o. (The insoluble material was largely tosic acid and had m 101-104°). [Pelletier *Chem Ind* 1034 *1953*.]
Also crystd from toluene/pet ether in the cold, from pet ether (b 40-60°) or benzene. Its soln in ethyl ether has been washed with aq 10% NaOH until colourless, then dried with Na$_2$SO$_4$ and crystd by cooling in powdered Dry-ice. It has also been purified by dissolving in benzene, washing with aq 5% NaOH, then dried with K$_2$CO$_3$ or MgSO$_4$, and distd under reduced pressure.

α-**Toluenethiol** see **benzylmercaptan**.

p-**Toluenethiol** *[106-45-6]* M 124.2, m 43.5-44°. Crystd from pet ether (b 40-70°).

Toluhydroquinone *[95-71-6]* M 124.1, m 128-129°. Crystd from EtOH.

o-**Toluic acid** *[118-90-1]* M 136.2, m 102-103°. Crystd from benzene (2.5ml/g) and dried in air.

m-**Toluic acid** *[99-04-7]* M 136.2, m 111-113°. Crystd from water.

p-**Toluic acid** *[99-94-5]* M 136.2, m 178.5-179.5^o. Crystd from water, water/EtOH (1:1), MeOH/water or benzene.

o-**Toluidine** *[95-53-4]* M 107.2, f.p. -16.3^o, b 80.1^o/10mm, 200.3^o/760mm, n 1.57246, n^{25} 1.56987, d 0.999. In general, methods similar to those for purifying aniline can be used, e.g. distn from zinc dust, at reduced pressure, under nitrogen. Berliner and May [*JACS* **49** 1007 *1927*] purified *via* the oxalate.

Twice-distd *o*-toluidine was dissolved in four times its volume of ethyl ether and the equivalent amount of oxalic acid needed to form the dioxalate was added as its soln in ethyl ether. (If *p*-toluidine is present, its oxalate pptes and can be removed by filtration.) Evapn of the ether soln gave crystals of *o*-toluidine dioxalate. They were filtered off, recrystd five times from water containing a small amount of oxalic acid (to prevent hydrolysis), then treated with dil aq Na_2CO_3 to liberate the amine which was separated, dried ($CaCl_2$) and distd under reduced pressure.

m-Toluidine *[108-44-1]* M 107.2, f.p. -30.4°, b 82.3°/10mm, 203.4°/760mm, n 1.56811, n^{25} 1.56570, d 0.989. It can be purified as for aniline, Twice-distd, *m*-toluidine was converted to the hydrochloride using a slight excess of HCl, and the salt was fractionally crystd from 25% EtOH (five times), and from distd water (twice), rejecting, in each case, the first material that crystd. The amine was regenerated and distd as for *o*-toluidine. [Berliner and May *JACS* **49** 1007 *1927*.]

p-Toluidine *[106-49-0]* M 107.2, m 44.8°, b 79.6°/10mm, 200.5°/760mm, n 1.5636, $n^{59.1}$ 1.5534, d 0.962. In general, methods similar to those for purifying aniline can be used. It can be separated from the *o*- and *m*-isomers by fractional crystn from its melt. *p*-Toluidine has been crystd from hot water (charcoal), EtOH, benzene, pet ether or EtOH/water (1:4), and dried in a vac desiccator. It can also be sublimed at 30° under vac. For further purification, use has been made of the oxalate, the sulphate and acetylation. The oxalate, formed as described for *o*-toluidine, was filtered, washed and recrystd three times from hot distd water. The base was regenerated with aq Na_2CO_3 and recrystd three times from distd water. [Berliner and May *JACS* **49** 1007 *1927*.] Alternatively, *p*-toluidine was converted to its acetyl derivative which, after repeated crystn from EtOH, was hydrolysed by refluxing (50g) in a mixture of 500ml of water and 115ml of conc H_2SO_4 until a clear soln was obtained. The amine sulphate was isolated, suspended in water, and NaOH was added. The free base was distd twice from zinc dust under vac. The *p*-toluidine was then recrystd from pet ether and dried in a vac desiccator or in a vac for 6hr at 40°. [Berliner and Berliner *JACS* **76** 6179 *1954*; Moore *et al, JACS* **108** 2257 *1986*.]

Toluidine Blue *[93-31-9]* M 305.8. Crystd from hot water (18ml/g) by adding one and a half volumes of alcohol and chilling on ice. Dried at 100° in an oven for 8-10hr.

p-**Toluidine hydrochloride** *[540-23-8]* M 143.6, m 245.9-246.1°. Crystd from MeOH containing a few drops of conc HCl. Dried under vac over paraffin chips.

2-*p*-Toluidinylnaphthalene-6-sulphonic acid *[7724-15-4]* M 313.9. Crystd twice from 2% aq KOH and dried under high vac for 4hr at room temperature. Crystd from water. Tested for purity by tlc on silica gel with isopropanol as solvent. The free acid was obtained by acidifying a satd aq soln.

o-**Tolunitrile** *[529-19-1]* M 117.2, b 205.2°, n 1.5279, d 0.992. Fractionally distd, washed with conc HCl or 50% H_2SO_4 at 60° until the smell of isonitrile had gone (this also removed any amines), then washed with satd $NaHCO_3$ and dil NaCl solns, then dried with K_2CO_3 and redistd.

m-**Tolunitrile** *[620-22-4]* M 117.2, b 209.5-210°/773mm, n 1.5250, d 0.986. Dried with $MgSO_4$, fractionally distd, then washed with aq acid to remove possible traces of amines, dried and redistd.

p-**Tolunitrile** *[104-85-8]* M 117.2, m 29.5°, b 104-106°/20mm. Melted, dried with $MgSO_4$, fractionally crystd from its melt, then fractionally distd under reduced pressure in a 6-in spinning band column. [Brown *JACS* **81** 3232 *1959*.] It can also be crystd from benzene/pet ether (b 40-60°).

p-**Toluquinone** see **methyl-1,4-benzoquinone**.

p-**Toluyl-*o*-benzoic acid** *[7148-03-0]* M 196.2, m 138-139°. Crystd from toluene.

p-**Tolylacetic acid** *[622-47-9]* M 150.2, m 90.8-91.3°. Crystd from water.

4-*o*-Tolylazo-*o*-toluidine see **2-amino-5-azotoluene**.

p-Tolyl carbinol *[589-18-4]* M 122.2, m 61°, b 116-118°/20mm, 217°/760mm. Crystd from pet ether (b 80-100°, 1g/ml). It can also be distd under reduced pressure.

Tolyl diphenyl phosphate *[26444-49-5]* M 340.3, n^{25} 1.5758. Vac distd, then percolated through a column of alumina. Finally, passed through a packed column maintained at 150° to remove traces of volatile impurities in a countercurrent stream of nitrogen at reduced pressure. [Dobry and Keller *JPC* **61** 1448 *1947*.]

p-Tolyl disulphide *[103-19-5]* M 246.4, m 45-46°. Purified by chromatography on alumina using hexane as eluent, then crystd from MeOH. [Kice and Bowers *JACS* **84** 2384 *1962*.]

p-Tolylsulphonylmethylditrosamide (Diazald) M 214.2, m 62°. Crystd from benzene/pet ether.

p-Tolyl urea *[622-51-5]* M 150.2, m 181°. Crystd from EtOH/water (1:1).

Tosylmethyl isocyanide *[36635-51-7]* M 195.2, m 114-115°. Recrystd from EtOH (charcoal) [Saito and Itano, JCSPT 1 *1986*].

trans-Traumatic acid *[6402-36-4]* M 228.3, m 165-166°. Crystd from EtOH or acetone.

α,α'-Trehalose (2H$_2$O) *[6138-23-4]* M 378.3, m 96.5-97.5°, 203° (anhydrous). Crystd (as the dihydrate) from aq EtOH. Dried at 13°.

TREN see tris(2-aminoethyl)amine.

1,2,3-Triaminopropane trihydrochloride *[free base 21291-99-6]* M 198.7, m 250°. Cryst from EtOH.

1,2,4-Triazole *[288-88-0]* M 69.1, m 121°, 260°. Crystd from EtOH or water [Barsczlaw *et al, JCSPT* 2025 *1986*]..

Tribenzylamine *[620-40-6]* M 287.4, m 93-94°. Crystd from abs EtOH or pet ether. Dried in a vac over P$_2$O$_5$ at room temperature.

2,4,6-Tribromoacetanilide *[607-93-2]* M 451.8, m 232°. Crystd from EtOH.

2,4,6-Tribromoaniline *[147-82-0]* M 329.8, m 120°. Crystd from MeOH.

sym-Tribromobenzene *[626-39-1]* M 314.8, m 122°. Crystd from glacial acetic acid/water (4:1), then washed with chilled EtOH and dried in air.

Tribromochloromethane *[594-15-0]* M 287.2, m 55°. Melted, washed with aq Na$_2$S$_2$O$_3$, dried with BaO and fractionally crystd from its melt.

2,4,6-Tribromophenol *[118-79-6]* M 330.8, m 94°. Crystd from EtOH or pet ether. Dried under vac over P$_2$O$_5$ at room temperature.

Tri-*n*-butylamine *[102-82-9]* M 185.4, b 68°/3mm, 120°/44mm, n 1.4294, d 0.7788. Purified by fractional distn from sodium under reduced pressure. Pegolotti and Young [*JACS* **83** 3251 *1961*] heated the amine overnight with an equal volume of acetic anhydride, in a steam bath. The amine layer was separated and heated with water for 2hr on the steam bath (to hydrolyse any remaining acetic anhydride). The soln was cooled, solid K$_2$CO$_3$ was added to neutralize any acetic acid that had been formed, and the amine was separated, dried (K$_2$CO$_3$) and distd at 44mm pressure. Davis and Nakshbendi [*JACS* **84** 2085 *1926*] treated the amine with one-eighth of its weight of benzenesulphonyl chloride in aq 15% NaOH at 0-5°. The mixture was shaken intermittently and allowed to warm to room temperature. After a day, the amine layer was washed with aq NaOH, then water and dried with KOH. (This treatment removes primary and secondary amines.) It was further dried with CaH_2 and distd under vac.

Tri-*n*-butylamine hydrobromide *[37026-85-0]* M 308.3, m 75.2-75-9°. Crystd from ethyl acetate.

Tri-*n*-butylammonium nitrate *[1941-27-1]* M 304.5. Crystd from mixtures of *n*-hexane and acetone (95:5). Dried over P_2O_5.

Tri-*n*-butylammonium perchlorate *[14999-66-7]* M 285.5. Recrystd from *n*-hexane.

sym-**Tri-*tert*-butylbenzene** *[1460-02-2]* M 246.4, m 73.4-73.9°. Crystd from EtOH.

2,4,6-Tri-*tert*-butylphenol *[732-26-3]* M 262.4, m 129-132°, 131°/1mm, 147°/10mm, 278°/760mm. Crystd from *n*-hexane or several times from 95% EtOH until the EtOH soln was colourless [Balasubramanian and Bruice *JACS* **108** 5495 *1986*]. It has also been purified by sublimation [Yuan and Bruice *JACS* **108** 1643 *1986*; Wong *et al, JACS* **109** 3428 *1987*]. Purification has been achieved by passage through a silica gel column followed by recrystn from *n*-hexane [Kajii *et al, JPC* **91** 2791 *1987*].

Tributyl phosphate *[126-73-8]* M 266.3, b 121-124°/3mm, n^{25} 1.4222, d 0.977. The main contaminants in commercial samples are organic pyrophosphates, mono- and di- butyl phosphates and butanol. It is purified by washing successively with 0.2M HNO_3 (three times), 0.2M NaOH (three times) and water (three times), then fractionally distd under vac. [Yoshida *J Inorg Nuclear Chem* **24** 1257 *1962*.] It has also been purified *via* its uranyl nitrate addition compound, obtained by saturating the crude phosphate with uranyl nitrate. This compound was crystd three times with *n*-hexane by cooling to -40°, and then decomposed by washing with Na_2CO_3 and water. Hexane was removed by steam distn and the water was then evapd under reduced pressure and the residue was distd under reduced pressure. [Siddall and Dukes *JACS* **81** 790 *1959*.]

Tricarballylic acid *[99-14-9]* M 176.1, m 166°. Crystd from ethyl ether.

Trichloroacetamide *[594-65-0]* M 162.4, m 139-141°, b 238-240°. Its xylene soln was dried with P_2O_5, then fractionally distd.

Trichloroacetanilide *[2563-97-5]* M 238.5, m 95°. Crystd from benzene.

Trichloroacetic acid *[76-03-9]* M 163.4, m 59.4-59.8°. Purified by fractional crystn from its melt, then crystd repeatedly from dry benzene and stored over conc H_2SO_4 in a vac desiccator. It can also be crystd from $CHCl_3$ or cyclohexane, and dried over P_2O_5 or $Mg(ClO_4)_2$ in a vac desiccator. Trichloroacetic acid can be fractionally distd under reduced pressure from $MgSO_4$. Layne, Jaffé and Zimmer [*JACS* **85** 435 *1963*] dried trichloroacetic acid in benzene by distilling off the benzene-water azeotrope, then crystd the acid from the remaining benzene soln. Manipulations were carried out under nitrogen. *[Use a well ventilated fumecupboard.]*

2,3,4-Trichloroaniline *[634-67-3]* M 196.5, m 67.5°, b 292°/774mm,
2,4,5-Trichloroaniline *[636-30-6]* M 196.5, m 96.5°, b 270°/760mm,
2,4,6-Trichloroaniline *[634-93-5]* M 196.5, m 78.5°, b 262°/746mm. Crystd from ligroin.

1,2,3-Trichlorobenzene *[87-61-6]* M 181.5, m 52.6°. Crystd from EtOH.

1,2,4-Trichlorobenzene *[120-82-1]* M 181.5, m 17°, b 210°. Separated from a mixture of isomers by washing with fuming H_2SO_4, then water, drying with $CaSO_4$ and slowly fractionally distilling. [Jensen, Marino and Brown *JACS* **81** 3303 *1959*.]

1,3,5-Tribromobenzene *[108-70-3]* M 181.5, m 64-65°. Recrystd from dry benzene or toluene.

1,1,1-Trichloro-2,2-bis(*p*-chlorophenyl)ethane (*p.p'*-DDT) *[50-29-3]* M 354.5, m 108.5-109°. Crystd from 95% EtOH, and checked by tlc.

3,4,5-Trichloro-*o*-cresol *[608-92-4]* M 211.5, m 77°,
2,3,5-Trichloro-*p*-cresol *[608-91-3]* M 211.5, m 66-67°. Crystd from pet ether.

1,1,1-Trichloroethane *[71-55-6]* M 133.4, f.p. -32.7°, b 74.0°, n 1.4385, d 1.337,
1,1,2-Trichloroethane *[79-00-5]* M 131.4, f.p. -36.3°, b 113.6°, n 1.472, d 1.435. Washed successively with conc HCl (or conc H_2SO_4), aq 10% K_2CO_3 (Na_2CO_3), aq 10% NaCl, dried with $CaCl_2$ or Na_2SO_4, and fractionally distd. It can contain up to 3% dioxane as preservative. This is removed by washing successively with 10% aq HCl, 10% aq $NaHCO_3$ and 10% aq NaCl; and distd over $CaCl_2$ before use.

Trichloroethylene *[79-01-6]* M 131.4, f.p. -88°, b 87.2°, n^{21} 1.4767, d 1.463. Undergoes decomposition in a similar way to $CHCl_3$, giving HCl, CO, $COCl_2$ and organic products. It reacts with KOH, NaOH and 90% H_2SO_4, and forms azeotropes with water, MeOH, EtOH, and acetic acid. It is purified by washing successively with 2M HCl, water and 2M K_2CO_3, then dried with K_2CO_3 and $CaCl_2$, and fractionally distd immediately before use. It has also been steam distd from 10% $Ca(OH)_2$ slurry, most of the water being removed from the distillate by cooling to -30° to -50° and filtering off the ice through chamois skin: the trichloroethylene was then fractionally distd at 250mm pressure and collected in a blackened container. [Carlisle and Levine *IEC* 24 1164 *1932*.]

2,4,5-Trichloro-1-nitrobenzene *[89-69-0]* M 226.5, m 57°. Crystd from EtOH.

3,4,6-Trichloro-2-nitrophenol *[82-62-2]* M 242.4, m 92-93°. Crystd from pet ether.

2,4,5-Trichlorophenol *[95-95-4]* M 197.5, m 67°. Crystd from EtOH or pet ether.

2,4,6-Trichlorophenol *[88-06-2]* M 197.5, m 67-68°. Crystd from benzene, EtOH or EtOH/water.

3,4,5-Trichlorophenol *[609-19-8]* M 197.5, m 100°. Crystd from pet ether/benzene mixture.

2,4,5-Trichlorophenoxyacetic acid *[93-76-5]* M 255.5, m 153°. Crystd from benzene.

1,1,1-Trichloro-2-(2,2,2-trichloro-1-hydroxyethoxy)-2-methylpropane see **chloralacetone chloroform.**

1,1,2-Trichlorotrifluoroethane *[76-13-1]* M 187.4, b 47.6°/760mm, n 1.360, d 1.576. Washed with water, then with weak alkali. Dried with $CaCl_2$ or H_2SO_4 and distd. [Locke *et al, JACS* 56 1726 *1934*.]

Tricyclohexylphosphine *[2622-14-2]* M 280.4, m 82-83°. Recrystd from EtOH [Boert *et al, JACS* 109 7781 *1987*].

Tricycloquinazoline *[195-84-6]* M 230.3, m 322-323°. Crystd repeatedly from toluene, followed by vac sublimation at 210° at a pressure of 0.15-0.3 Torr in subdued light.

Tridecanoic acid *[638-53-9]* M 214.4, m 44.5-45.5°, b 199-200°/24mm. Crystd from acetone.

7-Tridecanone *[462-18-0]* M 198.4, m 33°, b 255°/766mm. Crystd from EtOH.

Tri-*n*-dodecylammonium nitrate *[2305-34-2]* M 585.0. Crystd from *n*-hexane/acetone (95:5) and kept in a desiccator over P_2O_5.

Tri-*n*-dodecylammonium perchlorate *[5838-82-4]* M 622.4. Recrystd from *n*-hexane or acetone and kept in a desiccator over P_2O_5.

TRIEN see **triethylenetetramine**.

Triethanolamine hydrochloride *[637-39-8]* M 185.7, m 177°. Crystd from EtOH. Dried at 80°.

1,1,2-Triethoxyethane *[4819-77-6]* M 162.2, b 164°, n 1.401, d 0.897. Dried with Na_2SO_4, and distd.

Triethylsilane *[617-86-7]* M 116.3, b 107-108°. It was passed through neutral alumina before use [Randolph and Wrighton *JACS* **108** 3366 *1986*].

Triethylamine *[121-44-8]* M 101.2, b 89.4°, n 1.4005, d 0.7280. Dried with $CaSO_4$, $LiAlH_4$, Linde type 4A molecular sieves, CaH_2, KOH, or K_2CO_3, then distd, either alone or from BaO, sodium, P_2O_5 or CaH_2. It has also been distd from zinc dust, under nitrogen. To remove traces of primary and secondary amines, triethylamine has been refluxed with acetic anhydride, benzoic anhydride, phthalic anhydride, then distd, refluxed with CaH_2 (ammonia-free) or KOH (or dried with activated alumina), and again distd. Another purification involved refluxing for 2hr with p-toluenesulphonyl chloride, then distd. Grovenstein and Williams [*JACS* **83** 412 1961] treated triethylamine (500ml) with benzoyl chloride (30ml), filtered off the ppte, and refluxed the liquid for 1hr with a further 30ml of benzoyl chloride. After cooling, the liquid was filtered, distd, and allowed to stand for several hours with KOH pellets. It was then refluxed with, and distd from, stirred molten potassium. Triethylamine has been converted to its hydrochloride, crystd from EtOH (to **m** 254°), then liberated with aq NaOH, dried with solid KOH and distd from sodium under nitrogen.

Triethylammonium bromide *[4636-73-1]* M 229.1, m 248°. Equimolar portions of triethylamine and aqueous solutions of HBr in acetone were mixed. The pptd salt was washed with anhydrous acetone and dried in vac for 1-2hr. [Odinekov *et al, JCSFT 2* **80** 899 *1984*.] Recrystd from $CHCl_3$ or EtOH.

Triethylammonium chloride *[554-68-7]* M 137.7, m 257-260°(dec). Purified like the bromide above.

Triethylammonium iodide *[4636-73-1]* M 229.1, m 181°. Purified as for triethylammonium bromide, except the soln for pptn was precooled acetone at -10° and the ppte was twice recrystd from a cooled acetone/hexane mixture at -10°.

Triethylammonium trichloroacetate *[4113-06-8]* M 263.6. Equimolar solns of triethylamine and trichloroacetic acid in n-hexane were mixed at 10°. The solid so obtained was recrystd from $CHCl_3$/benzene mixture.

Triethylammonium trifluoroacetate *[454-49-9]* M 196.2. Purified as for the corresponding trichloroacetate. The salt was a colourless liquid at ambient temperature.

1,2,4-Triethylbenzene *[877-44-1]* M 162.3, b 96.8-97.1°/12.8mm, n 1.5015, d 0.8738,
1,3,5-Triethylbenzene *[102-25-0]* M 162.3, b 102-102.5°, n 1.4951, d 0.8631. For separation from a commercial mixture see Dillingham and Reid [*JACS* **60** 2606 *1938*].

Triethylenediamine [Dabco, TED, 1,4-diazabicyclo(2.2.2)octane] *[280-57-9]* M 112.2, m 156-157° **(sealed tube).** Crystd from 95% EtOH, pet ether or MeOH/ethyl ether (1:1). Dried under vac over $CaCl_2$ and BaO. It can be sublimed *in vacuo*. Also purified by removal of water during azeotropic distn of a benzene soln. It was then recrystd twice from anhydrous ethyl ether under argon, and stored under argon [Blackstock *et al, JOC* **52** 1451 *1987*].

Triethylene glycol *[112-27-6]* M 150.2, b 115-117°/0.1mm, 278°/760mm, n^{15} 1.4578, d^{15} 1.1274. Dried with $CaSO_4$ for 1 week, then repeatedly and very slowly fractionally distd under vac. Stored in a vac desiccator over P_2O_5. It is very hygroscopic.

Triethylene glycol dimethyl ether *[112-49-2]* M 178.2, b 225°, m 1.425, d 0.987. Refluxed with, and distd from sodium hydride or LiAlH$_4$.

Triethylenetetramine (TRIEN) *[112-24-3]* M 146.2, b 157°/20mm, n 1.497, d 0.971. Dried with sodium, then distd under vac. Further purification has been *via* the nitrate or the chloride. For example, Jonassen and Strickland [*JACS* **80** 312 *1958*] separated TRIEN from admixture with TREN (38%) by soln in EtOH, cooling to approximately 5° in an ice-bath and adding conc HCl dropwise from a burette, keeping the temperature below 10°, until all of the white crystalline ppte of TREN.HCl had formed and been removed. Further addition of HCl then pptd thick creamy white TRIEN > HCl which was crystd several times from hot water by adding an excess of cold EtOH. The crystals were finally washed with acetone, then ether and dried in a vac desiccator.

Triethylenetetramine tetrahydrochloride *[4961-10-4]* M 292.1, m 266-270°. Crystd repeatedly from hot water by pptn with cold EtOH or EtOH/HCl. Washed with acetone and abs EtOH and dried in a vac oven at 80°.

Tri-(2-ethylhexyl)phosphate (TEHP, tri-isooctylphosphate) *[25103-23-5]* M 434.6, b 219°/5mm, n 1.44464, d^{25} 0.92042. TEHP, in an equal volume of ethyl ether, was shaken with aq 5% HCl and the organic phase was filtered to remove traces of pyridine (used as a solvent during manufacture) as its hydrochloride. This layer was shaken with aq Na$_2$CO$_3$, then water, and the ether was distd off at room temperature. The ester was filtered, dried for 12hr at 100°/15mm, and again filtered, then shaken intermittently for 2 days with activated alumina (100g/L). It was decanted through a fine sintered-glass disc (with exclusion of moisture), and distd under vac. [French and Muggleton *JCS* 5064 *1957*.] Benzene can be used as a solvent (to give 0.4M soln) instead of ether.

Triethyloxonium fluoroborate *[368-39-8]* 190.0, m 92-93°(dec). Crystd from ethyl ether. *Very hygroscopic*, and should be handled in a dry box and stored at 0°. [*Org Synth* **46** 113 *1966*.] Pure material should give a clear and colourless soln in dichloromethane (1 in 50, w/v).

Triethyl phosphate *[78-40-0]* M 182.2, b 98-98.5°/8-10mm, 161°/188mm, n 1.409, d 1.069. Dried by refluxing with solid BaO and fractionally distd under reduced pressure. It is kept with sodium and distd. Stored in the receiver protected from light and moisture.

Triethyl phosphite *[554-70-1]* M 166.2, b 52°/12mm, n^{25} 1.4108, d^{25} 0.9610. Treated with sodium (to remove water and any dialkyl phosphonate), then decanted and distd under reduced pressure, with protection against moisture.

Triethyl silane *[617-86-7]* M 116.3, b 105-107°. n 1.414, d 0.734. Refluxed over molecular sieves, then distd.

Trifluoroacetic acid *[76-86-7]* M 114.0, f.p.-15.5°, b 72.4°, n 1.2850, d 1.494. The purification of trifluoroacetic acid, reported in earlier editions of this work, by refluxing over KMnO$_4$ for 24hr and slowly distilling has resulted in very **SERIOUS EXPLOSIONS** on various occasions, but not always. This apparently depends on the source and/or age of the acid. The method is NOT RECOMMENDED. Water can be removed by making 0.05% in trifluoroacetic anhydride (to diminish water content) and distd. [Conway and Novak *JPC* **81** 1459 *1977*.] It can be refluxed and distd from P$_2$O$_5$. It is further purified by fractional crystn by partial freezing and again distd.

Trifluoroacetic anhydride *[407-25-0]* M 210.0, b 38-40°/760mm, d 1.508. Purification by distilling over KMnO$_4$, as for the acid above is **EXTREMELY DANGEROUS** due to the possiblility of **EXPLOSION**. It is best purified by distilling from P$_2$O$_5$ slowly, and collecting the fraction boiling at 39.5°. Store in a dry atmosphere.

1,1,1-Trifluoro-2-bromoethane *[421-06-7]* M 163.0. Washed with water, dried (CaCl$_2$) and distd.

2,2,2-Trifluoroethanol *[75-89-8]* **M 100.0, b 72.4°/738mm, d 1.400.** Dried with $CaSO_4$ and a little $NaHCO_3$ (to remove traces of acid).

4-(Trifluoromethyl)acetophenone *[728-86-9]* **M 250.2, m 115-116°.** Purified by sublimation *in vacuo*.

3-Trifluoromethyl-4-nitrophenol *[88-30-2]* **M 162.1, m 81°.** Crystd from benzene or from pet ether/benzene mixture.

α,α,α-Trifluorotoluene *[98-08-8]* **M 144.1, b 102.5°. n^{30} 1.4100, d 1.190.** Purified by repeated treatment with boiling aq Na_2CO_3 (until no test for chloride ion was obtained), dried with K_2CO_3, then with P_2O_5, and fractionally distd.

Triglycyl glycine (tetraglycine) *[637-84-3]* **M 246.2, m 270-275°(dec).** Crystd from distd water (optionally, by the addition of EtOH).

Triglyme see **triethyleneglycol dimethyl ether.**

Trigonelline *[535-83-1]* **M 137.1, m 218°(dec).** Crystd (as monohydrate) from aq EtOH, then dried at 100°.

2,3,4-Trihydroxybenzoic acid *[610-02-6]* **M 170.1, m 207-208°,**
2,4,6-Trihydroxybenzoic acid *[83-30-7]* **M 170.1, m 205-212°(dec).** Crystd from water.

4',5,7-Trihydroxyflavone (apigenin) *[520-36-5]* **M 270.2, m 345-350°.** Crystd from aq pyridine.

3,4,5-Triiodobenzoic acid *[2338-20-7]* **M 499.8, m 289-290°.** Crystd from aq EtOH.

3,4,5-Triiodobenzyl chloride *[52273-54-8]* **M 504.3, m 138°.** Crystd from CCl_4/pet ether (charcoal).

3,3',5-Triiodo-S-thyronine *[6893-02-3]* **M 651.0, m 236-237°(dec), $[\alpha]_D^{29.5}$ +21.5° (EtOH/1M aq HCl, 2:1).** Likely impurities are as in *thyroxine*. Purified by dissolving in dil ammonia at room temperature, then crystd by addition of dil acetic acid to *p*H 6.

Triisoamyl phosphate *[919-62-0]* **M 308.4, b 143°/3mm,**
Triisobutyl phosphate *[126-71-6]* **M 266.3, b 119-129°/8-12mm, 192°/760mm, n 1.421, d 0.962.** Purified by repeated crystn, from hexane, of its addition compound with uranyl nitrate. (see *tributyl phosphate*.) [Siddall *JACS* **81** 4176 *1959*.]

Triisooctyl thiophosphate *[30108-39-5]* **M 450.6.** Purified by passage of its soln in CCl_4 through a column of activated alumina.

Triisopropyl phosphite *[116-17-6]* **M 208.2, b 58-59°/7mm, n^{25} 1.4082.** Distd from sodium, under vac, through a column with glass helices. (This removes any dialkyl phosphonate.)

1,2,3-Triketohydrindene hydrate see **ninhydrin.**

Trimellitic acid *[528-44-9]* **M 210.1, m 218-220°.** Crystd from acetic acid or aq EtOH.

Trimesitylphosphine *[23897-15-6]* **M 388.5, m 205-206°.** Recrystd from EtOH [Boert *et al*, *JACS* **109** 7781 *1987*].

Trimethallyl phosphate *[14019-81-9]* **M 260.3, b 134.5-140°/5mm, n^{25} 1.4454.** Purified as for triisoamyl phosphate.

1,2,3-Trimethoxybenzene *[634-36-6]* **M 168.2, m 45-46°,**

1,3,5-Trimethoxybenzene *[621-23-8]* M 168.2, m 53°. Sublimed under vac.

2-(Trimethoxyphenyl)ethylamine sulphate *see* **mescaline sulphate.**

Trimethylacetic acid see **pivalic acid.**

Trimethylamine *[75-50-3]* M 59.1, b 3.5°. Dried by passage of the gas through a tower filled with solid KOH. Water and impurities containing labile hydrogen were removed by treatment with freshly sublimed, ground, P_2O_5. Has been refluxed with acetic anhydride, and then distd through a tube packed with HgO and BaO. [Comyns *JCS* 1557 *1955*.] For more extensive purification, trimethylamine has been converted to the hydrochloride, crystd (see below), and regenerated by treating the hydrochloride with excess aq 50% KOH, the gas passing through a $CaSO_4$ column into a steel cylinder containing sodium ribbon. After 1-2 days, the cylinder was cooled at -78° and hydrogen and air were removed by pumping. [Day and Felsing *JACS* 72 1698 *1950*.] Trimethylamine has also been trap-to-trap distd and then freeze-pump-thaw degassed [Halpern *et al, JACS* 108 3907 *1986*].

Trimethylamine hydrochloride *[593-81-7]* M 95.7, m >280°(dec). Crystd from $CHCl_3$, EtOH or *n*-propanol, and dried under vac. It has also been crystd from benzene/MeOH, MeOH/ethyl ether and dried under vac over paraffin wax and H_2SO_4. Stood over P_2O_5. It is *hygroscopic*.

Trimethylamine hydroiodide *[20230-89-1]* M 186.0, m 263°. Crystd from MeOH.

1,2,4-Trimethylbenzene (pseudocumene) *[95-63-6]* M 120.2, m -43.8°, b 51.6°/10mm, 167-168°/760mm, n 1.5048, d 0.889. Refluxed over sodium and distd under reduced pressure.

2,4,6-Trimethylbenzoic acid (mesitoic acid) *[480-63-7]* M 164.2, m 155°. Crystd from water, ligroin or carbon tetrachloride [Ohwada *et al, JACS* 108 3029 *1986*].

Trimethyl-1,4-benzoquinone *[935-92-2]* M 150.1. Sublimed *in vacuo* before use.

Triton B see **trimethylbenzylammonium**

2,2,5-Trimethylhexane *[3522-94-9]* M 128.3, m -105.8°, b 124.1°, n 1.39971, n^{25} 1.39727, d 0.716. Extracted with conc H_2SO_4, washed with water, dried (type 4A molecular sieves), and fractionally distd.

Trimethyl-1,4-hydroquinone *[700-13-0]* M 152.2, m 173-174°. Recrystd from water, under anaerobic conditions.

1',3,3'-Trimethyl-6-nitrospiro[2H-benzopyran-2,2'-indoline] *[1498-88-0]* M 322.4, m 180°. Recrystd from abs EtOH [Hinnen *et al, Bull Soc Chim Fr* 2066 *1968*; Ramesh and Labes *JACS* 109 3228 *1987*].

Trimethylolethane see **2-hydroxymethyl-2-methylpropane-1,3-diol.**

Trimethylolpropane *[77-99-6]* M 134.2, m 57-59°. Crystd from acetone and ether.

2,2,3-Trimethylpentane *[564-02-3]* M 114.2, b 109.8°, n 1.40295, n^{25} 1.40064, d 0.7161. Purified by azeotropic distn with 2-methoxyethanol, which was subsequently washed out with water. The trimethylpentane was then dried and fractionally distd. [Forziati *et al, J Res Nat Bur Stand* 36 129 *1946.*]

2,2,4-Trimethylpentane (isooctane) *[540-84-1]* M 114.2, b 99.2°, n 1.39145, n^{25} 1.38898, d 0.693. Distd from sodium, passed through a column of silica gel or activated alumina (to remove traces of olefines), and again distd from sodium. Extracted repeatedly with conc H_2SO_4, then agitated with aq $KMnO_4$, washed with water, dried ($CaSO_4$) and distd. Purified by azeotropic distn with EtOH, which was subsequently washed out with water, and the trimethylpentane was dried and fractionally distd. [Forziati *et al, J Res Nat Bur Stand* 36 126 *1946*.] Also purified by fractional crystn.

2,3,5-Trimethylphenol *[697-82-5]* M 136.2, m 95-96°, b 233°/760mm. Crystd from water or pet ether.

2,4,5-Trimethylphenol *[496-78-6]* M 136.2, m 70.5-71.5°. Crystd from water.

2,4,6-Trimethylphenol *[527-60-6]* M 136.2, m 69°, b 220°/760mm. Crystd from water and sublimed *in vacuo.*

3,4,5-Trimethylphenol *[527-54-8]* M 136.2, m 107°, b 248-249°/760mm. Crystd from pet ether.

Trimethylphenylammonium benzenesulphonate *[16093-66-6]* M 293.3. Crystd repeatedly from MeOH (charcoal).

2,2,4-Trimethyl-6-phenyl-1,2-dihydroquinoline *[3562-69-4]* M 249.3, m 102°. Vac distd, then crystd from abs EtOH.

Trimethyl phosphite *[121-45-9]* M 124.1, b 22°/23mm, 111-112°/760mm, n 1.4095, d 1.053. Treated with sodium (to remove water and any dialkyl phosphonate), then decanted and distd with protection against moisture. It has also been treated with sodium wire for 24hr, then distd in an inert atmosphere onto activated molecular sieves [Connor *et al, JCSDT* 511 *1986*].

2,4,6-Trimethylpyridine (collidine) *[108-75-8]* M 121.1, b 60.7°/13mm, 170.4°/760mm, n 1.4981, n^{25} 1.4959, d^{25} 0.9100. Commercial samples may be grossly impure. Likely contaminants include 3,5-dimethylpyridine, 2,3,6-trimethylpyridine and water. Brown, Johnson and Podall [*JACS* 76 5556 *1954*] fractionally distd 2,4,6-trimethylpyridine under reduced pressure through a 40-cm Vigreux column and added to 430ml of the distillate slowly, with cooling to 0°, 45g of BF$_3$-ethyl etherate. The mixture was again distd, and an equal volume of dry benzene was added to the distillate. Dry HCl was passed into the soln, which was kept cold in an ice-bath, and the hydrochloride was filtered off. It was recrystd from abs EtOH (1.5ml/g) to m 286-287°(sealed tube). The free base was regenerated by treatment with aq NaOH, then extracted with benzene, dried (MgSO$_4$) and distd under reduced pressure. Sisler *et al* [*JACS* 75 446 *1953*] pptd trimethylpyridine as its phosphate from a soln of the base in MeOH by adding 85% H$_3$PO$_4$, shaking and cooling. The free base was regenerated as above. Garrett and Smythe [*JCS* 763 *1903*] purified the trimethylpyridine via the HgCl$_2$ complex.

Trimethylsulphonium iodide *[2181-42-2]* M 204.1, m 215-220°(dec). Crystd from EtOH.

1,3,7-Trimethyluric acid *[5415-44-1]* M 210.2, m 345°(dec),
1,3,9-Trimethyluric acid *[519-32-4]* M 210.2, m 347°. Crystd from water.

1,7,9-Trimethyluric acid *[55441-72-0]* M 210.2, m 345°. Crystd from water or EtOH, and sublimed *in vacuo.*

Trimyristin *[555-45-3]* M 723.2, m 56.5°. Crystd from ethyl ether.

Trineopentyl phosphate *[14540-59-1]* M 320.4. Crystd from hexane.

2,4,6-Trinitroanisole *[606-35-9]* M 243.1, m 68°. Crystd from EtOH or MeOH. Dried under vac.

1,3,6-Trinitrobenzene *[99-35-4]* M 213.1, m 122-123°. Crystd from glacial acetic acid, CHCl$_3$, CCl$_4$, EtOH aq EtOH or EtOH/benzene, after (optionally) heating with dil HNO$_3$. Air dried. Fused, and crystd under vac.

2,4,6-Trinitrobenzoic acid *[129-66-8]* **M 225.1, m 227-228°.** Crystd from distd water. Dried in a vac desiccator.

2,4,6-Trinitro-*m*-cresol *[602-99-3]* **M 243.1, m 107.0-107.5°.** Crystd successively from water, aq EtOH and benzene/cyclohexane, then dried at 80° for 2hr. [Davis and Paabo *J Res Nat Bur Stand* **64A** 533 *1960.*]

2,4,7-Trinitro-9-fluorenone *[129-79-3]* **M 315.2, m 176°.** Crystd from nitric acid/water (3:1), washed with water and dried under vac over P_2O_5, or recrystd from dry benzene.

2,4,6-Trinitroresorcinol *[82-71-3]* **M 245.1, m 177-178°.** Crystd from water containing HCl.

2,4,6-Trinitrotoluene (TNT) *[118-96-7]* **M 227.1, m 81.0-81.5°.** Crystd from benzene and EtOH. Then fused and allowed to cryst under vac. Gey, Dalbey and Van Dolah [*JACS* **78** 1803 *1956*] dissolved TNT in acetone and added cold water (1:2:15), the ppte was filtered, washed free from solvent and stirred with five parts of aq 8% Na_2SO_3 at 50-60° for 10min. It was filtered, washed with cold water until the effluent was colourless, and air dried. The product was dissolved in five parts of hot CCl_4, washed with warm water until the washings were colourless and TNT was recoverd by cooling and filtering. It was recrystd from 95% EtOH and carefully dried over H_2SO_4. The dry solid should not be heated without taking precautions for a possible **EXPLOSION.**

2,4,6-Trinitro-*m*-xylene *[632-92-8]* **M 241.2, m 182.2°.** Crystd from ethyl methyl ketone.

Tri-*n*-octylamine *[1116-76-3]* **M 353.7, b 365-367°/760mm, n 1.450, d 0.813.** It was converted to the amine hydrochloride etherate which was recrystd four times from ethyl ether at -30° (see below). Neutralisation of this salt regenerated the free amine. [Wilson and Wogman *JPC* **66** 1552 *1962.*] Distd at 1-2mm pressure.

Tri-*n*-octylammonium chloride *[1188-95-0]* **M 384.2.** Crystd from ethyl ether, then *n*-hexane (see above).

Tri-*n*-octylammonium perchlorate *[2861-99-6]* **M 454.2.** Crystd from *n*-hexane.

Tri-*n*-octylmethylammonium chloride see Aliquat 336.

Tri-*n*-octylphosphine oxide *[78-50-2]* **M 386.7, m 59.5-60°.** Mason, McCarty and Peppard [*J Inorg Nuclear Chem* **24** 967 *1962*] stirred an 0.1M soln in benzene with an equal volume of 6M HCl at 40° in a sealed flask for 48hr, then washed the benzene soln successively with water (twice), 5% aq Na_2CO_3 (three times) and water (six times). The benzene and water were then evapd under reduced pressure at room temperature. Zingaro and White [*J Inorg Nuclear Chem* **12** 315 *1960*] treated a pet ether soln with aq $KMnO_4$ (to oxidise any phosphinous acids to phosphinic acids), then with sodium oxalate, H_2SO_4 and HCl (to remove any manganese compounds). The pet ether soln was slurried with activated alumina (to remove phosphinic acids) and recrystd from pet ether or cyclohexane at -20°. It can also be crystd from EtOH.

1,3,5-Trioxane *[110-88-3]* **M 90.1, m 64°, b 114.5°/759mm.** Crystd from sodium-dried ethyl ether or water, and dried over $CaCl_2$. Purified by zone refining.

Tripalmitin *[555-44-2]* **M 807.4, m 66.4°.** Crystd from acetone, ethyl ether or EtOH.

Triphenylamine *[603-34-9]* **M 245.3, m 127.3-127.9°.** Crystd from EtOH or from benzene/abs EtOH, ethyl ether and pet ether. It was sublimed under vac and carefully dried in a vacuum line. Stored in the dark under nitrogen.

1,3,5-Triphenylbenzene *[612-71-5]* **M 306.4, m 173-175°.** Purified by chromatography on alumina using benzene or pet ether as eluents.

Triphenyl carbinol see **triphenylmethanol.**

Triphenylene *[58-72-0]* **M 228.3, m 198°.** Purified by zone refining.

1,2,3-Triphenylguanidine *[101-01-9]* **M 287.3, m 144°.** Crystd from EtOH or EtOH/water, and dried under vac.

Triphenylmethane *[519-73-3]* **M 244.3, m 92-93°.** Crystd from EtOH or benzene (with one molecule of benzene of crystn which is lost on exposure to air or by heating on a water bath). It can also be sublimed under vac. It can also be given a preliminary purification by refluxing with tin and glacial acetic acid, then filtered hot through a glass sinter disc, and pptd by addition of cold water.

Triphenylmethanol *[76-84-6]* **M 260.3, m 163°.** Crystd from EtOH, CCl_4 (4ml/g), benzene, hexane or pet ether (b 60-70°). Dried at 90°. [Ohwada *et al, JACS* **108** 3029 *1986.*]

Triphenylmethyl chloride (trityl chloride) *[76-83-5]* **M 278.9, m 111-112°.** Crystd from iso-octane. Also crystd from 5 parts of pet ether (b 90-100°) and 1 part of acetyl chloride using 1.8g of solvent per g of chloride. Dried in a desiccator over soda lime and paraffin wax. [Org Synth **Col Vol III** 841 *1955*; Moisel *et al, JACS* **108** 4706 *1986.*]

Triphenyl phosphate *[115-86-6]* **M 326.3, m 49.5-50°, b 245°/0.1mm.** Crystd from EtOH.

Triphenylphosphine *[603-35-0]* **M 362.3, m 80-81°.** Crystd from hexane, MeOH, ethyl ether, CH_2Cl_2/hexane or 95% EtOH. Dried at 65°/<1mm over $CaSO_4$ or P_2O_5. Chromatographed on alumina using (4:1) benzene/$CHCl_3$ as eluent. [Blau and Espenson *et al, JACS* **108** 1962 *1986*; Buchanan *et al, JACS* **108** 1537 *1986*; Randolph and Wrighton *JACS* **108** 3366 *1986*; Asali *et al, JACS* **109** 5386 *1987.*]

Triphenylphosphine oxide *[791-28-6]* **M 278.3, m 152.0°.** Crystd from abs EtOH. Dried *in vacuo.*

Triphenyl phosphite *[101-28-6]* **M 310.3, b 181-189°/1mm, d 1.183.** Its ethereal soln was washed succesively with aq 5% NaOH, distd water and satd aq NaCl, then dried with Na_2SO_4 and distd under vac after evaporating the ethyl ether.

Triphenyl silanol *[791-31-1]* **M 276.4, m 154-155°.** Recrystd by partial freezing from the melt to constant melting point.

2,3,5-Triphenyltetrazolium chloride *(TTC)* *[298-96-4]* **M 334.8, m 243°(dec).** Crystd from EtOH or $CHCl_3$, and dried at 105°.

Tri-*n*-propylamine *[102-69-2]* **M 143.3, b 156.5°, n 1.419, d 0.757.** Dried with KOH and fractionally distd. Also refluxed with toluene-*p*-sulphonyl chloride and with KOH, then fractionally distd. The distillate, after addn of 2% phenyl isocyanate, was redistd and the residue fractionally distd from sodium. [Takahashi *et al, JOC* **52** 2666 *1987.*]

2,2',2"-Tripyridine see **2,2':6',2"-terpyridyl.**

Tripyridyl triazine *[3682-35-7]* **M 312.3, m 245-248°.** Purified by repeated crystn from aq EtOH.

Tris-(2-aminoethyl)amine *[4097-89-6]* **M 146.2, b 114°/15mm, 263°/744mm, n 1.498, d 0.977.** For a separation from a mixture containing 62% TRIEN, see entry under triethylenetetramine. Also purified by conversion to the hydrochloride (see below), recrystn and regeneration of the free base [Xie and Hendrickson *JACS* **109** 6981 *1987*].

Tris-(2-aminoethyl)amine trihydrochloride *[14350-52-8]* M 255.7, m 300°(dec). Crystd several times by dissolving in a minimum of hot water and precipitating with excess cold EtOH. The ppte was washed with acetone, then ethyl ether and dried in a vac desiccator.

Tris-(2-biphenylyl) phosphate *[132-28-5]* M 554.6, m 115.5-117.5°. Crystd from MeOH containing a little acetone.

Tris-(1,2-dioxyphenyl)cyclotriphosphazine Recrystd from chlorobenzene, then sublimed (200°/0.1mm) [Meirovitch *JPC* **88** 1522 *1984*.]

Tris-(2-ethylhexyl) phosphate *[78-42-2]* M 744.0, n 1.443, d 0.924. Purified by partial crystn of its addition compound with uranyl nitrate. [Siddall *JACS* **81** 4176 *1959*.]

TRIS Buffer see **trishydroxymethylaminomethane.**

Tris-(hydroxymethylamino)methane (TRIS) *[77-86-1]* M 121.1, m 172°. Tris can ordinarily be obtained in highly pure form suitable for use as an acidimetric standard. If only impure material is available, it should be crystd from 20% EtOH. Dry in a vac desiccator over P_2O_5 or $CaCl_2$. Alternatively, it is dissolved in twice its weight of water at 55-60°, filtered, concd to half its volume and poured slowly, with stirring, into about twice the volume of EtOH. The crystals which separate on cooling to 3-4° are filtered off, washed with a little MeOH, air dried by suction, then finally ground and dried in a vac desiccator over P_2O_5. It has also been crystd from water, MeOH or aq MeOH, and vac dried at 80° for 2 days.

Tris-(hydroxymethylamino)methane hydrochloride *[1185-53-1]* M 157.6, m 149-150°(dec). Crystd from 50% EtOH, then from 70% EtOH. Tris-hydrochloride is also available commercially in a highly pure state. Otherwise, crystd from 50% EtOH, then 70% EtOH, and dried below 40° to avoid risk of decomposition.

1,1,1-Tris-(hydroxymethyl)ethane *[77-85-0]* M 120.2, m 200°. Dissolved in hot tetrahydrofuran, filtered and pptd with hexane. It has also been crystd from acetone/water (1:1). Dried in vac.

N-Tris-(hydroxymethyl)methyl-2-aminomethanesulphonic acid (TES) *[7365-44-8]* M 229.3, m 224-226°(dec). Crystd from hot EtOH containing a little water.

N-Tris-(hydroxymethyl)methylglycine (Tricine) *[5704-04-1]* M 179.2, m 186-188°(dec). Crystd from EtOH and water.

Tris-(hydroxymethyl)nitromethane see **2-(hydroxymethyl)-2-nitropropane-1,3-diol.**

1,3,5-Trithiane *[291-21-4]* M 138.3, m 220°(dec). Crystd from acetic acid.

Tri-p-tolyl phosphate *[20756-92-7]* M 368.4, b 232-234°, n 1.56703, d^{25} 1.16484. Dried with $CaCl_2$, then distd under vac and percolated through a column of alumina. Passage through a packed column at 150°, with a countercurrent stream of nitrogen under reduced pressure, removed residual traces of volatile impurities.

Tri-o-tolylphosphine *[6163-58-2]* M 304.4, m 129-130°. Crystd from EtOH [Boert *et al, JACS* **109** 7781 *1987*].

Trityl chloride see **triphenylmethyl chloride.**

Triuret *[556-99-0]* **M 146.1.** Crystd from aq ammonia.

Tropaeolin 00. Recrystd twice from water.

Tropaeolin 000. Purified by salting out from hot distd water using sodium acetate, then three times from distd water and twice from EtOH.

3-Tropanol *[120-29-6]* **M 141.2, m 63°, b 229°/760mm.** Distd in steam and crystd from ethyl ether. *Hygroscopic.*

*dl-***Tropic acid** *[529-64-6]* **M 166.2, m 118°.** Crystd from water or benzene.

Tropine see **3-tropanol.**

Tropolone *[533-75-5]* **M 122.1, m 49-50°, b 81-84°/0.1mm.** Crystd from hexane or pet ether and sublimed at 40°/4mm.

Tryptamine *[61-54-1]* **M 160.1, m 116°.** Crystd from benzene.

Tryptamine hydrochloride *[343-94-3]* **M 196.7, m 252-253°.** Crystd from EtOH/water.

L-Tryptophan *[73-22-3]* **M 204.3, m 278°,** $[\alpha]_D$ **-33.4° (EtOH),** $[\alpha]_{546}$ **-36° (c 1, H_2O).** Crystd from water/EtOH, washed with anhydrous ethyl ether and dried at room temperature under vac over P_2O_5.

Tryptophol [3-(2-hydroxyethyl)indole] *[526-55-6]* **M 161.2, m 59°.** Crystd from ethyl ether/pet ether.

(+)-Tubocurarine chloride (5H_2O) *[57-94-3]* **M 771.7, m 274-275°,** $[\alpha]_{546}$ **+235° (c 0.5, H_2O).** Crystd from water.

D(+)-Turanose *[5349-40-6]* **M 342.3, m 168-170°,** $[\alpha]_D$ **88° (c 4, H_2O).** Crystd from water by addition of EtOH.

Tyramine *[51-67-2]* **M 137.2, m 164-165°.** Crystd from benzene or EtOH.

Tyramine hydrochloride *[60-17-5]* **M 173.6, m 274-276°.** Crystd from EtOH by addition of ethyl ether, or from conc HCl.

Tyrocidine A *[1481-70-5]* **M 1268.8, m 240°(dec),** $[\alpha]_D^{25}$ **-115° (c 0.91, MeOH).** Crystd as hydrochloride from MeOH or EtOH and HCl. [Paladin and Craig *JACS* **76** 688 *1954*; King and Craig *JACS* **77** 6624 *1955*.]

L-Tyrosine *[60-18-4]* **M 181.2, m 290-295°(dec),** $[\alpha]_D^{25}$ **-10.0° (5M HCl).** Likely impurities are L-cysteine and the ammonium salt. Dissolved in dil ammonia, then crystd by adding dil acetic acid to *p*H 5. Also crystd from water or EtOH/water, and dried at room temperature under vac over P_2O_5.

Umbelliferone *[93-35-6]* M 162.2, m 225-228°. Crystd from water.

Undecan-1-ol *[112-42-5]* M 172.3, m 16.5°,
Undec-10-enoic acid *[112-38-9]* M 184.3, m 25-25.5°. Purified by repeated fractional crystn from its melt

Uracil *[26-22-8]* M 122.1, m 335°(dec). Crystd from water.

Uramil *[118-78-5]* M 143.1, m >400°(dec). Crystd from water.

Urea *[57-13-6]* M 60.1, m 132.7-132.9°. Crystd twice from conductivity water using centrifugal drainage and keeping the temperature below 60°. The crystals were dried under vac at 55° for 6hr. Levy and Margouls [*JACS* **84** 1345 *1962*] prepared a 9M soln in conductivity water (keeping the temperature below 25°) and, after filtering through a medium-porosity glass sinter, added an equal volume of abs EtOH. The mixture was set aside at -27° for 2-3 days and filtered cold. The ppte was washed with a small amount of EtOH and dried in air. Crystn from 70% EtOH between 40° and -9° has also been used. Ionic impurities such as ammonium isocyanate have been removed by treating the conc aq soln at 50° with Amberlite MB-1 cation- and anion-exchange resin, and allowing to crystallise. [Benesch, Lardy and Benesch *JBC* **216** 663 *1955*.] Also crystd from MeOH or EtOH, and dried under vac at room temperature.

Urea nitrate *[124-47-0]* M 123.1, m 152°(dec). Crystd from dil HNO_3.

Uric acid *[69-93-2]* M 168.1. Crystd from hot distd water.

Uridine *[58-96-8]* M 244.2, m 165°. $[\alpha]_D$ +4.0° (H_2O). Crystd from aq 75% MeOH.

Uridylic acid (di-Na salt) *[27821-45-0]* M 368.2, m 198.5°. Crystd from MeOH.

Urocanic acid *[104-98-3]* M 138.1, m 225°. Crystd from water and dried at 100°.

Ursodeoxycholic acid *[128-13-2]* M 392.5, m 203°, $[\alpha]_D$ +60° (c 0.2, EtOH). Crystd from EtOH.

(+)-Usnic acid *[7562-61-0]* M 344.3, m 204°, $[\alpha]_{546}$ +630° (c 0.7, $CHCl_3$). Crystd from acetone, MeOH or benzene.

Ustilagic acid *[8002-36-6]* M , m 146-147°. Crystd from ethyl ether.

trans-Vaccenic acid *[693-72-1]* M 282.5, m 43-44°. Crystd from acetone.

n-Valeraldehyde *[110-62-3]* M 86.1, b 103°, n^{25} 1.40233, d 0.811. Purified *via* the bisulphite derivative. [Birrell and Trotman-Dickinson *JCS* 2059 *1960*.]

n-Valeramide *[626-97-1]* M 101.1, m 115-116°. Crystd from EtOH.

Valeric acid *[109-52-4]* **M 102.1, b 186.4°, n 1.4080, d 0.938.** Water was removed from the acid by distn using a Vigreux column, until the boiling point reached 183°. A few crystals of $KMnO_4$ were added, and after refluxing, the distn was continued, [Andrews and Keefer *JACS* **83** 3708 *1961*.]

Valeronitrile *[110-59-8]* **M 83.1, b 142.3°, n^{15} 1.39913, n^{30} 1.39037, d 0.799.** Washed with half its volume of conc HCl (twice), then with satd aq $NaHCO_3$, dried with $MgSO_4$ and fractionally distd from P_2O_5.

L-Valine *[72-18-4]* **M 117.2, m 315°, $[\alpha]_D$ +266.7° (6M HCl).** Crystd from water by addition of EtOH.

Vanillin *[121-33-5]* **M 152.2, m 83°.** Crystd from water or aq EtOH.

Veratraldehyde *[120-14-9]* **M 166.2, m 42-43°.** Crystd from ethyl ether, pet ether, CCl_4 or toluene.

Veratric acid see **3,4-dimethyoxybenzoic acid.**

Veratrole see *o*-**dimethoxybenzene.**

Variamine Blue RT (salt) *[4477-28-5]* **M 293.3, λ_{max} 377 nm.** Dissolved 10g in 100ml of hot water. Sodium dithionite (0.4g) was added, followed by active carbon (1.5g) and filtered hot. To the colourless or slightly yellow filtrate a soln of satd NaCl was added and the mixture cooled. The needles were filtered off, washed with cold water, dried at room temperature, and stored in a dark bottle. [Erdey *Chem Analyst* **48** 106 *1959*.]

Vicine *[152-93-2]* **M 304.3.** Crystd from water or aq 85% EtOH, and dried at 135°.

Vinyl acetate *[108-05-4]* **M 86.1, b 72.3°, n 1.396, d 0.932.** Inhibitors such as hydroquinone, and other impurities, are removed by drying vinyl acetate with $CaCl_2$ and fractionally distilling under nitrogen, then refluxing briefly with a small amount of benzoyl peroxide and again distilling under nitrogen. Stored in the dark at 0°.

9-Vinylanthracene *[2444-68-0]* **M 204.3, m 65-67°, b 61-66°/10mm.** Purified by vacuum sublimation. Also by chromatography on silica gel with cyclohexane as eluent, and recrystd from EtOH [Werst *et al*, *JACS* **109** 32 *1987*].

Vinyl butoxyethyl ether *[4223-11-4]* **M 144.2.** Washed with aq 1% NaOH, dried with CaH_2, then refluxed with and distd from, sodium.

N-**Vinylcarbazole** *[484-13-5]* **M 193.3, m 66°.** Crystd repeatedly from MeOH in amber glassware. Vac sublimed.

Vinylene carbonate *[872-36-6]* **M 86.1, m 22°.** Purified by zone melting.

1-Vinylnaphthalene *[826-74-4]* **M 154.2, b 124-125°/15mm.** Fractionally distd under reduced pressure on a spinning-band column, dried with CaH_2 and again distd under vac. Stored in sealed ampoules in a freezer.

2-Vinylnaphthalene see **naphthylethylene.**

2-Vinylpyridine *[100-69-6]* **M 105.1, b 79-82°/29mm, n 1.550, d 0.974.** Steam distd, then dried with $MgSO_4$ and distd under vac.

Vinyl stearate *[111-63-7]* **M 310.5, m 35°, b 166°/1.5mm.** Vac distd under nitrogen, then crystd from acetone (3ml/g) or ethyl acetate at 0°.

Vioform see **5-chloro-7-iodo-8-hydroxyquinoline.**

Violanthrene *[81-31-2]* **M 428.5.** Purified by vac sublimation in a muffle furnace.

Viologen (*N,N'*-dimethyl-4,4'-dipyridyl dihydrochloride) *[27926-72-3]* **M 229.1, m >300°.** Purified by pptn on adding excess of acetone to a conc soln in aq MeOH. It has also been recrystd several times from MeOH and dried at 70° under vac for 24hr [Prasad *et al*, *JACS* **108** 5135 *1986*], and recrystd three times from MeOH/isopropanol [Stramel and Thomas *JCSFT* **82** 799 *1986*].

Violuric acid see **5-isonitrosobarbituric acid.**

Visnagin *[82-57-5]* **M 230.2, m 142-145°.** Crystd from water.

Vitamin-A acetate see **retinyl acetate.**

Vitamin-A alcohol see **retinol.**

Vitamin B$_{12}$ *[68-19-9]* **M 1355.4, darkens at 210-220°, but does not melt below 300°, $[\alpha]_{656}^{23}$ -59°.** Crystd from de-ionized water and dried under vac over Mg(ClO$_4$)$_2$.

Vitamin D$_2$ *[50-14-6]* **M 396.7, m 114-116°, $[\alpha]_{546}$ +122°** (c 4, EtOH),
Vitamin D$_3$ *[67-97-0]* **384.6, m 83-85°, $[\alpha]_{546}$ 126°** (c 2, EtOH). Converted into their 3,5-dinitrobenzoyl esters, and crystd repeatedly from acetone. The esters were then saponified and the free vitamins were isolated. [Laughland and Phillips *AC* **28** 817 *1956*.]

Vitamin K$_1$ *[84-80-0]* **M 450.7, m -20°, b 141-140/0.001mm, n^{25} 1.525, d$_{25}^{25}$ 0.967.** Crystd from acetone or EtOH at 170°.

dl-**Warfarin** *[81-81-2]* **M 308.3, m 161°.** Crystd from MeOH.

Xanthatin *[26791-73-1]* **M 114.5-115°.** Crystd from MeOH.

Xanthene *[92-83-1]* **182.2, m 100.5°, b 310-312°/760mm.** Crystd from benzene or EtOH.

9-Xanthenone see **xanthone.**

Xanthine *[69-89-6]* **M 152.1, m >300°(dec).** Pptd by the addition of conc ammonia to its soln in hot 2M HCl (after treatment with charcoal), then crystd from distd water.

Xanthone *[90-47-1]* **M 196.2, m 175.6-175.4°.** Crystd from EtOH (25ml/g) and dried at 100°. It has also been recrystd from *n*-hexane three times and sublimed *in vacuo*. [Saltiel *JACS* **108** 2674 *1986*.]

Xanthophyll see **lutein.**

Xanthopterin (H$_2$O) *[5979-01-1]* **M 197.2, m >410°.** Crystd by acidifying an ammoniacal soln, and collecting by centrifugation followed by washing with EtOH, ether and drying at 100° *in vacuo*.

Xanthorhamnin *[1324-63-6]* **M 770.7, m 195°, [α]$_D$ +3.75° (EtOH).** Crystd from a mixture of ethyl and isopropyl alcohols, air dried, then dried for several hours at 110°.

Xanthosine (2H$_2$O) *[5968-90-1]* **M 320.3, [α]$_D$ -53° (c 8, 0.3M NaOH).** Crystd from EtOH or water (as dihydrate).

Xanthydrol *[90-46-0]* **M 198.2, m 123-124°.** Crystd from EtOH and dried at 40-50°.

Xylene *[1330-20-7]* **M 106.1 (mixed isomers).** Usual impurites are ethylbenzene, paraffins, traces of sulphur compounds and water. It is not practicable to separate the *m*-, and *p*-isomers of xylene by fractional distn, although, with a sufficiently efficient still, *o*-xylene can be fractionally distd from a mixture of isomers. Purified (and dried) by fractional distn from LiAlH$_4$, P$_2$O$_5$, CaH$_2$ or sodium. This treatment can be preceded by shaking successively with conc H$_2$SO$_4$, water. aq 10% NaOH, water and mercury, and drying with CaCl$_2$ for several days. Xylene can be purified by azeotropic distn with 2-ethoxyethanol or 2-methoxyethanol, the distillate being washed with water to remove the alcohol, then dried and fractionally distd.

o-**Xylene** *[95-47-6]* **M 106.2, f.p. -25.2°, b 84°/14mm, 144.4°/760mm, n 1.50543, n^{25} 1.50292, d 0.88020, d^{25} 0.87596.** The general purification methods listed under xylene are applicable [Clarke and Taylor *JACS* **45** 831 *1923*] sulphonated *o*-xylene (4.4Kg) by stirring for 4hr with 2.5L of conc H$_2$SO$_4$ at 95°. After cooling, and separating the unsaponified material, the product was diluted with 3L of water and neutralised with 40% NaOH. On cooling, sodium *o*-xylene sulphonate separated and was recrystd from half its weight of water. [A further crop of crystals was obtained by concentrating the mother liquor to one-third of its volume.] The salt was dissolved in the minimum amount of cold water, then mixed with the same amount of cold water, and with the same volume of conc H$_2$SO$_4$ and heated to 110°. *o*-Xylene was regenerated and steam distd. It was then dried and redistd.

m-**Xylene** *[108-38-3]* **M 106, f.p. -47.9°, b 139.1°, n 1.49721, n^{25} 1.49464, d 0.86417, d^{25} 0.85990.** The general purification methods listed under *xylene* are applicable. The *o*- and *p*-isomers can be removed by their selective oxidation when a *m*-xylene sample containing them is boiled with dil HNO$_3$ (one part conc acid to three parts water). After washing with water and alkali, the product can be steam distd, then distd and purified by sulphonation. [Clarke and Taylor *JACS* **45** 831 *1923*.] *m*-Xylene is selectively sulphonated when a mixture of xylenes is refluxed with the theoretical amount of 50-70% H$_2$SO$_4$ at 85-95° under reduced pressure. By using a still resembling a Dean and Stark apparatus, water in the condensate can be progressively withdrawn while the xylene is returned to the reaction vessel. Subsequently, after cooling, then adding water, unreacted xylenes are distd off under reduced pressure. The *m*-xylenesulphonic acid is subsequently hydrolysed by steam distn up to 140°, the free *m*-xylene being washed, dried with silica gel and again distd. Stored over molecular sieves Linde type 4A.

p-**Xylene** *[106-42-3]* **M 106.2, f.p.13.3, b 138.3°, n 1.49581, n^{25} 1.49325, d 0.86105, d^{25} 0.85669.** The general purification methods listed for *xylene* are applicable. *p*-Xylene can readily be separated from its isomers by crystn from such solvents as MeOH, EtOH, isopropanol, acetone, butanone, toluene, pentane or pentene. It can be further purified by fractional crystn by partial freezing, and stored over sodium wire or molecular sieves Linde type 4A. [Stokes and French *JCSFT 1* **76** 537 *1980*].

Xylenol see **dimethylphenol.**

Xylenol Orange *[1611-35-4]* **M 758.6, m 210°(dec), ε_{578} 6.09 x 10^4 (pH 14), ε_{435} 2.62 x 10^4 (pH 3.1).** Generally contaminated with starting material (cresol red) and semixylenol orange. Purified by ion-exchange chromatography using DEAE-cellulose, eluting with 0.1M NaCl soln. Cresol Red, semixylenol orange and iminodiacetic acid bands elute first. [Sato, Yokoyama and Momoki *Anal Chim Acta* **94** 317 *1977*.].

Xylidine see **dimethylaniline**.

α-D-Xylose *[58-86-6]* **M 150.1, m 146-147°, [α]$_D$ -18.8° (c 4, H$_2$O).** Purified by slow crystn from aq 80% EtOH or EtOH, then dried at 60° under vac over P$_2$O$_5$. Stored in a vac desiccator over CaSO$_4$.

α-Yohimbine *[146-48-5]* M 354.5, m 278°(dec), [α]$_D$ +55.6° (c 2, EtOH). Crystd from EtOH, and dried to remove EtOH of crystn.

δ-Yohimbine see **ajmalicine**.

Zeaxanthin *[144-68-3]* M 568.9, m 215.5°, λ_{max} 275 (log ε 4.34), 453 (log ε 5.12), 480 (log ε 5.07) in EtOH. Yellow plates from MeOH.

CHAPTER 4

PURIFICATION OF INORGANIC AND METAL-ORGANIC CHEMICALS

The commonest method of purification of inorganic species is by recrystallisation, usually from water. However, especially with salts of weak acids or of cations other than the alkaline and alkaline earth metals, care must be taken to minimise the effect of hydrolysis. This can be achieved, for example, by recrystallising acetates in the presence of dilute acetic acid. Nevertheless, there are many inorganic chemicals that are too insoluble or are hydrolysed by water so that no general purification method can be given. It is convenient that many inorganic substances have large temperature coefficients for their solubility in water, but in other cases recrystallisation is still possible by partial solvent evaporation.

The abbreviations e.g. of molecular weight, melting point, etc., in this chapter follow the same arrangement as in Chapter 3, and the same precautions about the handling of substances are applicable here.

Acetarsol see *N*-Acetyl-4-hydroxy-*m*-arsanilic acid.

N-Acetyl-4-hydroxy-*m*-arsanilic acid *[97-44-9]* **M 275.1.** Crystd from water.

Alizarin Red S (sodium salt, H_2O) *[130-22-3]* **M 360.3.** Commercial samples contain large amounts of sodium and potassium chlorides and sulphates. It is purified by passing through a Sephadex G-10 column, followed by elution with water, then 50% aq EtOH [King and Pruden *Analyst* **93** 601 *1968*].

Alumina (neutral) *[1344-28-1]* **M 102.0 (anhyd.).** Stirred with hot 2M HNO_3, either on a steam bath for 12hr (changing the acid every hour) or three times for 30min, then washed with hot dist. water until the washings had *p*H 4, followed by three washings with hot MeOH. The product was dried at 270° [Angyal and Young *JACS* **81** 5251 *1959*]. For the preparation of alumina for chromatography see Chapter 1.

Aluminium ammonium sulphate ($10H_2O$) *[7784-26-1]* **M 453.3, m 93°.** Crystd from hot water by cooling in ice.

Aluminium bromide *[7727-15-3]* **M 266.7, m 97°, b 114°/10mm.** Refluxed and then distilled from pure aluminium chips in a stream of nitrogen into a flask containing more of the chips. It was then distd under vacuum into ampoules [Tipper and Walker *JCS* 1352 *1959*]. Anhydrous conditions are essential, and the white to very light brown solid distillate can be broken into lumps in a dry-box (under nitrogen). Fumes in moist air.

Aluminium caesium sulphate ($12H_2O$) *[14284-36-7]* **M 568.2.** Crystd from hot water (3ml/g).

Aluminium tri-*tert*-butoxide *[556-91-2]* **M 246.3.** Crystd from benzene and sublimed at 180°.

Aluminium chloride (anhydrous) *[7446-70-0]* **M 133.3.** Sublimed several times in an all glass system under nitrogen at 30-50mm pressure. Has also been sublimed in a stream of dry HCl and has been subjected to a preliminary sublimation through a section of granular aluminium metal [for manipulative details see Jensen *JACS* **79** 1226 *1957*]. Fumes in moist air.

Aluminium fluoride (anhydrous) *[7784-18-4]* **M 84.0, m 250°.** Technical material may contain up to 15% alumina, with minor impurities such as aluminium sulphate, cryolite, silica and iron oxide. Reagent grade AlF$_3$ (hydrated) contains only traces of impurities but its water content is very variable (may be up to 40%). It can be dried by calcining at 600-800° in a stream of dry air (some hydrolysis occurs), followed by vac. distn at low pressure in a graphite system, heated to approximately 925° (condenser at 900°) [Henry and Dreisbach *JACS* **81** 5274 *1959*]

Aluminium isopropoxide *[555-31-7]* **M 204.3, m 119°, b 94°/0.5mm, 135°/10mm.** Distd under vac. Hygroscopic.

Aluminium nitrate (9H$_2$O) *[7784-27-2]* **M 375.1.** Crystd from dil HNO$_3$, and dried by passing dry nitrogen through the crystals for several hours at 40°.

Aluminium potassium sulphate (12H$_2$O, alum) *[7784-24-9]* **M 474.4, m 92°.** Crystd from weak aq H$_2$SO$_4$ (*ca* 0.5ml/g).

Aluminium rubidium sulphate (12H$_2$O) *[7784-29-4]* **M 496.2.** Crystd from aq H$_2$SO$_4$ (*ca* 2.5ml/g).

Aluminium sulphate *[10043-01-3]* **M 342.2, m 765°(dec).** Crystd from hot dil H$_2$SO$_4$ (1ml/g) by cooling in ice.

Ammonia (gas) *[7664-41-7]* **M 17.0.** Major contaminants are water, oil and non-condensible gases. Most of these impurities are removed by passing the ammonia through a trap at -22° and condensing it at -176° under vac. Water is removed by distilling the ammonia into a tube containing a small lump of sodium. Also dried by passage through porous BaO, or over alumina followed by glass wool impregnated with sodium (prepared by soaking the glass wool in a solution of sodium in liquid ammonia, and evaporating off the ammonia). Can be rendered oxygen-free by passage through a soln of potassium in liquid ammonia.

Ammonia (liquid) *[7664-41-7]* **M 17.0, m -77.7°, b -33.4°, d 0.597.** Dried, and stored, with sodium in a steel cylinder, then distd and condensed by means of liquid air, the non-condensable gases being pumped off.

Ammonia (aqueous) *[7664-41-7]* **M 17.0 + H$_2$O, d 0.90 (satd, 27%w/v, 14.3 N)..** Obtained metal-free by saturating distd water, in a cooling bath, with ammonia (tank) gas. Alternatively, can use isothermal distn by placing a dish of conc aq ammonia and a dish of pure water in an empty desiccator and leaving for several days.

Ammonium acetate *[631-61-8]* **M 77.1, m 112-114°.** Crystd twice from anhydrous acetic acid, dried under vac for 24hr at 100° [Proll and Sutcliff *TFS* **57** 1078 *1961*].

Ammonium bisulphate *[7803-63-6]* **M 115.1°.** Crystd from water at room temperature (1ml/g) by adding EtOH and cooling.

Ammonium bromide *[12124-97-9]* **M 98.0, m 450°(sublimes).** Crystd from 95% EtOH.

Ammonium chloride *[12125-02-9]* **M 53.5.** Crystd several times from conductivity water (1.5ml/g) between 90° and 0°. Sublimes.

Ammonium chromate *[7788-98-9]* **M 152.1.** Crystd from weak aq ammonia (*ca* 2.5ml/g) by cooling from room temperature.

Ammonium dichromate *[7788-09-5]* **M 252.1, m 170°(dec).** Crystd from weak aq HCl (*ca* 1ml/g).

Ammonium dihydrogen arsenate *[13462-93-6]* **M 159.0.** Crystd from water (1ml/g).

Ammonium dihydrogen orthophosphate *[7722-76-1]* **M 115.0, m 190°.** Crystd from water (0.7ml/g) between 100° and 0°.

Ammonium ferric oxalate (3H$_2$O) *[13268-42-3]* **M 428.1.** Crystd from water (0.5ml/g) between 80° and 0°.

Ammonium ferric sulphate (12H$_2$O) *[7783-83-7]* **M 482.2.** Crystd from aq ethanol.

Ammonium ferrous sulphate (6H$_2$O) *[7783-85-9]* **M 392.1.** A soln in warm water (1.5ml/g) was cooled rapidly to 0°, and the resulting fine crystals were filtered at the pump, washed with cold distd water and pressed between sheets of filter paper to dry.

Ammonium fluorosilicate *[16919-19-0]* **M 178.1.** Crystd from water (2ml/g).

Ammonium hexachloroiridate (IV) *[1694-92-4]* **M 641.0.** Pptd several times from aq soln by satn with ammonium chloride. This removes any palladium and rhodium. Then washed with ice-cold water and dried over conc H$_2$SO$_4$ in a vac. desiccator. If osmium or ruthenium is present, it can be removed as the tetroxide by heating with conc HNO$_3$, followed by conc HClO$_4$, until most of the acid has been driven off. (This treatment is repeated). The near-dry residue is dissolved in a small amount of water and added to excess NaHCO$_3$ soln. and bromine water. On boiling, iridic (but not platinic) hydroxide is pptd. It is dissolved in HCl and pptd several times, then dissolved in HBr and treated with HNO$_3$ and HCl to convert the bromides to chlorides. Satn with ammonium chloride and cooling precipitates ammonium hexachloroiridate which is filtered off and purified as above [Woo and Yost *JACS* **53** 884 *1931*].

Ammonium hypophosphite *[7803-65-8]* **M 117.1.** Crystd from hot EtOH.

Ammonium iodate *[13446-09-8]* **M 192.9.** Crystd from water (8ml/g) between 100° and 0°.

Ammonium iodide *[12027-06-4]* **M 144.9.** Crystd from EtOH by addition of ethyl iodide. Very hygroscopic. Stored in the dark.

Ammonium magnesium chloride (6H$_2$O) *[60314-43-4]* **M 256.8.** Crystd from water (6ml/g) by partial evapn in a desiccator over KOH.

Ammonium magnesium sulphate (6H$_2$O) *[20861-69-2]* **M 360.6.** Crystd from water (1ml/g) between 100° and 0°.

Ammonium manganous sulphate (6H$_2$O) *[13566-22-8]* **M 391.3.** Crystd from water (2ml/g) by partial evapn in a desiccator.

Ammonium metavanadate *[7803-55-6]* **M 117.0, m 200°(dec).** Crystd from conductivity water (20ml/g).

Ammonium molybdate *[13106-76-8]* **M 196.0.** Crystd from water (2.5ml/g) by partial evapn in a desiccator.

Ammonium nickel sulphate (6H$_2$O) *[15699-18-0]* **M 395.0.** Crystd from water (3ml/g) between 90° and 0°.

Ammonium nitrate *[6484-52-2]* **M 80.0.** Crystd twice from distd water (1ml/g) by adding EtOH, or from warm water (0.5ml/g) by cooling in an ice-salt bath. Dried in air, then under vac.

Ammonium oxalate (H_2O) *[6009-70-7]* **M 142.1.** Crystd from water (10ml/g) between 50° and 0°.

Ammonium perchlorate *[7790-98-9]* **M 117.5.** Crystd twice from distd water (2.5ml/g) between 80° and 0°, and dried in a vac desiccator over P_2O_5. Drying at 110° might lead to slow decomposition to chloride. **POTENTIALLY EXPLOSIVE.**

Ammonium reineckate *[13573-16-5]* **M 345.5, m 270-273°(dec).** Crystd from water, between 30° and 0°, working by artificial light. Solns of reineckate decompose slowly at room temperature in the dark and more rapidly at higher temperatures or in diffuse sunlight.

Ammonium selenate *[7783-21-3]* **M 179.0.** Crystd from water at room temperature by adding EtOH and cooling.

Ammonium sulphamate *[7773-06-0]* **M 114.1, m 132-135°.** Crystd from water at room temperature (1ml/g) by adding EtOH and cooling.

Ammonium sulphate *[7783-20-2]* **M 132.1, m 230°(dec).** Crystd twice from hot water containing 0.2% EDTA to remove metal ions, then finally from distd water. Dried in a desiccator for two weeks over $Mg(ClO_4)_2$.

Ammonium tetrafluoroborate *[13826-53-0]* **M 104.8.** Crystd from conductivity water (1ml/g) between 100° and 0°.

Ammonium thiocyanate *[1762-95-4]* **M 76.1, m 138°(dec).** Crystd three times from dil. $HClO_4$, to give material optically transparent at wavelengths longer than 270nm. Has also been crystd from absolute MeOH and from acetonitrile.

Ammonium tungstate *[11120-25-5]* **M 283.9.** Crystd from warm water by adding EtOH and cooling.

n-Amylmercuric chloride *[544-15-0]* **M 307.2, m 110°.** Crystd from EtOH.

9,10-Anthraquinone-2,6-disulphonic acid (disodium salt) *[84-50-4]* **M 412.3, m >325°.** Crystd three times from water, in the dark [Moore *et al, JCSFT1* 82 745 1986].

9,10-Anthraquinone-2-sulphonic acid (sodium salt, H_2O) *[131-08-8]* **M 328.3.** Crystd from water using active charcoal.

Antimony trichloride *[10025-91-9]* **M 228.1, m 73°, b 283°.** Dried over P_2O_5 or by mixing with toluene or xylene and distilling (water is carried off with the organic solvent), then distd twice under dry nitrogen at 50mm, degassed and sublimed twice, under vac into ampoules. Can be crystd from CS_2. Deliquescent. Fumes in moist air.

Antimony trifluoride *[7783-56-4]* **M 178.8, m 292°.** Crystd from MeOH to remove oxide and oxyfluoride, then sublimed under vac in an aluminium cup on to a water-cooled copper condenser [Woolf *JCS* 279 *1955*].

Antimony triiodide *[7790-44-5]* **M 502.5, m 167°.** Sublimed under vac.

Antimony trioxide *[1309-64-4]* **M 291.5, m 656°.** Dissolved in minimum volume of dil HCl, filtered, and six volumes of water were added to ppte a basic antimonous chloride (free from Fe and Sb_2O_5). The ppte was redissolved in dil. HCl, and added slowly, with stirring, to a boiling soln (containing a slight excess) of Na_2CO_3. The oxide was filtered off, washed with hot water, then boiled and filtered, the process being repeated until the filtrate gave no test for chloride ions. The product was dried in a vac desiccator [Schuhmann *JACS* **46** 52 *1924*].

Argon *[7440-37-1]* M 39.95, b -185.6°. Rendered oxygen-free by passage over reduced copper at 450°, or by bubbling through alkaline pyrogallol and H_2SO_4, then dried with $CaSO_4$, $Mg(ClO_4)_2$, or Linde 5A molecular sieves. Other purification steps include passage through Ascarite (asbestos impregnated with sodium hydroxide), through finely divided uranium at about 800° and through a -78° cold trap. Alternatively, passed over CuO pellets at 300° to remove hydrogen and hydrocarbons, over Ca chips at 600° to remove oxygen and, finally, over titanium chips at 700° to remove nitrogen. Also purified by freeze-pump-thaw cycles and by passage over sputtered sodium [Arnold and Smith *JCSFT 2* **77** 861 *1981*].

o-**Arsanilic acid** *[2045-00-3]* M 216.1, m 153°,
p-**Arsanilic acid** *[98-50-0]* M 216.1, m 232°. Crystd from water or ethanol/ether.

Arsenazo I *[520-10-5]* M 614.3, ε 2.6 x 10⁴ at 500nm, pH 8.0. A satd aq soln of the free acid was slowly added to an equal vol. of conc HCl. The orange ppte was filtered, washed with acetonitrile and dried for 1-2hr at 110° [Fritz and Bradford *AC* **30** 1021 *1958*].

Arsenazo III *[1667-00-4]* M 776.4. Contaminants include monoazo derivatives, starting materials for synthesis and byproducts. Partially purified by pptn of the dye from aq alkali by addition of HCl. More thorough purification by taking a 2g sample in 15-25ml of 5% aq NH_3 and filter. Add 10ml HCl (1:1) to the filtrate to ppte the dye. Repeat procedure and dissolve solid dye (0.5g) in 7ml of a 1:1:1 mixture of *n*-propanol:conc NH_3:water at 50°. After cooling, filter soln and treat the filtrate on a cellulose column using 3:1:1 mixture of *n*-propanol:conc NH_3:water as eluent. Collect the blue band and evaporate to 10-15ml below 80°, then add 10ml conc HCl to ppte pure Arsenazo III. Wash with EtOH and air-dry. [Borak *et al, Talanta* **17** 215 *1970*.] The purity of the dye can be checked by paper chromatography using M HCl as eluent.

Arsenic *[7440-38-2]* M 74.9, m 816°. Heated under vac. at 350° to sublime oxides, then sealed in a Pyrex tube under vac. and sublimed at 600°, the arsenic condensing in the cooler parts of the tube. Stored under vac [Shih and Peretti *JACS* **75** 608 *1953*].

Arsenic tribromide *[82868-10-8]* M 394.6, m 89°/11mm, 221°/760mm. Distd under vac.

Arsenic trichloride *[60646-36-8]* M 181.3, b 130.0°. Refluxed with arsenic for 4hr, then fractionally distd. The middle fraction was stored with sodium wire for two days, then again distd [Lewis and Sowerby *JCS* 336 *1957*].

Arsenic triiodide *[50288-23-8]* M 455.6, m 146°. Crystd from acetone.

Arsenic III oxide *[1327-53-3]* M 197.8. Crystd from dil. HCl (1:2), washed, dried and sublimed. Analytical reagent grade material is suitable for use as an analytical standard after it has been dried by heating at 105° for 1-2hr or has been left in a desiccator for several hours over conc H_2SO_4.

3-(2-Arsenophenylazo)-4,5-dihydroxynaphthalene-2,7-disulphonic acid (trisodium salt) see **Arsenazo**.

Barium (metal) *[7440-39-3]* M 137.3. Cleaned by washing with ethyl ether to remove adhering paraffin, then filed in an argon-filled glove box, washed first with ethanol containing 2% conc HCl, then with dry ethanol. Dried under vac and stored under argon [Addison, Coldrey and Halstead *JCS* 3868 *1962*]. Has also been purified by double distn under 10mm argon pressure.

Barium acetate *[543-80-6]* **M 255.4.** Crystd twice from anhydrous acetic acid and dried under vac for 24hr at 100°.

Barium bromate *[13967-90-3]* **M 265.3.** Crystd from hot water (20ml/g).

Barium bromide (2H$_2$O) *[7791-28-8]* **M 333.2.** Crystd from water (1ml/g) by partial evapn in a desiccator.

Barium chlorate (H$_2$O) *[13477-00-4]* **M 322.3.** Crystd from water (1ml/g) between 100° and 0°.

Barium chloride (2H$_2$O) *[10326-27-9]* **M 244.3.** Twice crystd from water (2ml/g) and oven dried to constant weight.

Barium dithionate (2H$_2$O) *[13845-17-5]* **M 333.5.** Crystd from water.

Barium ferrocyanide (6H$_2$O) *[13821-06-2]* **M 594.8.** Crystd from hot water (100ml/g).

Barium formate *[541-43-5]* **M 277.4.** Crystd from warm water (4ml/g) by adding EtOH and cooling.

Barium hydroxide (8H$_2$O) *[12230-71-6]* **M 315.5, m 78°.** Crystd from water (1ml/g).

Barium hypophosphite (H$_2$O) *[14871-79-5]* **M 285.4.** Pptd from aq soln (3ml/g) by adding EtOH.

Barium iodate (H$_2$O) *[10567-69-8]* **M 505.2.** Crystd from a large volume of hot water by cooling.

Barium iodide (2H$_2$O) *[7787-33-9]* **M 427.2.** Crystd from water (0.5ml/g) by partial evapn in a desiccator.

Barium nitrate *[10022-31-8]* **M 261.4, m 593°(dec).** Crystd twice from water (4ml/g) and dried overnight at 110°.

Barium nitrite (H$_2$O) *[7787-38-4]* **M 247.4.** Crystd from water (1ml/g) by cooling in an ice-salt bath.

Barium perchlorate *[13465-95-7]* **M 336.2, m 505°.** Crystd twice from water.

Barium propionate (H$_2$O) *[5908-77-0]* **M 301.5.** Crystd from warm water (50ml/g) by adding EtOH and cooling.

Barium sulphate *[7722-43-7]* **M 233.4.** Washed five times by decantation with hot distilled water, dialysed against distd water for one week, then freeze-dried and oven dried at 105° for 12hr.

Barium tetrathionate *[82203-66-5]* **M 361.6.** Purified by dissolution in a small volume of water and pptd with EtOH below 5°. After drying the salt was stored in the dark at 0°.

Barium thiocyanate (2 H$_2$O) *[2092-17-3]* **M 289.6.** Crystd from water (2.5ml/g) by partial evapn in a desiccator.

Barium thiosulphate *[35112-53-9]* **M 249.5.** Very slightly sol in water. Washed repeatedly with chilled water and dried in air at 40°.

Benzenearsonic acid, see **Phenylarsonic acid.**

Benzenechromium tricarbonyl *[12082-08-5]* **M 214.1. m 163-166°.** Purified by sublimation *in vacuo.*

Benzenestibonic acid *[535-46-6]* **M 248.9, m >250°(dec).** Crystd from acetic acid, or from EtOH-CHCl$_3$ mixture by addition of water.

Beryllium acetate (basic) *[543-81-7]* **M 406.3, m 285-286°.** Crystd from chloroform.

Beryllium potassium fluoride *[7787-50-0]* **M 105.1.** Crystd from hot water (25ml/g).

Beryllium sulphate (4H$_2$O) *[7787-56-6]* **M 177.1.** Crystd from weak aq H$_2$SO$_4$.

2,2'-Bipyridineruthenous dichloride (6H$_2$O), see **Tris(2,2'-bipyridine)ruthenous dichloride.**

Bis(2,9-dimethyl-1,10-phenanthroline)copper(I)perchlorate *[54816-44-5]* **M 579.6.** Crystd from acetone.

Bis(ethyl)titanium(IV) chloride *[2247-00-9]* **M 177.0,**
Bis(ethyl)zirconium(IV) chloride *[92212-70-9]* **M 220.3.** Crystd from boiling toluene.

Bismuth *[7440-69-9]* **M 209.0, m 271-273°.** Melted in an atmosphere of dry helium and filtered through dry Pyrex wool to remove any bismuth oxide present [Mayer, Yosim and Topol *JPC* **64** 238 *1960*].

Bismuthiol I, potassium salt *[4628-94-8]* **M 226.4, m 275-276°(dec).** Usually contaminated with disulphide. Purified by crystn from EtOH.

Bismuth trichloride *[7787-60-2]* **M 315.3, m 233.6°.** Sublimed under high vac, or dried under a current of HCl gas, followed by fractional distn, once under HCl and once under argon.

Borax, see **Sodium borate.**

Boric acid *[10043-35-3]* **M 61.8.** Crystd three times from water (3ml/g) between 100° and 0°, after filtering through sintered glass. Dried to constant weight over metaboric acid in a desiccator.

Boron trichloride *[10294-34-5]* **M 117.2, b 0°/476mm.** Purified (from chlorine) by passage through two mercury-filled bubblers, then fractionally distd under vac. In a more extensive purification the nitrobenzene addition compound is formed by passage of the gas over nitrobenzene in a vac system at 10°. Volatile impurities are removed from the crystalline yellow solid by pumping at -20°, and the BCl$_3$ is recovered by warming the addition compound at 50°. Passage through a trap at -78° removes entrained nitrobenzene; the BCl$_3$ finally condensing in a trap at -112° [Brown and Holmes *JACS* **78** 2173 *1956*]. Also purified by condensing into a trap cooled in acetone/Dry-ice, where it was pumped for 15min to remove volatile impurities. It was then warmed, recondensed and again pumped.

Boron trifluoride *[7637-07-2]* **M 67.8, b 111.8°/300mm.** The usual impurities - bromine, BF$_5$, HF and non-volatile fluorides - are readily separated by distn. Brown and Johannesen [*JACS* **72** 2934 *1950*] passed BF$_3$ into benzonitrile at 0° until the latter was satd. Evacuation to 10^{-5}mm then removed all traces of SiF$_4$ and other gaseous impurities. [A small amount of the BF$_3$-benzonitrile addition compound sublimed and was collected in a U-tube cooled to -80°.] Pressure was raised to 20mm by admitting dry air, and the flask containing the BF$_3$ addition compound was warmed with hot water. The BF$_3$ evolved was passed through a -80° trap (to condense any benzonitrile) into a tube cooled in liquid air. The addition compound with anisole can also be used. For drying, BF$_3$ can be passed through H$_2$SO$_4$ saturated with boric oxide. Fumes in moist air.

Boron trifluoride diethyl etherate *[109-63-7]* **M 141.9, b 67°/43mm, b 126°/760mm, d 1.154, n 1.340.** Treated with a small quantity of ethyl ether (to remove an excess of this component), and then distd under reduced pressure, from CaH$_2$. Fumes in moist air.

Bromine *[7726-95-6]* **M 159.8, b 59°, d 3.102, n 1.661.** Refluxed with solid KBr and distd, dried by shaking with an equal volume of conc H_2SO_4, then distd. The H_2SO_4 treatment can be replaced by direct distn from BaO or P_2O_5. A more extensive purification [Hildenbrand *et al. JACS* **80** 4129 *1958*] is to reflux *ca* 1L of bromine for 1hr with a mixture of 16g of CrO_3 in 200ml of conc H_2SO_4 (to remove organic material). The bromine is distd into a clean dry, glass-stoppered bottle, and chlorine is removed by dissolving *ca* 25g of freshly fused CsBr in 500ml of the bromine and standing overnight. To remove HBr and water, the bromine was then distd back and forth through a train containing alternate tubes of MgO and P_2O_5. **HIGHLY TOXIC.**

Bromine pentafluoride *[7789-30-2]* **M.174.9, m -60.5°, b 41.3°, d^{25} 2.466.** Purified *via* its KF complex, as described for chlorine trifluoride. **HIGHLY TOXIC.**

Bromopyrogallol Red *[16574-43-9]* **M 576.2, ε 5.45 x 10^4 at 538nm (water pH 5.6-7.5).** Recrystd from aq alkaline soln (Na_2CO_3 or NaOH) by pptn on acidification [Suk *Coll Czech Chem Commun* **31** 3127 *1966*].

***n*-Butylmercuric chloride** *[543-63-5]* **M 293.1, m 130°.** Crystd from EtOH.

***n*-Butylstannoic acid** [PhSn(OH)$_3$] *[22719-01-3]* **M 208.8.** Purified by adding excess of KOH in $CHCl_3$ to remove *n*-BuSn(OH)Cl$_2$ and *n*-BuSn(OH)$_2$Cl, and isolated by acidification [Holmes *et al, JACS* **109** 1408 *1987*].

Cacodylic acid *[75-60-5]* **M 138.0, m 195-196°.** Crystd from warm EtOH (3ml/g) by cooling and filtering. Dried in vac desiccator over CaCl$_2$. Has also been twice recrystd from propan-2-ol. [Koller and Hawkridge *JACS* **107** 7412 *1985*.]

Cadmium *[7440-43-9]* **M 112.4.** Oxide has been removed by filtering the molten metal, under vac, through quartz wool.

Cadmium acetate *[543-90-8]* **M 230.5.** Crystd twice from anhydrous acetic acid and dried under vac for 24hr at 100°.

Cadmium bromide (4H$_2$O) *[7789-42-6]* **M 344.2.** Crystd from water (0.6ml/g) between 100° and 0°, and dried at 110°.

Cadmium chloride *[10108-64-2]* **M 183.3, m 568°.** Crystd from water (1ml/g) by addition of EtOH and cooling.

Cadmium fluoride *[7790-79-6]* **M 150.4, m >1000°.** Crystd by dissolving in water at room temperature (25ml/g) and heating to 60°.

Cadmium iodide *[7790-80-9]* **M 366.2, m 388°.** Crystd from ethanol (2ml/g) by partial evapn.

Cadmium lactate *[16039-55-7]* **M 290.6.** Crystd from water (10ml/g) by partial evapn in a desiccator.

Cadmium nitrate (4H$_2$O) *[10022-68-1]* **M 308.5.** Crystd from water (0.5ml/g) by cooling in ice-salt.

Cadmium potassium iodide *[13601-63-3]* **M 532.2.** Crystd from ethanol by partial evapn.

Cadmium salicylate *[19010-79-8]* **M 248.5,**

Cadmium sulphate *[10124-36-4]* **M 769.5.** Crystd from distd water by partial evapn in a desiccator.

Caesium bromide *[7787-69-1]* **M·212.8.** Crystd from water (0.8ml/g) by partial evapn in a desiccator.

Caesium carbonate *[534-17-8]* **M 325.8.** Crystd from ethanol (10ml/g) by partial evapn.

Caesium chloride *[7647-17-8]* **M 168.4.** Crystd from acetone-water, or from water (0.5ml/g) by cooling in $CaCl_2$/ice. Dried at 78° under vac.

Caesium chromate *[13454-78-9]* **M 381.8.** Crystd from water (1.4ml/g) by partial evapn in a desiccator.

Caesium fluoride *[13400-13-0]* **M 151.9.** Crystd from aq soln by adding ethanol.

Caesium iodide *[7789-17-5]* **M 259.8.** Crystd from warm water (1ml/g) by cooling to -5°.

Caesium nitrate *[7789-18-6]* **M 194.9.** Crystd from water (0.6ml/g) between 100° and 0°.

Caesium oleate *[31642-12-3]* **M 414.4.** Crystd from ethyl acetate, dried in an oven at 40° and stored over P_2O_5.

Caesium perchlorate *[13454-84-7* **M 232.4.** Crystd from water (4ml/g) between 100° and 0°.

Caesium perfluoro-octanoate *[17125-60-9]* **M 546.0.** Recrystd from a butanol-petroleum ether mixture, dried in an oven at 40° and stored over P_2O_5 under vac.

Caesium sulphate *[10294-54-9]* **M 361.9.** Crystd from water (0.5ml/g) by adding ethanol and cooling.

Calcium *[7440-70-2]* **M 40.1, m 845°.** Cleaned by washing with ether to remove adhering paraffin, then filed in an argon-filled glove box and washed with ethanol containing 2% of conc HCl. Then washed with dry ethanol, dried in a vacuum and stored under pure argon [Addison, Coldrey and Halstead, *JCS* 3868 *1962*].

Calcium acetate *[62-54-4]* **M 158.2.** Crystd from water (3ml/g) by partial evapn in a desiccator.

Calcium benzoate ($3H_2O$) *[2090-05-3]* **M 336.4.** Crystd from water (10m/g) between 90° and 0°.

Calcium bromide (H_2O) *[62648-72-0]* **M 217.9.** Crystd from ethanol or acetone.

Calcium butyrate *[5743-36-2]* **M 248.2.** Crystd from water (5ml/g) by partial evapn in a desiccator.

Calcium carbamate *[543-88-4]* **M 160.1.** Crystd from aq ethanol.

Calcium chloride ($2H_2O$) *[22691-02-7]* **M 147.0.** Crystd from ethanol.

Calcium dithionite *[13812-88-9]* **M 168.2.** Crystd from water, or water followed by acetone and dried in air at room temperature.

Calcium formate *[544-17-2]* **M 130.1.** Crystd from water (5ml/g) by partial evaporation in a desiccator.

Calcium hydroxide *[1305-62-0]* **M 74.1.** Heat analytical grade calcium carbonate at 1000° during 1 hr. Allow the resulting oxide to cool and add slowly to water. Heat the suspension to boiling, cool and filter through a sintered glass funnel of medium porosity (to remove soluble alkaline impurities). Dry the solid at 110° and crush to a uniformly fine powder.

Calcium iodate *[7789-80-2]* **M 389.9.** Crystd from water (100ml/g).

Calcium iodide (H_2O) *[71626-98-7]* **M 293.9.** Dissolved in acetone, which was then diluted and evapd. This drying process was repeated twice, then the CaI_2 was crystd from acetone-ethyl ether and stored over P_2O_5. Very hygroscopic when anhydrous [Cremlyn *et al. JCS* 528 *1958*].

Calcium isobutyrate *[533-90-4]* **M 248.2.** Crystd from water (3ml/g) by partial evapn in a desiccator.

Calcium lactate ($5H_2O$) *[814-80-2]* **M308.3.** Crystd from warm water (10ml/g) by cooling to $0°$.

Calcium nitrate ($4H_2O$) *[13477-34-4]* **M 236.1, m 45°.** Crystd four times from water (0.4ml/g) by cooling in a $CaCl_2$-ice freezing mixture. The tetrahydrate was dried over conc H_2SO_4 and stored over P_2O_5, to give the anhydrous salt.

Calcium nitrite ($2H_2O$) *[13780-06-8]* **M 150.1.** Crystd from hot water (1.4ml/g) by adding ethanol and cooling.

(+)-Calcium pantothenate (H_2O) *[63409-48-3]* **M 476.5,** $[\alpha]_D^{20}$ **+26.5°** (c 5, H_2O). Crystd from methanol.

Calcium permanganate ($4H_2O$) *[10118-76-0]* **M 350.0.** Crystd from water (3.3ml/g) by partial evapn in a desiccator.

Calcium propionate *[4075-81-4]* **M 186.2.** Crystd from water (2ml/g) by partial evapn in a desiccator.

Calcium salicylate ($2H_2O$) *[824-35-1]* **M 350.4.** Crystd from water (3ml/g) between $90°$ and $0°$.

Calcium thiosulphate **M 152.2.** Recrystd from water below $60°$, followed by drying with EtOH and Et_2O. Stored in a refrigerator. [Pethybridge and Taba *JCSFT 1* 78 1331 *1982*.]

(4-Carbamylphenylarsylenedithio)diacetic acid *[531-72-6]* **M 345.1.** Crystd from methanol or ethanol.

Carbon dioxide *[124-38-9]* **M 44.0, sublimes at -78.5°.** Passed over CuO wire at $800°$ to oxidise CO and other reducing impurities (such as H_2), then over copper dispersed on kieselguhr at $180°$ to remove oxygen. Drying at $-78°$ removed water vapour. Final purification was by vac distn at liquid nitrogen temperature to remove non-condensable gases [Anderson, Best and Dominey *JCS* 3498 *1962*]. Sulphur dioxide can be removed at $450°$ using silver wool combined with a plug of platinized quartz wool. Halogens are removed by using Mg, Zn or Cu, heated to $450°$.

Carbon disulphide, see entry in Chapter 3.

Carbon monoxide *[630-08-0]* **M 28.0, b -191.5°.** Iron carbonyl is a likely impurity in CO stored under pressure in steel tanks: It can be decomposed by passage of the gas through a hot porcelain tube at $350-400°$. Passage through alkaline pyrogallol soln removes oxygen (and CO_2). Removal of CO_2 and water are effected by passage through soda-lime followed by $Mg(ClO_4)_2$. Carbon monoxide can be condensed and distd at $-195°$.

Carbon tetrachloride, see entry in Chapter 3.

Carbonyl bromide *[593-95-3]* **M 187.8.** Purified by distn from Hg and from powdered Sb to remove free bromine, then vac distd to remove volatile SO_2 (the major impurity)[Carpenter *et al. JCSFT 2* 384 *1977*].

Carbonyl sulphide *[463-58-1]* **M 60.1.** Purified by scrubbing through three consecutive fritted washing flasks containing conc NaOH at 0°. Then freeze-pumped repeatedly and distd through a trap packed with glass wool and cooled to -130° (using an *n*-pentane slurry).

Celite 545 (diatomaceous earth) *[12279-49-1]*. Stood overnight in conc HCl after stirring well, then washed with distd water until neutral and free of chloride ions. Washed with methanol and dried at 50°.

Ceric ammonium nitrate *[16774-21-3]* **M 548.2.** Ceric ammonium nitrate (125g) is warmed with 100ml of dil. HNO_3 (1:3 v/v) and 40g of NH_4NO_3 until dissolved, and filtered off on a sintered-glass funnel. The solid which separates on cooling in ice is filtered off on a sintered funnel (at the pump) and air is sucked through the solid for 1-2 hr to remove most of the nitric acid. Finally, the solid is dried at 80-85°.

Cerous acetate *[537-00-8]* **M 317.3.** Crystd twice from anhydrous acetic acid, then pumped dry under vac at 100° for 8hr.

Chloramine-T *[127-65-1]* **M 227.6, m 168-170°.** Crystd from hot water (2ml/g). Dried in a desiccator over $CaCl_2$ (protect from sunlight).

Chlorine *[7782-50-5]* **M 70.9.** Passed in succession through aq $KMnO_4$, dil H_2SO_4, conc H_2SO_4, and a drying tower containing $Mg(ClO_4)_2$. Or, washed with water, dried over P_2O_5 and distd from bulb to bulb. **HIGHLY TOXIC.**

Chlorine trifluoride *[7790-91-2]* **M 92.5, b 12.1°.** Impurities include chloryl fluoride, chlorine dioxide and hydrogen fluoride. Passed first through two U-tubes containing NaF to remove HF, then through a series of traps in which the liquid is fractionally distd. Can be purified *via* the KF complex, $KClF_4$, formed by adding excess ClF_3 to solid KF in a stainless steel cylinder in a dry-box and shaking overnight. After pumping out the volatile materials, pure ClF_3 is obtained by heating the bomb to 100-150° and condensing the evolved gas in a -196° trap [Schack, Dubb and Quaglino *Chem Ind* 545 *1967*]. **HIGHLY TOXIC.**

Chloroform, see entry in Chapter 3.

p-**Chloromercuribenzoic acid** *[59-85-8]* **M 357.2, m >300°.** Its suspension in water is stirred with enough 1M NaOH to dissolve most of it: a small amount of insoluble matter is removed by centrifugation. The chloromercuribenzoic acid is then pptd. by adding 1M HCl and centrifuged off. The pptn is repeated twice. Finally, the ppte is washed three times with distd water (by centrifuging), then dried in a thin layer under vac over P_2O_5 [Boyer *JACS* 76 4331 *1954*].

Chlorosulphonic acid *[7790-94-5]* **M 116.5, b 151-152°/750mm, n 1.4929, d$_4$ 1.753.** Distd in an all-glass apparatus, taking the fraction boiling at 156-158°. Reacts **EXPLOSIVELY** with water.

Chromazurol S *[1667-99-8]* **M 605.3.** Purified by paper chromatography using *n*-butanol, acetic acid and water (7:3:1). First and second spots extracted.

Chromic chloride (anhydrous) *[10025-73-7]* **M 158.4.** Sublimed in a stream of dry HCl. Alternatively, the impure chromic chloride (100g) was added to 1L of 10% aq $K_2Cr_2O_7$ and several millilitres of conc HCl, and the mixture was brought to a gentle boil with constant stirring for 10 min. (This removed a reducing impurity.) The solid was separated, and washed by boiling with successive 1L lots of distd water until the wash water no longer gave a test for chloride ion, then dried at 110° [Poulsen and Garner *JACS* 81 2615 *1959*].

Chromium ammonium sulphate (12H$_2$O) *[34275-72-4]* **M 478.4.** Crystd from a satd aq soln at 55° by cooling slowly with rapid mechanical stirring. The resulting fine crystals were filtered on a Büchner funnel, partly dried on a porous plate, then equilibrated for several months in a vac desiccator over crude chromium ammonium sulphate (partially dehydrated by heating at 100° for several hours before use)[Johnson, Hu and Horton *JACS* 75 3922 *1953*].

Chromium potassium sulphate (12H$_2$O) *[7788-99-0]* **M 499.4.** Crystd from hot water (2ml/g) by cooling.

Chromium trioxide *[1333-82-0]* **M 100.0.** Crystd from water (0.5ml/g) between 100° and -5°, or from water/conc HNO$_3$ (1:5). Dried in a vac desiccator over NaOH pellets.

Chromium (III) tris-2,4-pentanedionate *[21679-31-2]* **M 349.3, m 210°.** Crystd three times from aq ethanol.

Chromyl chloride *[14977-61-8]* **M 154.9, b 115.7°, d 1.911.** Purified by distn under reduced pressure.

Cisplatin see *cis*-**diamminedichloroplatinum(II).**

Cobaltous acetate (4H$_2$O) *[6147-53-1]* **M 249.1.** Crystd several times as the tetrahydrate from 50% aq acetic acid. Converted to the anhydrous salt by drying at 80°/1mm for 60hr.

Cobaltous acetylacetonate *[14024-48-7]* **M 257.2, m 172°.** Crystd from acetone.

Cobaltous ammonium sulphate (6H$_2$O) *[13596-46-8]* **M 395.5.** Crystd from boiling water (2ml/g) by cooling. Washed with ethanol.

Cobaltous bromide (6H$_2$O) *[7789-43-7]* **M 326.9.** Crystd from water (1ml/g) by partial evapn in a desiccator.

Cobaltous chloride (6H$_2$O) *[7791-13-7]* **M 237.9.** A satd aq soln at room temperture was fractionally crystd by standing overnight. The first half of the material that crystd in this way was used in the next crystn. The process was repeated several times, water being removed in a dry-box using air filtered through glass wool and dried over CaCl$_2$ [Hutchinson *JACS* 76 1022 *1954*]. Has also been crystd from dil aq HCl.

Cobaltous nitrate (6H$_2$O) *[10026-22-9]* **M 291.0.** Crystd from water (1ml/g), or ethanol (1ml/g), by partial evapn.

Cobaltous perchlorate (6H$_2$O) *[13455-31-7]* **M 365.9.** Crystd from warm water (0.7ml/g) by cooling.

Cobaltous potassium sulphate *[13596-22-0]* **M 329.4.** Crystd from water (1ml/g) between 50° and 0°.

Cobaltous sulphate (7H$_2$O) *[10124-43-3]* **M 281.1.** Crystd three times from conductivity water (1.3ml/g) between 100° and 0°.

Cupric acetate (H$_2$O) *[6046-93-1]* **M 199.7.** Crystd twice from warm dil acetic acid solns (5ml/g) by cooling.

Cupric ammonium chloride (2H$_2$O) *[10534-87-9]* **M 277.5.** Crystd from weak aq HCl (1ml/g).

Cupric benzoate *[533-01-7]* **M 305.8.** Crystd from hot water.

Cupric bromide *[7789-45-9]* **M 223.4.** Crystd twice by dissolving in water (140ml/g), filtering to remove any Cu$_2$Br$_2$, and concentrating under vac at 30° until crystals appeared. The cupric bromide was then allowed to crystallise by leaving the soln in a vac desiccator containing P$_2$O$_5$ [Hope, Otter and Prue *JCS* 5226 *1960*].

Cupric chloride *[7447-39-4]* **M 134.4.** Crystd from hot dil aq HCl (0.6ml/g) by cooling in a CaCl$_2$-ice bath. Dehydrated by heating on a steam-bath under vac.

Cupric lactate (H$_2$O) *[814-81-3]* **M 295.7.** Crystd as the monohydrate from boiling water (3ml/g) by cooling.

Cupric nitrate (3H$_2$O) *[19004-19-4]* **M 241.6.** Crystd from weak aq HNO$_3$ (0.5ml/g) by cooling from room temperature. The anhydrous salt can be prepared by dissolving copper metal in a 1:1 mixture of liquid NO$_2$ and ethyl acetate and purified by sublimation [Evans *et al. JCSFT I* **75** 1023 *1979*].

Cupric oleate *[1120-44-1]* **M 626.5.** Crystd from ethyl ether.

Cupric perchlorate (6H$_2$O) *[13770-18-8]* **M 370.5.** Crystd from distd water.

Cupric phthalocyanine *[147-14-8]* **M 576.1.** Precipitated twice from conc H$_2$SO$_4$ by slow diln with water. Also purified by two or three sublimations in an argon flow at 300-400Pa.

Cupric sulphate *[7758-98-7]* **M 159.6.** After adding 0.02g of KOH to a litre of nearly satd aq soln, it was left for two weeks, then the ppte was filtered on to a fibreglass filter with pore diameter of 5-15 microns. The filtrate was heated to 90° and allowed to evaporate until some CuSO$_4$.5H$_2$O had crystd. The soln was then filtd hot and cooled rapidly to give crystals which were freed from mother liquor by filtering under suction [Geballe and Giauque *JACS* **74** 3513 *1952*]. Alternatively crystd from water (0.6ml/g) between 100° and 0°.

Cuprous bromide *[7787-70-4]* **M 286.9.** Purified as for cuprous iodide but using aq NaBr.

Cuprous chloride *[7758-89-6]* **M 99.0.** Dissolved in strong HCl, pptd by diln with water and filtered off. Washed with ethanol and ethyl ether, then dried and stored in a vac desiccator [Österlöf *Acta Chem Scand* **4** 375 *1950*]. Alternatively, to an aq. soln of CuCl$_2$.2H$_2$O was added, with stirring, an aq soln of anhydrous sodium sulphite. The colourless product was dried at 80° for 30min and storred under N$_2$. CuCl$_2$ can be purified by zone-refining [Hall *et al, JCSFT I* **79** 243 *1983*].

Cuprous iodide *[7681-65-4]* **M 190.4.** Freshly prpeared by dissolving an appropriate quantity of CuI in boiling satd aq. NaI over 30min. Pure CuI is obtained by cooling and diluting the soln with water, followed by filtering and washing sequentially with H$_2$O, EtOH, EtOAc and Et$_2$O, pentane, then drying *in vac* for 24 hr [Dieter, *JACS* **107** 4679 *1985*].

Cuprous thiocyanate *[18223-42-2]* **M 121.6.** Purified as for cuprous iodide but using aq NaSCN.

Decaborane *[17702-41-9]* **M 122.2, m 99.7-100°.** Purified by vac sublimation at 80°/0.1mm, followed by crystn from methylcyclohexane, methylene chloride, or dry olefin-free-*n*-pentane, the solvent being subsequently removed by storing the crystals in a vac desiccator containing CaCl$_2$.

Deuterium *[7782-39-0]* **M 4.** Passed over activated charcoal at -195° [MacIver and Tobin *JPC* **64** 451 *1960*]. Purified by diffusion through nickel [Pratt and Rogers, *JCSFT I* **92** 1589 *1976*].

Deuterium oxide *[7789-20-0]* **M 20, fp 3.8°/760mm, b 101.4°/760mm, d 1.105.** Distd from alkaline KMnO$_4$ [de Giovanni and Zamenhof *BJ* **92** 79 *1963*]. **NOTE that D$_2$O invariably contains tritiated water and will therefore be RADIOACTIVE; always check the radioactivity of D$_2$O in a scintillation counter before using.**

cis-Diamminedichloroplatinum(II) *[15663-27-1]* M 300.1. Recrystd from dimethylformamide and the purity checked by IR and UV-vis spectroscopy. [Raudaschl *et al, Inorg Chim Acta* **78** 143 *1983*].

Diammonium hydrogen orthophosphate *[7783-28-0]* M 132.1. Crystd from water (1ml/g) between 70° and 0°.

Dicobalt octacarbonyl *[15226-74-1]* M 341.9, m 51°. Orange-brown crystals by recrystn from *n*-hexane under a carbon monoxide atmosphere [Ojima *et al, JACS* **109** 7714 *1987*; see also Hileman in *Preparative Inorganic Reactions,* Jolly ed, vol 1 101 1987].

Diethyl aluminium chloride *[96-10-6]* M 120.6, m -75.5°, b 106.5-108°/24.5mm, d 0.96. Distd from excess dry NaCl (to remove ethyl aluminium dichloride) in a 50-cm column containing a heated nichrome spiral.

1,2-Dihydroxybenzene-3,5-disulphonic acid, disodium salt (TIRON) *[149-45-1]* M 332.2, ε 6.9 x 10^4 at 260nm, pH 10.8. Recrystd from water [Hamaguchi *et al, Anal Chim Acta* **9** 563 *1962*].

2,6-Dimethyl-1,10-phenanthrolinedisulphonic acid, disodium salt (H_2O) *[52698-84-7]* M 564.5. Inorganic salts and some coloured species can be removed by dissolving the crude material in the minimum volume of water and precipitating by adding EtOH. Purified reagent can be obtained by careful evapn of the filtrate.

Dinitrogen tetroxide, N_2O_4 *[10544-72-6]* M 92.0 m -11.2°, b 21.1°. Purified by oxidation at 0° in a stream of oxygen until the blue colour changed to red-brown. Distd from P_2O_5, then solidified on cooling in a deep-freeze (giving nearly colourless crystals). Oxygen can be removed by alternate freezing and melting.

Diphenylmercury *[587-85-9]* M 354.8, m 125.5-126°. Sublimed, then crystd from nitromethane or ethanol. If phenylmercuric halides are present they can be converted to phenylmercuric hydroxide which, being much more soluble, remains in the alcohol or benzene used for crystn. Thus, crude material (10g) is dissolved in warm ethanol (*ca* 150ml) and shaken with moist Ag_2O (*ca* 10g) for 30min, then heated under reflux for 30min and filtered hot. Concn of the filtrate by evapn gives diphenylmercury, which is recrystd from benzene [Blair, Bryce-Smith and Pengilly *JCS* 3174 *1959*].

4,7-Diphenyl-1,10-phenanthrolinedisulphonic acid, disodium salt *[52746-49-3]* M 536.5. Dissolve crude sample in the minimum volume of water and add EtOH to ppte the contaminants. Carefully evap the filtrate to obtain pure material.

Diphenylsilanediol *[947-42-2]* M 216.3, m 148°(dec). Crystd from $CHCl_3$-methyl ethyl ketone.

Disodium anthraquinone-2,6-disulphonate *[853-693-9]* M 412.3. Crystd from water.

Disodium calcium ethylenediaminetetraacetate *[62-33-9]* M 374.3. Dissolved in a small amount of water, filtered and pptd with excess EtOH. Dried at 80°.

Disodium dihydrogen ethylenediaminetetraacetic acid ($2H_2O$) *[6381-92-6]* M 372.2, m 248°(dec). Analytical reagent grade material can be used as primary standard after drying at 80°. Commercial grade material can be purified by crystn from water or by preparing a 10% aq. soln at room temperature, then adding ethanol slowly until a slight permanent ppte is formed, filtering, and adding an equal volume of ethanol. The ppte is filtered off on a sintered-glass funnel, is washed with acetone, followed by ethyl ether, and dried in air overnight to give the dihydrate. Drying at 80° for at least 24hr converts it to the anhydrous form.

Disodium 1,8-dihydroxynaphthalene-3,6-disulphonate (2H$_2$O) *[2808-22-0]* **M 400.3, m >300°.** Crystd from water.

Disodium ethylenebis[dithiocarbamate] *[142-59-6]* **M 436.5.** Crystd (as hexahydrate) from aq ethanol.

Disodium-β-glycerophosphate *[819-83-0]* **M 216.0, m 102-104°.** Crystd from water.

Disodium hydrogen orthophosphate (anhydrous) *[7558-79-4]* **M 142.0.** Crystd twice from warm water, by cooling. Air dried, then oven dried overnight at 130°. Hygroscopic: should be dried before use.

Disodium magnesium ethylenediaminetetraacetate *[14402-88-1]* **M 358.5.** Dissolved in a small amount of water, filtered and pptd with an excess of methanol. Dried at 80°.

Disodium naphthalene-1,5-disulphonate *[1655-29-4]* **M 332.3.** Recrystd from aq acetone [Okahata *et al, JACS* **108** 2863 *1986*].

Disodium phenylphosphate (2H$_2$O) *[3279-54-7]* **M 254.1.** Dissolved in a minimum amount of methanol, filtering off an insoluble residue of inorganic phosphate, then pptd by adding an equal volume of ethyl ether. Washed with ethyl ether and dried [Tsuboi *Biochim Biophys Acta* **8** 173 *1952*].

Disodium succinate *[150-90-3]* **M 162.1.** Crystd twice from water. Freed from other metal ions by passage of an 0.1M soln through a column of Dowex ion-exchange resin A-1, sodium form.

Di-*p*-tolylmercury *[50696-65-6]* **M 382.8, m 244-246°.** Crystd from xylene.

Eosin (as disodium salt) *[548-26-5]* **M 624.1.** Dissolved in water and pptd by addition of dil HCl. The ppte was washed with water, crystd from ethanol, then dissolved in the minimum amount of dil NaOH soln and evapd to dryness on a water-bath. The purified disodium salt was then crystd twice from ethanol [Parker and Hatchard *TFS* **57** 1894 *1961*].

Ethylarsonic acid *[507-32-4]* **M 154.0, m 99.5°.** Crystd from ethanol.

Ethylmercuric chloride *[107-27-7]* **M 265.1, m 193-194°.** Mercuric chloride can be removed by suspending ethylmercuric chloride in hot distd water, filtering with suction in a sintered-glass crucible and drying. Then crystd from ethanol and sublimed under reduced pressure. It can also be crystd from water.

Ethylmercuric iodide *[2440-42-8]* **M 356.6, m 186°.** Crystd once from water (50ml/g).

Europium (III) acetate (2H$_2$O) *[101953-41-7]* **M 383.1.** Recrystd several times from water [Ganapathy *et al, JACS* **108** 3159 *1986*].

Ferric acetylacetonate *[14024-18-1]* **M 353.2, m 179°.** Crystd from 95% ethanol and dried for 1hr at 120°.

Ferric chloride (anhydrous) *[7705-08-0]* **M 162.2, m >300°(dec).** Sublimed at 200° in an atmosphere of chlorine. Stored in a weighing bottle inside a desiccator.

Ferric chloride (6H$_2$O) *[10025-77-1]* **M 270.3.** An aq soln, satd at room temperature, was cooled to -20° for several hours. Pptn was slow, even with scratching and seeding, and it was generally necessary to stir overnight. The presence of free HCl retards the pptn [Linke *JPC* **60** 91 *1956*].

Ferric perchlorate (9H$_2$O) *[13537-24-1]* **M 516.3.** Crystd twice from conc HClO$_4$, the first time in the presence of a small amount of H$_2$O$_2$ to ensure that the iron is fully oxidised [Sullivan *JACS* **84** 4256 *1962*]. Extreme care should be taken with this preparation because it is potentially **DANGEROUS**.

Ferrocene *[102-54-5]* **M 186.0, m 173-174°.** Crystd from methanol or ethanol, and sublimed *in vacuo*. [Saltiel *et al, JACS* **109** 1209 *1987*.]

Ferrocenecarboxaldehyde *[12093-10-6]* **M 214.1, m 117-120°.** Purified by vac sublimation.

Ferrocenecarboxylic acid *[1271-42-7]* **M 230.1, m 210°(dec).** Crystd from aq. ethanol. [Matsue *et al, JACS* **107** 3411 *1985*.]

Ferrous chloride (4H$_2$O) *[13478-10-9]* **M 198.8.** A 550ml round-bottomed Pyrex flask was connected, *via* a glass tube fitted with a medium porosity sintered-glass disc, to a similar flask. To 240g of FeCl$_2$.4H$_2$O in the first flask was added conductivity water (200ml), 38% HCl (10ml), and pure electrolytic iron (8-10g). A stream of purified N$_2$ was passed through the assembly, escaping through a mercury trap. The salt was dissolved by heating which was continued until complete reduction had occurred. By inverting the apparatus and filtering (under N$_2$ pressure) through the sintered glass disc, unreacted iron was removed. After cooling and crystn, the unit was again inverted and the crystals of ferrous chloride were filtered free from mother liquor by applied N$_2$ pressure. Partial drying by overnight evacuation at room temperature gave a mixed hydrate which, on further evacuation on a water bath at 80°, lost water of hydration and its absorbed HCl (with vigorous effervescence) to give a white powder, FeCl$_2$.2H$_2$O [Gayer and Wootner *JACS* **78** 3944 *1956*].

Ferrous perchlorate (6H$_2$O) *[13933-23-8]* **M 362.9.** Crystd from HClO$_4$.

Ferrous sulphate (7H$_2$O) *[7782-63-0]* **M 278.0.** Crystd from 0.4M H$_2$SO$_4$.

Fluorine *[7782-41-4]* **M 38.0, b -129.2°.** Passed through a bed of NaF at 100° to remove HF and SiF$_4$. [For description of stills used in fractional distn, see Greenberg *et al. JPC* **65** 1168 *1961*; Stein, Rudzitis and Settle *Purification of Fluorine by Distillation, Argonne National Laboratory,* ANL-6364 1961 (from Office of Technical Services, US Dept of Commerce, Washington 25)]. **HIGHLY TOXIC**.

Fluoroboric acid *[16872-11-0]* **M 87.8.** Crystd several times from conductivity water.

Gallium *[7440-55-3]* M 69.7, m 29.8°.
Dissolved in dil HCl and extracted into Et$_2$O. Pptn with H$_2$S removed many metals, and a second extraction with Et$_2$O freed Ga more completely, except for Mo, Th(III) and Fe which were largely removed by pptn with NaOH. The soln was then electrolysed in 10% NaOH with a Pt anode and cathode (2-5A at 4-5V) to deposit Ga, In, Zn and Pb, from which Ga was obtained by fractional crystn of the melt [Hoffman *J Res Nat Bur Stand* **13** 665 *1934*]. Also purified by heating to boiling in 0.5-1M HCl, then heating to 40° in water and pouring the molten Ga with water under vac on to a glass filter (30-50 μ pore size), to remove any unmelted metals or oxide film. The Ga was then fractionally crystd from the melt under water.

Germanium *[7440-56-4]* **M 72.6.** Copper contamination on the surface and in the bulk of single crystals of Ge can be removed by immersion in molten alkali cyanide under N_2. The Ge was placed in dry cyanide powder in a graphite holder in a quartz or porcelain boat. The boat was then inserted into a heated furnace which, after a suitable time, was left to cool to room temperature. At 750°, a 1mm thickness requires about 1min, whereas 0.5cm needs about half hour. The boat was removed and the samples were taken out with plastic-coated tweezers, carefully rinsed in hot water and dried in air [Wang *JPC* **60** 45 *1956*].

Glass powder (100-300 mesh). Washed with 10% HNO_3, water and dried.

Graphite *[7782-42-5]*. Treated with hot 1:1 HCl. Filtered, washed, dried, powdered and heated in an evacuated quartz tube at 1000° until a high vacuum was obtained. Cooled and stored in an atmosphere of helium [Craig, Van Voorhis and Bartell *JPC* **60** 1225 *1956*].

Haematoporphyrin IX *[14459-29-1]* M 598.7. Recrystd from MeOH.

Helium *[7440-59-7]* **M 4.0.** Dried by passage through a column of Linde 5A molecular sieves and $CaSO_4$, then passed through an activated-charcoal trap cooled in liquid N_2, to adsorb N_2, argon, xenon and krypton. Passed over CuO pellets at 300° to remove hydrogen and hydrocarbons, over Ca chips at 600° to remove oxygen, and then over titanium chips at 700° to remove N_2 [Arnold and Smith *JCSFT 2* **77** 861 *1981*].

Hexachlorocyclotriphosphazene *[940-71-6]* **M 347.7, m 113-115°.** Recrystd from *n*-hexane [Winter and van de Grampel *JCSDT* 1269 *1986*].

Hexamminecobalt(III) chloride *[10534-89-1]* **M 267.5.** Crystd from warm water (8ml/g) by cooling.

Hexammineruthenium(III) chloride *[14282-91-8]* **M 309.6.** Crystd twice from 1M HCl.

Hydrazine (anhydrous) *[302-01-2]* **M 32.1, fp 1.5-2.0°, b 113-113.5°, n 1.470, d 1.91.** Hydrazine hydrate is dried by refluxing with an equal weight of NaOH pellets for 3hr, then distd from fresh NaOH or BaO in a current of dry N_2.

Hydrazine dihydrochloride *[5341-61-7]* **M 105.0.** Crystd from aq. EtOH and dried under vac over $CaSO_4$.

Hydrazine monohydrochloride *[2644-70-4]* **M 68.5, m 89°.** Prepd by dropwise addition of cold conc HCl to cold liquid hydrazine in equimolar amounts. The crystals were harvested from water and were twice recrystd from absolute MeOH and dried under vac. [Kovack *et al*, *JACS* **107** 7360 *1985*].

Hydriodic acid *[10034-85-2]* **M 127.9, b 127°, d 1.701.** Iodine can be removed from aq HI, probably as the amine hydrogen triiodide, by three successive extractions using a 4% soln of Amberlite LA-2 (a long-chain aliphatic amine) in CCl_4, toluene or pet ether (10ml per 100ml of acid). [Davidson and Jameson *Chem & Ind* 1686 *1963*.] Extraction with tributyl phosphate in $CHCl_3$ or other organic solvents is also suitable. Alternatively, a De-acidite FF anion-exchange resin column in the OH⁻-form using 2M NaOH, then into its I⁻-form by passing dil. KI soln, can be used. Passage of an HI soln under CO_2 through such a column removes polyiodide. The column can be regenerated with NaOH. [Irving and Wilson *Chem & Ind* 653 *1964*.] The earlier method was to reflux with red phosphorus and distil in a stream of N_2. The colourless product was stored in ampoules in the dark [Bradbury *JACS* **74** 2709 *1952*]. Fumes in moist air.

Hydrobromic acid *[10035-10-6]* **M 80.9.** A soln of aq HBr *ca* 48% (w/w, constant boiling) was distd twice with a little red phosphorus, and the middle half of the distillate was taken. (The azeotrope at 760mm contains 47.8% (w/w) HBr.) [Hetzer, Robinson and Bates *JPC* **66** 1423 *1962*.] Free bromine can be removed by Irvine and Wilson's method for HI (see above), except that the column is regenerated by washing with an ethanolic soln of aniline or styrene. Hydrobromic acid can also be purified by aerating with H_2S distilling and collecting the fraction boiling at 125-127°.

Hydrochloric acid *[7647-01-0]* **M 36.5, d 1.20.** Readily purified by fractional distn as constant boiling point acid, following diln with water. The constant-boiling fraction contains 1 mole of HCl in the following weights of distillate at the stated pressures: 179.555g (730mm), 179,766g (740mm), 179,979 (750mm), 180.193 (760mm), 180.407 (770mm) [Foulk and Hollingsworth *JACS* **45** 1220 *1923*].

Hydrofluoric acid *[7664-36-3]* **M 20.0, d 1.150.** Freed from lead (Pb *ca* 0.002ppm) by co-pptn with SrF_2, by addition of 10ml of 10% $SrCl_2$ soln per kilogram of the conc acid. After the ppte has settled, the supernatant is decanted through a filter in a hard-rubber or paraffined-glass vessel [Rosenqvist *Amer J Sci* **240** 358 *1942*. Pure aq HF solns (up to 25M) can be prepared by isothermal distn in polyethylene, polypropylene or platinum apparatus [Kwestroo and Visser *Analyst* **90** 297 *1965*]. **HIGHLY TOXIC.**

Hydrogen *[1333-75-0]* **M 2.0, m -259.1°, -252.9°.** Usually purified by passage through suitable absorption train. Carbon dioxide is removed with KOH pellets, soda-lime or NaOH pellets. Oxygen is removed with a "De-oxo" unit or by passage over Cu heated to 450-500°, Cu on kieselguhr at 250°. Passage over a mixture of MnO_2 and CuO (Hopcalite) oxidises any CO to CO_2 (which is removed as above). Hydrogen can be dried by passage through dried silica-alumina at -195°, through a dry-ice trap followed by a liquid-N_2 trap packed with glass wool, through $CaCl_2$ tubes, or through $Mg(ClO_4)_2$ or P_2O_5. Other purification steps include passage through a hot palladium thimble [Masson *JACS* **74** 4731 *1952*], through an activated-charcoal trap at -195°, and through non-absorbent cotton-wool filter or small glass spheres coated with a thin layer of silicone grease. *Potentially* **EXPLOSIVE** *in air.*

Hydrogen bromide (anhydrous) *[10035-10-6]* **M 80.9.** Dried by passage through $Mg(ClO_4)_2$ towers. This procedure is **hazardous**, see Stoss and Zimmermann [*Ind Eng Chem* **17** 70 *1939*]. Shaken with mercury, distd through a -78° trap and condensed at -195°/10⁻⁵mm. Fumes in moist air.

Hydrogen chloride *[7647-01-0]* **M 36.5.** Passed through conc H_2SO_4, then over activated charcoal and silica gel. Fumes in moist air. Hydrogen chloride in gas cylinder include ethylene, 1,1-dichloroethane and ethyl chloride. The latter rwo may be removed by fractionating the HCl through a trap cooled to -112°. Ethylene is difficult to remove. Fumes in moist air.

Hydrogen cyanide (anhydrous) *[74-90-8]* **M 27.0, b 25.7°.** Prepared from NaCN and H_2SO_4, and dried by passage through H_2SO_4 and over $CaCl_2$, then distd in a vacuum system and degassed at 77°K before use [Arnold and Smith *JCSFT 2* **77** 861 *1981*]. Cylinder HCN may contain stabilisers against explosive polymerisation, together with small amounts of H_3PO_4, H_2SO_4, SO_2, and water. It can be purified by distn over P_2O_5, then frozen in Pyrex bottles at Dry-ice temperature for storage. **HIGHLY POISONOUS.**

Hydrogen fluoride (anhydrous) *[7664-39-3]* **M 20.0, b 19.4°.** Can be purified by trap-to-trap distn, followed by drying over CoF_2 at room temperature and further distn. Alternatively, it can be absorbed on NaF to form $NaHF_2$ which is then heated under vac at 150° to remove volatile impurities. The HF is regenerated by heating at 300° and stored with CoF_3 in a nickel vessel, being distd as required. (Water content *ca* 0.01%.) To avoid contact with base metal, use can be made of nickel, polychlorotrifluoroethylene and gold-lined fittings [Hyman, Kilpatrick and Katz *JACS* **79** 3668 *1957*]. **HIGHLY TOXIC.**

Hydrogen iodide (anhydrous) *[10034-85-2]* **M 127.9, b -35.5°.** After removal of free iodine from aq HI (q.v.), the soln is frozen, then covered with P_2O_5 and allowed to melt under vac. The gas evolved is dried by passage through P_2O_5 on glass wool. It can be freed from iodine contamination by repeated fractional distn at low temperatures. Fumes in moist air.

Hydrogen peroxide *[7722-84-1]* **M 34.0, d 1.110.** The 30% material has been steam distd using distd water. Gross and Taylor [*JACS* **72** 2075 *1950*] made 90% H_2O_2 approx 0.001M in NaOH and then distd under its own vapour pressure, keeping the temperature below 40°, the receiver being cooled with a Dry-ice/isopropyl alcohol mush. The 98% material has been rendered anhydrous by repeated fractional crystn in all-quartz vessels. **EXPLOSIVE IN CONTACT WITH ORGANIC MATERIAL.**

Hydrogen sulphide *[7783-06-4]* **M 34.1, b -59.6°.** Washed, then passed through a train of flasks containing satd $Ba(OH)_2$ (two), water (two), and dil HCl [Goates *et al*, *JACS* **73** 707 *1951*]. **HIGHLY POISONOUS.**

Hydroxylamine *[7803-49-8]* **M 33.0, m 33.1°, b 56.5°/22mm.** Crystd from *n*-butanol at -10°, collected by vac filtratn and washed with cold ethyl ether.

Hydroxylamine hydrochloride *[5470-11-1]* **M 69.5, m 151°.** Crystd from aq. 75% ethanol or boiling methanol, and dried under vac over $CaSO_4$ or P_2O_5. Has also been dissolved in a minimum of water and satd with HCl; after three such crystns it was dried under vac over $CaCl_2$ and NaOH.

Hydroxylamine sulphate *[10039-54-0]* **M 164.1, m 170°(dec).** Crystd from boiling water (1.6ml/g) by cooling to 0°.

Hydroxynaphthol Blue, disodium salt, M 620.5. Crude material was treated with hot EtOH to remove soluble impurities, then dissolved in 20% aq MeOH and chromatographed on a cellulose powder column with propanol:EtOH:water (5:5:4) as eluent. The upper of three zones was eluted to give the pure dye which was pptd as the monosodium salt trihydrate by adding conc HCl to the concentrated eluate [Ito and Ueno *Analyst* **95** 583 *1970*].

4-Hydroxy-3-nitrobenzenearsonic acid *[121-19-7]* **M 263.0.** Crystd from water.

Hypophosphorous acid *[6303-21-5]* **M 66.0, m 26.5°.** Phosphorous acid is a common contaminant of commercial 50% hypophosphorous acid. Jenkins and Jones [*JACS* **74** 1353 *1952*] purified this material by evaporating about 600ml in a 1L flask at 40°, under reduced pressure (in N_2), to a volume of about 300ml. After the soln was cooled, it was transferred to a wide-mouthed Erlenmeyer flask which was stoppered and left in a Dry-ice/acetone bath for several hours to freeze (if necessary, with scratching of the wall). When the flask was then left at *ca* 5° for 12hr, about 30-40% of it liquefied, and again filtered. This process was repeated, then the solid was stored over $Mg(ClO_4)_2$ in a vac desiccator in the cold. Subsequent crystns from *n*-butanol by dissolving it at room temperature and then cooling in an ice-salt bath at -20° did not appear to purify it further.

Indium *[7440-74-6]* **M 114.8.** Before use, the metal surface can be cleaned with dil HNO_3, followed by a thorough washing with water and an alcohol rinse.

Indium sulphate *[13464-82-9]* **M 517.8.** Crystd from dil aq H_2SO_4.

Iodine *[7553-56-2]* **M 253.8, m 113.6°.** Usually purified by vac. sublimation. Preliminary purifications include grinding with 25% by weight of KI, blending with 10% BaO and subliming; subliming with CaO; grinding to a powder and treating with successive portions of water to remove dissolved salts, then drying; and crystn from benzene. Barrer and Wasilewski [*TFS* **57** 1140 *1961*] dissolved I_2 in conc KI and distd it, then steam distd three times, washing with distd water. Organic material was removed by sublimation in a current of oxygen over platinum at about 700°, the iodine being finally sublimed under vac.

Iodine monobromide *[7789-33-5]* **M 206.8, m 42°,**
Iodine monochloride *[7790-99-0]* **M 162.4, m 27.2°.** Purified by repeated fractional crystn from its melt.

Iodine pentafluoride *[7783-66-6]* **M 221.9, m -8.0°, b 97°.** Rogers *et al*, [*JACS* **76** 4843 *1954*] removed dissolved iodine from IF_5 by agitating with a mixture of dry air and ClF_3 in a fluorothene beaker using a magnetic stirrer. The mixture was transferred to a still and the more volatile impurities were pumped off as the pressure was reduced below 40mm. The still was gradually heated (kept at 40mm) to remove the ClF_3 before IF_5 distd. Stevens [*JOC* **26** 3451 *1961*] pumped IF_5 under vac. from its cylinder, trapping it at -78°, then allowing it to melt in a stream of dry N_2.

Iodine trichloride *[22520-96-3]* **M 233.3, m 33°, b 77°(dec).** Purified by sublimation at room temperature.

Iron (wire) *[7439-89-6]* **M 55.9, m 1535°.** Cleaned in conc HCl, rinsed in de-ionised water, then reagent grade acetone and dried under vac.

Iron pentacarbonyl *[13463-40-6]* **M 195.9, b 102.5°, n 1.520, d 1.490.** Distd under vac, the middle cut being redistd twice and stored in a bulb protected from light (*photosensitive*).

Lead (II) acetate *[301-04-2]* **M 325.3, m 280°.** Crystd twice from anhydrous acetic acid and dried under vac. for 24hr at 100°.

Lead (II) bromide *[10031-22-8]* **M 367.0, m 373°.** Crystd from water containing a few drops of HBr (25ml of water per gram $PbBr_2$) between 100° and 0°.
A neutral soln was evapd at 110° and the crystals that separated were collected by rapid filtration at 70°, and dried at 105° (to give the *monohydrate*). To prepare the anhydrous bromide, the hydrate is heated for several hours at 170° and then in a Pt boat at 200° in a stream of HBr and H_2. Finally fused [Clayton *et al*, *JCSFT I* **76** 2362 *1980*].

Lead (II) chloride *[7758-95-4]* **M 278.1, m 501°.** Crystd from distd water at 100° (33ml/g) after filtering through sintered-glass and adding a few drops of HCl, by cooling. After three crystns the solid was dried under vac or under anhydrous HCl vapour by heating slowly to 400°.

Lead (II) formate *[811-54-4]* **M 297.3.** Crystd from aq formic acid.

Lead (II) iodide *[10101-63-0]* **M 461.0, m 402°.** Crystd from a large volume of water.

Lead monoxide *[1317-36-8]* **M 223.2, m 886°.** Higher oxides were removed by heating under vac at 550° with subsequent cooling under vac. [Ray and Ogg *JACS* **78** 5994 *1956*].

Lead nitrate *[10099-74-8]* **M 331.2.** Pptd twice from hot (60°) conc aq soln by adding HNO_3. The ppte was sucked dry in a sintered-glass funnel, then transferred to a crystallising dish which was covered by a clock glass and left in an electric oven at 110° for several hours [Beck, Singh and Wynne-Jones *TFS* **55** 331 *1959*].

Lead (biscyclopentadienyl) *[1294-74-2]* **M 337.4.** Purified by vacuum sublimation. Handled and stored under N_2.

Lead tetraacetate *[546-67-8]* **M 443.4.** Dissolved in hot glacial acetic acid, any lead oxide being removed by filtration. White crystals of lead tetraacetate separated on cooling. Stored in a vac desiccator over P_2O_5 and KOH for 24hr before use.

Lissapol C (mainly sodium salt of cetyl oleyl alcohol sulphate) *[2425-51-6]*.
Lissapol LS (mainly sodium salt of anisidine sulphate) *[28903-20-0]*. Refluxed with 95% EtOH, then filtered to remove insoluble inorganic electrolytes. The alcohol soln was then concnd and the residue was poured into dry acetone. The ppte was filtered off, washed in acetone and dried under vac. [Biswas and Mukerji *JPC* **64** 1 *1960*].

Lithium (metal) *[7439-93-2]* **M 6.9.** After washing with pet. ether to remove storage oil, lithium was fused at 400° and then forced through a 10-micron stainless-steel filter with argon pressure. It was again melted in a dry-box, skimmed, and poured into an iron distn pot. After heating under vac to 500°, cooling and returning to the dry-box for a further cleaning of its surface, the lithium was distd at 600° using an all-iron distn apparatus [Gunn and Green *JACS* **80** 4782 *1958*].

Lithium acetate ($2H_2O$) *[6108-17-4]* **M 102.0, m 54-56°.** Crystd from ethanol (5ml/g) by partial evapn.

Lithium aluminium hydride *[16853-85-3]* **M 37.9, m 125°(dec).** Extracted with Et_2O, and, after filtering, the solvent was removed under vac. The residue was dried at 60° for 3hr, under high vac [Ruff *JACS* **83** 1788 *1961*]. **Ignites in the presence of a small amount of water.**

Lithium benzoate *[553-54-8]* **M 128.1.** Crystd from EtOH (13ml/g) by partial evapn.

Lithium borohydride *[16949-15-8]* **M 21.8.** Crystd from Et_2O, and pumped free of ether at 90-100° during 2hr [Schaeffer, Roscoe and Stewart *JACS* **78** 729 *1956*].

Lithium bromide *[7550-35-8]* **M 86.8, m 550°.** Crystd several times from water or EtOH, then dried under high vac for 2 days at room temperature, followed by drying at 100°.

Lithium carbonate *[554-13-2]* **M 73.9, m 618°.** Crystd from water. Its solubility decreases as the temperature is raised

Lithium chloride *[7447-47-8]* **M 42.4, m 600°.** Crystd from water (1ml/g) or MeOH and dried for several hours at 130°. Other metal ions can be removed by preliminary crystn from hot aq. 0.01M disodium EDTA. Has also been crystd from conc HCl, fused in an atmosphere of dry HCl gas, cooled under dry N_2 and pulverised in a dry-box. Kolthoff and Bruckenstein [*JACS* **74** 2529 *1952*] pptd with ammonium carbonate, washed with Li_2CO_3 five times by decantation and finally with suction, then dissolved in HCl. The LiCl soln was evapd slowly with continuous stirring in a large evaporating dish, the dry powder being stored (while still hot) in a desiccator over $CaCl_2$.

Lithium dodecylsulphate *[2044-56-6]* **M 272.3.** Recrystd twice from abs EtOH and dried under vac.

Lithium formate (H_2O) *[556-63-8]* **M 70.0.** Crystd from hot water (0.5ml/g) by chilling.

Lithium hydroxide (H_2O) *[1310-66-3]* **M 42.0.** Crystd from hot water (3ml/g) as the monohydrate. Dehydrated at 150° in a stream of CO_2-free air.

Lithium iodate *[13765-03-2]* **M 181.9.** Crystd from water and dried in a vac oven at 60°.

Lithium iodide *[10377-51-2]* **M 133.8.** Crystd from hot water (0.5ml/g) by cooling in $CaCl_2$-ice, or from acetone. Dried under vac over P_2O_5 for 1hr at 60° and then at 120°.

Lithium nitrate *[7790-69-4]* **M 68.9.** Crystd from water or EtOH. Dried at 180° for several days by repeated melting under vac. If it is crystd from water keeping the temperature above 70°, formation of trihydrate is avoided. The anhydrous salt is dried at 120° and stored in a vac desiccator over $CaSO_4$.

Lithium nitrite (H₂O) *[13568-33-7]* **M 71.0.** Crystd from water by cooling from room temperature.

Lithium picrate *[18390-55-1]* **M 221.0.** Recrystd three times from EtOH and dried under vac at 45° for 48hr [D'Aprano and Sesta *JPC* **91** 2415 *1987*]. The neccessary precautions should be taken in case of **EXPLOSION**.

Lithium perchlorate *[7791-03-9]* **M 106.4.** Crystd from water or 50% aq MeOH. Rendered anhydrous by heating the trihydrate at 170-180° in an air oven. It can then be recrystd from acetonitrile and again dried at 170°.

Lithium salicylate *[552-38-5]* **M 144.1.** Crystd from EtOH (2ml/g) by partial evapn.

Lithium sulphate (anhydrous) *[10377-48-7]* **M 109.9.** Crystd from water (4ml/g) by partial evapn.

Magnesium acetate *[16674-78-5]* **M 214.5, m 80°.** Crystd from anhydrous acetic acid, then dried under vac for 24hr at 100°

Magnesium benzoate (3H₂O) *[553-70-8]* **M 320.6.** Crystd from water (6ml/g) between 100° and 0°.

Magnesium bromide (anhydrous) *[7789-84-6]* **M 184.1.** Crystd from EtOH.

Magnesium chloride (6H₂O) *[7791-18-6]* **M 203.3.** Crystd from hot water (0.3ml/g) by cooling.

Magnesium dodecylsulphate *[3097-08-3]* **M 555.1.** Recrystd three times from EtOH and dried in vac.

Magnesium iodate (4H₂O) *[7790-32-1]* **M 446.2.** Crystd from water (5ml/g) between 100° and 0°.

Magnesium iodide *[10377-58-9]* **M 278.1.** Crystd from water (1.2ml/g) by partial evapn in a desiccator.

Magnesium lactate *[18917-93-6]* **M 113.4.** Crystd from water (6ml/g) between 100° to 0°.

Magnesium nitrate (6H₂O) *[13446-18-9]* **M 256.4.** Crystd from water (2.5ml/g) by partial evapn in a desiccator.

Magnesium perchlorate (2H₂O) *[10034-81-8]* **M 259.2.** Crystd from water. Coll, Nauman and West [*JACS* **81** 1284 *1959*] removed traces of unspecified contaminants by washing with small portions of Et₂O. **Explosive** *in contact with organic materials.*

Magnesium succinate *[556-32-1]* **M 141.4.** Crystd from water (0.5ml/g) between 100° and 0°.

Magnesium sulphate (anhydrous) *[7487-88-9]* **M 120.4.** Crystd from warm water (1ml/g) by cooling.

Manganous acetate (4H₂O) *[6156-78-1]* **M 245.1.** Crystd from water acidified with acetic acid.

Manganous chloride (4H₂O) *[13446-34-9]* **M 197.9.** Crystd from water (0.3ml/g) by cooling.

Manganous ethylenebis(dithiocarbamate) *[12427-38-2]* **M 265.3.** Crystd from EtOH.

Manganous lactate (3H$_2$O) *[51877-53-3]* **M 287.1.** Crystd from water.

Manganous sulphate (H$_2$O) *[10034-96-5]* **M 169.0.** Crystd from water (0.9ml/g) at 54-55° by evaporating about two-thirds of the water.

Mercuric acetate *[1600-27-7]* **M 318.7.** Crystd from glacial acetic acid.

Mercuric bromide *[7789-47-1]* **M 360.4, m 238.1°.** Crystd from hot satd ethanolic soln, dried and kept at 100° for several hours under vac, then sublimed.

Mercuric chloride *[7487-94-7]* **M 271.5.** Crystd twice from distd water, dried at 70° and sublimed under high vac.

Mercuric cyanide *[592-04-1]* **M 252.6.** Crystd from water.

Mercuric iodide *[7774-29-0]* **M 454.4, m 259°.** Crystd from MeOH or EtOH, and washed repeatedly with distd water. Has also been mixed thoroughly with excess 0.001M iodine soln, filtered, washed with cold distd water, rinsed with EtOH and Et$_2$O, and dried in air.

Mercuric oxide *[21908-53-2]* **M 216.6.** Dissolved in HClO$_4$ and pptd with NaOH soln.

Mercuric thiocyanate *[592-85-8]* **M 316.8, m 165°(dec).** Recrystd from water

Mercury *[7439-97-6]* **M 200.6, m -38.9°, b 356.7°, d 13.594.** After air had been bubbled through mercury for several hours to oxidise metallic impurities, it was filtered to remove coarser particles of oxide and dirt, then sprayed through a 4-ft column containing 10% HNO$_3$. It was washed with distd water, dried with filter paper and distd under vac.

Mercury(II) bis(cyclopentadienyl) *[18263-08-6]* **M 330.8.** Purified by low-temperature recrystn from Et$_2$O.

Mercury(II) trifluoroacetate *[13257-51-7]* **M 426.6.** Recrystd from trifluoroacetic anhydride/trifluoroacetic acid [Lan and Kochi *JACS* 108 6720 *1986*]. Very **TOXIC** and *hygroscopic*.

Metanil Yellow *[587-98-4]* **M 375.4.** Salted out from water three times with sodium acetate, then repeatedly extracted with EtOH [McGrew and Schneider, *JACS* 72 2547 *1950*].

Methylarsonic acid *[124-58-3]* **M 137.9, m 161°.** Crystd from absolute EtOH.

Methylmercuric chloride *[115-09-3]* **M 251.1, m 167°.** Crystd from absolute EtOH (20ml/g).

Methyl Orange see **Sodium** *p*-(*p*-dimethylaminobenzeneazo)-benzenesulphonate.

Methyl Thymol Blue, sodium salt *[1945-77-3]* **M 844.8, ε 1.89 x 10^4 at 435nm, pH 5.5.** Starting material for synthesis is Thymol Blue. Purified as for Xylenol Orange

Molybdenum hexacarbonyl *[13939-06-5]* **M 264.0, m 150°(dec), b 156°.** Sublimed in a vac before use [Connor *et al*, *JCSDT* 511 *1986*].

Molybdenum hexafluoride *[7783-77-9]* **M 209.9, b 35°/760mm.** Purified by low-temperature trap-to-trap distn over predried NaF. [Anderson and Winfield *JCSDT* 337 *1986*.]

Molybdenum trichloride *[13478-18-7]* **M 202.3.** Boiled with 12M HCl, washed with absolute EtOH and dried in a vac desiccator.

Molybdenum trioxide *[1313-27-5]* **M 143.9.** Crystd from water (50ml/g) between 70° and 0°.

Monocalcium phosphate (H_2O) *[7758-23-8]* **M 154.1.** Crystd from a near-satd soln in 50% aq reagent grade phosphoric acid at 100° by filtering through fritted glass and cooling to room temperature. The crystals were filtered off and this process was repeated three times using fresh acid. For the final crystn the soln was cooled slowly with constant stirring to give thin plate crystals that were filtered off on fritted glass, washed free of acid with anhydrous acetone and dried in a vac desiccator [Egan, Wakefield and Elmore, *JACS* **78** 1811 *1956*].

Neodymium oxide *[1313-97-9]* M 336.5.
Dissolved in $HClO_4$, pptd as the oxalate with doubly recrystd oxalic acid, washed free of soluble impurities, dried at room temperature and ignited in a platinum crucible at higher than 850° in a stream of oxygen [Tobias and Garrett *JACS* **80** 3532 *1958*].

Neon *[7440-01-9]* **M 20.2.** Passed through a ¼" copper coil packed with 60/80 mesh 13X molecular sieves which is cooled in liquid N_2, or through a column of Ascarite (NaOH-coated silica adsorbent).

Neopentoxy lithium *[3710-27-8]* **M 94.1.** Recrystd from hexane [Kress and Osborn *JACS* **109** 3953 *1987*].

Nickel bromide *[13462-86-9]* **M 218.5.** Crystd from dil HBr (0.5ml /g) by partial evapn in a desiccator.

Nickel chloride ($6H_2O$) *[7791-20-0]* **M 237.7.** Crystd from dil HCl.

Nickel nitrate ($6H_2O$) *[13478-00-7]* **M 290.8, m 57°.** Crystd from water (0.3ml/g) by partial evapn in a desiccator.

Nickel potassium sulphate see **potassium nickel sulphate.**

Nickel sulphate ($7H_2O$) *[1010-98-1]* **M 280.9.** Crystd from warm water (0.25ml/g) by cooling.

Nickel 5,10,15,20-tetraphenylporphyrin *[14172-92-0]* **M 671.4,** λ_{max} 414(525)nm. Purified by chromatography on neutral (Grade I) alumina, followed by recrystn from CH_2Cl_2/MeOH [Yamashita *JPC* **91** 3055 *1987*].

Nitric acid *[7697-37-2]* **M 63.0, m -42°, b 83°,** d_{25} **1.5027;** [Constant boiling acid has composition 68% HNO_3 + 32% H_2O, b 120.5°, d 1.41]. Obtained colourless (approx. 92%) by direct distn of fuming HNO_3 under reduced pressure at 40-50° with an air leak at the head of the fractionating column. Stored in a desiccator kept in a refrigerator. Nitrite-free HNO_3 can be obtained by vac distn from urea.

Nitric oxide *[10102-43-9]* **M 30.0, b -151.8°.** Bubbling through 10M NaOH removes NO_2. It can also be freed from NO_2 by passage through a column of Ascarite followed by a column of silica gel held at -197°K. The gas is dried with solid NaOH pellets or by passing through silica gel cooled at -78°, followed by fractional distn from a liquid N_2 trap. This purification does not eliminate nitrous oxide. Other gas scrubbers sometimes used include one containing conc H_2SO_4 and another containing mercury. It is freed from traces of N_2 by a freeze and thaw method. **TOXIC.**

p-Nitrobenzenediazonium fluoroborate *[456-27-9]* M 236.9. Crystd from water. **Can be EXPLOSIVE when dry.**

Nitrogen *[7727-37-9]* **M 28.0, b -195.8°.** Cylinder N_2 can be freed from oxygen by passage through Fieser's soln [which comprises 2g sodium anthraquinone-2-sulphonate and 15g sodium hydrosulphite dissolved in 100ml of 20% KOH (Fieser, *JACS* **46** 2639 *1924*)] followed by scrubbing with satd lead acetate soln (to remove any H_2S generated by the Fieser soln), conc H_2SO_4 (to remove moisture), then soda-lime (to remove any H_2SO_4 and CO_2). Alternatively, after passage through Fieser's soln, N_2 can be dried by washing with a soln of the metal ketyl from benzophenone and Na wire in absolute ethyl ether. [If ether vapour in N_2 is undesirable, the ketyl from liquid Na-K alloy under xylene can be used.] Another method for removing O_2 is to pass the nitrogen through a long tightly packed column of Cu turnings, the surface of which is constantly renewed by scrubbing it with ammonia (s.g. 0.880) soln. The gas is then passed through a column packed with glass beads moistened with conc H_2SO_4 (to remove ammonia), through a column of packed KOH pellets (to remove H_2SO_4 and to dry the N_2), and finally through a glass trap packed with chemically clean glass wool immersed in liquid N_2. Nitrogen has also been purified by passage over Cu wool at 723°K and Cu(II) oxide [prepared by heating $Cu(NO_3)_2.6H_2O$ at 903°K for 24hr] and then into a cold trap at 77°K.
A typical dry purification method consists of a mercury bubbler (as trap), followed by a small column of silver and gold turnings to remove any mercury vapour, towers containing anhydrous $CaSO_4$, dry molecular sieves or $Mg(ClO_4)_2$, a tube filled with fine Cu turnings and heated to 400° by an electric furnace, a tower containing soda-lime , and finally a plug of glass wool as filter. Variations include tubes of silica gel, traps containing activated charcoal cooled in a Dry-ice bath, copper on kieselguhr heated to 250°, and copper and iron filings at 400°.

Nitrophenolarsonic acid *[121-19-7]* **M 350.1.** Crystd from water.

Nitroso-R-salt see **1-Nitroso-2-naphthol-3,6-disulphonic acid, disodium salt, hydrate.**

1-Nitroso-2-naphthol-3,6-disulphonic acid, disodium salt, hydrate *[525-05-3]* **M 377.3, m >300°.** Purified by dissolution in aq alkali and pptn by addition of HCl.

Nitrosyl chloride *[2696-92-6]* **M 65.5, b -5.5°.** Fractionally distd at atmospheric pressure in an all-glass, low temperature still, taking the fraction boiling at -4° and storing it in sealed tubes.

Nitrous oxide *[10024-97-2]* **M 44.0, b -88.5°.** Washed with conc alkaline pyrogallol soln, to remove O_2, CO_2, and NO_2, then dried by passage through columns of P_2O_5 or Drierite, and collected in a dry trap cooled in liquid N_2. Further purified by freeze-pump-thaw and distn cycles under vac [Ryan and Freeman *JPC* **81** 1455 *1977*].

Orange II *[633-96-5]* M 350.3.
Extracted with a small volume of water, then crystd by dissolving in boiling water, cooling to *ca* 80°, adding two volumes of EtOH and cooling. When cold, the ppte is filtered off, washed with a little EtOH and dried in air. It can be salted out from aq. soln with sodium acetate, then repeatedly extracted with EtOH. Meggy and Sims [*JCS* 2940 *1956*], after crystallising the sodium salt twice from water, dissolved it in cold water (11ml/g) and conc HCl added to ppte the dye acid which was separated by centrifugation, redissolved and again pptd with acid. After washing the ppte three times with 0.5M acid it was dried over NaOH, recrystd twice from absolute EtOH, washed with a little Et_2O, dried over NaOH and stored over conc H_2SO_4 in the dark.

Orange RO *[5850-86-2]* **M 364.4.** Salted out three times with sodium acetate, then repeatedly extracted with EtOH.

Oxygen *[7782-44-7]* **M 32.00, m -218.4°, b -182.96°, d⁻¹⁸³ 1.149, d⁻²⁵²·⁵ 1.426.** Purified by passage over finely divided platinum at 673°K and Cu(II) oxide (see under nitrogen) at 973°, then condensed in liquid N₂-cooled trap. **HIGHLY EXPLOSIVE in contact with organic matter.**

Perchloric acid *[7601-90-3]* **M 100.5, d 1.665.** The 72% acid has been purified by double distn from silver oxide under vacuum: this frees the acid from metal contamination. Anhydrous acid can be obtained by adding gradually 400-500ml of oleum (20% fuming H₂SO₄) to 100-120ml of 72% HClO₄ in a reaction flask cooled in an ice-bath. The pressure is reduced to 1mm (or less), with the reaction mixture at 20-25°. The temperature is gradually raised during 2hr to 85°, the distillate being collected in a receiver cooled in Dry-ice. For further details of the distn apparatus see Smith *JACS* **75** 184 *1953*]. **HIGHLY EXPLOSIVE, a strong protective screen should be used at all times.**

Phenylarsonic acid *[98-05-5]* **M 202.2, m 155-158°(dec).** Crystd from water (3ml/g) between 90° and 0°.

Phenylmercuric hydroxide *[100-57-2]* **M 294.7, m 195-203°.** Crystd from dil aq NaOH.

Phenylmercuric nitrate *[8003-05-2]* **M 634.4, m 178-188°.** Crystd from water.

Phosgene *[75-44-5]* **M 98.9, b 8.2°/756mm.** Dried with Linde 4A molecular sieves, degassed and distd under vac. **HIGHLY TOXIC, should not be inhaled.**

Phosphonitrilic chloride (tetramer) *[1832-07-1]* **M (115.9)₄.** Purified by zone melting, then crystd from pet. ether (b 40-60°) or *n*-hexane. [van der Huizen *et al, JCSDT* 1317 *1986*.]

Phosphonitrilic chloride (trimer) *[940-71-6]* **M (115.9)₃, m 112.8°.** Purified by zone melting, by crystn from pet.ether, *n*-hexane or benzene, and by sublimation. [van der Huizen *et al, JCSDT* 1311 *1986*.]

Phosphoric acid *[7664-38-2]* **M 98.0, m 42.3°.** Pyrophosphate can be removed from phosphoric acid by diluting with distd water and refluxing overnight. By cooling to 11° and seeding with crystals obtained by cooling a few millilitres in a Dry-ice/acetone bath, 85% orthophosphoric acid crystallises as H₃PO₄.½H₂O. The crystals are separated using a sintered glass filter.

Phosphorus (red) *[7723-14-0]* **M 31.0, m 590°/43atm, ignites at 200°, d 2.34.** Boiled for 15min with distd water, allowed to settle and washed several times with boiling water. Transferred to a Büchner funnel, washed with hot water until the washings are neutral, then dried at 100° and stored in a desiccator.

Phosphorus (white) *[7723-14-0]* **M 31.0, m 590, d 1.82.** Purified by melting under dil H₂SO₄-dichromate mixture and allowed to stand for several days in the dark at room temperature. It remains liquid, and the initial milky appearance due to insoluble, oxidisable material gradually disappears. The phosporus can then be distilled under vac. in the dark [Holmes *TFS* **58** 1916 *1962*]. Other methods include extraction with dry CS₂ followed by evapn of the solvent, or washing with 6M HNO₃, then water, and drying under vac. **POISONOUS.**

Phosphorus oxychloride *[10025-87-3]* **M 153.3, b 105.5°, n 1.461, d 1.675.** Distd under reduced pressure to separate from the bulk of the HCl and the phosphoric acid, the middle fraction being distd into ampoules containing a little purified mercury. These ampoules are sealed and stored in the dark for a 4-6 weeks with occasional shaking to facilitate reaction of any free chloride with the mercury. The POCl₃ is then again fractionally distd and stored in sealed ampoules in the dark until used [Herber

JACS **82** 792 *1960*]. Lewis and Sowerby [*JCS* 336 *1957*] refluxed their distd POCl$_3$ with sodium wire for 4hr, then removed the sodium and again distd.

Phosphorus pentabromide *[7789-69-7]* **M 430.6, m <100°, b 106°(dec).** Dissolved in pure nitrobenzene at 60°, filtering off any insoluble residue on to sintered glass, then crystd by cooling. Washed with dry Et$_2$O and removed the ether in a current of dry N$_2$. (All manipulations should be performed in a dry-box.) [Harris and Payne, *JCS* 3732 *1958*.] Fumes in moist air because of hydrolysis. **TOXIC.**

Phosphorus pentachloride *[10026-13-8]* **M 208.2, m 179-180°(sublimes).** Sublimed at 160-170° in an atmosphere of chlorine. The excess chlorine was then displaced by dry N$_2$ gas. All subsequent manipulations were performed in a dry-box [Downs and Johnson *JACS* **77** 2098 *1955*]. Fumes in moist air.

Phosphorus pentasulphide *[1314-80-3]* **M 444.5, m 277-283°.** Purified by extraction and crystn with CS$_2$, using a Soxhlet extractor. Liberates H$_2$S in moist air.

Phosphorus pentoxide *[1314-56-3]* **M 141.9.** Sublimed at 250° under vac into glass ampoules. Fumes in moist air and reacts violently with water.

Phosphorus sesquisulphide P$_4$S$_3$ *[1314-85-8]* **M 220.1, m 172°.** Extracted with CS$_2$, filtered and evapd to dryness. Placed in water, and steam was passed through for an hour. The water was then removed, the solid was dried, followed by crystn from CS$_2$ [Rogers and Gross *JACS* **74** 5294 *1952*].

Phosphorus trichloride *[7719-12-2]* **M 137.3, b 76°, n 1.515, d 1.575.** Heated under reflux to expel dissolved HCl, then distd. It has been further purified by vac fractionation several times through a -45° trap into a receiver at -78°.

12-Phosphotungstic acid *[12501-23-4]* **M 2880.2.** A few drops of conc HNO$_3$ were added to 100g of phosphotungstic acid dissolved in 75ml of water, in a separating funnel, and the soln was extracted with ethyl ether. The lowest of the three layers, which contained a phosphotungstic acid-ether complex, was separated, washed several times with 2M HCl, then with water and again extracted with ether. Evapn of the ether, under vac. with mild heating on a water bath gave crystals which were dried under vac and ground [Matijevic and Kerker, *JACS* **81** 1307 *1959*].

Phthalocyanine *[574-93-6]* **M 514.6.** Purified by sublimation (two to three times) in an argon flow at 300-400Pa. Similarly for the Cu(II), Ni(II), Pb(II), VO(II) and Zn(II) phthalocyanine complexes.

Poly(sodium 4-styrenesulphonate) [-CH$_2$CH(C$_6$H$_4$SO$_3$Na)-] *[25704-18-1]*. Recrystd from EtOH.

Potassium (metal) *[7440-09-07]* **M 39.1, m 62.3°.** Oil was removed from the surface of the metal by immersion in *n*-hexane and pure Et$_2$O for long periods. The surface oxide was next removed by scraping under ether, and the potassium was melted under vac. It was then allowed to flow through metal constricitons into tubes that could be sealed, followed by distn under vac in the absence of mercury vapour (see Sodium). **EXPLOSIVE IN WATER.**

Potassium acetate *[127-08-2]* **M 98.2.** Crystd three times from water-ethanol (1:1) dried to constant weight in a vac oven, or crystd from anhydrous acetic acid and pumped dry under vac for 30hr at 100°.

Potassium 4-aminobenzoate *[138-84-1]* **M 175.2.** Crystd from EtOH.

Potassium antimonyltartrate (½H$_2$O) *[28300-74-5]* **M 333.9, [α]$_D$ +141° (c 2,H$_2$O).** Crystd from water (3ml/g) between 100° and 0°. Dried at 100°.

Potassium benzoate *[582-25-2]* **M 160.2.** Crystd from water (1ml/g) between 100° and 0°.

Potassium bicarbonate *[298-14-6]* **M 100.1.** Crystd from water at 65-70° (1.25ml/g) by filtering, then cooling to 15°. During all operations, CO_2 is passed through the stirred mixture. The crystals, sucked dry at the pump, are washed with distd water, dried in air and then over H_2SO_4 in an atmosphere of CO_2.

Potassium biiodate *[13455-24-8]* **M 389.9.** Crystd three times from hot water (3ml/g), stirred continuously during each cooling. After drying at 100° for several hours, the crystals are suitable for use in volumetric analysis.

Potassium bisulphate *[7646-93-7]* **M 136.2, m 214°.** Crystd from water (1ml/g) between 100° and 0°.

Potassium borohydride *[13762-51-1]* **M 53.9.** Crystd from liquid ammonia.

Potassium bromate *[7758-01-2]* **M 167.0.** Crystd from distd water (2ml/g) between 100° and 0°. To remove bromide contamination, a 5% soln in distd water, cooled to 10°, has been bubbled with gaseous chlorine for 2hr, then filtered and extracted with reagent grade CCl_4 until colourless and odourless. After evaporating the aq phase to about half its volume, it was cooled again slowly to about 10°. The crystalline $KBrO_3$ was separated, washed with 95% EtOH and vac dried [Boyd, Cobble and Wexler *JACS* **74** 237 *1952*]. Another way to remove Br⁻ ions was by stirring several times in MeOH and then dried at 150° [Field and Boyd *JPC* **89** 3767 *1985*].

Potassium bromide *[7758-02-3]* **M 119.0.** Crystd from distd water (1ml/g) between 100° and 0°. Washed with 95% EtOH, followed by Et_2O. Dried in air, then heated at 115° for 1hr, pulverised and heated in a vac oven at 130° for 4hr. Has also been crystd from aq 30% EtOH, or EtOH, and dried over P_2O_5 under vac before heating in an oven.

Potassium carbonate *[584-08-7]* **M 138.2.** Crystd from water beteen 100° and 0°.

Potassium chlorate *[3811-04-9]* **M 122.6.** Crystd from water (1.8ml/g) between 100° and 0°, and the crystals are filtered on to sintered glass.

Potassium chloride *[7447-40-7]* **M 74.6.** Dissolved in conductivity water, filtered, and satd with chlorine (generated from A.R. HCl and $KMnO_4$). Excess chlorine was boiled off, and the KCl was pptd by HCl (generated by dropping conc A.R. HCl into conc H_2SO_4). The ppte was washed with water, dissolved in conductivity water at 90-95°, and crystd by cooling to about -5°. The crystals were drained at the centrifuge, dried in a vac desiccator at room temperature, then fused in a platinum dish under N_2, cooled and stored in desiccator. Potassium chloride has also been sublimed in a stream of prepurified N_2 gas and collected by electrostatic discharge [Craig and McIntosh *Canad J Chem* **30** 448 *1952*].

Potassium chromate *[7784-00-6]* **M 194.2.** Crystd from conductivity water (0.6g/ml at 20°), and dried between 135° and 170°.

Potassium cobalticyanide *[13963-58-1]* **M 332.4.** Crystd from water to remove traces of HCN.

Potassium cyanate *[590-28-3]* **M 81.1.** Common impurities include ammonia and bicarbonate ion (from hydrolysis). Purified by preparing a satd aq soln at 50°, neutralising with acetic acid, filtering, adding two volumes of EtOH and keeping for 3-4hr in an ice bath. (More EtOH can lead to co-pptn of $KHCO_3$.) Filtered, washed with EtOH and dried rapidly in a vac desiccator (P_2O_5). The process is repeated [Vanderzee and Meyers *JCS* **65** 153 *1961*].

Potassium cyanide *[151-50-8]* **M 65.1.** A satd soln in water-ethanol (1:3) at 60° was filtered and cooled to room temperature. Absolute EtOH was added, with stirring, until crystn ceased. The soln was again allowed to cool to room temperature (during 2-3hr) then the crystals were filtered off, washed with absolute EtOH, and dried, first at 70-80° for 2-3hr, then at 105° for 2hr [Brown, Adisesh and Taylor *JPC* **66** 2426 *1962*]. Also purified by vac melting and zone refining. **HIGHLY POISONOUS.**

Potassium dichromate *[7778-50-9]* **M 294.2.** Crystd from water (1ml/g) between 100° and 0° and dried under vac at 156°.

Potassium dihydrogen citrate *[866-83-1]* **M 230.2.** Crystd from water. Dried at 80°.

Potassium dihydrogen phosphate *[7778-77-0]* **M 136.1.** Dissolved in boiling distd water (2ml/g), kept on a boiling water-bath for several hours, then filtered through paper pulp to remove any turbidity. Cooled rapidly with constant stirring, and the crystals were separated on to hardened filter paper, using suction, washed twice with ice-cold water, once with 50% EtOH, and dried at 105°. Alternative crystns are from water, then 50% EtOH, and again water, or from conc aq soln by addition of EtOH. Freed from traces of Cu by extracting its aq soln with diphenylthiocarbazone in CCl_4, followed by repeated extraction with CCl_4 to remove traces of diphenylthiocarbazone.

Potassium dithionate *[13455-20-4]* **M 238.3.** Crystd from water (1.5ml/g) between 100° and 0°.

Potassium ethylxanthate *[140-89-6]* **M 160.3, m > 215°(decomp).** Crystd from absolute EtOH, ligroin-ethanol or acetone by addition of Et_2O. Washed with ether, then dried in a desiccator.

Potassium ferricyanide *[13746-66-2]* **M 329.3.** Crystd repeatedly from hot water (1.3ml/g). Dried under vac in a desiccator.

Potassium ferrocyanide (3H_2O) *[14459-95-1]* **M 422.4.** Crystd repeatedly from distd water, never heating above 60°. Prepared anhydrous by drying at 110° over P_2O_5 in a vac desiccator. To obtain the trihydrate, it is necessary to equilibrate in a desiccator over satd aq soln of sucrose and NaCl. Can also be pptd from a satd soln at 0° by adding an equal volume of cold 95% EtOH, standing for several hours, then centrifuging and washing with cold 95% EtOH. Finally sucked air dry with a water-pump. The anhydrous salt can be obtained by drying in a platinum boat at 90° in a slow stream of N_2 [Loftfield and Swift *JACS* **60** 3083 *1938*]

Potassium fluoroborate *[14075-53-7]* **M 152.9.** Crystd several times from conductivity water (15ml/g).

Potassium fluorosilicate *[16871-90-2]* **M 220.3.** Crystd several times from conductivity water (100ml/g) between 100° and 0°.

Potassium hexachloroiridate (IV) *[16920-56-2]* **M 483.1.** Crystd from hot aq soln containing a few drops of HNO_3.

Potassium hexachloroosmate (IV) *[16871-60-6]* **M 481.1.** Crystd from hot dil aq HCl.

Potassium hexachloroplatinate (IV) *[16921-30-5]* **M 486.0.** Crystd from water (20ml/g) between 100° and 0°.

Potassium hexacyanochromate (III) (3H_2O) *[13601-11-1]* **M 418.5.** Crystd from water.

Potassium hexafluorophosphate *[17084-13-8]* **M 184.1.** Crystd from alkaline aq soln, using polyethylene vessels, or from 95% EtOH, and dried in a vac desiccator over KOH.

Potassium hydrogen fluoride *[7789-29-9]* **M 78.1.**
Potassium hydrogen D-glucarate *[18404-47-2]* **M 248.2, m 188°(dec).** Crystd from water.

Potassium hydrogen malate *[4675-64-3]* **M 172.2.** A satd aq soln at 60° was decolorised with activated charcoal, and filtered. The filtrate was cooled in water-ice bath and the salt was pptd by addition of EtOH. After being crystd five times from ethanol-water mixtures, it was dried overnight at 130° in air [Eden and Bates *J Res Nat Bur Stand* **62** 161 *1959*].

Potassium hydrogen oxalate (½H$_2$O) *[127-95-7]* **M 137.1.** Crystd from water by dissolving 20g in 100ml water at 60° containing 4g of potassium oxalate, filtering and allowing to cool to 25°. The crystals, after washing three or four times with water, are allowed to dry in air.

Potassium hydrogen phthalate *[877-24-7]* **M 204.2.** Crystd first from a dil aq soln of K$_2$CO$_3$, then water (3ml/g) between 100° and 0. Before being used as a standard in volumetric analysis, analytical grade potassium hydrogen phthalate should be dried at 120° for 2hr, then allowed to cool in a desiccator.

Potassium hydrogen saccharate see **potassium hydrogen D-glucarate.**

Potassium hydrogen *d*-tartrate *[868-14-4]* **M 188.2, [α]$_{546}$ +37.5° (c 10, M NaOH).** Crystd from water (17ml/g) between 100° and 0°. Dried at 110°.

Potassium hydroxide (solution) *[1310-58-3]* **M 56.1.** Its carbonate content can be reduced by rinsing KOH sticks rapidly with water prior to dissolving them in boiled out distd water. Alternatively, a slight excess of satd BaCl$_2$ or Ba(OH)$_2$ can be added to the soln which, after shaking well, is left so that the BaCO$_3$ ppte will separate out. Davies and Nancollas [Nature **165** 237 *1950*] rendered KOH solns carbonate free by ion exchange using a column of Amberlite IR-100 in the OH⁻ form.

Potassium iodate *[7758-05-6]* **M 214.0.** Crystd twice from distd water (3ml/g) between 100° and 0°, dried for 2hr at 140° and cooled in a desiccator. Analytical reagent grade material dried in this way is suitable for use as an analytical standard.

Potassium iodide *[7681-11-0]* **M 166.0.** Crystd from distd water (0.5ml/g) by filtering the near-boiling soln and cooling. To minimise oxidation to iodine, the crystn can be carried out under N$_2$ and the salt is dried under vac over P$_2$O$_5$ at 70-100°. Before drying, the crystals can be washed with EtOH or with acetone followed by pet ether. Has also been crystd from water/ethanol.

Potassium isoamyl xanthate *[61792-26-5]* **M 202.4.** Crystd twice from acetone-ethyl ether. Dried in a desiccator for two days and stored under refrigeration.

Potassium laurate *[10124-65-9]* **M 338.4.** Recrystd three times from EtOH [Neto and Helene *JPC* **91** 1466 *1987*].

Potassium metaperiodate see **potassium periodate.**

Potassium nickel sulphate (6H$_2$O) *[13842-46-1]* **M 437.1.** Crystd from water (1.7ml/g) between 75° and 0°.

Potassium nitrate *[7757-79-1]* **M 101.1, m 334°.** Crystd from hot water (0.5ml/g) by cooling (*cf* potassium nitrite below). Dried for 12hr under vac. at 70°.

Potassium nitrite *[7758-09-0]* **M 85.1, m 350°(dec).** A satd soln at 0° can be warmed and partially evapd under vac., the crystals so obtained being filtered from the warm soln. (This procedure is designed to reduce the level of nitrate impurity and is based on the effects of temperature on solubility. The solubility of KNO$_3$ in water is 13g/100ml at 0°, 247g/100ml at 100°; for KNO$_2$ the corresponding figures are 280g/100ml and 413g/100ml.)

Potassium oleate *[143-18-0]* **M 320.6.** Crystd from EtOH (1ml/g).

Potassium oxalate *[6487-48-5]* **M 184.2.** Crystd from hot water.

Potassium perchlorate *[7778-74-7]* **M 138.6.** Crystd from boiling water (5ml/g) by cooling. Dried under vac at 105°.

Potassium periodate *[779-21-8]* **M 230.0.** Crystd from distd water.

Potassium permanganate *[7722-64-7]* **M 158.0.** Crystd from hot water (4ml/g at 65°), then dried in a vac. desiccator over $CaSO_4$. Phillips and Taylor [*JCS* 4242 *1962*] cooled an aq. soln of $KMnO_4$, satd at 60°, to room temperature in the dark, and filtered through a No.4 porosity sintered-glass filter funnel. The soln was allowed to evaporate in air in the dark for 12hr, and the supernatant liquid was decanted from the crystals, which were dried as quickly as possible with filter paper.

Potassium peroxydisulphate *[7727-21-1]* **M 270.3.** Crystd twice from distd water (10ml/g) and dried at 50° in a vac desiccator.

Potassium perrhenate *[10466-65-6]* **M 289.3.** Crystd from water (7ml/g), then fused in a platinum crucible in air at 750°.

Potassium persulphate see **potassium peroxydisulphate.**

Potassium *p*-phenolsulphonate *[30145-40-5]* **M 212.3.** Crystd several times from distd water at 90°, after treatment with charcoal, by cooling to *ca* 10°. Dried at 90-100°.

Potassium picrate *[573-83-1]* **M 267.2.** Crystd from water or 95% EtOH, and dried at room temperature in vac. It is soluble in 200 parts of cold water and 4 parts of boiling water. **THE DRY SOLID EXPLODES WHEN STRUCK OR HEATED.**

Potassium propionate *[327-62-8]* **M 112.2.** Crystd from water (30ml/g) or 95% EtOH.

Potassium reineckate *[34430-73-4]* **M 357.5.** Crystd from KNO_3 soln, then from warm water [Adamson *JACS* **80** 3183 *1958*].

Potassium selenocyanate *[3425-46-5]* **M 144.1.** Dissolved in acetone, filtered and pptd by adding Et_2O.

Potassium sodium tartrate ($4H_2O$) *[304-59-6]* **M 282.3.** Crystd from distd water (1.5ml/g) by cooling to 0°.

Potassium sulphate *[7778-80-5]* **M 174.3.** Crystd from distd water (4ml/g at 20°; 8ml/g at 100°) between 100° and 0°.

Potassium *d*-tartrate (½H_2O) *[921-53-9]* **M 235.3.** Crystd from distd water (solubility: 0.4ml/g at 100°; 0.7ml/g at 14°).

Potassium tetrachloroplatinate(II) *[10025-99-7]* **M 415.1.** Crystd from aq 0.75M HCl (20ml/g) between 100° and 0°. Washed with ice-cold water and dried.

Potassium tetraphenylborate *[3244-41-5]* **M 358.3.** Pptd from a soln of KCl acidified with dil HCl, then crystd twice from acetone, washed thoroughly with water and dried at 110° [Findeis and de Vries *AC* **28** 1899 *1956*].

Potassium tetroxalate ($2H_2O$) *[127-96-8]* **M 358.3.** Crystd from water below 50°. Dried below 60° at atmospheric pressure.

Potassium thiocyanate *[333-20-0]* **M 97.2, m 172°.** Crystd from water if much chloride ion is present in the salt, otherwise from EtOH or MeOH (optionally by addition of Et_2O). Filtered on to a Büchner funnel without paper, and dried in a desiccator at room temperature before being heated for 1hr at 150°, with a final 10-20min at 200° to remove the last traces of solvent [Kolthoff and Lingane *JACS* **57** 126 *1935*]. Stored in the dark.

Potassium thiosulphate *[10294-66-3]* **M 190.3.** Crystd from warm water (0.5ml/g) by cooling in an ice-salt mixture.

Potassium tungstate (*ortho* $2H_2O$) *[37349-36-3]* **M 362.1.** Crystd from hot water (0.7ml/g).

Praseodymium acetate *[6192-12-7]* **M 318.1.** Recrystd from water several times [Ganapathyl *JACS* **109** 3159 *1986*].

Praseodymium trichloride ($6H_2O$) *[10361-79-2]* **M 355.4.** Its 1M soln in 6M HCl was passed twice through a Dowex-1 anion-exchange column. The eluate was evapd in a vac desiccator to about half its volume and allowed to crystallise [Katzin and Gulyas *JPC* **66** 494 *1962*].

Praseodymium oxide (Pr_6O_{11}) *[12036-32-7]* **M 1021.4.** Dissolved in acid, pptd as the oxalate and ignited at 650°.

3-(2-Pyridyl)-5,6-diphenyl-1,2,4-triazine-*p,p'*-disulphonic acid, monosodium salt (H_2O) *[63451-29-6]* **M 510.5.** Purified by recrystn from water or by dissolving in the minimum volume of water, followed by addn of EtOH to ppte the pure salt.

Pyrocatechol Violet *[115-41-3]* **M 386.4, ε 1.4 x 10^4 at 445nm in acetate buffer pH 5.2-5.4.** Recrystd from glacial acetic acid. Very hygroscopic. [Mustafin *et al*, *Zh Analit Khim* **22** 1808 *1967*.]

Pyrogallol Red *[32638-38-3]* **M 418.4, m >300°(dec), ε 4.3 x 10^4 at 542nm, pH 7.9-8.6.** Recrystd from aq alkaline soln (Na_2CO_3 or NaOH) by pptn on acidification [Suk *Collect Czech Chem Commun* **31** 3127 *1966*].

Reinecke salt see ammonium reineckate.

Rubidium bromide *[7789-39-1]* **M 165.4.** Crystd from near-boiling water (0.5ml/g) by cooling to 0°.

Rubidium chlorate *[13446-71-4]* **M 168.9.** Crystd from water (1.6ml/g) by cooling from 100°.

Rubidium chloride *[7791-11-9]* **M 120.9.** Crystd from water (0.7ml/g) by cooling to 0° from 100°.

Rubidium nitrate *[13126-12-0]* **M 147.5.** Crystd from hot water (0.25ml/g) by cooling to room temperature.

Rubidium perchlorate *[13510-42-4]* **M 184.9.** Crystd from hot water (1.6ml/g) by cooling to 0°.

Rubidium sulphate *[7488-54-2]* **M 267.0.** Crystd from water (1.2ml/g) between 100° and 0°.

Ruthenium dioxide *[12157-25-6]* **M 133.1.** Freed from nitrates by boiling in distd water and filtering. A more complete purification is based on fusion in a KOH-KNO_3 mix to form the soluble ruthenate and perruthenate salts. The melt is dissolved in water, and filtered, then acetone is added to reduce the ruthenates to the insoluble hydrate oxide which, after making a slurry with paper pulp, is filtered and ignited in air to form the anhydrous oxide [Campbell, Ortner and Anderson *AC* **33** 58 *1961*].

Selenious acid *[7783-00-8]* M 129.0. Crystd from water.

Selenium *[7782-49-2]* **M 79.0, m 217.4°.** Dissolved in small portions in hot conc HNO_3 (2ml/g) filtered and evapd to dryness to give selenious acid which was then dissolved in conc HCl. Passage of SO_2 into the soln pptd selenium (but not tellurium) which was filtered off and washed with conc HCl. This purification process was repeated. The selenium was then converted twice to the selenocyanate by treating with a 10% excess of 3M aq. KCN, heating for half an hour on a sand-bath and filtering. Addition of an equal weight of chopped ice to the cold soln, followed by an excess of cold, conc HCl, with stirring (in a well ventilated fume hood because HCN is evolved) pptd selenium powder, which, after washing with water until colourless, and then with MeOH, was heated in an oven at 105°, then by fusion for 2hr under vac. It was cooled, crushed and stored in a desiccator [Tideswell and McCullough *JACS* **78** 3036 *1956*].

Selenium dioxide *[7446-08-4]* **M 111.0, m 340°.** Purified by sublimation, or by soln in HNO_3, pptn of selenium which, after standing for several hours or boiling, is filtered off, then re-oxidised by HNO_3 and cautiously evapd to dryness below 200°. The dioxide is dissolved in water and again evapd to dryness.

Silica *[7631-86-9].* Purification of silica for high technology applications uses isopiestic vapour distn from conc volatile acids and is absorbed in high purity water. The impurities remain behind. Preliminary cleaning to remove surface contaminants uses dip etching in HF or a mixture of HCl, H_2O_2 and deionised water [Phelan and Powell *Analyst* **109** 1299 *1984*].

Silica gel *[63231-67-4].* Before use as a drying agent, silica gel is heated in an oven, then cooled in a desiccator. Conditions in the literature range from heating at 110° for 15hr to 250° for 2-3hr. Silica gel has been purified by washing with hot acid (in one case successively with aqua regia, conc HNO_3, then conc HCl; in another case digested overnight with hot conc H_2SO_4), followed by exhaustive washing with distd water (one week in a Soxhlet apparatus has also been used), and prolonged oven drying. Alternatively, silica gel has been extracted with acetone until all soluble material was removed, then dried in a current of air, washed with distd water and oven dried. Silica gel has also been washed successively with water, M HCl, water, and acetone, then activated at 110° for 15hr.

Silicon tetrachloride *[10026-04-7]* **M 169.9, m -70°, b 57.6°, d 1.483.** Distd under vac and stored in sealed ampoules under N_2. Very sensitive to moisture.

12-Silicotungstic acid (tungstosilicic acid; $H_4SiW_{12}O_{40}$) *[12027-38-2]* **M 2914.5.** Extracted with ethyl ether from a soln acidified with HCl. The ethyl ether was evapd under vac, and the free acid was crystd twice [Matijevic and Kerker *JPC* **62** 1271 *1958*].

Silver (metal) *[7440-22-4]* **M 107.9, m 961.9°, b 2212°, d 10.5.** For purification by electrolysis, see Craig *et al.* [*J Res Nat Bur Stand* **64A** 381 *1960*].

Silver acetate *[563-63-3]* **M 166.9.** Shaken with acetic acid for three days, the process being repeated with fresh acid, the solid then being dried in a vac oven at 40° for 48hr. Has also been crystd from water containing a trace of acetic acid, and dried in air.

Silver bromate *[7785-23-1]* **M 235.8.** Crystd from hot water (80ml/g).

Silver bromide *[7785-23-1]* **M 187.8, m 432°.** Purified from Fe, Mn, Ni and Zn by zone melting in a quartz vessel under vac.

Silver chlorate *[7783-92-8]* **M 191.3.** Recrystd three times from water (10ml/g at 15°; 2ml/g at 80°).

Silver chloride *[7783-90-6]* **M 143.3, m 455°.** Purified by recrystn from conc NH_3 soln.

Silver diethyldithiocarbamate *[1470-61-7]* **M 512.3.** Purified by recrystn from pyridine. Stored in a desiccator in a cool and dark place.

Silver iodate *[7783-97-3]* **M 282.8.** Washed with warm dil. HNO_3, then water and dried at 100°, or crystd from ammoniacal soln by adding HNO_3, filtering, washing with water and drying at 100°.

Silver nitrate *[7761-88-8]* **M 169.9, m 212°.** Purified by recrystn from hot water (solubility of $AgNO_3$ in water is 992g/100ml at 100° and 122g/100ml at 0°). It has also been purified by crystn from hot conductivity water by slow addition of freshly distd EtOH.
    CAUTION: avoid using EtOH for washing the ppte; and of concentrating the filtrate to obtain **further crops of AgNO$_3$ owing to the risk of an EXPLOSION (as has been reported to us) caused by the presence of silver fulminate. When using EtOH in the purification the apparatus should be enveloped in a strong protective shield.** [Tully, *News Ed (Am Chem Soc)* **19** 3092 *1941*; Garin and Henderson *J Chem Educ* **47** 741 *1970*; Bretherick, *Handbook of Reactive Chemical Hazards* 4th edn, Butterworths, London, 1985, pp 13-14.]
    Before being used as a standard in volumetric analysis, analytical reagent grade $AgNO_3$ should be finely powdered, dried at 120° for 2hr, then cooled in a desiccator.
    Recovery of silver residues as $AgNO_3$ **[use protective shield during the whole of this procedure]** can be achieved by washing with hot water and adding 16M HNO_3 to dissolve the solid. Filter through glass wool and concentrate the filtrate on a steam bath until pptn commences. Cool the soln in an ice-bath and filter the pptd $AgNO_3$. Dry at 120° for 2hr, then cool in a desiccator in vac. Store over P_2O_5 in vac in the dark. *AVOID contact with hands due to formation of black stains.*

Silver nitrite *[7783-99-5]* **M 153.9, m 141°(dec).** Crystd from hot conductivity water (70ml/g) in the dark. Dried in the dark under vac.

Silver(I) oxide *[20667-12-3]* **M 231.7.** Leached with hot water in a Soxhlet apparatus for several hours to remove any entrained electrolytes.

Silver perchlorate (H$_2$O) *[7783-93-9]* **M 207.3.** Refluxed with benzene (6ml/g) in a flask fitted with a Dean and Stark trap until all the water was removed azeotropically (*ca* 4hr). The soln was cooled and diltd with dry pentane (4ml/g of $AgClO_4$). The pptd $AgClO_4$ was filtered off and dried in a desiccator over P_2O_5 at 1mm for 24hr [Radell, Connolly and Raymond *JACS* **83** 3958 *1961*]. It has also been crystd from perchloric acid. [**Caution** *due to explosive nature in the presence of organic matter.*]

Silver sulphate *[10294-26-5]* **M 311.8.** Crystd by dissolving in hot conc H_2SO_4 containing a trace of HNO_3, cooling and diluting with water. The ppte was filtered off, washed and dried at 120°.

Silver trifluoromethanesulphonate *[2923-28-6]* **M 256.9.** Recrystd twice from hot CCl_4 [Alo *et al, JCSPT* 808 *1986*].

Sodium (metal) *[7440-23-5]* **M 23.0, m 97.5°, d 0.97.** The metal was placed on a coarse grade of sintered-glass filter, melted under vac. and forced through the filter using argon. The Pyrex apparatus was then re-evacuated and sealed off below the filter, so that the sodium could be distd at 460° through a side arm and condenser into a receiver bulb which was then sealed off [Gunn and Green *JACS* **80** 4782 *1958*]. **EXPLODES and IGNITES in water.**

Sodium acetate *[127-09-3]* **M 82.0, m 324°.** Crystd from acetic acid and pumped under vac for 10hr at 120°. Alternatively, crystd from aq EtOH, as the trihydrate. This material can be converted to the anhydrous salt by heating slowly in a porcelain, nickel or iron dish, so that the salt liquefies. Steam is evolved and the mass again solidifies. Heating is now increased so that the salt melts again. (NB: if it is heated too strongly, the salt chars.) After several minutes, the salt is allowed to solidify and cooled to a convenient temperature before being powdered and bottled (water content should now less than 0.02%).

Sodium alginate *[9005-38-3]*. Freed from heavy metal impurities by treatment with ion-exchange resins (Na$^+$-form), or with a dil soln of the sodium salt of EDTA. Also dissolved in 0.1M NaCl, centrifuged and fractionally pptd by gradual addition of EtOH or 4M NaCl. The resulting gels were centrifuged off, washed with aq EtOH or acetone, and dried under vac. [Büchner, Cooper and Wassermann *JCS* 3974 *1961*].

Sodium *n*-alkylsulphates. Recrystd from EtOH/acetone [Hashimoto and Thomas *JACS* **107** 4655 *1985.*]

Sodium 4-aminobenzoate *[555-06-6]* **M 159.1.** Crystd from water.

Sodium 4-aminosalicylate *[133-10-8]* **M 175.1.** Crystd from water at room temperature (2ml/g) by adding acetone and cooling.

Sodium ammonium hydrogen phosphate *[13011-54-6]* **M 209.1.** Crystd from hot water (1ml/g).

Sodium amylpenicillin *[575-47-3]* **M 350.4.** Crystd from moist acetone or moist ethyl acetate.

Sodium anthraquinone-1,5-disulphonate (H_2O) *[853-35-0]* **M 412.3.** Separated from insoluble impurities by contnuous extraction with water. Crystd twice from hot water and dried under vac.

Sodium anthraquinone-1-sulphonate (H_2O) *[107439-61-2]* **M 328.3.** Crystd from hot water (4ml/g) after treatment with active charcoal, or from water by addition of EtOH. Dried under vac over $CaCl_2$, or in an oven at 70°. Stored in the dark.

Sodium anthraquinone-2-sulphonate (H_2O) *[131-08-8]* **M 328.3.** Recrystd from MeOH [Costa and Bookfield *JCSFT 1* **82** 991 *1986*]

Sodium antimonyl tartrate *[34521-09-0]* **M 308.8.** Crystd from water.

Sodium arsenate (7H_2O) *[10048-95-0]* **M 312.0.** Crystd from water (2ml/g).

Sodium azide *[26628-22-8]* **M 65,0.** Crystd from hot water or from water by the addition of absolute EtOH or acetone. Also purified by repeated crystn from an aq soln satd at 90° by cooling it to 10°, and adding an equal volume of EtOH. The crystals were washed with acetone and the azide dried at room temperature under vac for several hours in an Abderhalden pistol. [Das *et al, JCSFT I* **78** 3485 *1982.*] **HIGHLY POISONOUS.**

Sodium barbitone *[144-02-5]* **M 150.1.** Crystd from water (3ml/g) by adding an equal volume of EtOH and cooling to 5°. Dried under vac over P_2O_5.

Sodium benzenesulphinate *[873-55-2]* **M 164.2, m >300°.** Recrystd from EtOH and dried at 120°for 4hr under reduced pressure [Kornblum and Wade *JOC* **52** 5301 *1987*].

Sodium benzenesulphonate *[144-42-4]* **M 150.1.** Crystd from EtOH or aq 70-100% MeOH, and dried under vac at 80-100°.

Sodium benzoate *[532-32-1]* **M 144.1.** Crystd from EtOH (12ml/g).

Sodium benzylpenicillin *[69-57-8]* **M 356.4.** Crystd from methanol-ethyl acetate.

Sodium bicarbonate *[144-55-8]* **M 84.0.** Crystd from hot water (6ml/g). The solid should not be heated above 40° due to the formation of carbonate.

Sodium bisulphite *[7631-90-5]* **M 104.1.** Crystd from hot water (1ml/g). Dried at 100° under vac for 4hr.

Sodium borate (borax) *[1330-43-4]* **M 201.2.** Most of the water of hydration was removed from the decahydrate by evacuation at 25° for three days, followed by heating to 100° and evacuation with a high-speed diffusion pump. The dried sample was then heated gradually to fusion (above 966°), and allowed to cool gradually to 200° before being tranferred to a desiccator containing P_2O_5 [Grenier and Westrum *JACS* **78** 6226 *1956*].

Sodium borate (decahydrate, hydrated borax) *[1303-96-4]* **M 381.2.** Crystd from water (3.3ml/g) keeping below 55° to avoid formation of the pentahydrate. Filtered at the pump, washed with water and equilibrated for several days in a desiccator containing an aq soln satd with respect to sucrose and NaCl. Borax can be prepared more quickly (but its water content is somewhat variable) by washing the recrystd material at the pump with water, followed by 95% EtOH, then Et_2O, and air dried at room temperature for 12-18hr on a clock glass.

Sodium borohydride *[16940-66-2]* **M 37.8.** After adding $NaBH_4$ (10g) to freshly distd diglyme (120ml) in a dry three-necked flask fitted with a stirrer, nitrogen inlet and outlet, the mixture was stirred for 30min at 50° until almost all of the solid had dissolved. Stirring was stopped, and, after the solid had settled, the supernatant liquid was forced under N_2 pressure through a sintered-glass filter into a dry flask. [The residue was centrifuged to obtain more of the soln which was added to the bulk.] The soln was cooled slowly to 0° and then decanted from the white needles that separated. The crystals were dried by pumping for 4hr to give anhydrous $NaBH_4$. Alternatively, after the filtration at 50° the soln was heated at 80° for 2hr to give a white ppte of substantially anhydrous $NaBH_4$ which was collected on a sintered-glass filter under N_2, then pumped at 60° for 2hr [Brown, Mead and Subba Rao *JACS* **77** 6209 *1955*].

$NaBH_4$ has also been crystd from isopropylamine by dissolving it in the solvent at reflux, cooling, filtering and allowing the soln to stand in a filter flask connected to a Dry-ice/acetone trap. After most of the solvent was passed over into the cold trap, crystals were removed with forceps, washed with dry ethyl ether and dried under vac. [Kim and Itoh *JPC* **91** 126 *1987*.] Somewhat less pure crystals were obtained more rapidly by using Soxhlet extraction with only a small amount of solvent and extracting for about 8hr. The crystals that formed in the flask were filtered off, then washed and dried as before. [Stockmayer, Rice and Stephenson *JACS* **77** 1980 *1955*.] Other solvents used for crystn include water and liquid ammonia.

Sodium bromate *[7789-38-0]* **M 150.9.** Crystd from hot water (1.1ml/g) to decrease contamination by NaBr, bromine and hypobromide. [Noszticzius *et al*, *JACS* **107** 2314 *1985*.]

Sodium bromide *[7647-15-6]* **M 102.9.** Crystd from water (0.86ml/g) between 50° and 0°, and dried at 140° under vac (this purification may not eliminate chloride ion).

Sodium 4-bromobenzenesulphonate *[5015-75-8]* **M 258.7.** Crystd from MeOH, EtOH or distd water.

Sodium butyrate *[156-54-7]* **M 110.1.** Prepared by neutralisation of the acid and recrystn from EtOH.

Sodium cacodylate ($3H_2O$) *[124-65-2]* **M 214.0.** Crystd from aq EtOH.

Sodium carbonate *[497-19-8]* **M 106.0.** Crystd from water as the decahydrate which was redissolved in water to give a near-satd soln. By bubbling CO_2, $NaHCO_3$ was ppted. It was filtered, washed and ignited for 2hr at 280° [MacLaren and Swinehart *JACS* **73** 1822 *1951*.] Before being used as a volumetric standard, analytical grade material should be dried by heating at 260-270° for ½hr and allowed to cool in a desiccator.
For preparation of primary standard sodium carbonate, see *PAC* **25** 459 *1969*.

Sodium carboxymethylcellulose *[9004-32-4]*. Dialysed for 48hr against distd water.

Sodium cetyl sulphate *[1120-01-0]* **M 344.5.** Crystd from MeOH.

Sodium chlorate *[7775-09-9]* **M 106.4.** Crystd from hot water (0.5ml/g).

Sodium chloride *[7647-14-5]* **M 58.4.** Crystd from satd aq soln (2.7ml/g) by passing in HCl gas, or by adding EtOH or acetone. Can be freed from bromide and iodide impurities by adding chlorine water to an aq. soln and boiling for some time to expel free bromine and iodine. Traces of iron can be removed by prolonged boiling of solid NaCl in 6M HCl, the crystals then being washed with EtOH and dried at *ca* 100°. Sodium chloride has been purified by sublimation in a stream of pre-purified N_2 and collected by electrostatic discharge [Ross and Winkler *JACS* **76** 2637 *1954*]. For use as a primary analytical standard, analytical reagent grade NaCl should be finely ground, dried in an electric furnace at 500-600° in a platinum crucible, and allowed to cool in a desiccator. For most purposes, however, drying at 110-120° is satisfactory.

Sodium chlorite *[7758-19-2]* **M 90.4.** Crystd from hot water and stored in a cool place. Has also been crystd from MeOH by countercurrent extraction with liquid ammonia [Curti and Locchi *AC* **29** 534 *1957*]. Major impurity is chloride ion; can be recrystd from 0.001M NaOH.

Sodium 4-chlorobenzenesulphonate *[5138-90-9]* **M 214.6,**
Sodium 4-chloro-*m*-toluenesulphonate *[5138-92-1]* **M 228.7.** Crystd twice from MeOH and dried under vac.

Sodium chromate ($4H_2O$) *[10034-82-9]* **M 234.0.** Crystd from hot water (0.8ml/g).

dl-**Sodium creatinephosphate** ($4H_2O$) *[922-32-7]* **M 327.1.** Crystd from water-ethanol.

Sodium cyanate *[917-61-3]* **M 65.0.** Crystd from water.

Sodium *p*-cymenesulphonate *[77060-21-0]* **M 236.3.** Dissolved in water, filtered and evapd to dryness. Crystd twice from absolute EtOH and dried at 110°.

Sodium decanoate (caproate) *[1002-62-6]* **M 194.2.** Neutralised by adding a slight excess of the free acid, recovering the excess acid by Et_2O extraction. The salt is crystd from soln by adding pure acetone, repeating the steps several times, then dried in an oven at *ca* 110° [Chaudhury and Awuwallia *TFS* **77** 3119 *1981*].

Sodium 1-decanesulphonate *[13419-61-9]* **M 244.33.** Recrystd from absolute EtOH and dried over silica gel.

Sodium *n*-decylsulphate *[142-87-0]* **M 239.3.** Rigorously purified by continuous Et_2O extraction of a 1% aq. soln for two weeks.

Sodium deoxycholate (H_2O) *[302-95-4]* **M 432.6,** $[\alpha]_D$ +48° (c 1, EtOH). Crystd from EtOH and dried in an oven at 100°. The solution is freed from soluble components by repeated extraction with acid-washed charcoal.

Sodium 2,5-dichlorobenzenesulphonate *[5138-93-2]* **M 249.0.** Crystd from MeOH, and dried under vac.

Sodium 5,5-diethylbarbiturate see **sodium barbitone.**

Sodium diethyldithiocarbamate (3H$_2$O) *[20624-25-3]* **M 225.3, m 94-96°(anhydr).** Recrystd from water.

Sodium di(ethylhexyl)sulphosuccinate (Aerosol-OT) *[577-11-7]* **M 444.6.** Dissolved in MeOH and inorganic salts which pptd were filtered off. Water was added and the soln was extracted several times with hexane. The residue was evapd to one fifth its original volume, benzene was added and azeotropic distn. was continued until no water remained. Solvent was then evaptd. The white solid was crushed and dried in vac over P$_2$O$_5$ for 48hr [El Seoud and Fendler *JCSFT I* **71** 452 *1975*]

Sodium diethyloxalacetate *[88330-76-1]* **M 210.2.** Extracted several times with boiling Et$_2$O (until the solvent remained colourless) and then the residue was dried in air.

Sodium dihydrogen orthophosphate (2H$_2$O) *[7558-80-7]* **M 156.0.** Crystd from warm water (0.5ml/g) by chilling.

Sodium 2,4-dihydroxyphenylazobenzene-4'-sulphonate *[30117-38-5]* **M 304.2.** Crystd from absolute EtOH.

Sodium 2,2'-dihydroxy-1-naphthaleneazobenzene-5'-sulphonate *[2092-55-9]* **M 354.3.** Purified by pptn of the free acid from aq soln using conc HCl, washing and extracting with EtOH in a Soxhlet extractor. The acid pptd on evapn of the EtOH and was reconverted to the sodium salt.

Sodium *p*-(*p*-dimethylaminobenzeneazo)-benzenesulphonate *[23398-40-5]* **M 327.3.** Crystd from water.

Sodium *p*-dimethylaminoazobenzene-*o*'-carboxylate *[845-10-3]* **M 219.2,**
Sodium *p*-dimethylaminoazobenzene-*p*'-carboxylate *[845-46-5]* **M 219.2.** Pptd from aq soln as the free acid which was crystd from 95% EtOH, then reconverted to the sodium salt.

Sodium 2,4-dimethylbenzenesulphonate *[827-21-4]* **M 208.2,**
Sodium 2,5-dimethylbenzenesulphonate *[827-19-0]* **M 208.2.** Crystd from MeOH and dried under vac.

Sodium *N,N*-dimethylsulphanilate *[2244-40-8]* **M 223.2, m >300°.** Crystd from water.

Sodium dithionite (2H$_2$O) *[7631-94-9]* **M 242.1.** Crystd from hot water (1.1ml/g) by cooling.

Sodium dodecanoate *[629-25-4]* **M 200.3.** Neutralised by adding a slight excess of dodecanoic acid, removing it by ether extraction. The salt crystd from the aq soln by adding pure acetone, repeating the process several times (see sodium decanoate).

Sodium 1-dodecanesulphonate *[2386-53-0]* **M 272.4.** Twice recrystd from EtOH.

Sodium dodecylbenzenesulphonate *[25155-30-0]* **M 348.5.** Recrystd from propan-2-ol.

Sodium dodecylsulphate (SDS, sodium laurylsulphate) *[151-21-3]* **M 288.4, m 204-207°.** Purified by Soxhlet extraction with pet ether for 24hr, followed by dissolution in acetone:MeOH:H$_2$O 90:5:5(v/v) and recrystn [Politi *et al, JPC* **89** 2345 *1985*]. Also purified by two recrystns from abs. EtOH, aq 95% EtOH, MeOH, isopropanol or a 1:1 mixture of EtOH:isopropanol to remove dodecanol, and dried under vac. [Ramesh and Labes *JACS* **109** 3228 *1987*.] Also purified by foaming [see Cockbain and McMullen *TFS* **47** 322 *1951*] or by liquid-liquid extraction [see Harrold *J Colloid Sci* **15** 280 *1960*].

Dried over silica gel. For DNA work it should be dissolved in excess MeOH passed through an activated charcoal column and evaporated until it crystallises out.
Also purified by dissolving in hot 95% EtOH (14ml/g), filtering and cooling, then drying in a vac. desiccator. Alternatively, it was crystd from H_2O, vac dried, washed with anhydrous Et_2O, vac dried. These operations were repeated five times [Maritato *JPC* **89** 1341 *1985*; Lennox and McClelland *JACS* **108** 3771 *1986*; Dressik *JACS* **108** 7567 *1986*].

Sodium ethylmercurithiosalicylate *[14737-80-5]* **M 404.8.** Crystd from ethanol-ethyl ether

Sodium ethylsulphate *[546-74-7]* **M 166.1.** Recrystd three times from MeOH-Et_2O and vac dried.

Sodium ferricyanide (H_2O) *[14217-21-1]* **M 298.9.** Crystd from hot water (1.5ml/g) or by pptn from 95% EtOH.

Sodium ferrocyanide ($10H_2O$) *[13601-19-9]* **M 484.1.** Crystd from hot water (0.7ml/g), until free of ferricyanide as shown by absence of Prussian Blue formation with ferrous sulphate soln.

Sodium fluoride *[7681-49-4]* **M 42.0.** Crystd from water by partial evapn in a vac desiccator. Or, dissolved in water, and *ca* half of it pptd by addition of EtOH. Ppte was dried in an air oven at 130° for one day, and then stored in a desiccator over KOH.

Sodium fluoroborate *[13755-29-8]* **M 109.8.** Crystd from hot water (50ml/g) by cooling to 0°. Alternatively, purified from insoluble material by dissolving in a minimum amount of water, then fluoride ion was removed by adding conc lanthanum nitrate in excess. After removing lanthanum fluoride by centrifugation, the supernatant was passed through a cation-exchange column (Dowex 50, Na^+-form) to remove any remaining lanthanum [Anbar and Guttman *JPC* **64** 1896 *1960*].

Sodium fluorosilicate *[16893-85-9]* **M 188.1.** Crystd from hot water (40ml/g) by cooling.

Sodium formate (anhydrous) *[141-53-7]* **M 68.0.** A satd aq soln at 90° (0.8ml water/g) was filtered and allowed to cool slowly. (The final temperature was above 30° to prevent formation of the hydrate.) After two such crystns the crystals were dried in an oven at 130°, then under high vac. [Westrum, Chang and Levitin *JPC* **64** 1553 *1960*; Roecker and Meyer *JACS* **108** 4066 *1986*]. The salt has also been recrystd twice from 1mM DTPA, then twice from water [Bielski and Thomas *JACS* **109** 7761 *1987*].

Sodium glycochenodeoxycholate *[16564-43-5]* **M 472.6,**
Sodium glycocholate *[863-57-0]* **M 488.6.** Dissolved in EtOH, filtered and concentrated to crystn, and recrystd from a little EtOH.

Sodium glycollate ($2H_2O$) *[2053-21-6]* **M 98.0.** Pptd from aq soln by EtOH, and air dried.

Sodium hexadecylsulphate *[1120-01-0]* **M 323.5.** Recrystd from absolute EtOH [Abu Hamdiyyah and Rahman *JPC* **91** 1531 *1987*].

Sodium hexafluorophosphate *[21324-39-0]* **M 167.9.** Recrystd from acetonitrile and vac dried for 2 days at room temperature. It is an **irritant** and is *hygroscopic*. [Delville *et al*, *JACS* **109** 7293 *1987*.]

Sodium hydrogen diglycollate *[50795-24-9]* **M 156.1.** Crystd from hot water (7.5ml/g) by cooling to 0° with constant stirring, the crystals being filtered off on to a sintered-glass funnel and dried at 110° overnight.

Sodium hydrogen oxalate ($2H_2O$) *[1186-49-8]* **M 130.0.** Crystd from hot water (5ml/g) by cooling.

Sodium hydrogen succinate *[2922-54-5]* **M 140.0.** Crystd from water and dried at 110°.

Sodium hydrogen *d*-tartrate *[526-94-3]* **M 190.1,** $[\alpha]_{546}$ **+26°** (c 1, H$_2$O). Crystd from warm water (10ml/g) by cooling to 0°.

Sodium hydroxide (anhydrous) *[1310-73-2]* **M 40.0.** Common impurities are water and sodium carbonate. Sodium hydroxide can be purified by dissolving 100g in 1L of pure EtOH, filtering the soln under vac through a fine sintered-glass disc to remove insoluble carbonates and halides. (This and subsequent operations should be performed in a dry, CO$_2$-free box.) The soln is concentrated under vac, using mild heating, to give a thick slurry of the mono-alcoholate which is transferred to a coarse sintered-glass disc and pumped free of mother liquor. After washing the crystals several times with purified alcohol to remove traces of water, they are vac dried, with mild heating, for about 30hr to decompose the alcoholate, leaving a fine white crystalline powder [Kelly and Snyder *JACS* **73** 4114 *1951*].

Sodium hydroxide solutions (*caustic*). Carbonate ion can be removed by passage through an anion-exchange column (such as Amberlite IRA-400; OH⁻-form). The column should be freshly prepared from the chloride form by slow prior passage of sodium hydroxide soln until the effluent gives no test for chloride ions. After use, the column can be regenerated by washing with dil HCl, then water. Similarly, other metal ions are removed when a 1M (or more dilute) NaOH soln is passed through a column of Dowex ion-exchange A-1 resin in its Na⁺-form.
Alternatively, carbonate contamination can be reduced by rinsing analytical reagent quality sticks of NaOH rapidly with water, then dissolving them in distd water, or by preparing a conc aq soln of NaOH and drawing off the clear supernatant liquid. (Insoluble Na$_2$CO$_3$ is left behind.) Carbonate contamination can be reduced by adding a slight excess of conc BaCl$_2$ or Ba(OH)$_2$ to a sodium hydroxide soln, shaking well and allowing the BaCO$_3$ ppte to settle. If the presence of barium in the soln is unacceptable, an electrolytic purification can be used. For example, sodium amalgam is prepared by the electrolysis of 3L of 30% NaOH with 500ml of pure mercury for cathode, and a platinum anode, passing 15 Faradays at 4A, in a thick-walled polyethylene bottle. The bottle is then fitted with inlet and outlet tubes, the spent soln being flushed out by CO$_2$-free N$_2$. The amalgam is then washed thoroughly with a large volume of deionised water (with the electrolysis current switched on to minimize loss of sodium). Finally, a clean steel rod is placed in contact in the soln with the amalgam (to facilitate hydrogen evolution), reaction being allowed to proceed until a suitable concn is reached, before being transferred to a storage vessel and diluted as required [Marsh and Stokes *Aust J Chem* **17** 740 *1964*].

Sodium 2-hydroxy-4-methoxybenzophenone-5-sulphonate *[6628-37-1]* **M 330.3.** Crystd from MeOH and dried under vac.

Sodium *p*-hydroxyphenylazobenzene-*p*'-sulphonate *[2623-31-1]* **M 288.2.** Crystd from 95% EtOH.

Sodium iodate *[7681-55-2]* **M 197.9.** Crystd from water (3ml/g) by cooling.

Sodium iodide *[7681-82-5]* **M 149.9.** Crystd from water/ethanol soln and dried for 12hr under vac, at 70°. Alternatively, dissolved in acetone, filtered and cooled to -20°, the resulting yellow crystals being filtered off and heated in a vac oven at 70° for 6hr to remove acetone. The NaI was then crystd from very dil. NaOH, dried under vac, and stored in a vac desiccator [Verdin *TFS* **57** 484 *1961*].

Sodium isopropylxanthate *[140-93-2]* **M 158.2.** Crystd from ligroin/ethanol.

Sodium laurate *[629-25-4]* **M 222.0.** Crystd from MeOH.

Sodium mandelate *[114-21-6]* **M 174.1.** Crystd from 95% EtOH.

Sodium metanilate *[1126-34-7]* **M 195.2,**
Sodium metaperiodate (NaIO$_4$) *[7790-28-5]* **M 213.9.** Crystd from hot water.

Sodium metasilicate (5H$_2$O) *[6834-92-0]* **M 212.1.** Crystd from aq 5% NaOH soln.

Sodium 3-methyl-1-butanesulphonate *[5343-41-9]* **M 174.1.** Crystd from 90% MeOH.

Sodium molybdate (2H$_2$O) *[10102-40-6]* **M 241.9.** Crystd from hot water (1ml/g) by cooling to 0°.

Sodium monensin *[22373-78-0]* **M 693.8.** Recrystd from EtOH-H$_2$O [Cox *et al, JACS* **107** 4297 *1985*].

Sodium 1-naphthalenesulphonate *[85-47-2]* **M 230.2.** Recrystd from water or aq acetone [Okadata *et al, JACS* **108** 2863 *1986*].

Sodium 2-naphthalenesulphonate *[532-02-5]* **M 230.2.** Crystd from hot 10% aq NaOH or water, and dried in a steam oven.

Sodium 2-naphthylamine-5,7-disulphonate *[79004-97-0]* **M 235.4.** Crystd from water (charcoal) and dried in a steam oven.

Sodium nitrate *[7631-99-4]* **M 85.0.** Crystd from hot water (0.6ml/g) by cooling to 0°, or from conc aq soln by addition of MeOH. Dried under vac at 140°.

Sodium nitrite *[7632-00-0]* **M 69.0, m 271°.** Crystd from hot water (0.7ml/g) by cooling to 0°, or from its own melt. Dried over P$_2$O$_5$.

Sodium 1-octanesulphonate *[5324-84-5]* **M 216.2.** Recrystd from absolute EtOH.

Sodium oleate *[143-19-1]* **M 304.4, m 233-235°.** Crystd from EtOH and dried in an oven at 100°.

Sodium oxalate *[62-76-0]* **M 134.0.** Crystd from hot water (16ml/g) by cooling to 0°. Before use as a volumetric standard, analytical grade quality sodium oxalate should be dried for 2hr at 120° and allowed to cool in a desiccator.

Sodium palmitate *[408-35-5]* **M 278.4, m 285-201°.** Crystd from EtOH and dried in an oven.

Sodium perchlorate (anhydrous) *[7601-89-0]* **M 122.4.** Because its solubility in water is high (2.1g/ml at 15°) and it has a rather low temperature coefficient of solubility, sodium perchlorate is usually crystd from acetone, MeOH, water-ethanol or dioxane-water (33g dissolved in 36ml of water and 200ml of dioxane). After filtering and crystallising, the solid is dried under vac at 140-150° to remove solvent of crystn. Basic impurities can be removed by crystn from hot acetic acid, followed by heating at 150°. If NaClO$_4$ is pptd from distd water by adding HClO$_4$ to the chilled soln, the ppte contains some free acid.

Sodium *p*-phenolsulphonate (2H$_2$O) *[825-90-1]* **M 232.2.** Crystd from hot water (1ml/g) by cooling to 0°, or from MeOH, and dried in vac.

Sodium phenoxide *[139-02-6]* **M 116.1.** Washed with Et$_2$O, then heated under vac to 200° to remove any free phenol.

Sodium phenylacetate *[114-70-5]* **M 158.1.** Its aq soln was evapd to crystn on a steam bath, the crystals being washed with absolute EtOH and dried under vac at 80°.

Sodium *o*-phenylphenolate (4H$_2$O) *[132-27-4]* **M 264.3.** Crystd from acetone and dried under vac at room temperature.

Sodium phosphoamidate *[3076-34-4]* **M 119.0.** Dissolved in water below 10°, and acetic acid added dropwise to *p*H 4.0 so that the monosodium salt was pptd. The ppte was washed with water and Et$_2$O, then air dried. Addition of one equivalent of NaOH to the soln gave the sodium salt, the soln being adjusted to *p*H 6.0 before use [Rose and Heald *BJ* **81** 339 *1961*].

Sodium phytate (H₂O) *[14306-25-3]* M 857.9. Crystd from water.

Sodium piperazine-N,N'-bis(2-ethanesulphonate) (H₂O) *[76836-02-7]* M 364.3. Crystd from water and EtOH.

Sodium polyacrylate (NaPAA) *[9003-04-7]*. Commercial polyacrylamide was neutralised with an aq soln of NaOH and the polymer pptd with acetone. The ppte was redissolved in a small amount of water and freeze-dried. The polymer was repeatedly washed with EtOH and water to remove traces of low molecular weight material, and finally dried in vac at 60° [Vink *JCSFT 1* **75** 1207 *1979*]. Also dialysed overnight against distd water, then freeze-dried.

Sodium poly(α-L-glutamate). It was washed with acetone, dried, dissolved in water and pptd with isopropanol at 5°. Impurities and low mol.wt. fractions were removed by dialysis of the aq soln for 50hr, followed by ultrafiltration through a filter impermeable to polymers of mol. wts greater the 10⁴. The polymer was recovered by freeze-drying. [Mori *et al*, *JCSFT 1* 2583 *1978.*]

Sodium pyrophosphate (10H₂O) *[13472-36-1]* M 446.1. Crystd from warm water and air dried at room temperature.

Sodium selenate *[13410-01-0]* M 188.9,
Sodium selenite *[10102-18-8]* M 172.9. Crystd from water.

Sodium silicate solution *[1344-09-8]*. Purified by contact filtration with activated charcoal.

Sodium succinate *[150-90-3]* M 162.1. Crystd from hot water (1.2ml/g) by cooling to 0°. Dried at 125°.

Sodium sulphanilate *[515-74-2]* M 195.2. Crystd from water.

Sodium sulphate (10H₂O) *[7727-73-3]* M 322.2. Crystd from water at 30° (1.1ml/g) by cooling to 0°. Sodium sulphate becomes anhydrous at 32°.

Sodium sulphide (9H₂O) *[1313-84-4]* M 240.2. Some purification of the hydrated salt can be achieved by selecting large crystals and removing the surface layer (contaminated with oxidation products) by washing with distd water. Other metal ions can be removed from Na₂S solns by passage through a column of Dowex ion-exchange A-1 resin, Na⁺-form. The hydrated salt can be rendered anhydrous by heating in a stream of H₂ or N₂ until water is no longer evolved. (The resulting cake should not be heated to fusion because it is readily oxidised.) Recrystd from distd water [Anderson and Azowlay *JCSDT* 469 *1986*].

Sodium sulphite *[7757-83-7]* M 126.0. Crystd from warm water (0.5ml/g) by cooling to 0°. Purified by repeated crystns from deoxygenated water inside a glove-box, finally drying under vac. [Rhee and Dasgupta *JPC* **89** 1799 *1985*].

Sodium R-tartrate *[868-18-8]* M 230.1. Crystd from warm dil aq NaOH by cooling.

Sodium taurocholate *[145-42-6]* M 555.7. Purified by recrystn and gel chromatography using Sephadex LH-20.

Sodium tetradecylsulphate *[1191-50-0]* M 316.4. Recrystd from absolute EtOH [Abu Hamdiyyah and Rahman *JPC* **91** 1531 *1987*].

Sodium tetrafluoroborate *[13755-29-8]* **M 109.8.** Recrystd from anhydrous MeOH and dried in a vac at 70° for 16hr. It is affected by moisture. [Delville *et al, JACS* **109** 7293 *1987*.]

Sodium tetrametaphosphate *[13396-41-3]* **M 429.9.** Crystd twice from water at room temperature by adding EtOH (300g of $Na_4P_4O_{12}$,H_2O, 2L of water, and 1L of EtOH), washed first with 20% EtOH then with 50% EtOH and air dried [Quimby *JPC* **58** 603 *1954*].

Sodium tetraphenylborate *[143-66-8]* **M 342.2.** Crystd from acetone-hexane or $CHCl_3$, or from Et_2O-cyclohexane (3:2) by warming the soln to ppte the compound. Dried in vac at 80°. Dissolved in acetone and added to an excess of toluene. After a slight milkiness developed on standing, the mixture was filtered. The clear filtrate was evapd at room temperature to a small bulk and again filtered. The filtrate was then warmed to 50-60°, giving clear dissolution of crystals. After standing at this temperature for 10min the mixture was filtered rapidly through a pre-heated Büchner funnel, and the crystals were collected and dried in a vac desiccator at room temperature for 3 days [Abraham *et al, JCSFT 1* **80** 489 *1984*]. If the product gives a turbid aq soln, the turbidity can be removed by treating with freshly prepared alumina gel.

Sodium thioantimonate (Na_3SbS_4.$9H_2O$) *[10101-91-4]* **M 481.1.** Crystd from warm water (2ml/g) by cooling to 0°.

Sodium thiocyanate *[540-72-7]* **M 81.1.** Crystd from water, acetonitrile or from MeOH using Et_2O for washing, then dried at 130°, or dried under vac at 60° for 2 days. [Strasser *et al, JACS* **107** 789 *1985*; Szezygiel *et al, JACS* **91** 1252 *1987*.] (The latter purification removes material reacting with iodine.) Sodium thiocyanate solns can be freed from traces of iron by repeated batch extractions with Et_2O.

Sodium thioglycollate *[367-51-1]* **M 114.1.** Crystd from 60% EtOH (charcoal).

Sodium thiosulphate ($5H_2O$) *[10102-17-7]* **M 248.2,** (anhydrous) *[7772-98-7]*. Crystd from EtOH-H_2O solns or from water (0.3ml/g) below 60° by cooling to 0°, and dried at 35° over P_2O_5 under vac.

Sodium *p*-toluenesulphinate *[824-79-3]* **M 178.2.** Crystd from water (to constant UV spectrum), and dried under vac. or extracted with hot benzene, then dissolved in EtOH-H_2O and heated with decolorising charcoal. The soln was filtered and cooled to give crystals of the dihydrate.

Sodium *p*-toluenesulphonate *[657-84-1]* **M 194.2.** Dissolved in distd water, filtered to remove insoluble impurities and evapd to dryness. Then crystd from MeOH or EtOH, and dried at 110°. Its solubility in EtOH is not high (maximum 2.5%) so that Soxhlet extraction with EtOH may be preferable. Sodium *p*-toluenesulphonate has also been crystd from Et_2O and dried under vac at 50°.

Sodium trifluoroacetate *[2923-18-4]* **M 136.0, m 206-210°(dec).** Pptd from EtOH by adding dioxane, then crystd several times from hot absolute EtOH. Dried at 120-130°/1mm.

Sodium 2,2',4-trihydroxyazobenzene-5'-sulphonate *[3564-26-9]* **M 320.2.** Purified by precipitating the free acid from aq soln using conc HCl, then washing and extracting with EtOH in a Soxhlet extractor. Evapn of the EtOH left the purified acid.

Sodium trimetaphosphate ($6H_2O$) *[7785-84-4]* **M 320.2.** Pptd from an aq soln at 40° by adding EtOH. Air dried.

Sodium 2,4,6-trimethylbenzenesulphonate *[6148-75-0]* **M 222.1.** Crystd twice from MeOH and dried under vac.

Sodium triphosphate see **sodium tripolyphosphate.**

Sodium tripolyphosphate *[7758-29-4]* **M 367.9.** Purified by repeated pptn from aq soln by slow additn of MeOH and air dried. Also a soln of anhydrous sodium tripolyphosphate (840g) in water (3.8L) was filtered, MeOH (1.4L) was added with vigorous stirring to ppte $Na_5P_3O_{10}.6H_2O$. The ppte was collected on a filter, air dried by suction, then left to dry in air overnight. It was crystd twice more in this way, using a 13% aq soln (w/w), and leachimg the crystals with 200ml portions of water [Watters, Loughran and Lambert *JACS* **78** 4855 *1956*]. Similarly, EtOH can be added to ppte the salt from a filtered 12-15% aq soln, the final soln containing *ca* 25% EtOH (v/v). Air drying should be at a relative humidity of 40-60%. Heat and vac drying should be avoided. [Quimby *JPC* **58** 603 *1954*.]

Sodium tungstate (2H$_2$O) *[10213-10-2]* **M 329.9.** Crystd from hot water (0.8ml/g) by cooling to 0°.

Sodium *m*-xylenesulphonate *[30587-85-0]* **M 208.2,**
Sodium *p*-xylenesulphonate *[827-19-0]* **M 208.2.** Dissolved in distd water, filtered, then evapd to dryness. Crystd twice form absolute EtOH and dried at 110°.

Stannic chloride *[7646-78-8]* **M 260.5, d 2.215.** Refluxed with clean mercury or P_2O_5 for several hours, then distd under (reduced) N_2 pressure into a receiver containing P_2O_5. Finally redistd. Alternatively, distd from Sn metal under vac in an all-glass system and sealed off in large ampoules. Fumes in moist air.

Stannic iodide (SnI$_4$) *[7790-47-8]* **M 626.3, m 144°.** Crystd from anhydrous CHCl$_3$, dried under vac. and stored in a vac desiccator.

Stannic oxide (SnO$_2$) *[18282-10-5]* **M 150.7.** Refluxed repeatedly with fresh HCl until the acid showed no tinge of yellow. The oxide was then dried at 110°.

Stannous (biscyclopentadienyl) *[26078-96-6]* **M 248.9.** Purified by vacuum sublimation. Handled and stored under dry N_2. The related thallium and indium compounds are similarly prepared.

Stannous chloride (anhydrous) *[7772-99-8]* **M 189.6.** Analytical reagent grade stannous chloride dihydrate is dehydrated by adding slowly to vigorously stirred, redistd acetic anhydride (120g salt per 100g of anhydride). (In a fume cupboard.) After *ca* an hour, the anhydrous SnCl$_2$ is filtered on to a sintered-glass or Büchner funnel, washed free from acetic acid wth dry Et$_2$O (2 x 30ml), and dried under vac. It is stored in a sealed container. [Stephen *JCS* 2786 *1930*.]

Strontium acetate *[543-94-2]* **M 205.7.** Crystd from acetic acid, then dried under vac for 24hr at 100°.

Strontium bromide *[10476-81-0]* **M 247.4.** Crystd from water (0.5ml/g).

Strontium chloride (6H$_2$O) *[1025-70-4]* **M 266.6, m 114°.** Crystd from warm water (0.5ml/g) by cooling to 0°.

Strontium chromate *[7789-06-2]* **M 203.6.** Crystd from water (40ml/g) by cooling.

Strontium hydroxide (8H$_2$O) *[18480-07-4]* **M 265.8.** Crystd from hot water (2.2ml/g) by cooling to 0°.

Strontium lactate (3H$_2$O) *[29870-99-5]* **M 319.8.** Crystd from aq EtOH.

Strontium nitrate *[10042-76-9]* **M 211.6.** Crystd from hot water (0.5ml/g) by cooling to 0°.

Strontium oxalate (H$_2$O) *[814-95-9]* **M 193.6.** Crystd from hot water (20ml/g) by cooling.

Strontium salicylate *[526-26-1]* **M 224.7.** Crystd from hot water (4ml/g) or EtOH.

Strontium tartrate *[868-19-9]* **M 237.7.** Crystd from hot water.

Strontium thiosalicylate (5H$_2$O) *[15123-90-7]* **M 289.8.** Crystd from hot water (2ml/g) by cooling to 0°.

Sulphamic acid *[5329-14-6]* **M 97.1, m 205°(dec).** Crystd from water at 70° (300ml per 25g), after filtering, by cooling a little and discarding the first batch of crystals (about 25g) before standing in an ice-salt mixture for 20min. The crystals were filtered by suction, washed with a small quantity of ice water, then twice with cold EtOH and finally with Et$_2$O. Air dried for 1hr, then stored in a desiccator over Mg(ClO$_4$)$_2$ [Butler, Smith and Audrieth *IECAE* **10** 690 *1938*].
For preparation of primary standard material see *PAC* **25** 459 *1969*.

Sulphamide *[7803-58-9]* **M 96.1, m 91.5°.** Crystd from absolute EtOH.

Sulphur *[7704-34-9]* **M 32.1, m between 112.8° and 120°, depending on form.** Murphy, Clabaugh and Gilchrist [*J Res Nat Bur Stand* **64A** 355 *1960*] have obtained sulphur of about 99.999 moles per cent purity by the following procedure: Roll sulphur was melted and filtered through a coarse-porosity glass filter funnel into a 2L round-bottomed Pyrex flask with two necks. Conc H$_2$SO$_4$ (300ml) was added to the sulphur (2.5Kg), and the mixture was heated to 150°, stirring continuously for 2hr. Over the next 6hr, conc HNO$_3$ was added in about 2ml portions at 10-15min intervals to the heated mixture. It was then allowed to cool to room temperature and the acid was poured off. The sulphur was rinsed several times with distd water, then remelted, cooled, and rinsed several times with distd water again, this process being repeated four or five times to remove most of the acid entrapped in the sulphur. An air-cooled reflux tube (*ca* 40cm long) was attached to one of the necks of the flask, and a gas delivery tube (the lower end about 1in above the bottom of the flask) was inserted into the other. While the sulphur was boiled under reflux, a stream of helium or N$_2$ was passed through to remove any water, HNO$_3$ or H$_2$SO$_4$, as vapour. After 4hr, the sulphur was cooled so that the reflux tube could be replaced by a bent air-cooled condenser. The sulphur was then distd, rejecting the first and the final 100ml portions, and transferred in 200ml portions to 400ml glass cylinder ampoules (which were placed on their sides during solidification). After adding about 80ml of water, displacing the air with N$_2$, and sealing the ampoule was cooled, and the water was titrated with 0.02M NaOH, the process being repeated until the acid content was negligible. Finally, entrapped water was removed by alternate evacuation to 10mm Hg and refilling with N$_2$ while the sulphur was kept molten.
Other purifications include crystn from CS$_2$ (which is less satisfactory becuase the sulphur retains appreciable amounts of organic material), benzene or benzene/acetone, followed by melting and degassing. Has also been boiled with 1% MgO, then decanted, and dried under vac. at 40° for 2 days over P$_2$O$_5$. [For purification of S$_6$, "recryst. S$_8$" and 'Bacon-Fanelli sulphur" see Bartlett, Cox and Davis *JACS* **83** 103, 109 *1961*.]

Sulphur dichloride *[10545-99-0]* **M 103.0, m -78°, b 59°/760mm(dec), d 1.621.** Twice dist in the presence of a small amount of PCl$_3$ through a 12in Vigreux column, the fraction boiling between 55-61° being redistd (in the presence of PCl$_3$), and the fraction distilling between 58-61° retained. (The PCl$_3$ is added to inhibit the decompn of SCl$_2$ into S$_2$Cl$_2$ and Cl$_2$. The SCl$_2$ must be used as quickly as possible after distn; within 1hr at room temperature the sample contains 4% S$_2$Cl$_2$. On long standing this reaches 16-18%.

Sulphur dioxide *[7446-09-5]* **M 64.1, b -10°.** Dried by bubbling through conc H$_2$SO$_4$ and by passage over P$_2$O$_5$, mist being removed by passage through a glass-wool plug. Frozen with liquid air and pumped to a high vac to remove dissolved gases.

Sulphuric acid *[7664-93-9]* **M 98.1, d 1.83.** Sulphuric acid, and also 30% fuming H$_2$SO$_4$, can be distd in an all-Pyrex system, optionally from potassium persulphate. Also purified by fractional crystn of the monohydrate from the liquid.

Sulphur monochloride *[10025-67-9]* **M 135.0, b 138°/760mm, n 1.658, d 1.682.** Purified by distn below 60° from a mixture containing sulphur (2%) and activated charcoal (1%), under reduced pressure (e.g. 50mm). Stored in the dark in a refrigerator.

Telluric acid *[11120-48-2]* **M 229.6.** Crystd once from nitric acid, then repeatedly from hot water (0.4ml/g).

Tellurium *[13494-80-9]* **M 127.6, m 450°.** Purified by zone refining and repeated sublimation to an impurity of less than 1 part in 10^8 (except for surface contamination by TeO_2). [Machol and Westrum *JACS* **80** 2950 *1958*.] Tellurium is volatile at 500°/0.2mm. Also purified by electrodeposition [Mathers and Turner *Trans Amer Electrochem Soc* **54** 293 *1928*].

Tellurium dioxide *[7446-07-3]* **M 159.6.** Dissolved in 5M NaOH, filtered and pptd by adding 10M HNO_3 to the filtrate until the soln was acid to phenolphthalein. After decanting the supernatant, the ppte was washed five times with distd water, then dried for 24hr at 110° [Horner and Leonhard *JACS* **74** 3694 *1952*.]

Terbium oxide *[12037-01-3]* **M 747.7.** Dissolved in acid, pptd as its oxalate and ignited at 650°.

Tetraethyl lead *[78-00-2]* **M 323.5.** Its more volatile comtaminants can be removed by exposure to a low pressure (by continuous pumping) for 1hr at 0°. Purified by stirring with an equal volume of H_2SO_4 (s.g. 1.40), keeping the temperature below 30°, repeating this process until the acid layer is colourless. It is then washed with dil Na_2CO_3 and distd water, dried with $CaCl_2$ and fractionally distd at low pressure under H_2 or N_2 [Calingaert *Chem Rev* **2** 43 *1926*].

Tetramethylammonium triphenylborofluoride *[437-11-6]* **M 392.2.** Crystd from acetone or acetone/ethanol.

Tetramethylsilane *[75-76-3]* **M 88.2, b 26.3°, n 1.359, d 0.639.** Distd from conc H_2SO_4 (after shaking with it) or $LiAlH_4$, through a 5ft vacuum-jacketted column packed with glass helices into an ice-cooled condenser, then percolated through silica gel to remove traces of halide.

Tetraphenylarsonium chloride *[507-28-8]* **M 418.8, m 261-263°.** A neutralised aq soln was evapd to dryness. The residue was extracted into absolute EtOH, evapd to a small volume and pptd by addition of absolute Et_2O. It was again dissolved in a small volume of absolute EtOH or ethyl acetate and repptd with Et_2O. Alternatively purified by adding conc HCl to ppte the chloride dihydrate. Redissolved in water, neutralised with Na_2CO_3 and evapd to dryness. The residue was extracted with $CHCl_3$ and finally crystd from CH_2Cl_2 or EtOH by adding Et_2O. If the aqueous layer is somewhat turbid treat with Celite and filter through filter paper.

Tetraphenylarsonium iodide *[7422-32-4]* **510.2,**
Tetraphenylarsonium perchlorate *[3084-10-4]* **M 482.8.** Crystd from MeOH.

Tetraphenylboron potassium *[3244-41-5]* **M 358.2.** Recrystd from acetone or water.

Tetraphenylsilane *[1048-08-4]* **M 336.4, m 231-233°.** Crystd from benzene.

Tetraphenyltin *[595-90-4]* **M 427.1, m 226°.** Crystd from $CHCl_3$, xylene or benzene/cyclohexane, and dried at 75°/20mm.

Tetrasodium pyrene-1,3,6,8-tetrasulphonate *[59570-10-0]* **M 610.5.** Recrystd from aq acetone [Okahata *et al, JACS* 108 2863 *1986*].

Thallous bromide *[7789-40-4]* **M 284.3, m 460°.** Thallous bromide (20g) was refluxed for 2-3hr with water (200ml) containing 3ml of 47% HBr. It was then washed until acid-free, heated to 300° for 2-3hr and stored in brown bottles.

Thallous carbonate *[6533-73-9]* **M 468.7, m 268-270°.** Crystd from hot water (4ml/g) by cooling.

Thallous chlorate *[13453-30-0]* **M 287.8.** Crystd from hot water (2ml/g) by cooling.

Thallous chloride *[7791-12-0]* **M 239.8, m 429.9°.** Crystd from 1% HCl and washed until acid-free. Or crystd from hot water (50ml/g), then dried at 140° and stored in brown bottles. Also purified by subliming in vac, followed by treatment with dry HCl gas and filtering while molten. (Soluble in 260 parts of cold water and 70 parts of boiling water.)

Thallous hydroxide *[12026-06-1]* **M 221.4.** Crystd from hot water (0.6ml/g) by cooling.

Thallous iodide *[7790-30-9]* **M 331.3.** Refluxed for 2-3hr with water containing HI, then washed until acid-free, and dried at 120°. Stored in brown bottles.

Thallous nitrate *[10102-45-1]* **M 266.4.** Crystd from warm water (1ml/g) by cooling to 0°.

Thallous perchlorate *[13453-40-2]* **M 303.8.** Crystd from hot water (0.6ml/g) by cooling. Dried under vac for 12hr at 100° (protect from possible **explosion**).

Thallous sulphate *[7446-18-6]* **M 504.8, m 633°.** Crystd from hot water (7ml/g) by cooling, then dried under vac over P_2O_5.

Thionyl chloride *[7719-09-7]* **M 119.0, b 77°, d 1.636.** Crude $SOCl_2$ can be freed from sulphuryl chloride, sulphur monochloride and sulphur dichloride by refluxing with sulphur and then fractionally distilling twice. [The $SOCl_2$ is converted to SO_2 and sulphur chlorides. The S_2Cl_2 (b 135.6°) is left in the residue, whereas SCl_2 (b 59°) passes over in the forerun.] The usual purification is to distil from quinoline (50g $SOCl_2$ to 10g quinoline) to remove acid impurities, followed by distn from boiled linseed oil (50g $SOCl_2$ to 20g of oil). Precautions must be taken to exclude moisture.
Thionyl chloride for use in organic syntheses can be prepared by distn of technical $SOCl_2$ in the presence of diterpene (12g/250m $SOCl_2$), avoiding overheating. Further purification is achieved by redistn from linseed oil (1-2%) [Rigby *Chem & Ind* 1508 *1969*]. Gas chromatographically pure material is obtained by distn from 10% (w/w) triphenyl phosphite [Friedman and Wetter *JCS (A)* 36 *1967*; Larsen *et al, JACS* 108 6950 *1986*].

Thorium chloride *[1002-08-1]* **M 373.8.** Freed from anionic impurities by passing a 2M soln of $ThCl_4$ in 3M HCl through a Dowex-1 anion-resin column. The eluate was partially evapd to give crystals which were filtered off, washed with Et_2O and stored in a desiccator over H_2SO_4 to dry. Alternatively, a satd soln of $ThCl_4$ in 6M HCl was filtered through quartz wool and extracted twice with ethyl, or isopropyl, ether (to remove iron), then evapd to a small volume on a hot plate. (Excess silica pptd, and was filtered off. The filtrate was cooled to 0° and satd with dry HCl gas.) It was shaken with an equal volume of Et_2O, agitating with HCl gas, until the mixture becomes homogeneous. On standing, $ThCl_4.8H_2O$ pptd and was filtered off, washed with Et_2O and dried [Kremer *JACS* 64 1009 *1942*].

Thorium sulphate (4H$_2$O) *[10381-37-0]* **M 496.2.** Crystd from water.

Thyroxine sodium salt (5H$_2$O) *[1491-91-4]* **M 888.9,** $[\alpha]_{546}$ **+20°** (c 2, 1M HCl + EtOH, 1:4). Crystd from absolute EtOH and dried for 8hr at 30°/1mm.

Tin (powder) *[7440-31-5]* **M 118.7.** The powder was added to about twice its weight of 10% aq NaOH and shaken vigorously for 10min. (This removed oxide film and stearic acid or similar material sometimes added for pulverisation.) It was then filtered, washed with water until the washings were no longer alkaline to litmus, rinsed with MeOH and air dried. [Sisido, Takeda and Kinugama *JACS* **83** 538 *1961*.]

Titanium tetrachloride *[7550-31-0]* **M 189.7, b 136.4°, d 1.730.** Refluxed with mercury or a small amount of pure copper turnings to remove the last traces of light colour [due to FeCl$_3$ and V(IV)Cl$_4$], then distd under N$_2$ in an all-glass system, taking precautions to exclude moisture. Clabaugh, Leslie and Gilchrist [*J Res Nat Bur Stand* **55** 261 *1955*] removed organic material by adding aluminium chloride hexahydrate as a slurry with an equal amount of water (the slurry being *ca* one-fiftieth the weight of TiCl$_4$), refluxing for 2-6hr while bubbling in chlorine, which was subsequently removed by passing a stream of clean dry air. The TiCl$_4$ was then distd, refluxed with copper and again distd, taking precautions to exclude moisture. Volatile impurities were then removed using a technique of freezing, pumping and melting.

Tiron see **1,2-dihydroxybenzene-3,5-disulphonic acid, disodium salt.**

Titanyl sulphate (TiOSO$_4$.2H$_2$O) *[1325-74-6]* **M 160.0.** Dissolved in water, filtered and crystd three times from boiling 45% H$_2$SO$_4$, washing with EtOH to remove excess acid, then with Et$_2$O. Air dried for several hours, then oven dried at 105-110°. [Hixson and Fredrickson *IEC* **37** 678 *1945*.]

Tri-*n*-butyl borate *[688-74-4]* **M 230.2, b 232.4°, n 1.4092, d 0.857.** The chief impurities are *n*-butyl alcohol and boric acid (from hydrolysis). It must be handled in a dry-box, and can readily be purified by fractional distn, under reduced pressure.

Trichloroborane see **boron trichloride.**

B-Trichloroborazine *[933-18-6]* **M 183.1, m 87°, b 88-92°/21mm.** Purified by distn from mineral oil.

Triethylaluminium *[97-93-8]* **M 114.2, b 129-131°/55mm.** Distd under vac in a 50cm column containing a heated nichrome spiral, taking the fraction 112-114°/27mm.

Triethylborane *[97-94-9]* **M 146.0, b 118.6°, n 1.378, d 0.678.** Distd at 56-57°/220mm.

Triethyl borate *[150-46-9]* **M 146.0, b 118°, n 1.378, d 0.864.** Dried with sodium, then distd.

Triethyltin hydroxide *[994-32-1]* **M 222.9.** Treated with HCl, followed by KOH, and filtered to remove diethyltin oxide [Prince *JCS* 1783 *1959*].

Tri-*n*-hexylborane *[1188-92-7]* **M 265.3.** Treated with hex-1-ene and 10% anhydrous Et$_2$O for 6hr at gentle reflux under N$_2$, then vac distd through an 18in glass helices-packed column under N$_2$ taking the fraction b 130°/2.1mm to 137°/1.5mm. The distillate still contained some di-*n*-hexylborane [Mirviss *JACS* **83** 3051 *1961*].

Trihydroxy-*n*-butylstannane see ***n*-butylstannoic acid.**

Trimethyl borate *[121-43-7]* **M 103.9, b 65°, n 1.359, d 0.933.** Dried with sodium, then distd.

Triphenylantimony *[603-36-1]* **M 353.1, m 52-54°.** Recrystd from acetonitrile [Hayes *et al, JACS* **107** 1346 *1985*].

Triphenylarsine *[603-32-7]* **M 306.2, m 60-62°.** Recrystd from EtOH or aq EtOH [Dahlinger *et al, JCSDT* 2145 *1986*; Boert *et al, JACS* **109** 7781 *1987*].

Triphenylborane *[960-71-4]* **M 242.1. m 142-142.5°, b 203°/15mm.** Crystd three times from benzene under N_2.

Triphenylchlorosilane *[76-86-8]* **M 294.9, 97-99°.** Likely impurities are tetraphenylsilane, small amounts of hexaphenyldisiloxane and traces of triphenylsilanol. Purified by dtstn at 2mm, then crystd from EtOH-free $CHCl_3$, and from pet ether (b 30-60°) by cooling in a Dry-ice/acetone bath.

Triphenylchlorostannane *[639-58-7]* **M 385.5, m 104°.** Crystd repeatedly from pet ether (b 30-60°) or EtOH, then sublimed in vac.

Triphenylsilanol *[791-31-1]* **M 276.4, m 151-153°.** Crystd from benzene or Et_2O-pet ether (1:1).

Triphenyltin hydroxide *[76-87-9]* **M 367.0.** West, Baney and Powell [*JACS* **82** 6269 *1960*] purified a sample which was grossly contaminated with tetraphenyltin and diphenyltin oxide by dissolving it in EtOH, most of the impurities remaining behind as an insoluble residue. Evapn of the EtOH gave the crude hydroxide which was converted to triphenyltin chloride by grinding in a mortar under 12M HCl, then evaporating the acid soln. The chloride, after crystn from EtOH, had m 104-105°. It was dissolved in Et_2O and converted to the hydroxide by stirring with excess aq ammonia. The ether layer was separated, dried, and evapd to give triphenyltin hydroxide which, after crystn from EtOH and drying under vac, was in the form of white crystals (m 119-120°), which retained some cloudiness in the melt above 120°. The hydroxide retains water (0.1-0.5 moles of water per mole) tenaciously.

Tri-*n*-propyl borate *[688-71-1]* **M 140.1.** Dried with sodium and then distd.

Tris(2,2'-bipyridine)ruthenium(II) dichloride ($6H_2O$) *[14323-06-9]* **M 748.6.** Recrystd from water then from MeOH [Ikezawa *et al, JACS* **108** 1589 *1986*].

Trisodium citrate ($2H_2O$) *[68-04-2]* **M 294.1.** Crystd from warm water by cooling to 0°.

Trisodium 8-hydroxy-1,3,6-pyrenetrisulphonate *[6358-69-6]* **M 488.8.** Purified by chromatography with an alumina column, and eluted with propan-1-ol-water (3:1, v/v). Recrystd from aq. acetone (5:95, v/v) using decolorising charcoal.

Trisodium 1,3,6-naphthalenetrisulphonate *[5182-30-9]* **M 434.2.** The free acid was obtained by passage through an ion-exchnge column and converted to the lanthanum salt by treatment with La_2O_3. This salt was crystd twice from hot water. [The much lower solubility of $La_2(SO_4)_3$ and its retrograde temperature dependence allows a good separation from sulphate impurity.] The lanthanum salt was then passed through an appropriate ion-exchange column to obtain the free acid, the sodium or potassium salt. (The sodium salt is hygroscopic.) [Atkinson, Yokoi and Hallada *JACS* **83** 1570 *1961*.] Also recrystd from aq acetone [Okahata *et al, JACS* **108** 2863 *1986*].

Trisodium orthophosphate ($12H_2O$) *[10101-89-0]* **M 380.1.** Crystd from warm dil aq NaOH (1ml/g) by cooling to 0°.

Tris(2,4-pentandionate)aluminium *[13963-57-0]* **M 323.3, m 194°.** Recrystd twice from benzene.

Tritium *[10028-17-8]* **M 6.0.** Purified from hydrocarbons and ^{3}He by diffusion through the wall of a hot nickel tube [Landecker and Gray *Rev Sci Inst* **25** 1151 *1954*]. **RADIOACTIVE.**

Tungsten (rod) *[7440-33-7]* **M 183.6.** Cleaned with conc NaOH soln, rubbed with very fine emery paper until its surface was bright, washed with previously boiled and cooled conductivity water and dried with filter paper.

Tungsten hexacarbonyl *[14040-11-0]* **M 351.9, d 2.650.** Sublimed *in vacuo* before use [Connoe *et al, JCSDT* 511 *1986*].

Uranium hexafluoride *[7783-81-5]* **M 352.0, b 0°/17.4mm, 56.2°/765mm, m 64.8°.**
Purified by fractional distn to remove HF. Also purified by low temperature trap-to-trap distn over pre-dried NaF [Anderson and Winfield *JCSDT* 337 *1986*].

Uranium trioxide *[1344-58-7]* **M 286.0.** The oxide was dissolved in $HClO_4$ (to give a uranium content of 5%), and the soln was adjusted to $pH2$ by addition of dil ammonia. Dropwise addition of 30% H_2O_2, with rapid stirring, pptd U(VI) peroxide, the pH being held constant during the pptn, by addition of small amounts of the ammonia soln. (The H_2O_2 was added until further quantities caused no change in pH.) After stirring for 1hr, the slurry was filtered through coarse filter paper in a Büchner funnel and washed with 1% H_2O_2 acidified to pH 2 with $HClO_4$, then heated at 350° for three days in a large platinum dish [Baes *JPC* 60 878 *1956*].

Uranyl nitrate (6H$_2$O) *[10102-06-4]* **M 502.1, m 60.2°, b 118°.** Crystd from water by cooling to -5°, taking only the middle fraction of the solid which separated. Dried as the hexahydrate over 35-40% H_2SO_4 in a vac desiccator.

Vanadium (metal) *[7440-62-2]* **M 50.9.** Cleaned by rapid exposure consecutively to HNO_3, HCl, HF, de-ionised water and reagent grade acetone, then dried in a vac desiccator.

Vanadyl acetylacetonate *[3153-26-2]* **M 265.2, m 256-259°.** Crystd from acetone.

Water *[7732-18-5]* **M 18.0, b 100°.** Conductivity water (specific conductance *ca* 10^{-7} mho) can be obtained by distilling water in a steam-heated tin-lined still, then, after adding 0.25% of solid NaOH and 0.05% of $KMnO_4$, distilling once more from an electrically heated Barnstead-type still, taking the middle fractn into a Jena glass bottle. During these operations suitable traps must be used to protect against entry of CO_2 and NH_3. Water only a little less satisfactory for conductivity measurements (but containing traces of organic material) can be obtained by passing ordinary distd water through a mixed bed ion-exchange column containing, for example, Amberlite resins IR 120 (cation exchange) and IRA 400 (anion exchange), or Amberlite MB-1. This treatment is also a convenient one for removing traces of heavy metals. (The metals Cu, Zn, Pb, Cd and Hg can be tested for by adding pure conc ammonia to 10ml of sample and shaking vigorously with 1.2ml 0.001% dithizone in CCl_4. Less than 0.1μg of metal ion will impart a faint colour to the CCl_4 layer.) For almost all laboratory purposes, simple distn yields water of adequate purity, and most of the volatile contaminants such as ammonia and CO_2 are removed if the first fraction of distillate is discarded.

Xylenol Orange (sodium salt) *[1611-35-4]*. See entry in Chapter 3.

Zinc (dust) *[7440-66-6]* **M 65.4.** Commercial zinc dust (1.2Kg) was stirred with 2% HCl (3L) for 1min, (then the acid was removed by filtratn), and washed in a 4L beaker with a 3L portion of 2% HCl, three 1L portions of distd water, two 2L portions of 95% EtOH, and finally with 2L of absolute Et_2O. (The wash solns were removed each time by filtn.) The material was then dried thoroughly and if necessary, any lumps were broken up in a mortar.

Zinc (metal) *[7440-66-6]* **M 65.4, m 420°, d 7.141.** Fused under vac, cooled, then washed with acid to remove the oxide.

Zinc acetate ($2H_2O$) *[5970-45-6]* **M 219.5.** Crystd (in poor yield) from hot water or, better, from EtOH.

Zinc acetonylacetate *[14024-63-6]* **M 263.6.** Crystd from hot 95% EtOH.

Zinc bromide *[7699-45-8]* **M 225.2.** Heated to 300° under vac. (2×10^{-2}mm) for 1hr, then sublimed.

Zinc caprylate *[557-09-5]* **M 351.8.** Crystd from EtOH.

Zinc chloride *[7646-85-7]* **M 136.3, m 283°.** The anhydrous material can be sublimed under a stream of dry HCl, followed by heating to 400° in a stream of dry N_2. Also purified by refluxing (50g) in dioxane (400ml) with 5g zinc dust, filtering hot and cooling to ppte $ZnCl_2$. Crystd from dioxane and stored in a desiccator over P_2O_5. It has also been dried by refluxing in thionyl chloride. [Weberg *et al*, *JACS* **108** 6242 *1986*]. Hygroscopic: *minimal exposure to the atmosphere is necessary*.

Zinc diethyldithiocarbamate *[14324-55-1]* **M 561.7,**
Zinc dimethyldithiocarbamate *[137-30-4]* **M 305.8, m 248-250°,**
Zinc ethylenebis[dithiocarbamate] *[12122-67-7]* **M 249.7.**
Crystd several times from hot toluene or from hot $CHCl_3$ by addition of EtOH.

Zinc formate ($2H_2O$) *[557-41-5]* **M 191.4.** Crystd from water (3ml/g).

Zinc iodide *[10139-47-6]* **M 319.2.** Heated to 300° under vac (2×10^{-2}mm) for 1hr, then sublimed.

Zinc *RS*-lactate ($3H_2O$) *[554-05-2]* **M 297.5.** Crystd from water (6ml/g).

Zincon *[135-52-4]* **M 462.4.** Main impurities are inorganic salts which can be removed by treatment with dil acetic acid. Organic contaminants are removed by refluxing with ether.

Zinc perchlorate ($6H_2O$) *[13637-61-1]* **M 372.4, m 105-107°.** Crystd from water.

Zinc phenol-*o*-sulphonate ($8H_2O$) *[127-82-2]* **M 555.8.** Crystd from warm water by cooling to 0°.

Zinc phthalocyanine *[14320-04-8]* **M 580.9.** Purified by repeated sublimation in a flow of oxygen-free N_2.

Zinc sulphate ($7H_2O$) *[7446-20-0]* **M 287.5.** Crystd from aq EtOH.

Zinc 5,10,15,20-tetraphenylporphyrin *[14074-80-7]* **M 678.1,** λ_{max} **418(556)nm.** Purified by chromatography on neutral (Grade I) alumina, followed by recrystn from CH_2Cl_2/MeOH [Yamashita *et al*, *JPC* **91** 3055 *1987*].

Zirconium tetrachloride *[10026-11-6]* **M 233.0, m 300°(sublimes).** Crystd repeatedly from conc HCl.

Zirconyl chloride (6H$_2$O) *[7699-43-6]* **M 286.2.** Crystd repeatedly from 8M HCl as ZrOCl$_2$.8H$_2$O. (The product was not free from hafnium.)

Zirconyl chloride (8H$_2$O) *[13520-92-8]* **M 322.3.** Recrystd several times from water [Ferragina *et al*, *JCSDT 265 1986*].

CHAPTER 5

GENERAL METHODS FOR THE PURIFICATION OF CLASSES OF COMPOUNDS

Chapters 3 and 4 list a large number of individual compounds, with a brief indication of how each one may be purified. For substances that are not included in these chapters the following procedures may prove helpful.

If the laboratory worker does not know of a reference to the preparation of a commercially available substance, he may be able to make a reasonable guess at the method used from published laboratory syntheses. This information, in turn, can simplify the necessary purification steps by suggesting probable contaminants.

Impurities that amount to a few per cent can be observed by measuring some of the spectroscopic and physical properties. Two volumes on the NMR spectra [C.J.Pouchert, **The Aldrich Library of NMR Spectra, Vols 1 and 2**, 2nd ed, Aldrich Chemical Co,. Inc, Milwaukee, Wl, 1983], and one on the infrared spectra [C.J.Pouchert, **The Aldrich Library of FT-IR Spectra**, 2nd ed, Aldrich Chemical Co., Milwaukee, Wl, 1985], and computer software [**FT-IR Peak-search Data Base and Software**, for Apple IIE, IIC and II Plus computers; and for IBM PC computers, Nicholet Instruments, Madison, Wl, 1984] contain data for all the compounds in the Aldrich catalogue and are extremely useful for identifying compounds and impurities. They are also very useful, because of the vast amount of data, to deduce the spectra of new compounds and to identify them if they are present as impurities. A combination of UV, IR and NMR spectra, colour, liquid, gas and tlc chromatographic properties and spot tests can give one a very good idea of what impurities are present. Purification methods can then be devised to remove these impurities, and a monitoring method will have already been established.

Physical methods of purification depend largely on the melting and boiling points of the materials.. For gases and low-boiling liquids use is commonly made of the *freeze-pump-thaw* (see p. 19) procedure. Gas chromatography is also useful, especially for low-boiling point liquids. Liquids are usually purified by refluxing with drying agents, acids or bases, reducing agents, charcoal, etc., followed by fractional distillation under reduced pressure. For solids, general methods include fractional freezing of the melted material, taking the middle fraction. A related procedure is zone refining. Another procedure is sublimation of the solid under reduced pressure. The other commonly used method for purifying solids is by crystallisation from a solution in a suitable solvent, by cooling with or without the prior addition of a solvent in which the solute is not very soluble.

Purification becomes meaningful only insofar as adequate tests of purity are applied: the higher the degree of purity that is sought, the more stringent must these tests be. If the material is an organic solid, its melting point should first be taken and compared with the recorded value. Also, as part of this preliminary examination, the sample might be examined by thin layer (or paper) chromatography (see E.Demole, **Chromatographic Reviews**, 4 26 *1962*) in several different solvent systems and in high enough concentrations to facilitate the detection of minor components. On the other hand, if the substance is a liquid, its boiling point should be measured. If, further, it is a high boiling liquid, its chromatographic behaviour should be examined. Liquids, especially volatile ones, can be studied very satisfactorily by gas chromatography, preferably using at least two different stationary phases.

Application of these tests at successive steps will give a good indication of whether or not the purification is satisfactory and will also show when adequate purification has been achieved.

The nature of the procedure will depend to a large extent on the quantity of purified material that is required. For example, for small quantities (50-250mg) of a pure liquid, preparative gas chromatography is probably the best method. Two passes through a suitable column may well be sufficient. Similarly, for small amounts (100-500mg) of an organic solid, column chromatography is likely to be very satisfactory., the eluate being collected as a number of separate fractions (*ca* 5-10ml) which are examined by UV, FT-IR or NMR spectroscopy, tlc or by some other appropriate analytical technique. (For information on suitable adsorbents and eluents the texts referred to in the bibliography at the end of Chapters 1 and 2 should be consulted.) Preparative thin layer chromatography or HPLC can be used successfully for purifying up to 500mg of solid.

Where larger quantities (upwards of 1g) are required, most of the impurities should be removed by preliminary treatments, such as solvent extraction, liquid-liquid partition, or conversion to a derivative (preferably solid, see Chapter 2) which can be purified by crystallisation or fractional distillation before being reconverted to the starting material. The substance is then crystallised or distilled. If the final amounts must be in excess of 25g, preparation of a derivative is sometimes omitted because of the cost involved. In all of the above cases, purification is likely to be more laborious if the impurity is an isomer or a derivative with closely similar physical properties.

In the general methods of purification described below, it is assumed that the impurities belong essentially to a class of compounds different from the one being purified. They are suggested for use in cases where substances are not listed in Chapters 3 and 4. In such cases, the experimenter is advised to employ them in conjunction with information given in these chapters for the purification of suitable analogues. Also, for a wider range of drying agents, solvents for extraction and solvents for recrystallisation, the reader is referred to Chapter 1.

CLASSES OF COMPOUNDS

Acetals. These are generally diethyl or dimethyl acetal derivatives of aldehydes. They are more stable to alkali than to acids. Their common impurities are the corresponding alcohol, aldehyde and water. Drying with sodium wire removes alcohols and water, and polymerizes aldehydes so that, after decantation, the acetal can be fractionally distilled. In cases where the use of sodium is too drastic, aldehydes can be removed by shaking with alkaline hydrogen peroxide solution and the acetal is dried with sodium carbonate or potassium carbonate. Residual water and alcohols (up to *n*-propyl) can be removed with Linde type 4A molecular sieves. The acetal is then filtered and fractionally distilled. Solid acetals (i.e. acetals of high molecular weight aldehydes) are generally low-melting and can be recrystallised from low-boiling petroleum ether, toluene or a mixture of both.

Acids. (a) Carboxylic: Liquid carboxylic acids are first freed from neutral and basic impurities by dissolving them in aqueous alkali and extracting with ethyl ether. (The *p*H of the solution should be at least three units above the pK_a of the acid). The aqueous phase is then acidified to a *p*H at least three units below the pK_a of the acid and again extracted with ether. The extract is dried with magnesium sulphate or sodium sulphate and the ether is distilled off. The acid is fractionally distilled through an efficient column. It can be further purified by conversion to its methyl or ethyl ester (see p. 62) which is then fractionally distilled. Hydrolysis yields the original acid which is again purified as above.
Acids that are solids can be purified in this way, except that distillation is replaced by repeated crystallisation (preferable from at least two different solvents such as water, alcohol or aqueous alcohol, toluene, toluene/petroleum ether or acetic acid.) Water-insoluble acids can be partially purified by dissolution in *N* sodium hydroxide solution and precipitation with dilute mineral acid. If the acid is required to be free from sodium ions, then it is better to dissolve the acid in hot *N* ammonia, heat to *ca* 80°, adding slightly more than an equal volume of *N* formic acid and allowing to cool slowly for crystallisation.

The separation and purification of naturally occurring fatty acids, based on distillation, salt solubility and low temperature crystallisation, are described by K.S.Markley (ed), **Fatty Acids**, 2nd edn, part 3, Chap. 20, Interscience, New York. 1964.
Aromatic carboxylic acids can be purified by conversion to their sodium salts, recrystallisation from hot water, and reconversion to the free acids.

(b) Sulphonic: The low solubility of sulphonic acids in organic solvents and their high solubility in water makes necessary a treatment different from that for carboxylic acids. Sulphonic acids are strong, they have the tendency to hydrate, and many of them contain water of crystallisation. The lower-melting and liquid acids can generally be purified with only slight decomposition by fractional distillation, preferably under reduced pressure. A common impurity is sulphuric acid, but this can be removed by recrystallisation from concentrated aqueous solutions. The wet acid can be dried by azeotropic removal of water with toluene, followed by distillation. The higher-melting acids, or acids that melt with decomposition, can be recrystallised from water or, occasionally, from ethanol.

(c) Sulphinic: These acids are less stable, less soluble and less acidic than the corresponding sulphonic acids. The common impurities are the respective sulphonyl chlorides from which they have been prepared, and the thiolsulphonates (neutral) and sulphonic acids into which they decompose. The first two of these can be removed by solvent extraction from an alkaline solution of the acid. On acidification of an alkaline solution, the sulphinic acid crystallises out leaving the sulphonic acid behind. The lower molecular weight members are isolated as their metal (e.g. ferric) salts, but the higher members can be crystallised from water (made slightly acidic), or alcohol.

Acid chlorides. The corresponding acid and hydrogen chloride are the most likely impurities. Usually these can be removed by efficient fractional distillation. Where acid chlorides are not readily hydrolysed (e.g. aryl sulphonyl chlorides) the compound can be freed from contaminants by dissolving in a suitable solvent such as alcohol-free chloroform, dry toluene or petroleum ether and shaking with dilute sodium bicarbonate solution. The organic phase is then washed with water, dried with sodium sulphate or magnesium sulphate, and distilled. This procedure is hazardous with readily hydrolysable acid chlorides such as acetyl chloride and benzoyl chloride. Solid acid chlorides are satisfactorily crystallised from toluene, toluene-petroleum ether, petroleum ethers, alcohol-free chloroform/toluene, and, occasionally, from dry ethyl ether. Hydroxylic or basic solvents should be strictly avoided. *All operations should be carried out in a fume cupboard because of the irritant nature of these compounds.*

Alcohols. (a) Monohydric: The common impurities in alcohols are aldehydes or ketones, and water. [*Ethanol* in Chapter 3 is typical.] Aldehydes and ketones can be removed by adding a small amount of sodium metal and refluxing for 2 hours, followed by distillation. Water can be removed in a similar way but it is preferable to use magnesium metal instead of sodium because it forms a more insoluble hydroxide, thereby shifting the equilibrium more completely from metal alkoxide to metal hydroxide. The magnesium should be activated with iodine (or a small amount of methyl iodide), and the water content should be low, otherwise the magnesium will be deactivated.
Acidic materials can be removed by treatment with anhydrous Na_2CO_3, followed by a suitable drying agent, such as calcium hydride, and fractional distillation, using gas chromatography to establish the purity of the product [Ballinger and Long, *JACS* **82** 795 *1960*]. Alternatively, the alcohol can be refluxed with freshly ignited CaO for 4 hours and then fractionally distilled [McCurdy and Laidler, *Canad J Chem* **41** 1867 *1963*].
With higher-boiling alcohols it is advantageous to add some freshly prepared magnesium ethoxide solution (only slightly more than required to remove the water), followed by fractional distillation. Alternatively, in such cases, water can be removed by azeotropic distillation with toluene. Higher-melting alcohols can be purified by crystallisation from methanol or ethanol, toluene/petroleum ether or petroleum ethers. Sublimation in vacuum, molecular distillation and gas chromatography are also useful means of purification. For purification *via* derivatives, see p. 60.

(b) Polyhydric: These alcohols are more soluble in water than are the monohydric ones. Liquids can be freed from water by shaking with type 4A Linde molecular sieves and can safely be distilled only under high vacuum. Carbohydrate alcohols can be crystallised from strong aqueous solution or, preferably, from mixed solvents such as ethanol/petroleum ether or dimethyl formamide/toluene. Crystallisation usually requires seeding and is extremely slow. Further purification can be effected by conversion to the acetyl derivatives which are much less soluble in water and which can readily be recrystallised, e.g. from ethanol. Hydrolysis of the acetyl derivatives, followed by removal of acetate and metal ions by ion-exchange chromatography, gives the purified material. On no account should solutions of carbohydrates be concentrated above 40° because of darkening and formation of *caramel*. Ion exchange, charcoal or cellulose column chromatography has been used for the purification and separation of carbohydrates [see Heftman p. 54].

Aldehydes. Common impurities found in aldehydes are the corresponding alcohols, aldols and water from self-condensation, and the corresponding acids formed by autoxidation. Acids can be removed by shaking with aqueous 10% sodium bicarbonate solution. The organic liquid is then washed with water. It is dried with sodium sulphate or magnesium sulphate and then fractionally distilled. Water soluble aldehydes must be dissolved in a suitable solvent such as ethyl ether before being washed in this way. Further purification can be effected *via* the bisulphite derivative (see p. 60) or the Schiff base formed with aniline or benzidine. Solid aldehydes can be dissolved in ethyl ether and purified as above. Alternatively, they can be steam distilled, then sublimed and crystallised from toluene or petroleum ether.

Amides. Amides are stable compounds. The lower-melting members (such as acetamide) can be readily purified by fractional distillation. Most amides are solids which have low solubilities in water. They can be recrystallised from large quantities of water, ethanol, ethanol/ether, aqueous ethanol, chloroform/toluene, chloroform or acetic acid. The likely impurities are the parent acids or the alkyl esters from which they have been made. The former can be removed by thorough washing with aqueous ammonia followed by recrystallisation, whereas elimination of the latter is by trituration or recrystallisation from an organic solvent. Amides can be freed from solvent or water by drying below their melting points. These purifications can also be used for sulphonamides and acid hydrazides.

Amines. The common impurities found in amines are nitro compounds (if prepared by reduction), the corresponding halides (if prepared from them) and the corresponding carbamate salts. Amines are dissolved in aqueous acid, the *p*H of the solution being at least three units below the *p*K$_a$ value of the base to ensure almost complete formation of the cation. They are extracted with ethyl ether to remove neutral impurities and to decompose the carbamate salts. The solution is then made strongly alkaline and the amines that separate are extracted into a suitable solvent (ether or toluene) or steam distilled. The latter process removes coloured impurities. Note that chloroform cannot be used as a solvent for primary amines because, in the presence of alkali, poisonous carbylamines are formed. However, chloroform is a useful solvent for the extraction of heterocyclic bases. In this case it has the added advantage that while the extract is being freed from the chloroform most of the moisture is removed with the solvent.
Alternatively, the amine may be dissolved in a suitable solvent (e.g. toluene) and dry HCl gas is passed through the solution to precipitate the amine hydrochloride. This is purified by recrystallisation from a suitable solvent mixture (e.g. ethanol/ethyl ether). The free amine can be regenerated by adding sodium hydroxide and isolated as above.
Liquid amines can be further purified *via* their acetyl or benzoyl derivatives (see p. 62). Solid amines can be recrystallised from water, alcohol, toluene or toluene-petroleum ether. *Care should be taken in handling large quantities of amines because their vapours are harmful and they are readily absorbed through the skin.*

Amino acids. Because of their zwitterionic nature, amino acids are soluble in water. Their solubility in organic solvents rises as the fat-soluble portion of the molecule increases. The likeliest impurities are traces of salts, heavy metal ions, proteins and other amino acids. Purification of these is usually easy, by recrystallisation from water or ethanol/water mixtures. The amino acid is dissolved in the boiling solvent, decolorised if necessary by boiling with 1g of acid-washed charcoal/100g amino acid, then filtered hot, chilled, and stood for several hours to crystallise. The crystals are filtered off, washed with ethanol, then ether, and dried.

Amino acids have high melting or decomposition points and are best examined for purity by paper or thin layer chromatography. The spots are developed with ninhydrin (see Lederer and Lederer, p. 54). Customary methods for the purification of small quantities of amino acids obtained from natural sources (i.e. 1-5g) are ion-exchange chromatography (see p. 23) or countercurrent distribution (see p. 32). For general treatment of amino acids see Greenstein and Winitz [**The Amino Acids**, Vols 1-3, J.Wiley & Sons, New York 1961].

A useful source of details such as likely impurities, stability and tests for homogeneity of amino acids is **Specifications and Criteria for Biochemical Compounds**, 3rd edn, 1972, National Academy of Sciences, USA].

Anhydrides.   The corresponding acids, resulting from hydrolysis, are the most likely impurities. Distillation from phosphorus pentoxide, followed by fractional distillation, is usually satisfactory. With high boiling or solid anhydrides, another method involves refluxing for ½-1 hr with acetic anhydride, followed by fractional distillation. Acetic acid distils first, then acetic anhydride and finally the desired anhydride. Where the anhydride is a solid, removal of acetic acid and acetic anhydride at atmospheric pressure is followed by heating under vacuum. The solid anhydride is then either crystallised as for acid chlorides or (in some cases) sublimed in a vacuum. A preliminary purification when large quantities of acid are present in a solid anhydride (such as phthalic anhydride) can sometimes be achieved by preferential solvent extraction of the (usually) more soluble anhydride from the acid (e.g. with chloroform in the case of phthalic anhydride). *All operations with liquid anhydrides should be carried out in a fume cupboard because of their lacrimatory properties.*

Carotenoids.   These usually are decomposed by light, air and solvents, so that degradation products are probable impurities. Chromatography and adsorption spectra permit the ready detection of coloured impurities, and separations are possible using solvent distribution, chromatography or crystallisation. Thus, in partition between immiscible solvents, xanthophyll remains in 90% methanol while carotenes pass into the petroleum ether phase.   For small amounts of material, thin-layer or paper chromatography may be used, while column chromatography is suitable for larger amounts. Colourless impurities may be detected by IR, NMR or mass spectrometry.   The more common separation procedures are described by P.Karrer and E.Jucker in **Carotenoids**, E.A.Braude (translator), Elsevier, New York, 1950.

Purity can be assayed by chromatography (on thin-layer plates, Kieselguhr paper or columns), by UV or NMR procedures.

Esters.   The most common impurities are the corresponding acid and hydroxy compound (i.e. alcohol or phenol), and water. A liquid ester from a carboxylic acid is washed with $2N$ sodium carbonate or sodium hydroxide to remove acid material, then shaken with calcium chloride to remove ethyl or methyl alcohols (if it is a methyl or ethyl ester). It is dried with potassium carbonate or magnesium sulphate, and distilled. Fractional distillation then removes residual traces of hydroxy compounds. This method does not apply to esters of inorganic acids (e.g. dimethyl sulphate) which are more readily hydrolysed in aqueous solution when heat is generated in the neutralisation of the excess acid. In such cases, several fractional distillations, preferably under vacuum, are usually sufficient.

Solid esters are easily crystallisable materials. It is important to note that esters of alcohols must be recrystallised either from non-hydroxylic solvents (e.g. toluene) or from the alcohol from which the ester is derived. Thus methyl esters should be crystallised from methanol or methanol/toluene, but not from ethanol, n-butanol or other alcohols, in order to avoid alcohol exchange and contamination of the ester with a second ester. Useful solvents for crystallisation are the corresponding alcohols or aqueous alcohols, toluene, toluene/petroleum ether, and chloroform (ethanol-free)/toluene. Carboxylic acid esters derived from phenols are more difficult to hydrolyse and exchange, hence any alcoholic solvent can be used freely. Sulphonic acid esters of phenols are even more resistant to hydrolysis: they can safely be crystallised not only from the above solvents but also from acetic acid, aqueous acetic acid or boiling n-butanol.

Fully esterified phosphoric acid and phosphonic acids differ only in detail from the above mentioned esters. Their major contaminants are alcohols or phenols, phosphoric or phosphonic acids (from hydrolysis), and (occasionally) basic material, such as pyridine, which is used in their manufacture. Water-insoluble esters are washed thoroughly and successively with dilute acid (e.g. 0.2N sulphuric acid), water, 0.2N sodium hydroxide and water. After drying with calcium chloride they are fractionally distilled. Water-soluble esters should first be dissolved in a suitable organic solvent and, in the washing process, water should be replaced by saturated aqueous sodium chloride. Some esters (e.g. phosphate and phosphonate esters) can be further purified through their uranyl adducts (see p. 63). Traces of water or hydroxy compounds can be removed by percolation through, or shaking with, activated alumina (about 100g/L of liquid solution), followed by filtration and fractional distillation in a vacuum. For high molecular weight esters (which cannot be distilled without some decomposition) it is advisable to carry out distillation at as low a pressure as possible. Solid esters can be crystallised from toluene or petroleum ether. Alcohols can be used for recrystallising phosphoric or phosphonic esters of phenols.

Ethers. The purification of ethyl ether (see Chapter 3) is typical of liquid ethers. The most common contaminants are the alcohols or hydroxy compounds from which the ethers are prepared, their oxidation products (e.g. aldehydes), peroxides and water. Peroxides, aldehydes and alcohols can be removed by shaking with alkaline potassium permanganate solution for several hours, followed by washing with water, concentrated sulphuric acid, then water. After drying with calcium chloride, the ether is distilled. It is then dried with sodium or with lithium aluminium hydride, redistilled and given a final fractional distillation. The drying process should be repeated if necessary.
Alternatively, methods for removing peroxides include leaving the ether to stand in contact with iron filings or copper powder, shaking with a solution of ferrous sulphate acidified with sulphuric acid, shaking with a copper-zinc couple [Fierz-David, **Chimia 1** 246 *1947*], passage through a column of activated alumina, and refluxing with phenothiazine. Cerium(III) hydroxide has also been used [Ramsey and Aldridge, *JACS* **77** 2561 *1955*].
A simple test for ether peroxides is to add 10ml of the ether to a stoppered cylinder containing 1ml of freshly prepared 10% solution of potassium iodide containing a drop of starch indicator. No colour should develop during one minute. Alternatively, a 1% solution of ferrous ammonium sulphate, 0.1M in sulphuric acid and 0.01M in potassium thiocyanate should not increase appreciably in red colour when shaken with two volumes of the ether.
As a safety precaution against **EXPLOSION** (in case the purification has been insufficiently thorough) at least a quarter of the total volume of ether should remain in the distilling flask when the distillation is discontinued. To minimize peroxide formation, ethers should be stored in dark bottles and, if they are liquids, they should be left in contact with type 4A Linde molecular sieves, in a cold place, over sodium amalgam. The rate of formation of peroxides depends on storage conditions and is accelerated by heat, light, air and moisture. The formation of peroxides is inhibited in the presence of diphenylamine, di-*tert*-butylphenol, or other antioxidant as stabilizer.
Ethers that are solids (e.g. phenyl ethers) can be steam distilled from an alkaline solution which will hold back any phenolic impurity. After the distillate is made alkaline with sodium carbonate, the insoluble ether is collected either by extraction (e.g. with chloroform, ethyl ether or toluene) or by filtration. It is then crystallised from alcohols, alcohol/petroleum ether, petroleum ether, toluene or mixtures of these solvents, sublimed in a vacuum and recrystallised.

Halides. Aliphatic halides are likely to be contaminated with halogen acids and the alcohols from which they have been prepared, whereas in aromatic halides the impurities are usually aromatic hydrocarbons, amines or phenols. In both groups the halogen atom is less reactive than it is in acid chlorides. Purification is by shaking with concentrated hydrochloric acid, followed by washing successively with water, 5% sodium carbonate or bicarbonate, and water. After drying with calcium chloride, the halide is distilled and then fractionally distilled using an efficient column. For a solid halide the above purification is carried out by dissolving it in a suitable solvent such as toluene. Solid halides can also be purified by chromatography using an alumina column and eluting with toluene or petroleum ether. They can be crystallised from toluene, petroleum ethers, toluene/petroleum ether or toluene/chloroform/petroleum ether. Care should be taken when handling organic halogen compounds because of their **TOXICITY.**

Liquid aliphatic halides are obtained alcohol-free by distillation from phosphorus pentoxide. They are stored in dark bottles to prevent oxidation and, in some cases, the formation of phosgene.

A general method for purifying *chlorohydrocarbons* uses repeated shaking with concentrated sulphuric acid until no further colour develops in the acid, then washing with a solution of sodium bicarbonate, followed by water. After drying with calcium chloride, the chlorohydrocarbon is fractionally distilled. Finally, it is fractionally crystallised to constant boiling point [Barton and Howlett *JCS* 155 *1949*].

Hydrocarbons. Gaseous hydrocarbons are best freed from water and gaseous impurities by passage through suitable adsorbents and (if olefinic material is to be removed) oxidants such as alkaline potassium permanganate solution, followed by fractional cooling (see p. 42 for cooling baths) and fractional distillation at low temperature. To effect these purifications and also to store the gaseous sample, a vacuum line is necessary.

Impurities in hydrocarbons can be characterised and evaluated by gas chromatography and mass spectrometry. The total amount of impurities present can be estimated from the thermometric freezing curve.

Liquid aliphatic hydrocarbons are freed from aromatic impurities by shaking with concentrated sulphuric acid whereby the aromatic compounds are sulphonated. Shaking is carried out until the sulphuric acid layer remains colourless for several hours. The hydrocarbon is then freed from the sulphuric acid and the sulphonic acids by separating the two phases and washing the organic layer successively with water, 2N sodium hydroxide, and water. It is dried with $CaCl_2$ or Na_2SO_4, and then distilled. The distillate is dried with sodium wire, P_2O_5, or metallic hydrides, or passage through a dry silica gel column, or preferably, and more safely, with molecular sieves (see p. 32) before being finally fractionally distilled through an efficient column. If the hydrocarbon is contaminated with olefinic impurities, shaking with aqueous alkaline permanganate is necessary prior to the above purification. Alicyclic and paraffinic hydrocarbons can be freed from water, non-hydrocarbon and aromatic impurities by passage through a silica gel column before the final fractional distillation. This may also remove isomers. (For the use of a chromatographic method to separate mixtures of aromatic, paraffinic and alicyclic hydrocarbons, see Meir, *J Res Nat Bur Stand* 34 453 *1945*). Another method of removing branched-chain and unsaturated hydrocarbons from straight-chain hydrocarbons depends on the much faster reaction of the former with chlorosulphonic acid.

Isomeric materials which have closely similar physical properties can be serious contaminants in hydrocarbons. With aromatic hydrocarbons, e.g. xylenes and alkyl benzenes, advantage is taken of differences in ease of sulphonation. If the required compound is sulphonated more readily, the sulphonic acid is isolated, crystallised (e.g. from water), and decomposed by passing superheated steam through the flask containing the acid. The sulphonic acid undergoes hydrolysis and the liberated hydrocarbon distils with the steam. It is separated from the distillate, dried, distilled and then fractionally distilled. For small quantities (10-100mg), vapour phase chromatography is the most satisfactory method for obtaining a pure sample (for column packings see pp. 28, 51).

Azeotropic distillation with methanol or 2-ethoxyethanol has been used to obtain highly purified saturated hydrocarbons and aromatic hydrocarbons such as xylenes and isopropylbenzenes.

Carbonyl-containing impurities can be removed from hydrocarbons (and other oxygen-lacking solvents such as $CHCl_3$ and CCl_4) by passage through a column of Celite 545 (100g) mixed with concentrated sulphuric acid (60ml). After first adding some solvent and about 10g of granular Na_2SO_4, the column is packed with the mixture and a final 7-8cm of Na_2SO_4 is added at the top [Hornstein and Crowe, *Anal.Chem.* 34 1037 *1962*]. Alternatively, Celite impregnated with 2,4-dinitrophenylhydrazine can be used.

With solid hydrocarbons such as naphthalene, preliminary purification by sublimation in vacuum (or high vacuum if the substance is high melting), is followed by zone refining and finally by chromatography (e.g. on alumina) using low-boiling liquid hydrocarbon eluents. These solids can be recrystallised from alcohols, alcohol/petroleum ether or from liquid hydrocarbons (e.g. toluene) and dried below their melting points. Aromatic hydrocarbons that have been purified by zone melting include anthracene, biphenyl, fluoranthrene, naphthalene, perylene, phenanthrene, pyrene and terphenyl, among others.

Olefinic hydrocarbons have a very strong tendency to polymerise and commercially available materials are generally stabilized, e.g. with hydroquinone. When distilling compounds such as vinylpyridine or styrene, the stabilizer remains behind and the purified olefinic material is more prone to polymerization. The most common impurities are higher-boiling dimeric or polymeric compounds. Vacuum distillation in a nitrogen atmosphere not only separates monomeric from polymeric materials but in some cases also depolymerizes the impurities. The distillation flask should be charged with a polymerization inhibitor and the purified material should be used immediately or stored in the dark and mixed with a small amount of stabilizer (e.g. 0.1% of hydroquinone).

Imides. Imides (e.g. phthalimide) can be purified by conversion to their potassium salts by reaction in ethanol with ethanolic potassium hydroxide. The imides are regenerated when the salts are hydrolysed with dilute acid. Like amides, imides readily crystallise from alcohols and, in some cases (e.g. quinolinic imide), from glacial acetic acid.

Imino compounds. These substances contain the -C=NH group and, because they are strong, unstable bases, they are kept as their more stable salts, such as the hydrochlorides. (The free base usually hydrolyses to the corresponding oxo compound and ammonia.) Like amine hydrochlorides, the salts are purified by solution in alcohol containing a few drops of hydrochloric acid. After treatment with charcoal, and filtering, dry ethyl ether (or petroleum ether if ethanol is used) is added until crystallisation sets in. The salts are dried and kept in a vacuum desiccator.

Ketones. Ketones are more stable to oxidation than aldehydes and can be purified from oxidisable impurities by refluxing with potassium permanganate until the colour persists, followed by shaking with sodium carbonate (to remove acidic impurities) and distilling. Traces of water can be removed with type 4A Linde molecular sieves. Ketones which are solids can be purified by crystallisation from alcohol, toluene, or petroleum ether, and are usually sufficiently volatile for sublimation in vacuum. Ketones can be further purified *via* their bisulphite, semicarbazone or oxime derivatives (see p. 63). The bisulphite addition compounds are formed only by aldehydes and methyl ketones but they are readily hydrolysed in dilute acid or alkali.

Nitriles. *All purifications should be carried out in an efficient fume cupboard because of the toxic nature of these compounds.*
Nitriles are usually prepared either by reacting the corresponding halide or diazonium salts with a cyanide salt or by dehydrating an amide. Hence, possible contaminants are the respective halide or alcohol (from hydrolysis), phenolic compounds, amines or amides. Small quantities of phenols can be removed by chromatography on alumina. More commonly, purification of liquid nitriles or solutions of solid nitriles in a solvent such as ethyl ether is by shaking with dilute aqueous sodium hydroxide, followed by washing successively with water, dilute acid and water. After drying with sodium sulphate, the solvent is distilled off. Liquid nitriles are best distilled from a small amount of P_2O_5 which, besides removing water, dehydrates any amide to the nitrile. About one fifth of the nitrile should remain in the distilling flask at the end of the distillation (*the residue may contain some inorganic cyanide*). This purification also removes alcohols and phenols. Solid nitriles can be recrystallised from ethanol, toluene or petroleum ether, or a mixture of these solvents. They can also be sublimed under vacuum. Preliminary purification by steam distillation is usually possible.
Strong alkali or heating with dilute acids may lead to hydrolysis of the nitrile, and should be avoided.

Nitro compounds. Aliphatic nitro compounds are acidic. They are freed from alcohols or alkyl halides by standing for a day with concentrated sulphuric acid, then washed with water, dried with magnesium sulphate followed by calcium sulphate and distilled. The principal impurities are isomeric or homologous nitro compounds. In cases where the nitro compound was originally prepared by vapour phase nitration of the aliphatic hydrocarbon, fractional distillation should separate the nitro compound from the corresponding hydrocarbon. Fractional crystallisation is more effective than fractional distillation [Coetzee and Cunnungham *JACS* **87** 2529 *1965*].
The impurities present in aromatic nitro compounds depend on the aromatic portion of the molecule. Thus, benzene, phenols or anilines are probable impurities in nitrobenzene, nitrophenols and nitroanilines, respectively. Purification should be carried out accordingly. Isomeric compounds are likely to remain as impurities after the preliminary purifications to remove basic and acidic

contaminants. For example, o-nitrophenol may be found in samples of p-nitrophenol. Usually, the o-nitro compounds are more steam volatile than the p-nitro isomers, and can be separated in this way. Polynitro impurities in mononitro compounds can be readily removed because of their relatively lower solubilities in solvents. With acidic or basic nitro compounds which cannot be separated in the above manner, advantage may be taken of their differences in pK_a values. The compounds can thus be purified by preliminary extractions with several sets of aqueous buffers of known pH (see for example Table 19, p. 52) from a solution of the substance in a suitable solvent such as ethyl ether. This method is more satisfactory and less laborious the larger the difference between the pK_a value of the impurity and the desired compound.   Heterocyclic nitro compounds require similar treatment to the nitroanilines. Neutral nitro compounds can be steam distilled.

Phenols. Because phenols are weak acids, they can be freed from neutral impurities by dissolution in aqueous N sodium hydroxide and extraction with a solvent such as ethyl ether, or by steam distillation to remove the non-acidic material. The phenol is recovered by acidification of the aqueous phase with 20% sulphuric acid, and either extracted with ether or steam distilled. In the second case the phenol is extracted from the steam distillate after saturating it with sodium chloride. A solvent is necessary when large quantities of liquid phenols are purified. The phenol is fractionated by distillation under reduced pressure, preferably in an atmosphere of nitrogen to minimize oxidation. Solid phenols can be crystallised from toluene, petroleum ether or a mixture of these solvents, and can be sublimed under vacuum. Purification can also be effected by fractional crystallisation or zone refining. For further purification of phenols *via* their acetyl or benzoyl derivatives, see p. 63.

Quinones. These are neutral compounds which are usually coloured. They can be separated from acidic or basic impurities by extraction of their solutions in organic solvents with aqueous basic or acidic solutions, respectively. Their colour is a useful property in their purification by chromatography through an alumina column with, e.g. toluene as eluent. They are volatile enough for vacuum sublimation, although with high-melting quinones a very high vacuum is necessary. p-Quinones are stable compounds and can be recrystallised from water, ethanol, aqueous ethanol, toluene, petroleum ether or glacial acetic acid. o-Quinones, on the other hand, are readily oxidised. They should be handled in an inert atmosphere, preferably in the absence of light.

Salts (organic). (a) **With metal ions:** Water-soluble salts are best purified by preparing a concentrated aqueous solution to which, after decolorising with charcoal and filtering, ethanol or acetone is added so that the salts crystallise. They are collected, washed with aqueous ethanol or aqueous acetone, and dried. In some cases. water-soluble salts can be recystallised satisfactorily from alcohols. Water-insoluble salts are purified by Soxhlet extraction, first with organic solvents and then with water, to remove soluble contaminants. The purified salt is recovered from the thimble.

(b) **With organic ions:** Organic salts (e.g. trimethylammonium benzoate) are usually purified by recrystallisation from polar solvents (e.g. water, ethanol or dimethyl formamide). If the salt is too soluble in a polar solvent, its concentrated solution should be treated dropwise with a miscible nonpolar solvent (see p. 41) until crystallisation begins.

(c) **Sodium alkane disulphonates:** Purified from sulphites by boiling with aq HBr. Purified from sulphates by adding $BaBr_2$. Sodium alkane disulphonates are finally pptd by addition of MeOH. [Pethybridge and Taba *JCSFT I* **78** 1331 *1982*.]

Sulphur compounds. (a) **Disulphides** can be purified by extracting acidic and basic impurities with aqueous base or acid, respectively. However, they are somewhat sensitive to strong alkali which slowly cleaves the disulphide bond. The lower-melting members can be fractionally distilled under vacuum. The high members can be recrystallised from alcohol, toluene or glacial acetic acid.

(b) **Sulphones** are neutral and extremely stable compounds that can be distilled without decomposition. They are freed from acidic and basic impurities in the same way as disulphides. The low molecular weight members are quite soluble in water but the higher members can be recrystallised from water, ethanol, aqueous ethanol or glacial acetic acid.

(c) **Sulphoxides** are odourless, rather unstable compounds, and should be distilled under vacuum in an inert atmosphere. They are water-soluble but can be extracted from aqueous solution with a solvent such as ethyl ether.

(d) **Thioethers** are neutral stable compounds that can be freed from acidic and basic impurities as described for disulphides. They can be crystallised from organic solvents and distil without decomposition.

(e) **Thiols** are stronger acids than the corresponding hydroxy compounds but can be purified in a similar manner. However, care must be exercised in handling thiols to avoid their oxidation to disulphides. For this reason, purification is best carried out in an inert atmosphere in the absence of oxidising agents. Similarly, thiols should be stored out of contact with air. They can be distilled without change, and the higher-melting thiols (which are usually more stable) can be crystallised, e.g. from water or dilute alcohol. They oxidise readily in alkaline solution but can be separated from the disulphide which is insoluble in this medium. They should be stored in the dark below 0°. *All operations with thiols should be carried out in an efficient fume cupboard because of their unpleasant odour and their toxicity.*

(f) **Thiolsulphonates** (disulphoxides) are neutral and are somewhat light-sensitive compounds. Their most common impurities are sulphonyl chlorides (neutral) or the sulphinic acid or disulphide from which they are usually derived. The first can be removed by partial freezing or crystallisation, the second by shaking with dilute alkali, and the third by recrystallisation because of the higher solubility of the disulphide in solvents. Thiolsulphonates decompose slowly in dilute, or rapidly in strong, alkali to form disulphides and sulphonic acids. Thiolsulphonates also decompose on distillation but they can be steam distilled. The solid members can be recrystallised from water, alcohols or glacial acetic acid.

CHAPTER 6

PURIFICATION OF BIOCHEMICALS AND RELATED PRODUCTS

Biochemicals are chemical species produced by living organisms. They range widely in size, from simple molecules such as formic acid and glucose to macromolecules such as proteins and nucleic acids. Their *in vitro* synthesis is often impossibly difficult and in such cases they are available (if at all) only as commercial tissue extracts which have been subjected to purification procedures of widely varying stringency. The desired chemical may be, initially, only a minor constituent of the tissue that is the starting material, and which may vary considerably in its composition and complexity.

As a preliminary step the tissue might be separated into phases [e.g. whole egg into white and yolk, blood into plasma (or serum) and red cells], and the desired phase may be homogenised. Subsequent treatment usually comprises filtration, solvent extraction, salt fractionation, ultracentrifugation, chromatographic purification, gel filtration and dialysis. Fractional precipitation with ammonium sulphate gives crude protein species. Purification is finally judged by the formation of a single band of macromolecule (e.g. protein) on electrophoresis and/or analytical ultracentrifugation. Although these generally provide good evidence of high purity, none-the-less it does not follow that one band under one set of experimental conditions is an indication of homogeneity.

During the past 20 or 30 years a wide range of methods for purifying substances of biological origin have become available. For small molecules reference should be made to Chapters 1 and 2. The more important methods used for large molecules comprise:

1. *Centrifugation.* In addition to centrifugation for sedimenting proteins after ammonium sulphate precipitation in dilute aqueous buffer, this technique has been used for fractionation of large molecules in a denser medium or a medium of varying density. By layering sugar solutions of increasing densities in a centrifuge tube, proteins can be separated in a sugar-density gradient by centrifugation. Smaller DNA molecules (e.g. plasmid DNA) can be separated from RNA or nuclear DNA by centrifugation in aqueous caesium chloride (*ca* 0.975g/ml of buffer) for a long time (e.g. 40hr at 40,000 rpm). The plasmid DNA band appears at about the middle of the centrifuge tube, and is revealed by the fluorescent pink band formed by the binding of DNA to ethidium bromide which is added to the CsCl buffer. *Analytical centrifugation*, which is performed under specific conditions in an analytical centrifuge is very useful for studying purity and the molecular weight of macromolecules.

2. *Gel filtration* with polyacrylamide (mol wt exclusion limit from 3000 to 300,000) and agarose gel (mol wt exclusion limit 0.5 to 150 x 10^6). In this technique high-molecular weight substances are too large to fit into the gel microapertures and pass rapidly through the matrix (with the void volume), whereas low molecular weight species enter these apertures and are held there for longer periods of time, being retarded by the column material in the equilibria, relative to the larger molecules.

3. *Dry gels* and *crushed beads* are also useful in the gel filtration process. Selective retention of water and inorganic salts by the gels or beads results in increased concentration and purity of the protein fraction which moves with the void volume.

4. *Ion exchange* matrices are microreticular polymers containing carboxylic acid (e.g. Bio-Rad 70) or phosphoric acid (Pharmacia Mono-P) exchange functional groups for weak acidic cation exhangers, sulphonic acid groups (Dowex 50W) for strong acidic cation exchangers, diethylaminoethyl (DEAE) groups for weakly basic anion exchangers and quaternary ammonium (QEAE) groups for strong anion exchangers. These have been used extensively for the fractionation of peptides, proteins and enzymes. The use of *p*H buffers controls the strength with which the large molecules are bound to the support in the chromatographic process. Careful standardisation of experimental conditions and similarly the very uniform size distribution of Mono beads has led to high resolution in the purification of protein solutions. MonoQ (Pharmacia) is a useful strong anion exchanger; and MonoS (Pharmacia) is a useful strong cation exchanger whereas MonoP is a weak cation exchanger. Chelex 100 binds strongly and removes metal ions from macromolecules.

5. *Hydroxylapatite* is used for the later stages of purification of enzymes. It consists essentially of hydrated calcium phosphate which has been precipitated in a specific manner. It combines the characteristics of gel and ionic chromatography. Crystalline hydroxylapatite is a structurally organised, highly polar material which, in aqueous solution (in buffers) strongly adsorbs macromolecules such as proteins and nucleic acids, permitting their separation by virtue of the interaction with charged phosphate groups and calcium ions, as well as by physical adsorption. The procedure therefore is not entirely ion-exchange in nature. Chromatographic separations of singly and doubly stranded DNA are readily achievable whereas there is negligible adsorption of low molecular weight species.

6. *Affinity chromatography* is a chromatographic technique whereby the adsorbant has a particular affinity for one of the ingredients of the mixture to be purified. For example the adsorbant can be prepared by chemically binding an inhibitor of a specific enzyme (which is present in a complex mixture) to a matrix (e.g. Sepharose). When the mixture of impure enzyme is passed through the column containing the adsorbant, only the specific enzyme binds to the column. After adequate washing, the pure enzyme can be removed from the column by either increasing the salt concentration (e.g. NaCl) in the eluting buffer or adding the inhibitor to the eluting buffer. The salt or inhibitor can then be removed by dialysis or ultrafiltration (see below).

7. In the *Isoelectric focusing* of large charged molecules on polyacrylamide or agarose gels, slabs of these are prepared in buffer mixtures (e.g. ampholines) which have various *p*H ranges. When a voltage is applied for some time the buffers arrange themselves on the slabs in respective areas according to their *p*H ranges (prefocusing). Then the macromolecules are applied near one electrode and allowed to migrate in the electric field until they reach the *p*H area similar to their isoelectric points and focus at that position. This technique can also be used in a chromatographic mode, *chromatofocusing*, whereby a gel in a column is run (also under HPLC conditions) in the presence of ampholines (narrow or wide *p*H ranges as required) and the macromolecules are then run through in a buffer. The bands are eluted according to their isoelectric points. Isoelectric focusing standards are available which can be used in a preliminary run in order to calibrate the effluent from the column, or alternatively the *p*H of the effluent is recorded using a glass electrode designed for the purpose.

8. *High performance liquid chromatography* (HPLC) is liquid chromatography in which the eluting liquid is sent through the column containing the packing (materials as in 2-6 above, which can withstand higher than atmospheric pressures) under pressure. On a routine basis this has been found useful for purifying proteins (including enzymes) and polypeptides after enzymic digestion of proteins or chemical cleavage (e.g. with CNBr) prior to sequencing (using reverse-phase columns such as μ-Bondapak C18). Moderate pressures (50-300psi) have been found most satisfactory for large molecules (FPLC). [See Scopes *Anal Biochem*

114 8 *1981*; *High Performance Liquid Chromatography and Its Application to Protein Chemistry,* Hearn in **Advances in Chromatography 20** 7 1982.]

9. *Ultrafiltration* using a filter (e.g. Millipore) can remove water and low-molecular weight substances without the application of heat. Filters with a variety of molecular weight exclusion limits not only allow the concentration of a particular macromolecule, but also the removal (by washing during filtration) of smaller molecular weight contaminants (e.g. salts, inhibitors or cofactors). This procedure has been useful for changing the buffer in which the macromolecule is present (e.g. from Tris-Cl to ammonium carbonate), and for desalting. Ultrafiltration can be carried out in a stirrer cell (Amicon) in which the buffer containing the macromolecule (particularly protein) is pressed through the filter, with stirring, under argon or nitrogen pressure (e.g. 20-60psi). During this filtration process the buffer can be changed. This is rapid (e.g. 2L of solution can be concentrated to a few mls in 1 to 2 hr depending on pressure and filter). A similar application uses a filter in a specially designed tube (Centricon tubes, Amicon) and the filtration occurs under centrifugal force in a centrifuge (4-6000rpm/0°/40min). The macromolecule (usually DNA) then rests on the filter and can be washed on the filter by centrifugation. The macromolecule is recovered by inverting the filter, placing a conical receiver tube on the same side where the macromolecule rests, filling the other side of the filter tube with eluting solution (usually a very small volume e.g. 100 μl), and during further centrifugation this solution passes through the filter and collects the macromolecule from the underside into the conical receiver tube.

10. *Partial precipitation* of a protein in solution can often be achieved by controlled addition of a strong salt solution, e.g ammonium sulphate. This is commonly the first step in the purification process. Its simplicity is offset by possible denaturation of the desired protein and the (sometimes gross) contamination with other proteins. It should therefore be carried out by careful addition of small aliquots of the powdered salt or concentrated solution (below 4°, with gentle stirring) and allowing the salt to be evenly distributed in the solution before adding another small aliquot. Under carefully controlled conditions and using almost pure protein it is sometimes possible to obtain the protein in crystalline form suitable for X-ray analysis.

Other details of the above will be found in Chapters 1 and 2 which also contain relevant references.

Several illustrations of the usefulness of the above methods are given in the *Methods in Enzymology* series (Academic Press) in which 1000-fold purifications or more, have been readily achieved. In applying these sensitive methods to macromolecules, reagent purity is essential. It is disconcerting, therefore, to find that some commercial samples of the widely used affinity chromatography ligand Cibacron Blue F3GA contained this dye only as a minor constituent. The major component appeared to be the dichlorotriazinyl precursor of this dye. Commercial samples of Procion Blue and Procion Blue MX-R were also highly heterogeneous [Hanggi and Cadd *Anal Biochem* **149** 91 *1985*]. Variations in composition of sample dyes can well account for differences in results reported by different workers. The purity of substances of biological origin should therefore be checked by one or more of the methods given above. Water of high purity should be used in all operations. Double glass distd water or water purified by a MilliQ filtration system (see Chapter 2) is most satisfactory.

This chapter lists some representative examples of biochemicals and their origins, a brief indication of key techniques used in their purification, and literature references where further details may be found. Simpler low molecular weight compounds, particularly those that may have been prepared by chemical syntheses, e.g. acetic acid, glycine, will be found in Chapter 3.

Journal title abbreviations are as in Chapter 3.

Abrin A and Abrin B. Toxic proteins from seeds of *Abras precatorius*. Purified by successive chromatography on DEAE-Sephadex A-50, carboxymethylcellulose, and DEAE-cellulose. [Wei *et al, JBC* **249** 3061 *1974*.]

B-D-N-Acetylhexosaminidase A and B. From human placenta. Purified by Sephadex G-200 filtration and DEAE-cellulose column chromatography. Hexosaminidase A was further purified by DEAE-cellulose column chromatography, followed by an ECTEOLA-cellulose column, Sephadex-200 filtration, electrofocusing and Sephadex G-200 filtration. Hexosaminidase B was purified by a CM-cellulose column, electrofocusing and Sephadex G-200 filtration. [Srivastava *et al, JBC* **249** 2034 *1974*.]

p-**Acetylphenyl phosphate, potassium salt.** Purified by dissolving in the minimum volume of hot water (60°) and adding EtOH, with stirring, then left at 0° for 1hr. Crystals were filtered off and recrystd from water until free of Cl^- and SO_4^{2-} ions. Dried in vac over P_2O_5 at room temperature. [Milsom *et al, BJ* **128** 331 *1972*.]

(-)-S-Adenosyl-L-methionine chloride. Purified by ion exchange on Amberlite IRC-150, and eluting with 0.1-4M HCl. [Stolowitz and Minch *JACS* **103** 6015 *1981*.]

ADP-Ribosyl transferase. From human placenta. Purified by making an affinity absorbent for ADP-ribosyltransferase by coupling 3-aminobenzamide to Sepharose 4B. [Burtscher *et al, Anal Biochem* **152** 285 *1986*.]

Agglutin. From peanuts (*Arachis hypogaea*). Purified by affinity chromatography on Sepharose-ε-aminocaproyl-β-D-galactopyranosylamine. [Lotan *et al, JBC* **250** 8518 *1974*.]

Alamethicin. From *Tricoderma viridae*. Recrystd from MeOH. [Panday *et al, JACS* **99** 8469 *1977*.]

Albumin (bovine serum) Purified by soln in conductivity water and passage at 2-4° through two ion-exchange columns, each containing a 2:1 mixture of anionic and cationic resins (Amberlite IR-120, H-form; Amberlite IRA-400, OH-form). This treatment removed ions and lipoid impurities. Care was taken to exclude CO_2, and the soln was stored at -15°. [Möller, van Os and Overbeek *TFS* **57** 312 *1961*.] More complete lipid removal was achieved by lyophilizing the de-ionised soln, covering the dried albumin (human serum) with a mixture of 5% glacial acetic acid (v/v) in iso-octane (previously dried with Na_2SO_4) and allowing to stand at 0° (without agitation) for upwards of 6 hr before decanting and discarding the extraction mixture, washing with iso-octane, re-extracting, and finally washing twice with iso-octane. The purified albumin was dried under vac for several hours, then dialyzed against water for 12-24hr at room temperature, lyophilized, and stored at -10°C [Goodman *Science* **125** 1296 *1957*].

Alkaline phosphatase. From rat *osteosarcoma*. Purified by acetone pptn, followed by chromatography on DEAE-cellulose, Sephacryl S-200, and hydroxylapatite. [Nair *et al, ABB* **254** 18 *1987*.]

Amethopterin. Commonest impurities are 10-methyl pteroylglutamic acid, 4-amino-10-methylpteroylglutamic acid, aminopterin and pteroylglutamic acid. Purified by chromatography on Dowex-1 acetate, followed by filtration through a mixture of cellulose and charcoal [Momle *Biochem Preps* **8** 20 *1961*].

7-Amino-4-(trifluoromethyl)coumarin, m 222°. Purified by column chromatography on a C18 column, eluted with acetonitrile/0.01M aq HCl (1:1), and crystd from isopropanol. Alternatively, it is eluted from a silica gel column with CH_2Cl_2, or by extracting a CH_2Cl_2 solution (4g/L) with 1M aq NaOH (3 x 0.1L), followed by drying (MgSO$_4$), filtration and evapn. [Bissell *JOC* **45** 2283 *1980*].

3-Aminopyridine adenine dinucleotide. Purified by ion exchange chromatography [Fisher *et al, JBC* **248** 4293 *1973*].

Amylose *[9005-82-7] (for use in iodine complex formation).* Amylopectin was removed from impure amylose by dispersing in aq 15% pyridine at 80-90° (concn 0.6-0.7%) and leaving the soln stand at 44-45° for 7days. The ppte was redispersed and recrystd during 5 days. After a further dispersion in 15% pyridine, it was cooled to 45°, allowed to stand at this temperature for 12hr, then cooled to 25° and left for a further 10hr. The combined ppte was dispersed in warm water, pptd with EtOH, washed with abs EtOH, and vac dried [Foster and Paschall *JACS* **75** 1181 *1953*].

Angiotensin. From rat brain. Purified using extraction, affinity chromatography and HPLC [Hermann *et al, Anal Biochem* **159** 295 *1986*].

Angiotensinogen. From human blood serum. Purified by chromatography on Blue Sepharose, Phenyl-Sepharose, hydroxylapatite and immobilised 5-hydroxytryptamine [Campbell *et al, BJ* **243** 121 *1987*].

Aureomycin *[57-62-5]* M 478.5, m 172-174°(dec), $[\alpha]_D^{23}$ -275° (MeOH). Dehydrated by azeotropic distn of its soln with toluene. On cooling anhydrous material crysts out and is recrystd from benzene, then dried under vac at 100° over paraffin wax. (If it is crystd from MeOH, it contains MeOH which is not removed on drying.) [Stephens *et al JACS* **76** 3568 *1954*.]

Aureomycin hydrochloride *[64-72-2]* M 514.0, m 234-236°(dec), $[\alpha]_D^{25}$ -23.5° (H_2O). Purified by dissolving 1g rapidly in 20ml of hot water, cooling rapidly to 40°, treating with 0.1ml of 2M HCl, and chilling in an ice-bath. The process is repeated twice [Stephens *et al JACS* **76** 3568 *1954*].

Avidin. From egg white. Purified by chromatography of an ammonium acetate soln on CM-cellulose [Green *BJ* **101** 774 *1966*]. Also purified by affinity chromatography on 2-iminobiotin-6-aminohexyl-Sepharose 4B [Orr *JBC* **256** 761 *1981*].

Azurin. From *Pseudomonas aeruginosa.* Purified by gel chromatography on G-25 Sephadex with 5mM phosphate *p*H 7 buffer as eluent [Cho *et al, JPC* **91** 3690 *1987*].

L-Canavanine sulphate. From jackbean. Recrystd from water by adding 95% EtOH [Hunt and Thompson *Biochem Preps* **13** 416 *1971*].

Carboxypeptidase A. From bovine pancreas. Purified by DEAE-cellulose chromatography, activation with trypsin and dialysed against 0.1M NaCl, yielding crystals [Cox *et al, Biochemistry* **3** 44 *1964*].

Cathepsin B. From human liver. Purified by affinity chromatography on the semicarbazone of Gly-Phe-glycinal-linked to Sepharose 4B, with elution by 2,2'-dipyridyl disulphide [Rich *et al, BJ* **235** 731 *1986*].

Cathepsin D. From bovine spleen. Purified on a CM column after ammonium sulphate fractionation and dialysis, then starch-gel electrophoresis and by ultracentrifugal analysis. Finally chromatographed on a DEAE column [Press *et al, BJ* **74** 501 *1960*].

Ceruloplasmin. From human blood plasma. Purified by precipitation with polyethylene glycol 4000, batchwise adsorption and elution from QAE-Sephadex, and gradient elution from DEAE-Sepharose CL-6B. Ceruloplasmin was purified 1640-fold. Homogeneous on anionic polyacrylamide gel, electrophoresis (PAGE), SDS-PAGE, isoelectric focusing and low speed equilibrium centrifugation. [Oestnuizen *Anal Biochem* **146** 1 *1985*].

γ-**Chymotrypsin** (*EC* 3.4.21.1) *[9004-07-3]*. Crystd twice from four-tenths satd ammonium sulphate soln, then dissolved in 1mM HCl and dialysed against 1mM HCl at 2-4°. The soln was stored at 2° [Lang, Frieden and Grunwald *JACS* **80** 4923 *1958*].

Citric acid cycle components. From rat heart mitochondria. Resolved by anion-exchange chromatography [LaNoue *et al*, *JBC* **245** 102 *1970*].

Cocarboxylase *[532-40-1]* **M 416.8, m 195°(dec).** Crystd from EtOH slightly acidified with HCl.

Collagenase. From human polymorphonuclear leukocytes. Purified by using *N*-ethylmaleimide to activate the enzyme, and wheat germ agglutinin-agarose affinity chromatography [Callaway *et al*, *Biochemistry* **25** 4757 *1986*].

Convallatoxin *[508-75-8]* **M 550.6, m 238-239°.** Crystd from ethyl acetate.

Copper-zinc-superoxide dismutase. From blood cell haemolysis. Purified by DEAE-Sepharose and copper chelate affinity chromatography. The preparation was homogeneous by SDS-PAGE, analytical gel filtration chromatography and by isoelectric focusing [Weselake *et al*, *Anal Biochem* **155** 193 *1986*].

Corticotropin. Extract separated by ion-exchange on CM-cellulose, desalted, evapd and lyophilized. Then run on gel filtration (Sephadex G-50) [Lande *et al*, *Biochem Preps* **13** 45 *1971*].

Colicin E$_1$. From *E.coli*. Purified by salt extraction of extracellular-bound colicin followed by salt fractionation and ion-exchange chromatography on a DEAE-Sephadex column, and then on CM-Sephadex column chromatography [Schwartz and Helinski *JBC* **246** 6318 *1971*].

α-**Cyclodextrin** (H_2O) *[10016-20-3]* **M 972.9, m >280°(dec), [α]**$_{546}$ **+175° (c 10, H_2O).** Recrystd from 60% aq EtOH, then twice from water, and dried for 12hr in vac at 80°. Also purified by pptn from water with 1,1,2-trichloroethylene. The ppte was isolated, washed and resuspended in water. This was boiled to steam distil the trichloroethylene. The soln was freeze-dried to recover the cyclodextrin. [Armstrong *et al*, *JACS* **108** 1418 *1986*.]

β-**Cyclodextrin** (H_2O) *[7585-39-9]* **M 1135.0, m >300°(dec), [α]**$_{546}$ **+170° (c 10, H_2O).** Recrystd from water and dried for 12hr in vac at 110°, or 24hr in vac at 70°. The purity was assessed by tlc on cellulose with a fluorescent indicator. [Taguchi, *JACS* **108** 2705 *1986*; Tabushi *et al*, *JACS* **108** 4514 *1986*; Orstam and Ross *JPC* **91** 2739 *1987*.]

Cytochalasin B. From dehydrated mould matte. Purified by MeOH extractn, reverse phase C18 silica gel batch extraction, selective elution with 1:1 v/v hexane/tetrahydrofuran, crystn, subjected to tlc and recrystallised [Lipski *et al*, *Anal Biochem* **161** 332 *1987*].

Cytochrome c. From horse heart. Purified by chromatography on CM-cellulose (CM-52 Whatman) [Brautigan *et al*, *Methods in Enzymology* **53D** 131 *1978*].
From *Rhodospirillum rubrum*. ($\varepsilon_{270}/\varepsilon_{551}$ 0.967). Purified by chromatography on a column of CM-Whatman cellulose [Paleus and Tuppy *Acta Chem Scand* **13** 641 *1959*].

Cytochrome c oxidase. From bovine heart mitochondria. Purified by selective solubilisation with Triton X-100 and subsequently with lauryl maltoside: finally by sucrose gradient centrifugation [Li *et al*, *BJ* **242** 417 *1978*].

Deoxyribonucleic acid. From plasmids. Purified by two buoyant density ultracentrifuging using ethidium bromide-CsCl. The ethidium bromide was extracted and the DNA was dialysed against buffered EDTA. Lyophilized. [Marmur and Doty *J Mol Biol* **5** 109 *1962*; Guerry *et al, J Bacteriol* **116** 1064 *1973*.]

Dermatan sulphate (condroitin sulphate B). From pig skin. Purified by digestion with papain and hyaluronidase, and fractionation using aqueous EtOH. [Gifonelli and Roden *Biochem Preps* **12** 1 *1968*.]

Dextran. Soln keeps indefnitely at room temperature if 0.2ml of Roccal (10% alkyldimethylbenzylammonium chloride) or 2mg phenyl mercuric acetate added per 100 ml solution. [Scott and Melvin *Anal Biochem* **25** 1656 *1953*.]

Di- and tri-carboxylic acids. Resolution by anion-exchange chromatography. [Bengtsson and Samuelson *Anal Chim Acta* **44** 217b *1969*.]

Dihydrofolate reductase. From *Mycobacterium phlei*. Purified by ammonium sulphate pptn, then fractionated on Sephadex G-75 column, applied to a Blue Sepharose column and eluted with 1 nM dihydrofolate. [Al Rubeai and Dole *BJ* **235** 301 *1986*.]

Dihydropteridine reductase. (EC 1.6.99.7). From sheep liver. Purified by fractionation with ammonium sulphate, dialysed versus tris buffer, adsorbed and eluted from hydroxylapatite gel. Then run through a DEAE-cellulose column and also subjectd to Sephadex G-100 filtration. [Craine *et al*, *JBC* **247** 6082 *1972*.]

Dihydropteridine reductase (EC 1.6.99.7). From human liver. Purified to homogeneity on a naphthoquinone affinity adsorbent, followed by DEAE-Sephadex and CM-Sephadex chromatography. [Firgaira, Cotton and Danks, *BJ* **197** 31 *1981*.] [For other dihydropteridine reductases see Armarego *et al, Medicinal Research Reviews* **4**(3) 267 *1984*.]

3,4-Dihydroxyphenylalanine-containing proteins. Boronate affinity chromatography is used in the selective binding of proteins containing 3,4-dihydroxyphenylalanine to a *m*-phenylboronate agarose column, eluting with 1 M ammonium acetate at *p*H 10. [Hankus *et al, Anal Biochem* **150** 187 *1986*.]

Dipeptidyl aminopeptidase. From rat brain. Purified adout 2000-fold by column chromatography on CM-cellulose, hydroxylapatite and Gly-Pro AH-Sepharose. [Imai *et al, J Biochem (Tokyo)* **93** 431 *1983*.]

Dolichol. From pig liver. Cryst 6 times from pet ether/EtOH at -20°C. Ran as entity on a paper chromatogram on paper impregnated with paraffin, with acetone as the mobile phase. [Burgos *BJ* **88** 470 *1963*.]

Flavin adenine dinucleotide (diNa, 2H$_2$O salt) *[146-14-5]* M 865.6, [α]$_{546}$ -54° (c 1, H$_2$O). Small quantities, purified by paper chromatography using *tert*-butyl alcohols/water, cutting out the main spot and eluting with water. Larger amounts can be pptd from water as the uranyl complex by adding a slight excess of uranyl acetate to a soln at *p*H 6.0, dropwise and with stirring. The soln is stood overnight in the cold, and the ppte is centrifuged off, washed with small portions of cold EtOH, then with cold, peroxide-free ethyl ether. It is dried in the dark under vac over P$_2$O$_5$ at 50-60°. The uranyl complex is suspended in water and, after adding sufficient 0.01M NaOH to adjust the *p*H to 7, the ppte of uranyl hydroxide is removed by centrifugation [Huennekens and Felton, in *Methods in Enzymology*, Colowick and Kaplan, Academic Press, N Y, **Vol 3** 954 *1957*]. It can also be crystd from water. Should be kept in the dark. More recently it was purified by elution from a DEAE-cellulose (Whatman DE 23) column with 0.1M phosphate buffer pH 7, and the purity was checked by tlc. [Holt and Cotton, *JACS* **109** 1841 *1987*.]

Flavin mononucleotide (Na, $2H_2O$ salt) *[130-40-5]* **M 514.4**. Purified by paper chromatography using *tert*-butanol-water, cutting out the main spot and eluting with water. Also purified by adsorption onto an apo-flavodoxin column, followed by elution and freeze drying [Mayhew and Strating *Eur J Biochem* **59** 539 *1976*].

Ferritin. From human placenta. Purified by homogenisation in water and precipitating with ammonium sulphate repeating cycle of ultracentrifuging and molecular sieve chromatography through Sephadex 4B column. Isoelectric focusing revealed a broad spectrum of impurities which were separated by ion-exchange chromatography on Sephadex A-25 and stepwise elution. [Konijn *et al, Anal Biochem* **144** 423 *1985*.]

Fibrinogen (plasminogen-deficient). From blood plasma. Anticoagulated blood was centrifuged and the plasma was frozen and washed with saline solution. Treated with charcoal and freeze-thawed. Dialysed versus Tris/NaCl buffer. [Maxwell and Nikel *Biochem Preps* **12** 16 *1968*.]

Folic acid *[75708-92-8]* **M 441.4, m** >250°(dec) $[\alpha]_D^{25}$ +23° (c 0.5, 0.1N NaOH). Crystd from hot water after extraction with butanol see Blakley [*BJ* **65** 331 *1957*] and Kalifa, Furrer, Bieri and Viscontini [*Helv Chim Acta* **61** 2739 *1978*].

Galactal. Recryst from ethyl acetate. [Distler and Jourdian *JBC* **248** 6772 *1973*.]

Glucose oxidase. From *Aspergillus niger*. Purified by dialysis against deionized water at 6° for 48hr, and by molecular exclusion chromatography with Sephadex G25 at room temperature. [Holt and Cotton *JACS* **109** 1841 *1987*.]

β-Galatosidase. From bovine testes. Purified 600-fold by ammonium sulphate precipitation, acetone fractionation and affinity chromatography on agarose substituted with terminal thio-β-galactopyranosyl residues. [Distlern and Jourdian *JBC* **248** 6772 *1973*.]

Glutathione S-transferase. From human liver. Purified by affinity chromatography using a column prepared by coupling glutathione to epoxy-saturated Sepharose. [Simons and Vander Jag *Anal Biochem* **52** 334 *1977*.]

Glycogen synthase. From bovine heart. Purified by precipitation of the enzyme in the presence of added glycogen by polyethylene glycol, chromatography on DEAE-Sephacel and high speed centrifugation through a sucrose-containing buffer. [Dickey-Dunkirk and Kollilea *Anal Biochem* **146** 199 *1985*.]

N-**Guanyltyramine.** Purified on cation exchange columns. [Mekalanos *et al, JBC* **254** 5849 *1979*.]

Haemoglobin A. From normal human blood. Purified from blood using CM-32 cellulose column chromatography. [Matsukawa *et al, JACS* **107** 1108 *1985*.]

Heparin. From pig intestinal mucosa. Most likely contaminants are mucopolysaccharides including heparin sulphate and dermatan sulphate. Purified by pptn with cetylpyridinium chloride from saturated solutions of high ionic strength. [Cifonelli and Roden *Biochem Preps* **12** 12 *1968*.]

Heparin (sodium salt) *[9041-08-1]*. Dissolved in 0.1M NaCl (1g/100ml) and pptd by addition of EtOH (150ml).

Histones. From S4A mouse lymphoma. Purification used a macroprocess column, heptafluorobutyric acid as solubilising and ion-pairing agent and an acetonitrile gradient. [McCroskey *et al, Anal Biochem* **163** 427 *1987*.]

Hyaluronidase. Purified by chromatography on DEAE-cellulose prior to use. [Distler and Jourdain *JBC* **248** 6772 *1973*.]

3-Hydroxy butyrate dehydrogenase,
3-Hydroxy malate dehydrogenase. From *Rhodopseudomonas spheroides*. Purified by two sequential chromatography steps on two triazine dye-Sepharose matrices. [Scavan *et al, BJ* **203** 699 *1982*]

Interleukin. From human source. Purified using lyophilisation and desalting on Bio-Rad P-6DC desalting gel, then two steps of HPLC, first with hydroxylapatite, followed by a TSK-125 size exclusion column. [Kock and Luger *J Chromat* **296** 293 *1984*.]

Lactate dehydrogenase. From dogfish muscle. 40-Fold purification by affinity chromatography using Sepharose 4B coupled to 8-(6-aminohexyl)amino-5'-AMP or -NAD$^+$. [Lees *et al, Arch Biochem Biophys* **163** 561 *1974*.]

Lactoferrin. From human whey. Purified by direct adsorption on cellulose phosphate by batch extraction, then eluted by a stepped salt and *p*H gradient. [Foley and Bates *Anal Biochem* **162** 296 *1987*.]

Lecithin. For purification of commercial egg lecithin see Pangborn [*JBC* **188** 471 *1951*].

Lecithin. From hen egg white. Purified by solvent extraction and chromatography on alumina. Suspended in distd water and kept frozen until used [Lee and Hunt *JACS* **106** 7411 *1984*, Singleton *et al, J Am Oil Chemists Soc* **42** 53 *1965*].

Lectins. From seeds of *Robinia pseudoacacia*. Purified by pptn with ammonium sulphate and dialysis; then chromatographed on DE-52 DEAE-cellulose anion exchanger, hydroxylapatite and Sephacryl S-200. [Wantyghem *et al, BJ* **237** 483 *1986*.]

Lipoprotein lipase. From bovine skimmed milk. Purified by affinity chromatography on heparin-Sepharose [Shirai *et al, Biochim Biophys Acta* **665** 504 *1981*].

Lipoproteins. From human plasma. Individual human plasma lipid peaks were removed from plasma by ultracentrifugation,then separated and purified by agarose-column chromatography. Fractions were characterised immunologically, chemically, electrophoretically and by electron microscopy. [Rudel *et al, BJ* **13** 89 *1974*.]

Lipoteichoic acids. From gram-positive bacteria. Extracted by hot phenol/water from disrupted cells. Nucleic acids that were also extracted were removed by treatment with nucleases. Nucleic resistant acids, protein, polysaccharides and teichoic acid were separated from lipoteichoic acid by anion-exchange chromatography on DEAE-Sephacel or hydrophobic interaction on octyl-Sepharose [Fischer et al, *Eur J Biochem* **133** 523 *1983*].

Magnesium protoporphyrin dimethyl ester. Crude product
dissolved in as little hot dry benzene as possible and left overnight at room temperature to cryst. [Fuhrhof and Graniek *Biochem Preps* **13** 55 *1971*.]

α-Melanotropin,
β-Melanotropin. Extract separated by ion-exchange on carboxyymethyl cellulose, desalted, evapd and lyophilised, then run on gel filtration (Sephadex G-25). [Lande *et al, Biochem Preps* **13** 45 *1971*.]

Metallothionein. From rabbit liver. Purified by precipitation to give Zn- and Cd-containing protein fractions and running on a Sephadex G-75 column, then isoelectric focussing to give two protein peaks [Nordberg *et al, BJ* **126** 491 *1972*].

Myoglobin. From sperm whale muscle. Purified by CM-cellulose chromatography [Anres and Atassi *Biochemistry* **12** 942 *1980*.

Nicotinamide adenine dinucleotide (Diphosphopyridine nucleotide, NAD,
DPN) *[53-84-9]* M 663.4. Purified by chromatography on Dowex-1 ion-exchange resin. The column was prepared by washing with 3M HCl until free of material absorbing at 260nm, then with water, 2M sodium formate until free of chloride ions and, finally, with water. NAD, as an 0.2% soln in water, adjusted with NaOH to *p*H 8, was adsorbed on the column, washed with water, and eluted with 0.1M formic acid. Fractions with strong absorption at 360nm were combined, acidified to pH 2.0 with 2M HCl, and cold acetone (*ca* 5L/g of NAD) was added slowly and with constant agitation. It was left overnight in the cold, then the ppte was collected in a centrifuge, washed with pure acetone and dried under vac over $CaCl_2$ and paraffin wax shavings [Kornberg in *Methods in Enzymology* Eds Colowick and Kaplan, Academic Press, New York **Vol 3** p 876 *1957*]. Purified by anion exchange chromatography [Dalziel and Dickinson *Biochem Preps* **11** 84 *1966*.] The purity is checked by reduction to NADH (with EtOH and yeast alcohol dehydrogenase) which has ε_{340nm} 6220 $M^{-1}cm^{-1}$.

Nicotinamide adenine dinucleotide phosphate (NADP). Purified by anion-exchange chromatography [Dalziel and Dickinson *BJ* **95** 311 *1965*].

Novobiocin *[303-81-1]* **M 612.6, two forms m 152-156°** and **m 174-178°,** $\lambda_{max.}$ **330nm (acid EtOH), 305nm (alk EtOH).** Crystd from EtOH and stored in the dark. Then sodium salt can be crystd from MeOH, then dried at 60°/0.5mm. [Sensi, Gallo and Chiesa, *AC* **29** 1611 *1957*.]

5'-Nucleotidase. From Electric ray (*Torpedo sp*). Purified by dissolving in Triton X-100 and deoxycholate, and by affinity chromatography on concanavalin A-Sepharose and AMP-Sepharose [Grondal and Zimmerman *BJ* **245** 805 *1987*].

Orosomucoid.
From human plasma. Purified by passage through a carboxymethyl cellulose column [Aronson *et al*, *JBC* **243** 4564 *1968*].

R-Pantothenic acid
[867-81-2] M 241.2, m 122-124°, $[\alpha]_D^{25}$ +27° (c 5, H_2O). Crystd from EtOH.

Papain *(EC 3.4.22.2) [9001-73-4]*. A suspension of 50g of papain (freshly ground in a mortar) in 200ml of cold water was stirred at 4° for 4hr, then filtered through a Whatman No 1 filter paper. The clear yellow filtrate was cooled in an ice-bath while a rapid stream of H_2S was passed through it for 3hr, and the suspension was centrifuged at 2000rpm for 20min. Sufficient cold MeOH was added slowly and with stirring to the supernatant to give a final MeOH conc of 70 vol %. The ppte, collected by centrifugation for 20min at 2000rpm, was dissolved in 200ml of cold water, the soln was satd with H_2S, centrifuged, and the enzyme again pptd with MeOH. The process was repeated four times. [Bennett and Niemann *JACS* **72** 1798 *1950*.] Papain has also been purified by affinity chromatography on a column of Gly-Gly-Tyr-Arg-agarose [Stewart *et al*, *JACS* **109** 3480 *1986*.]

Pectic acid M $(176.1)_n$, $[\alpha]_D$ +250° (c 1, 0.1M NaOH). Citrus pectic acid (500g) was refluxed for 18hr with 1.5L of 70% EtOH and the suspension was filtered hot. The residue was washed with hot 70% EtOH and finally with ether. It was dried in a current of air, ground and dried for 18hr at 80° under vac. [Morell and Link *JBC* **100** 385 *1933*.] It can be further purified by dispersing in water and adding just enough dil NaOH to dissolve the pectic acid, then passing the soln through columns of cation- and anion-exchange resins [Williams and Johnson *IECAE* **16** 23 *1944*], and precipitating with two volumes of 95% EtOH containing 0.01% HCl. The ppte is worked with 95% EtOH, then ethyl ether, dried and ground.

Pectin *[9000-69-5]* M 25000-50000. Dissolved in hot water to give a 1% soln, then cooled, and made about 0.05M in HCl by addition of conc HCl, and pptd by pouring slowly and with vigorous stirring into two volumes of 95% EtOH. After standing for several hours, the pectin is filtered on to nylon cloth, then redispersed in 95% EtOH and stood overnight. The ppte is filtered off, washed with EtOH/ethyl ether, then ethyl ether and air dried.

Pepsin (EC 3.4.23.1) *[9001-75-6]*. Rechromatographed on a column of Amberlite CG-50 using a *p*H gradient prior to use. Crystd from EtOH. [Richmond *et al*, *Biochim Biophys Acta* **29** 453 *1958*; Huang and Tang, *JBC* **244** 1085 *1969*.]

Pertussis toxin. From *Bordetella pertussis*. Purified by stepwise elution from 3 columns comprising Blue Sepharose, Phenyl Sepharose and hydroxylapatite, and SDS-polyacrylamide gel electrophoresis [Svoboda *et al*, *Anal Biochem* **159** 402 *1986*].

Phospholipids. For the removal of ionic contaminants from raw zwitterionic phospholipids, most lipids were purified twice by mixed-bed ionic exchange (Amberlite AB-2) of methanolic solutions. (About 1 g of lipid in 10 ml of MeOH). With both runs the first 1 ml of the eluate was discarded. The main fraction of the solution was evaporated at 40°C under dry N and recryst three times from *n*-pentane. The resulting white powder was dried for about 4hr at 50° under reduced pressure and stored ar 3°. Some samples were purified by mixed-bed ion exchange of aqueous suspensions of the crystal/liquid crystal phase. [Kaatze *et al*, *JPC* **89** 2565 *1985*.]

Phosphoproteins (various). Purified by adsorbing onto an iminodiacetic acid substituted agarose column to which was bound ferric ions. This chelate complex acted as a selective immobilised metal affinity adsorbent for phosphoproteins. [Muszyfiska *et al*, *Biochemistry* **25** 6850 *1986*.]

Phosphoribosyl pyrophosphate synthetase. From human erythrocytes. Purified 5100-fold by elution from DEAE-cellulose, fractionation with ammonium sulphate, filtration on Sepharose 4B and ultrafiltration. [Fox and Kelley *JBC* **246** 5739 *1971*.]

Pituitary Growth Factor. From human pituitary gland. Purified by heparin and copper affinity chromatography, followed by carboxymethyl cellulose (Whatman 52). [Rowe *et al*, *Biochemistry* **25** 6421 *1986*.]

Polyethylene glycol. May be contaminated with aldehydes and peroxides. Methods are given for removing interfering species. [Ray and Purathingal *Anal Biochem* **146** 307 *1985*.]

Porphyrin a. From ox heart. A method is described for the extraction and purification of porphyrin a. [Morell *et al*, *BJ* **78** 793 *1961*.]

Protamine kinase. From rainbow trout testes. Partial purification by hydroxylapatite chromatography, followed by biospecific chromatography on nucleotide coupled Sepharose 4B (the nucleotide was 8-(6-aminohexyl)amine coupled cyclic-AMP.) [Jergil *et al*, *BJ* **139** 441 *1974*.]

Protease nexin. From cultured human fibroblasts. Purified by affinity binding of protease nexin to dextran sulphate-Sepharose. [Farrell *et al*, *BJ* **237** 707 *1986*.]

Proteoglycans. From cultured human muscle cells. Separated by ion-exchange HPLC using a Bio-gel TSKDEAE 5-PW analytical column. [Harper *et al*, *Anal Biochem* **159** 150 *1986*.]

Prothrombin. From equine blood plasma. Purified by chromatography on Sephadex G-200 or IRC-50. [Miller *Biochem Preps* **14** 49 *1971*.]

Pyridoxal hydrochloride *[65-12-5]* **M 203.6, m 176-180°(dec).** Dissolved in water, the pH was adjusted to 6 with NaOH and set aside overnight to crystallise. The crystals were washed with cold water, dried in a vac desiccator over P_2O_5 and stored in a brown bottle at room temperature. [Fleck and Alberty *JPC* **66** 1678 *1962*.]

Pyridoxamine hydrochloride *[5103-96-8]* **M 241.2, m 226-227°(dec).** Crystd from hot MeOH.

Pyridoxine hydrochloride (vitamin B_6)   *[58-56-0]* **M 205.7, m 209-210°(dec).** Crystd from EtOH/acetone.

Prymnesin. Toxic protein from phytoflagellate *Pyrymnesium parvum*. Purified by column chromatography, differential soln and pptn in solvent mixtures and differential partition between diphasic mixtures. The product has at least 6 components as observed by tlc. [Ulitzur and Shilo *BBA* **301** 350 *1970*.]

Pyruvate kinase isoenzymes. From *Salmonella typhimurium*. Purified by ammonium sulphate fractionation and gel filtration, ion-exchange and affinity chromatography. [Garcia-Olalla and Garrido-Pertierra *BJ* **241** 573 *1987*.]

Recombinant human interleukin-2. Purified by reverse phase HPLC.
[Weir and Sparks *BJ* **245** 85 *1987*.]

Renal dipeptidase. From porcine kidney cortex. Purified by homogenising the tissue, extracting with Triton X-100, elimination of insoluble material, and ion-exchange, size exclusion and affinity chromatography. [Hitchcock *et al, Anal Biochem* **163** 219 *1987*.]

Reverse transcriptase. From avian or murine RNA tumour viruses. Purified by solubilising the virus with non-ionic detergent. Lysed virions were adsorbed on DEAE-cellulose or DEAE-Sephadex columns and enzyme elution with a salt gradient, then chromatographed on a phosphocellulose column and enzyme activity eluted in a salt gradient. Purified from other virus proteins by affinity chromatography on a pyran-Sepharose column. [Verna *Biochim Biophys Acta* **473** 1 *1977*.]

Riboflavin *[83-88-5]* M 376.4, m 295-300°(dec), $[\alpha]_D$ -9.8° (H_2O), -125° (c 5, 0.05N NaOH). Crystd from 2M acetic acid, then extracted with $CHCl_3$ to remove lumichrome impurity. [Smith and Metzler *JACS* **85** 3285 *1963*.] Has also been crystd from water.

Riboflavin-5'-phosphate (Na salt, $2H_2O$) *[130-40-5]* M 514.4. Crystd from acidic aq soln.

Ribonuclease. From human plasma. Purified by ammonium sulphate fractionation, followed by PC cellulose chromatography and affinity chromatography (using Sepharose 4B to which $(G)_n$ was covalently bonded). [Schmukler *et al, JBC* **250** 2206 *1975*.]

Ribonucleic acid (RNA). Martin *et al* [*BJ* **89** 327 *1963*] dissolved RNA (5g) in 90ml of 0.1mM EDTA, then homogenised with 90ml of 90% (w/v) phenol in water using a Teflon pestle. The suspension was stirred vigorously for 1hr at room temperature, then centrifuged for 1hr at 0° at 25000rpm. The lower (phenol) layer was extracted four times with 0.1mM EDTA and the aq layers were combined, then made 2% (w/v) with respect to potassium acetate and 70% (v/v) with respect to EtOH. After standing overnight at -20°, the ppte was centrifuged down, dissolved in 50ml of 0.1mM EDTA, made 0.3M in NaCl and left 3 days at 0°. The purified RNA was then centrifuged down at 10000xg for 30min, dissolved in 100ml of 0.1mM EDTA, dialysed at 4° against water, and freeze-dried. It was stored at -20° in a desiccator. Michelson [*JCS* 1371 *1959*] dissolved 10g of RNA in water, added 2M ammonia to adjust the *p*H to 7, then dialysed in Visking tubing against five volumes of water for 24hr. The process was repeated three times, then the material after dialysis was treated with 2M HCl and EtOH to ppte the RNA which was collected, washed with EtOH, ether and dried.

Ricin. Toxin from Castor bean (*Ricinus communis*). Crude ricin, obtained by aqueous extraction and ammonium sulphate pptn, was chromatographed on a galactosyl-Sepharose column with sequential elution of pure ricin. The second peak was due to ricin agglutinin. [Simmons and Russell *Anal Biochem* **146** 206 *1985*.]

Saccharides. Resolved by anion-exchange chromatography. [Walberg and Kando *Anal Biochem* **37** 320 *1970*.]

Spirilloxanthin *[34255-08-8]* M 596.9, m 216-218°, λ_{max} 463, 493, 528 nm, $\varepsilon_{1cm}^{1\%}$ 2680 (493 nm), in pet ether (b 40-70°). Crystd from $CHCl_3$/pet ether, acetone/pet ether, benzene/pet ether or benzene. Purified by chromatography on a column of $CaCO_3/Ca(OH)_2$ mixture or deactivated alumina. [Polgar *et al, Arch Biochem Biophys* **5** 243 *1944*.] Stored in the dark in an inert atmosphere, at -20°.

Starch *[9005-84-9]* M (162.1)n. Defatted by Soxhlet extraction with ethyl ether or 95% EtOH. For fractionation of starch into "amylose" and "amylopectin" fractions, see Lansky, Kooi and Schoch [*JACS* **71** 4066 *1949*].

Subtilisin. From *Bacillus subtilis*. Purified by affinity chromatography using 4-(4-aminophenylazo)phenylarsonic acid complex to activated CH-Sepharose 4B. [Chandraskaren and Dhar *Anal Biochem* **150** 141 *1985*.]

Syrexin. From bovine liver. Purified by ammonium sulphate pptn, then by *p*H step elution from chromatofocusing media in the absence of ampholytes. [Scott *et al, Anal Biochem* **149** 163 *1985*.]

Thrombin. From bovine blood plasma. Purified by chromatography on a DEAE-cellulose
column, while eluting with 0.1M NaCl, pH 7.0, followed by chromatography on Sephadex G-200. Final preparation was free from plasminogen and plasmin. [Yin and Wessler *JBC* **243** 112 *1968*.
Thrombin from bovine blood was purified by chromatography using *p*-chlorobenzylamino-ε-aminocaproyl agarose, and gel filtration through Sephadex G-25. [Thompson and Davie *Biochim Biophys Acta* **250** 210 *1971*.]
Thrombin from various species was purified by precipitaion of impurities with rivanol. [Miller *Nature* **184** 450 *1959*.]

Tissue inhibitor of metalloproteins. From human blood plasma. Purified by immuno-affinity chromatography and gel filtration. [Cawstin *et al, BJ* **238** 677 *1986*.]

Transferrin. From human serum. Purified by affinity chromatography on phenyl-boronate agarose followed by DEAE-Sephacel chromatography. The product is free from haemopexin. [Cook *et al, Anal Biochem* **149** 349 *1985*.]

Trehalase. From kidney cortex. Purified by solubilising in Triton X-100 and sodium deoxycholate, and submitting to gel filtration, ion-exchange chromatography, conA-Sepharose chromatography, phenyl-Sepharose CL-4B hydrophobic interaction chromatography, Tris-Sepharose 6B affinity and hydrolyapatite chromatography. Activity was increased 3000-fold. [Yoneyama *Arch Biochem Biophys* **255** 168 *1987*.]

T4-RNA ligase. From bacteriophage-infected *E.coli*. Purified by differential centrifugation and separation on a Sephadex A-25 column, then through hydroxylapatite and DEAE-glycerol using Aff-Gel Blue to remove DNAase activity. (Greater than 90% of the protein in the enzyme preparation migrated as a single band on gradient polyacrylamide gels containing SDS during electrophoresis.) [McCoy *et al, Biochim Biophys Acta* **562** 149 *1979*.]

Ubiquinol-cytochrome c reductase. From beef heart mitochondria.
Purified by solubilising the crude enzyme with Triton X-100, followed by hydroxylapatite and gel chromatography. [Engel *et al, Biochim Biophys Acta* **592** 211 *1980*.]

Uracil, uridine and uridine nucleotides. Resolved by ion-exchange chromatography AG1 (Cl⁻ form). [Lindsay *et al, Anal Biochem* **24** 506 *1968*.]

Uridine diphosphoglucose pyrophosphorylase. From rabbit skeletal muscle. Purified by two hydrophobic chromatographic steps and gel filtration. [Bergamini *et al, Anal Biochem* **143** 35 *1984*.]

Veratridine. An alkaloid neurotoxin purified from veratrine. [McKinney *et al*, *Anal Biochem* **153** 33 *1986*.]

Xylanase. From *Streptomyces lividans*. Purified by anion exchange chromatography and HPLC. [Morosoli *et al*, *BJ* **239** 587 *1986*.]

INDEX

For individual organic chemicals, listed alphabetically, see Chapter 3, beginning on Page 65; for inorganic and metal-organic chemicals, see Chapter 4, beginning on Page 310, and for biochemicals see Chapter 6, beginning on Page 372.

A

Abderhalden pistol, 16
Acetals, drying agent for, 46
    purification of, 363
Acid anhydrides, purification of, 366
Acid chlorides, purification of, 364
Acids, drying agents for, 46
    purification of, 363
Acyl halides, drying agents for, 46
Adsorbents, for chromatography, 20, 46
Alcohols, drying agents for, 46
    purification of, 364
    purification *via* derivatives, 60
Aldehydes, drying agents for, 46
    purification of, 365
    purification using ion-exchange, 24
    purification *via* derivatives, 60
Aliphatic acids, purification of, 365
Aliphatic halides, purification of, 367
Alkyl halides, drying agents for, 46
Alumina, 20
    activation of, 20
    as drying agent, 17
    for chromatography, 20
    grades of, 21
    preparation of neutral, 21
Aluminium amalgam, as drying agent, 19
    preparation of, 174
Aluminium methoxide, preparation of, 217
Amberlite IRA-400,
    preparation of OH-form, 286
Amides, purification of, 365
Amines, drying agents for, 46
    Purification as *N*-acetyl derivatives, 62
    purification as *N*-tosyl derivatives, 62
    purification as double salts, 61
    purification as picrates, 61

    purification as salts, 61
    purification of, 365
Amino acids, purification of, 365
    purification by ion-exchange, 24
Aminosugars, purification by ion-exchange, 24
Apparatus, cleaning, 4
Aromatic halides, purification of, 365
Aromatic hydrocarbons, purification *via* derivatives, 62
Aryl halides, drying agents for, 46
Ascarite, 314
Azeotropic mixtures, 11
    separation of, 12

B

Barium oxide, as drying agent, 18
Barium perchlorate, as drying agent, 18
Boiling point, variation with pressure, 4, 5, 35, 36
Boric anhydride, as drying agent, 18
    preparation of, 18
Bruun column, 7
Bubble-cap column, 7
Buffer solutions, 52
Bumping, prevention in distillation, 6

C

Calcium chloride, as drying agent, 18, 59
Calcium hydride, 18
Calcium oxide, as drying agent, 18
Calcium sulphate, as drying agent, 18
Carboxylic acids, purification of, 363
    purification *via* derivatives, 62
Carotenoids, purification of, 366
Catharometer, 29

Nucleotides, purified by ion-
    exchange, 24

O

Oldershaw column, 7
Organic acids, purified by ion-
    exchange, 24
Organic bases, purified by ion-
    exchange, 24

P

Packing for distillation columns, *see*
    distillation columns, 7, 8
Paper chromatography, 29
Paraformaldehyde, preparation of,
    184
Partition chromtography, 22
Peptides, purification by ion-
    exchange, 24
Peroxides, detection of, 368
    removal of, 368
Phenols, purification of, 370
    purification *via* acetates, 63
    purification *via* benzoates, 63
N-Phenyl-2-naphthylamine as
    contaminant, 4
Phosphate esters, purification by ion-
    exchange, 24
    purification *via* derivatives, 63
Phosphonate esters, purification *via*
    derivatives, 63
Phosphorus pentoxide, as drying
    agent, 18
Plasticisers as impurities, 3
Potbielniak column, 7
Potassium, as drying agent, 18
Potassium borohydride, 59
Potassium carbonate, as drying agent,
    18
Potassium hydroxide, as drying agent,
    18
Precipitation, 58
    collectors in, 58
Purification, by extraction, 21
    via derivatives, 60
Purines, purified by ion-exchange, 24
Purity, criteria, 1, 2
Pyrimidines, purified by ion-
    exchange, 24

Q

Quinones, purification of, 370

R

Raoult's Law, 5
Raschig rings, 8
Reagents, removal of metals from,
    57
Recrystallisation, 12
    from the melt, 15
    from mixed solvents, 15
    solvents for,
Reflux ratio, 6
Resins, ion-exchange, 23

S

Safety, in the laboratory, 3, 33, 34
Salts (organic), purification of, 370
Schlenk techniques, 10
Sephadex, 25, 26, 48
Sepharose, 25, 26, 48
Silica gel, activation of, 21
    as drying agent, 19
    for chromatography, 21
    purification, 4
    self-indicating, 19
Skellysolves, composition of, 276
Sephadex, ion-exchange. 24, 25, 48
Sodium, as a drying agent, 19
Sodium benzophenone ketyl, 17
Sodium borohydride, 59
Sodium D line, 2
Sodium hydroxide, as drying agent,
    19
Sodium-potassium alloy, as drying
    agent, 19
Sodium sulphate, as drying agent, 19
Solvent extraction, 31
Solidification from the melt, 15
Solvents, boiling points of, 40
    eluting ability of, 46
    immiscible, 51
    impurities in, 3
    miscible, 41
    removal of, 16